Studies in Systems, Decision and Control

Volume 628

The series "Studies in Systems, Decision and Control" (SSDC) covers both new developments and advances, as well as the state of the art, in the various areas of broadly perceived systems, decision making and control–quickly, up to date and with a high quality. The intent is to cover the theory, applications, and perspectives on the state of the art and future developments relevant to systems, decision making, control, complex processes and related areas, as embedded in the fields of engineering, computer science, physics, economics, social and life sciences, as well as the paradigms and methodologies behind them. The series contains monographs, textbooks, lecture notes and edited volumes in systems, decision making and control spanning the areas of Cyber-Physical Systems, Autonomous Systems, Sensor Networks, Control Systems, Energy Systems, Automotive Systems, Biological Systems, Vehicular Networking and Connected Vehicles, Aerospace Systems, Automation, Manufacturing, Smart Grids, Nonlinear Systems, Power Systems, Robotics, Social Systems, Economic Systems and other. Of particular value to both the contributors and the readership are the short publication timeframe and the world-wide distribution and exposure which enable both a wide and rapid dissemination of research output.

Indexed by SCOPUS, DBLP, WTI Frankfurt eG, zbMATH, SCImago.

All books published in the series are submitted for consideration in Web of Science.

Witold Pedrycz · Gilberto Rivera ·
Rosa Ma. Rodríguez · Salvador Ibarra-Martínez
Editors

Decision-Making in Computational Intelligence-Based Systems

New Approaches, Methods, and Applications

Springer

Editors
Witold Pedrycz
Department of Electrical and Computer Engineering
University of Alberta
Edmonton, AB, Canada

Gilberto Rivera
Departamento de Ingeniería Eléctrica y Computación
Universidad Autónoma de Ciudad Juárez
Ciudad Juárez, Chihuahua, Mexico

Rosa Ma. Rodríguez
Departamento de Informática
University of Jaén
Jaén, Spain

Salvador Ibarra-Martínez
Facultad de Ingeniería
Universidad Autónoma de Tamaulipas
Tampico, Tamaulipas, Mexico

ISSN 2198-4182 ISSN 2198-4190 (electronic)
Studies in Systems, Decision and Control
ISBN 978-3-032-04055-8 ISBN 978-3-032-04056-5 (eBook)
https://doi.org/10.1007/978-3-032-04056-5

This Springer imprint is published by the registered company Springer Nature Switzerland AG
The registered company address is: Gewerbestrasse 11, 6330 Cham, Switzerland

Preface

Decision-Making in Computational Intelligence-Based Systems—New Approaches, Methods, and Applications emerges as a timely and essential contribution to the growing body of research at the intersection of computational intelligence, decision science, and data-driven innovation. In an era marked by increasing complexity and uncertainty, the ability to model, analyze, and optimize decision-making processes through intelligent systems has become not only desirable but imperative. This volume offers a comprehensive perspective on the latest methodologies, theoretical advances, and practical applications shaping the future of decision support systems grounded in computational intelligence.

This book is the result of a collaborative endeavor led by the Eurekas Community, an international and interdisciplinary network of researchers committed to advancing the fields of artificial intelligence, analytics, and smart decision-making. Since its inception, Eurekas has been dedicated to fostering innovation through knowledge exchange, peer-reviewed dissemination, and transdisciplinary dialogue.

The 21 chapters presented in this volume reflect the contributions of experts from diverse institutions and regions, showcasing the depth and breadth of ongoing research in the field.

The structure of the book is organized into three thematic parts that reflect the rich and multifaceted nature of decision-making in computational intelligence-based systems:

- First Part. *Prescriptive Analytics-Oriented Approaches* focuses on frameworks and algorithms designed to support and enhance decision-making through optimization and prescriptive modeling. Topics include the use of Shapley values for group recommendations, multiple criteria ordinal classification, hybrid methods for electric vehicle routing, and coevolutionary algorithms, among others.
- Second Part. *Soft Computing Models* highlights the role of fuzzy logic, neural networks, hybrid systems, and learning-based approaches in uncertain or imprecise environments. Applications span across healthcare, cybersecurity, robotic control, and energy systems, underlining the flexibility and effectiveness of soft computing in complex decision contexts.

- Third Part. *Challenging Case Studies and Applications* provides grounded, real-world scenarios that demonstrate how computational intelligence techniques can be deployed to solve concrete problems. These include simulations in airport operations, crime pattern detection, clinical engineering decision support, and time series analysis in forecasting competitions.

Each chapter in this book underwent a rigorous peer-review process to ensure scholarly quality and relevance. The diversity of perspectives presented—ranging from foundational models to applied studies—offers a panoramic view of how computational intelligence is reshaping modern decision-making.

We extend our sincere gratitude to all the contributing authors for their valuable work and commitment to excellence and to the Eurekas Community for their continuous support in promoting collaborative research efforts. Our appreciation also goes to the editorial and production teams at Springer for facilitating the publication of this volume within the *Studies in Systems, Decision and Control* series.

We hope that *Decision-Making in Computational Intelligence-Based Systems* will serve as both a reference and an inspiration to researchers, practitioners, and students seeking to understand and contribute to the next generation of intelligent decision-support systems.

Edmonton, Canada — Dr. Witold Pedrycz
Ciudad Juárez, Mexico — Dr. Gilberto Rivera
Jaén, Spain — Dr. Rosa Ma. Rodríguez
Tampico, Mexico — Dr. Salvador Ibarra-Martínez

Contents

Soft Computing Models

Challenging Case Studies and Applications

Prescriptive Analytics-Oriented Approaches

Using Shapley Additive Explanations for Supporting Individual and Group Recommendation: Survey and New Perspectives

Raciel Yera, Bapi Dutta, and Luis Martínez

Abstract Recommender systems are essential tools that assist users in navigating vast information spaces by filtering and ranking options based on their preferences and behaviors. While traditional recommender systems focus on individual users, group recommender systems extend this concept to scenarios where items are consumed collectively. These systems must aggregate diverse user preferences and resolve conflicts to generate recommendations that satisfy the group as a whole. Various aggregation techniques, including social choice-based methods, are employed to achieve this goal. Additionally, contextual factors, such as group size and decision-making styles, play a relevant role in shaping recommendations. Beyond enhancing accuracy, trustworthiness has become an important concern in recommender systems, aligning with European AI guidelines that emphasize transparency, accountability, and fairness. Explainability is central to fostering trust, enabling users to understand the rationale behind recommendations. This chapter explores post-hoc explainability approaches, which offer flexible explanations that can be integrated into various recommendation frameworks. Specifically, the chapter explores the use of SHapley Additive exPlanations for recommendation explanations. Despite some research efforts in this direction, a systematic analysis is lacking. This contribution aims to bridge that gap by examining SHAP's applicability in recommender systems, offering through a PRISMA-based systematic review, insights into its potential for enhancing transparency and user trust. A general framework for explaining individual and group recommendation using Shapley Additive Explanations is provided. Furthermore, a case study is developed to show how the proposal can effectively accomplish its goal.

R. Yera (✉) · B. Dutta · L. Martínez
University of Jaén, Jaén, Spain
e-mail: ryera@ujaen.es

B. Dutta
e-mail: bdutta@ujaen.es

L. Martínez
e-mail: martin@ujaen.es

W. Pedrycz et al. (eds.), *Decision-Making in Computational Intelligence-Based Systems*,
Studies in Systems, Decision and Control 628,
https://doi.org/10.1007/978-3-032-04056-5_1

Keywords Recommender system · Shapley additive explanations · Group recommendation · Survey · Trustworthy artificial intelligence

1 Introduction

Recommender systems are intelligent tools designed to assist users in discovering items or information of interest by filtering and ranking options based on their preferences or behavior [1]. These systems operate using various paradigms: content-based recommendation focuses on item attributes to match user preferences; collaborative recommendation leverages the preferences of similar users; knowledge-based recommendation relies on domain knowledge to suggest items based on explicit needs; demographic recommendation tailors suggestions based on user demographics; and hybrid recommendation combines multiple approaches to enhance accuracy. Recommender systems find applications across diverse domains, including e-commerce (personalized product suggestions), e-learning (customized learning resources), e-services (service recommendations), e-tourism (travel planning), and e-health (personalized health advice) [24]. By delivering relevant and timely suggestions, these systems improve user satisfaction, decision-making, and engagement in both personal and professional contexts.

Being traditional recommender systems usually focused on individuals, group recommendation systems extend such concept by focusing on scenarios where items are consumed collectively by a group rather than an individual [11]. These systems are designed to be focused on the preferences, needs, and dynamics of multiple users to suggest items that satisfy the group as a whole. This is particularly relevant in domains where decisions are usually collaborative, such as selecting travel packages for families or friends, choosing movies or TV shows for group viewing, or planning group dining experiences.

Group recommendation are centered on aggregating diverse preferences and resolving conflicts to find a solution acceptable to all members [10]. The main used techniques include preference aggregation and recommendation aggregation methods. Such aggregation is usually modeled through the use of social choice-based approaches such as additive, least misery, approval voting, or most pleasure aggregation [11]. Transversally, contextual factors, such as group size, relationships, and decision-making styles, also drive the generated recommendations [28].

Beyond accuracy, trustworthiness in recommender systems is essential for ensuring user acceptance, satisfaction, and long-term engagement, especially in light of the European AI guidelines emphasizing ethical and transparent AI [42]. These guidelines highlight the importance of accountability, fairness, and transparency in AI systems, pointing out recommender systems as AI-based tools, and the necessity of operating in a way that users can understand, trust, and rely on.

The current contribution concerns explainability in recommender systems, which is the key concept managed by the research community for accomplishing the transparency goal [48]. It refers to the ability of the system to provide clear, understandable reasons behind its suggestions, allowing users to comprehend why certain items are recommended. Zhang et al. [48] point out different types of explanations in recommender systems:

1. **Intrinsic recommendation explanations**, built as component of specific recommendation approaches, providing insights directly tied to its internal logic, such as highlighting shared features or remarkable directions that directly lead to the system's output.
2. **Post-hoc recommendation explanations**, focused on explaining the output of the recommendation approach, instead of being focused on its inner logic.

Particularly, the current contribution is centered on post-hoc explanations, which are flexible approaches that can be coupled with any recommendation framework and are focused on finding some common sense reasoning that would justify the delivery of some specific recommendation [48]. Due to their flexibility, they have attracted the attention of the research community over the last few years.

Peake and Wang [27] present one of the first approaches focused on this goal, focused on mining local association rules representing the relationship between items, that are used for building the subsequent explanations of the recommended items. In parallel, Quijano-Sánchez et al. [31] present a methodology for explaining group recommendations by exploiting social relationships between group members. Ariza-Casabona et al. [2] recently presented a comparative analysis between current text-based explainable recommender systems. Furthermore, Yera and Martínez [46] have also developed a local rules-based approach that works over item features to build factual and counterfactual recommendations for individuals and groups.

The development of explanation interfaces for supporting recommendations has been also a relevant research effort covered by several authors [37]. Graphical interfaces such as bar charts, connected graphs, or line charts; have been used for representing feature weights, feature contributions, neighborhood information etc., and therefore justify the recommendation. Over the last few years, visual explanation paradigms such as LIME or SHAP, have been also brought to the recommender systems scenario.

The aim of the current book chapter is to explore the use of SHAP for supporting recommendation explanations. While some research efforts have been made in this direction, there remains an important gap in systematic analysis and contributions that would facilitate the structured and consistent application of this approach both in individual and group recommendation systems. Particularly, Sect. 2 introduces the necessary background, including individual and group recommendation and post-hoc explanation in recommender systems. Section 3 presents the systematic review methodology on the use of SHAP for supporting recommender systems. Section 4 discusses a general framework for explaining individual and group recommendation using SHAP, being one of the first efforts on covering these research task in the group recommendation scenario. Section 5 develops a case study for applying the proposed framework in a real scenario. Section 6 concludes the contribution.

2 Background

This section will be focused on providing a background on individual and group recommendations, which is necessary for introducing the subsequent stages of this contribution. Furthermore, a brief overview on post-hoc explanations in recommender systems is also provided.

2.1 *Individual and Group Recommendation*

Two main paradigms have driven the development of individual recommendations since 90s [1]. These have been: (1) the content-based recommendation approach, focused on recommending items similar to those consumed in the past by the active user, and (2) the collaborative filtering approach, focused on recommending items that have been consumed in the past by other users with a rating behavior similar to the current one.

Content-based recommendation methods, as described by Adomavicius and Tuzhilin [1], operate on the principle of "Give me more of what I've already enjoyed." These systems are defined by three key components: a framework to represent the items that can be suggested, a mechanism to build a user profile capturing their preferences, and a method to match items against the user's profile to decide which ones to recommend.

The recommended items are typically characterized by a set of features or attributes, each with defined possible values. A user's level of preference for certain items is assessed, and machine learning algorithms are employed to develop a profile based on those attribute values. This enables the system to identify, among the remaining items, those that offer the highest relevance or utility to the user. Consequently, the two central challenges in these systems are how the items s are represented ($Content(s)$) and how the user's profile c is accurately learned ($ContentBasedProfile(c)$). The ultimate goal is to find the utility function for the user c, of item s:

$$u(c, s) = score(ContentBasedProfile(c), Content(s)) \tag{1}$$

Collaborative filtering operates on the principle of leveraging collective preferences to recommend movies to a specific user. This method identifies, either explicitly or implicitly, patterns in users' ratings that align with those of the current user and employs this shared information to generate recommendations. Collaborative filtering techniques can be broadly categorized into two main types: (1) memory-based approaches [13], which focus on directly identifying similar user groups, or "neighborhoods" for the active user and utilizing the preferences of these closest neighbors to make suggestions, and (2) model-based approaches [20], which involve creating intermediate models that encapsulate the preferences of the overall set of users, for leading to the recommendation generation. Summarizing, the goal of collaborative filtering would be the calculation of the current rating $r(c, s)$, based on the aggregation of the ratings associated with the same item, provided by other users c', similar to the current one c.

$$r(c, s) = agg_{c' \in C} r(c', s) \tag{2}$$

These methods have gained significant popularity in several recommendation scenarios, due to their ability to deliver precise suggestions by relying solely on user ratings, without requiring additional information. This distinguishes them from content-based methods, which depend on detailed item attributes to perform effectively.

Beyond the recommendation generation for individuals, there are several contexts with items that are usually consumed by groups, such as travel packages, TV programs, restaurants, music, etc.; and that therefore need different recommendation algorithms that suggest items that fit the necessity and preferences of a set of users, and not an individual one.

Two alternative approaches drive the conception of a group recommender system:

- *Recommendation aggregation approach*: Here each member of the group receives individual recommendations using a method tailored for the individual scenarios, and after that, such individual recommendations are aggregated to obtain the group recommendations.
- *Preference aggregation approach*: In this case the aggregation is done over the individual preferences of the group members, to build a pseudo-user profile that represents the preferences of the group. The group recommendations are delivered to this pseudo-user profile, using an individual recommendation method.

The aggregation of the user preferences or the generated recommendations has been usually done with the use of social choice-based aggregation approaches [11], such as the following ones:

- **Additive**: It is focused on adding the preferences associated with each individual in the group, for composing the group preferences.

$$r_{G,i} = \sum_{u \in G} r_{u,i} \tag{3}$$

- **Least Misery**: It is focused on identifying the preference associated with the group, as the minimum preference linked to any of their members.

$$r_{G,i} = \text{Min}_{u \in G} r_{u,i} \tag{4}$$

- **Approval Voting**: It identifies the group preferences, as the amount of users whose individual preferences are over some predefined threshold.

$$r_{G,i} = |r : r_{u,i} > \theta, u \in G| \tag{5}$$

- **Most Pleasure**: It identifies the group preferences, as that associated with the member with the highest preference value.

$$r_{G,i} = \text{Max}_{u \in G} r_{u,i} \tag{6}$$

Taking as base these foundations, in the last few years several authors have been focused on developing more sophisticated approaches.

Dara et al. [10] conducted a comprehensive survey on the latest advancements in group recommender systems, covering various domains and analyzing different systems based on their aggregation strategies and user preference modeling. Yalcin et al. [41] introduced two innovative aggregation methods that combine additive utilitarian and approval voting approaches to highlight popular items agreed upon by group members. Similarly, Contreras et al. [8] proposed a collaborative framework grounded in the social dynamics within web-based conversational group recommender systems. Their model enables the system to implicitly identify different roles within the group, such as collaborative members and leaders. Lastly, Ismailoglu [18] applied the Expectation-Maximization (EM) algorithm to merge group members' preferences, estimating group ratings and evaluating the expertise levels of individual members in group recommendation scenarios.

2.2 Post-hoc Explanations in Recommender Systems

This section will cover a brief overview of recent advances in post-hoc explanations in recommender systems. To accomplish this aim, the related works will be categorized into four main groups, recognizing that their objectives may be overlapped in some cases: (1) explanations supported by traditional data mining techniques, (2) explanations supported by alternative, simpler and transparent recommendation models, (3) explanations supported by generative models, usually with the aim of text generation, and (4) visual explanations supported by established model-agnostic machine learning approaches.

Explanations supported by traditional data mining techniques: Peake and Wang [27] developed a representative approach belonging to this category, that use the discovery of association rule mining for matching preferred and recommended items. Specifically, given a set of recommended items, the approach is focused on discovering rules with the shape $A \rightarrow B$ where the consequent B of the rule is represented by recommended items, which recommendation would be justified by the preferences of the users over items in A, in case of it applies for the active user. This work exploits the concept of local explanations, by mining the rules just using data of the neighborhood of the current user.

Across the recent years, this approach has been used for explaining recommendation in specific scenarios such as cooking recipes [42], as well as incorporated as key explanation approach in several recommender systems libraries [7].

Furthermore, Yera and Martinez [46] has recently explored the use of rules that connect item features with preferred items, for performing factual and counterfactual explanations in individual and group recommendations. The explanation generation is supported by the path from the root to the leaves, of decision trees built taking as base the information of items similar to the recommended one. Herein the group recommendation explanations are built by aggregating the explanations generated for the individuals.

Explanations supported by alternative, simpler and transparent recommendation models: Some previous works have explained black-box recommendation approaches through the use of simpler and transparent recommendation models, employed as a proxy for reaching the same output of the main black-box approach.

This is the goal of the framework proposed by Schmaryahu et al. [34] (Fig. 1), where initially the black-box recommendation algorithm delivers recommendations, which are attempted to be explained by several explanation algorithms, using additional data sources. The explanation algorithms are based on basic content-based and collaborative filtering approaches, such as item-item content-based, user-item content-based, item-item overview, item-item collaborative filtering, and user-user collaborative filtering.

In a different direction, Tran et al. [38] and Barile et al. [5] have systematized the use of several social choice-based explainable aggregation approaches, such as additive utilitarian, least misery, approval voting, most pleasure, fairness-based, and so on. Furthermore, they have shown that the generation of explanations focused on illustrating the internal reasoning behind these aggregation approaches could positively impact the perceived group satisfaction, consensus, and fairness. Additionally, the effect across several group configurations, such as divergent, minority, or coalitional, was investigated. Recently, Waterschoot et al. [40] evaluate the objective and subjective user understanding of social-choice based explainable aggregation approaches.

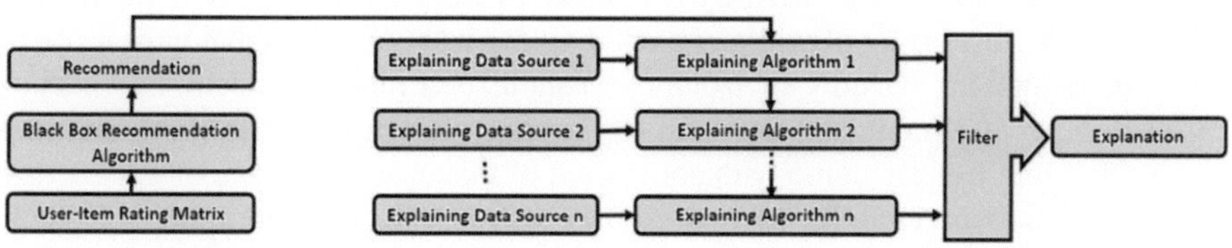

Fig. 1 General explanation approach based on simpler models

Explanations supported by generative models, usually with the aim of text generation. With the rapid growth of LLMs-based models, some authors have used their benefits for generating text-based explanations of the output of recommender systems. This research line has the particularity of working over previous user reviews to generate textual explanations. Even though this could be a desired capability for most e-commerce environments, it may also imply a limited application in other recommendation scenarios.

Ariza et al. [2] have recently presented an analysis and benchmarking of recent research focused on this aim. Herein, related works have been classified as text generation-based, text extraction-based, or hybrid.

Generation-based models compose the explanations word by word in an autoregressive way, taking into account past user-item interactions. The primary shortcoming of these models is their lack of controllability in their generation process, which produces repetitive and shallow explanations. In this family of works, PETER [22] was one of the first explainable models that took advantage of a transformer-like architecture to accomplish multiple tasks such as rating prediction, context prediction, and explanation generation. Focused on a more specialized goal, SEQUER [3] extends PETER through the inclusion of user interaction sequences and explorations of different masking alternatives to improve the representation capabilities. More recently, POD [23] introduced a transformer-like model that uses a pre-trained multitask LLM, that is tuned to perform top N recommendations, next-item recommendation, and explanation generation.

Extraction-based models build the explanation by selecting the top-k explanation sentences for a candidate set that is composed of user-item review sentences. In this group of works, Pugoy and Kao [30] have applied unsupervised text clustering over review sentences encoded with a BERT-based approach, for creating user and item profiles, which can be used to generate rating predictions and their subsequent explanation. Furthermore, Wang et al. [39] construct a heterogeneous graph of user-item-attribute nodes taking as starting point the user and item reviews, which is used later to rank the candidate sentences and attributes presented in the graph, supported by a graph neural network solution. Overall, these works are limited by the diversity and completeness of the set of review sentences and then can be affected by the cold start problem when the set of reviews does not have sufficient coverage. Hybrid architectures have been also proposed, by following a retrieval-augmented generation viewpoint, in order to combine the generation-based and extraction-based approaches.

Visual explanations supported by established model-agnostic machine learning approaches. In this group of works we will explore a set of works that have brought to the recommender systems environment, established model-agnostic approaches that have been systematically used for explaining the output of machine learning models, and that usually come with some kind of visual information.

Formerly, Ribeiro et al. [32] proposed the creation of Local Interpretable Model-agnostic Explanation models (LIME), which leverage sparse linear models to approximate the behavior of a black-box classifier in the vicinity of a specific instance. These linear models can then identify the sample features that significantly influence the predicted outcome. Building on the LIME framework, Nóbrega and Marinho [25] developed LIME-RS, a variant tailored for recommender systems, designed to pinpoint the top-n item attributes that most effectively elucidate individual recommendations generated by a factorization machine. Additionally, Chanson et al. [6] introduced LIRE, an enhanced version of LIME for recommender systems, which incorporates optimized sampling around the recommendation instance to construct a more accurate local surrogate model.

Taking an alternative approach, SHapley Additive exPlanation values (SHAP) have also emerged as a post-hoc explanation framework for recommender systems [16]. This method emphasizes calculating the average marginal impact of each feature value on the model's prediction across all possible permutations. Recently, Zhong and Negre [49] have used the SHAP framework as a tool for identifying possible candidates for composing counterfactual explanations.

In contrast to text-based explanation approaches, these methods are able to generate explanations using just ratings and attribute values. However, in a different way in relation to the previous group of works based on traditional data mining techniques or simpler recommendation methods, they have been explored with a lesser extent.

Particularly, we will be interested on SHAP values, which can be used both at the global and local level. Furthermore, in contrast to LIME-based approaches, they do not suffer from stability concerns [49]. The next section will be focused on discussing the direct antecedents on the use of SHAP for supporting explanations recommender systems at individual and group level.

3 Antecedents of the Use of SHAP for Supporting Recommender Systems

This section is focused on identifying key works across the literature, focused on explaining the recommender system output based on SHAP. With this purpose in mind, the Preferred Reporting Items for Systematic reviews and Meta-Analyses methodology (PRISMA) will be driven. Additionally, the identified works will be screened in detail.

3.1 Systematic Review Methodology

This section outlines the methodology used to perform a systematic review of the literature on food recommender systems for diabetic patients. The review was carried out according to the PRISMA 2020 guidelines (Preferred Reporting Items for Systematic Reviews and Meta-Analyses) [26]. To achieve this, we followed the steps specified in the PRISMA 2020 checklist throughout the development of the survey. The main stages of this checklist are described in detail in this section, being those also used by previous works also in the domain in recommendation systems, and therefore generalized for systematic reviews in this context. These are: (1) eligibility criteria, (2) information sources and search strategy, (3) the studies selection, (4) data extraction and data items, (5) quality evaluation, and (6) the analysis of the obtained results.

(1) *Eligibility criteria*. Our systematic review on the use of SHAP for supporting individual and group recommendation, is guided by the following inclusion and exclusion criteria:

- Inclusion criteria:
  - Papers proposing recommendation algorithms and applications.
- Exclusion criteria:
  - Papers incorporating methods that perform some kind of prediction/suggestions, but that do not match with the well-defined recommendation problem scenario.
  - Papers limited to identifying matches between feature values.
  - Papers presenting only general reports without a clear exposition of novel approaches.

(2) *Information sources and search strategy*. A literature search was performed on February 2025, using Web of Science Core Collection and Scopus as information sources (see Fig. 1). Both sources have been widely used in previous reviews focused on similar goals [44, 45]. Herein it is important to take into account that most relevant content published by major editorials like Elsevier, Springer, IEEE, Wiley, ACM, etc., are already covered in these two main information sources, and therefore we found it not necessary to perform the search directly at the editorial websites to avoid redundancy. Furthermore, inside Web of Science Core Collection we also searched through the Emerging Source Citation Index database (ESCI), as a source of emerging literature that ensure diversity and inclusiveness of our systematic review process. Two main queries were considered at this stage:

- "SHAP" AND "recommender systems"
- "SHAP" AND "recommendation systems"

(3) *Studies Selection*. Relevant papers were identified based on the search output, by analyzing title, abstract and keywords. This process was developed by two of the reviewers (R.Y. and B.P.), taking into account the inclusion and exclusion criteria. The third reviewer (L.M.) was involved in the scenarios where an additional and more experienced analysis was necessary to decide whether include or not include some specific outputs; and for ensuring that a consensus would be always reached.

(4) *Data Extraction and Data Items*. For polishing the established criteria, further information on the selected papers were retrieved and tabulated in CSV files. Specifically, authors, year, journal, artificial intelligence-based methodology, and used explanation-centered methodology. Disagreements on information extracted were also solved, with the moderation of the third reviewer (L.M.).

(5) *Quality evaluation*. The checklist is composed of the following categories and criteria: For quality evaluation of the identified manuscripts, it was used the checklist previously used by Ataei and Litchfield [4], and by Yera et al. [44], which are subsequently informed by those proposed by CASP for assessing qualitative research [9], and by Kitchenham et al. [19] concerning systematic reviews in software engineering (Fig. 2).

1. Minimum threshold quality:
    a. Does the study report empirical research or is it merely a 'lesson learnt' report based on an expert opinion?
    b. The objectives and aims of the study are clearly communicated, including the reasoning for why the study was undertaken.
    c. Does the study provide adequate information regarding the context in which the research was carried out?
2. Rigor:
    a. Is the research design appropriate to address the objective of the research?
    b. Is there any data collection method used and is it appropriate?
3. Credibility:
    a. Does the study report findings in a clear and unbiased manner?
4. Relevance:
    a. Does the study provide value for practice or research?

In the current work, each retrieved work was evaluated through yes/no answers for each of these seven criteria. As previous works [4, 44], works with positive responses under 75% of the number of questions, were excluded from the analysis.

(6) *Analysis of the obtained results*. As presented at the PRISMA flow diagram (Fig. 1), the application of the previous stages leads to the reaching of 165 records, being 69 from Web of Science Core Collection and 96 from Scopus. As expected, after duplicated removal 56 records were suppressed, leaving to 109 records to perform a more detailed screening. This screening identified a larger number of works (n =92) to be excluded as even though they pass the initial filtering on studies selection, were at the intersection with other knowledge domains and/or not focused on recommendation explanation based on SHAP. After this filtering, 17 contributions were retrieved for a detailed study. Herein, we identified nine of them as not essentially focused on explaining the recommendation output (i.e. just slightly facing features dependency), and four of them were overall papers

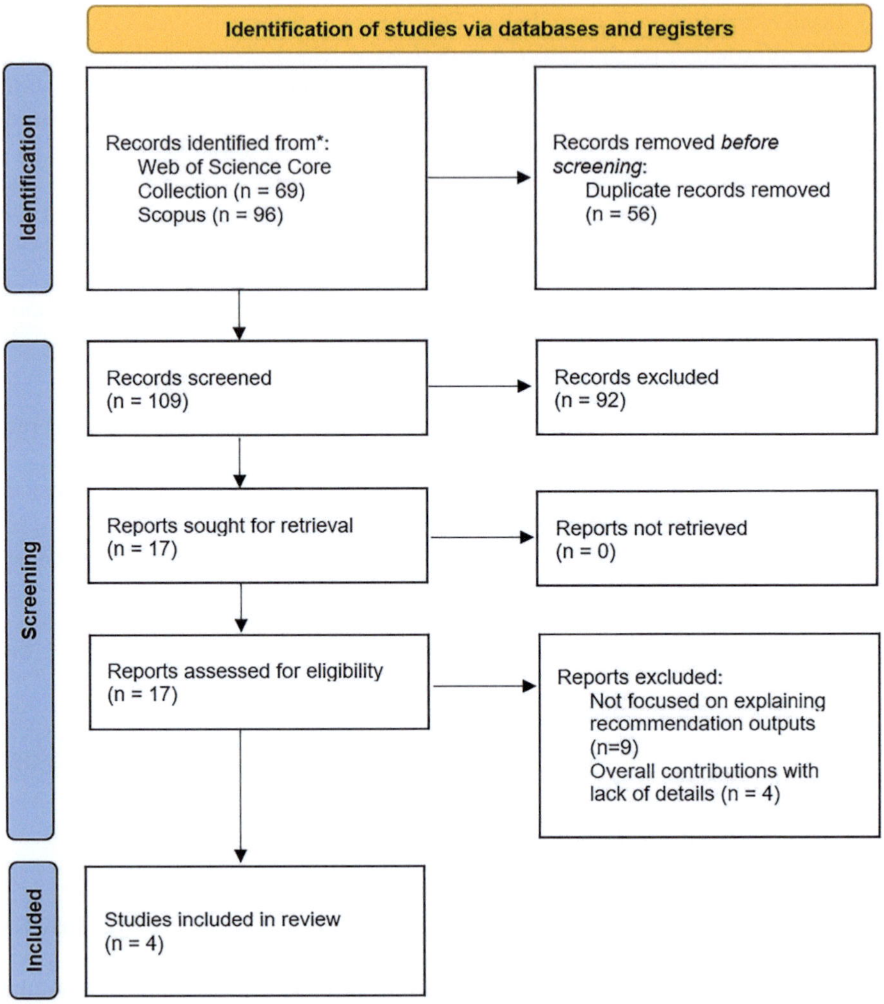

Fig. 2 Systematic flow diagram representing the inclusion of studies according to the PRISMA 2020 declaration

lacking of details. After disregarding these papers, we finally included in the review four papers that will be analyzed in detail in the next subsections.

3.2 Detailed Analysis of the Identified Works

This subsection will provide a detailed analysis of the four identified works, highly focused on exploring the use of SHAP for explaining recommendations.

3.2.1 Product Feature Recommendation

Guo et al. [16] are focused on selecting the feature products that have the best trade-off between price and functionality. Particularly, the authors formulate the unique feature product recommendation problem as a price-driven supervised learning problem to discover the features that explain best the price of a product in a given category. They use the machine learning models coupled with the Shapley value model-agnostic method to understand, rank, and recommend, the most important features.

In this work, the authors use as reference three machine learning models (linear regression, LightGBM, and CatBoost), that handle categorical values, and are able to learn the feature importance taking as reference the price of the products. This information is provided inside a Shapley values-based framework for obtaining the final ranking feature for each category. Figure 3 illustrates how each feature such as Product Weight, Product width or Chunk size, contributes to a higher or lower price.

Furthermore, Fig. 4 illustrates all the Shapley values for each specific feature, showing how the values of each attribute impact the final price of the item. For evaluation, the authors use domain experts to identify key product features, which are then contrasted with the features identified by their approach, using metrics such as NDCG, Precision, Recall, and Coverage. They also report the use of online A/B testing in real scenarios, for product feature recommendations and explainable recommendations using feature-based messages.

3.2.2 Interpretable Neural Network TV Program Recommendation Suing SHAP

Yin et al. [47] analyze the interpretability of a convolutional neural TV program recommendation model, that uses additional information to learn the program features efficiently. Particularly, they first learn program representations to subsequently generate user representations using convolutional neural networks. These two representations are provided to the recommendation module, for generating the recommendations. Herein, the output of the model for each user (to be explained), is the numerical label for each program having a binary shape, to indicate whether a program is watched or not. The model for program recommendation is then focused on predicting the probability of the user, of viewing each corresponding program.

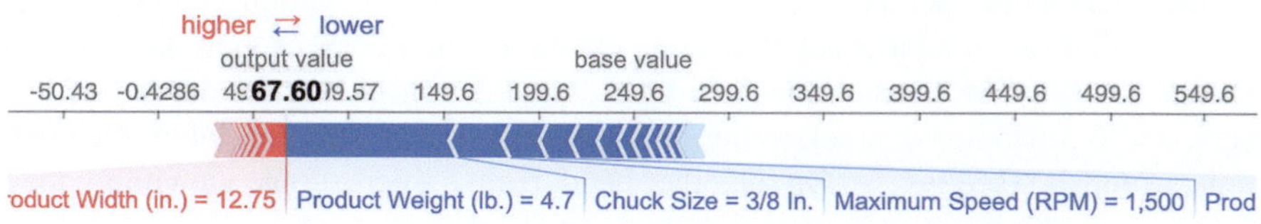

Fig. 3 Feature contributions of an item in the dataset, obtained by Guo et al. [16]

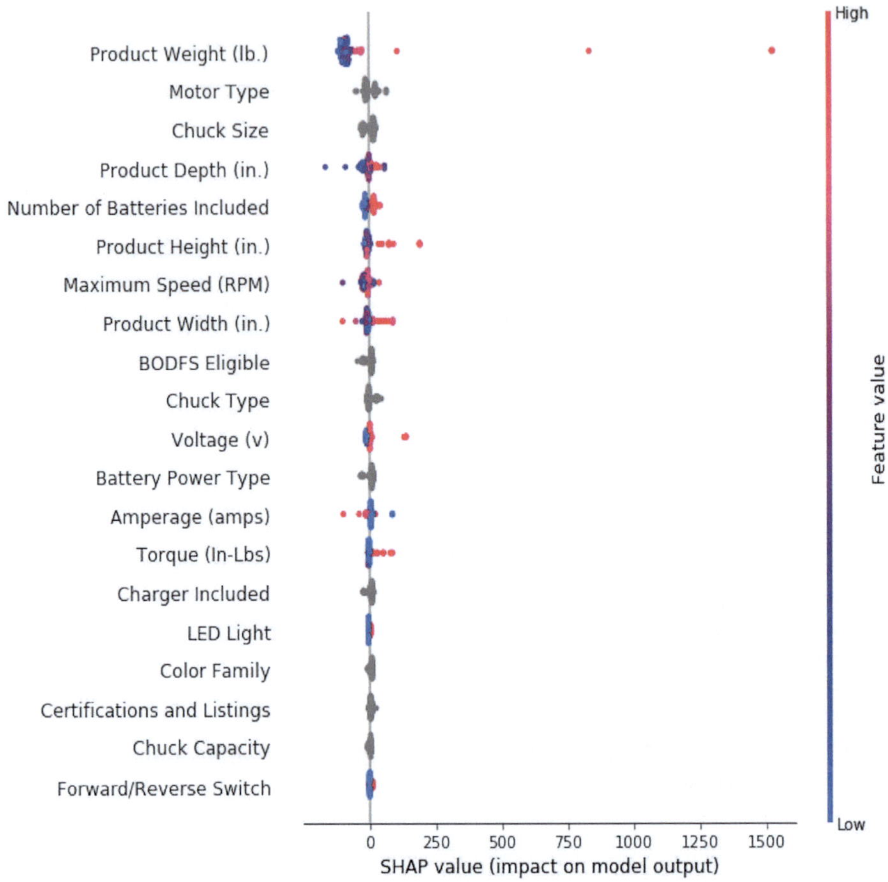

Fig. 4 Shapley values used for feature ranking, obtained by Guo et al. [16]

SHAP is used for generating visual interpretations of the generated outputs, to illustrate the role of each program feature in the prediction of user interests. Specifically, they use DeepLIFT algorithm [35], specifically designed for neural networks, to approximately obtain the SHAP values.

Using this tool, the contribution provides interpretations from the macroscopic and the microscopic perspectives. From the macroscopic viewpoint, it analyses the feature attribution of a large set of training examples (first 3000 of their dataset) and explains the first 500 predictions done by the model, concerning the dataset of 110 users, and 6 attributes characterizing programs (i.e. actor, director, label, channel, name, synopsis) (see Fig. 5). On the other hand, from a microscopic viewpoint, they examine how the magnitude of each feature affects different users when they watch it. As an example, Fig. 6 presents a sample focused on explaining the predictions made for a single user, where the name of the program negatively influences whether

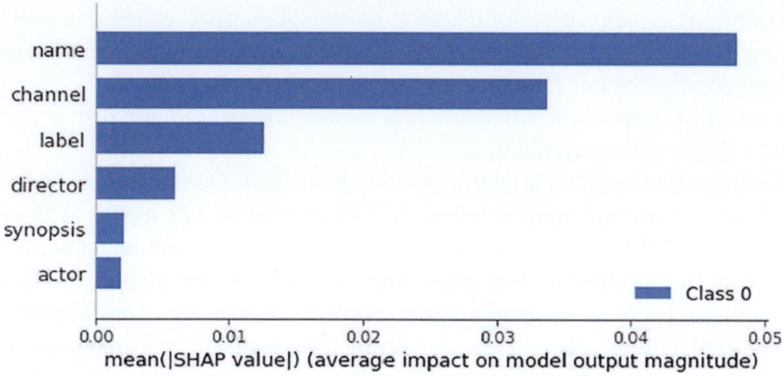

Fig. 5 Impact of each TV program feature in the overall set of prediction, obtained by Yin et al. [47]

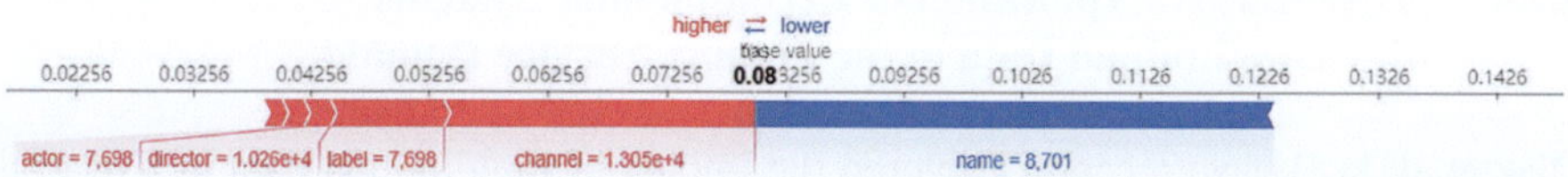

Fig. 6 Impact of each TV program feature at the individual user stage, obtained by Yin et al. [47]

it is watched or not. On the other hand, channel and label have a positive impact on the prediction of preference or non-preference.

3.2.3 SHAP-Enhanced Counterfactual Recommendation Explanations

Zhong and Negre [49] were focused on generating counterfactual recommendation explanations, using SHAP as an auxiliary tool. Taking into account the traditional top-N recommendation scenario, where items are characterized by a set of features, they consider two types of explanations:

- *Instance-level explanations.* Herein, counterfactual explanations consist of identifying the minimum set of feature values to be changed, in order to put the active item out of the recommendation list.
- *List-level explanations.* Subsequently, they are defined as the minimum set of feature values to be changed, in order to change the set of items to be recommended.

To accomplish the initial goal, the authors propose a greedy approach which is initially focused on identifying the features with a larger contribution to the recommendation output, based on the SHAP framework. For each feature in this set, the original value is modified towards a new one, generating a new recommendation list for this alternative set of features. If the item of interest is no longer on the recommendation list, then this alternative set of features could be considered as the counterfactual explanation.

The previous strategy, used for instance-level explanations, can be generalized for generating list-level explanations. In this case, it is necessary to define a similarity score between the former recommendation list and the new one after changing the feature values. If such similarity exceeds a specified threshold, the new explanation could be considered counterfactual.

The authors evaluated their approach using the LDOS-CoMoDa dataset, focusing on the context-aware recommendation. It is composed of 121 users, 1197 movies, 2296 ratings and 12 context factors. Such context factors include time, daytype, season, location, weather, mood, etc., which are used as features in this scenario. The overall proposal is evaluated using criteria such as the time consumed for searching explanations, the average length of the counterfactual explanations, and the percentage of recommendations explained.

3.2.4 SHAP-Based Exploration of Recommender Systems with Longitudinal Data in the Finance Service Domain

Diaz et al. [12] have recently explored the impact of longitudinal data in a recommendation environment, comparing traditional collaborative filtering approaches, multi-label classification algorithms, and a deep learning approach based on gated recurrent units. They are specifically focused on a dataset containing product ownership data for 35 distinct financial services and their recency, frequency, and monetary information. Demographic data is also available from 300000+ clients.

Moreover, SHAP is deployed to better understand the output of the best performing models. With this purpose in mind, the products are grouped into different categories such as credits, insurance, savings, and services. The absolute SHAP values for the features belonging to each group are summed to find their global importance associated with each category.

For each different category, different sets of relevant features were identified. For example, the recommendation of credit products was primarily influenced by features such as the recency and frequency of Savings products, as well as the frequency, monetary, and recency features for Insurance products (Fig. 7).

4 Towards a General Framework for Explaining Individual and Group Recommendations Using Shapley Additive Explanations

The previously analyzed works suggested the presence of a still incipient development towards the task of taking advantage of SHAP for explaining recommender system outputs. In the following section, we discuss a general framework that could be used for systematizing the research and development tasks towards this effort.

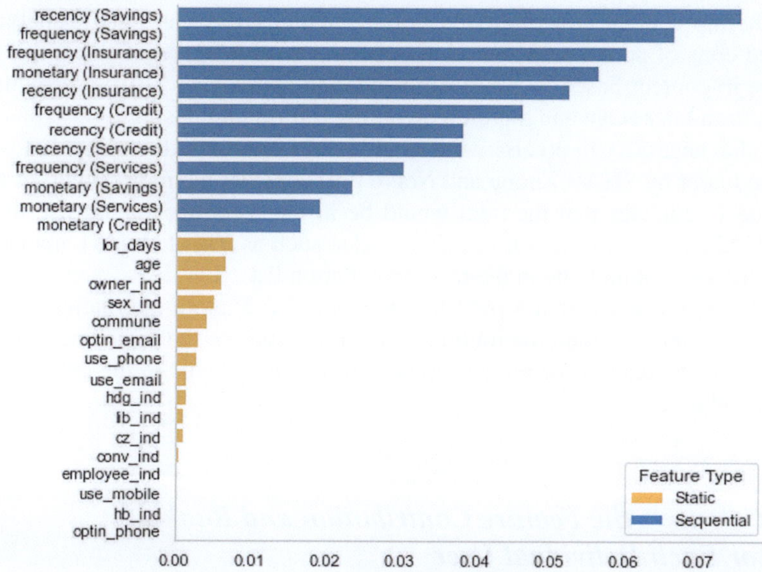

Fig. 7 Mean associated SHAP values for credit products (Diaz et al. [12])

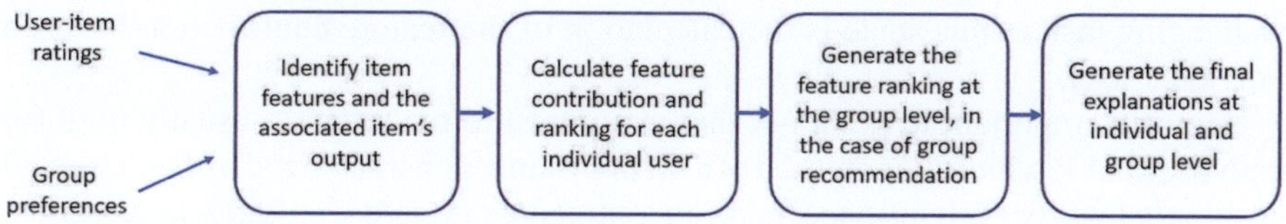

Fig. 8 Our framework for systematic SHAP-based explanations in recommender systems

Figure 8 presented this general framework, which would be composed of the steps discussed below, and that concerns individual and group recommendations.

4.1 *Identify the Item Features and Their Associated Output for Each Individual User, in the Recommendation Model*

At first, it is necessary to define the information input that will be used for explanation generation. Based on the works analyzed in the previous section, there is a consensus of previous works on the use of item features for supporting the subsequent explanations generation. In this case, it is necessary to precise whether taking into account the nature of the feature (categorical or continuous), it would be necessary some kind of discretization procedures for transforming them in a way that could be managed by the traditional SHAP approaches.

Furthermore, in the case of group recommendations, it is necessary to analyze the pros and cons of performing some kind of feature value aggregations at this stage (as done in content-based group recommendation approaches [29]), or leaving this process for a later stage and therefore avoiding information losing.

It is also necessary to precise what would be the output of the model that will be explained later by SHAP. Zhong and Negre [49] consider a binary output, in which the value 1 indicates that the item would be in the recommendation list, and the value 0 indicates that it does not. Other works, such as Yin et al. [47], manage the probability of an item to be in the recommendation list.

Finally, we stress that it would be interesting to explore alternative ways for representing items beyond the traditional features, such as the preferences of some key users in the dataset, or the preferences of certain groups in the case of group recommendation.

4.2 Calculate the Feature Contribution and Ranking for Each Individual User

Once defined the shape of the item features and the output of the model, the main challenging task at this stage is the calculation of the feature contribution for each individual user.

Here it is important to point out that in contrast to the scenario usually used for applying SHAP, which is composed of a set of instances characterized by features, and an output/class for each instance, in the recommender system scenario is necessary to also consider the presence of a user that would provided the output (rating), associated to the current instance (item). This fact leads us to consider the application of SHAP at two different levels: the global level considers an important part of all the available dataset, and the user level focuses on explaining just the feature contributions based on the active user preferences. The approach suggested by Yin et al. [47], focused on "macroscopic" or "microscopic" viewpoints of the SHAP explanations, could be an appropriate starting point for supporting this issue. Anyway, taking into account the sparsity of recommender systems datasets, it is important to analyze their feasibility, specifically for the user-level case.

Another important design decision to be regarded here is the appropriateness of the specific SHAP implementation to be used, for the recommender systems environment. In practice, it is expected that most of the features that characterize items will be categorical [7, 28, 29]. Therefore, it would be necessary to exploit previous SHAP implementations tailored to categorical feature scenarios [21], and even the use of some classifier such as CatBoost [17], developed for such environment.

4.3 Generate the Feature Ranking at the Group Level

In addition to the use of SHAP at the global and individual levels, defining SHAP-based explanations to justify the recommendations delivered to a specific group of users in the context of group recommendation is a challenging task.

To cover this goal, according to our viewpoint the first stage would be to define whether the aggregation of the group member information will be performed at the first stage of identification of features (e.g. the profiling stage), or at this current stage by obtaining the feature ranking based on the individual ranking linked to each of the group members'. In the latter case, some aggregation operators need to be formalized, which can be inspired by the proposed feature aggregation schemes for content-based group recommendation [14, 29].

Finally, hybrid approaches focused on simultaneously presenting explanations at the individual and group levels could also be considered.

4.4 Generate the Final Explanations at Individual and Group Level

In the last stage, generating the final explanations at both the individual and group levels is not a trivial task. Since the expected output of SHAP-based approaches is typically standardized (see Figs. 3, 4, 5, 6 and 7), some authors, such as Zhong and Negre [49], have pointed out that these outputs could be difficult for some users to interpret.

Particularly, Zhong and Negre [49] used the output of SHAP to generate counter-factual explanations in the form of a set of <attribute, value> pairs, similar to Yera and Martínez [46]. Subsequently, text-based handcrafted templates could be defined and populated with these pairs [15] to generate effective and plausible explanations.

The previous four stages have referred several open problems to be cover towards systematizing the use of SHAP-based explanations in recommender systems. Beyond this, it is also important to highlight the necessity of boosting the development and use of robust evaluation framework for measuring the real contribution of such developments, both from the offline analysis and the user studies perspective.

Offline analysis: The effectiveness of post-hoc explanations have been usually characterized by metrics that measure the fidelity of the explanation mechanism, i.e. the percentage of recommended items for which the proposed approach was able to find some explanation [42, 46]. Text-based explanation approaches, like those discussed in the background section, have also incorporated feature-oriented metrics, such as feature coverage rate (i.e. the percentage of features referred to in the generated explanations), that could be also important to consider in the SHAP environment.

In a different direction, experts-mediated offline evaluation, such as the one developed by Guo et al. [16], would also be an interesting direction to explore. In this approach, the authors ask domain experts to identify the most relevant features, which are then contrasted with the results of SHAP.

User studies: The development of user studies will answer important questions towards the appropriateness of SHAP-associated explanations in individual and group recommendations. In such a context, it is necessary to empirically determine the effect of such explanations in well-established aims regarding explanatory interfaces, such as transparency, trust, effectiveness, efficiency, and satisfaction [36].

Furthermore, for the case of group recommendation, tailored criteria such as perceived consensus, perceived fairness and group satisfaction [5], need to be covered.

Also, Rong et al. [33] have recently pointed out additional goals fairly explored in recommender systems explanations, such as understanding and usability, which should be specifically overseen in the case of SHAP-based explanations.

Beyond these issues, it is important to take into account that while SHAP provides detailed interpretability, its use in large-scale recommendation systems leads to important computational costs. The complexity of SHAP-based methods increases with the number of features and model evaluations, making it challenging to deploy in real-time, high-dimensional contexts. These drawbacks particularly appear when dealing with sparse user-item interaction matrices, where the lack of data increases computational needs without guaranteeing trustable explanations. Moreover, the high dimensionality usually associated to recommender models also increments these challenges. Research efforts on the scalability of SHAP and its relevance in sparse contexts is then important to balance interpretability with efficiency in practical scenarios.

5 Case Study

This section discusses the initial steps that would be necessary to deploy the general framework discussed in the previous section, for providing SHAP-based explanations of the recommendation outputs.

The selected scenario will be the group recommendation approach and dataset previously introduced by Perez-Almaguer et al. [28], for group restaurant recommendation in Havana City. Herein, the proposed group recommendation approach includes a hybridization between preference aggregation-based and recommendation-based approaches, as well as data imputation, feature weighting, and automatic selection of the aggregation methods [43]. Below, we subsequently discuss how the different steps of our framework introduced in the previous section, would be tailored to the current context.

1. *Identify item features and associated output for individuals*
   The referred work characterizes restaurants according to five distinctive attributes based on TripAdvisor as information source, as described in Perez-Almaguer et al. [28]. These attributes and their respective possible values are:

   a. Prize: cheap, average, expensive.
   b. Food and service quality: low, medium, high.
   c. Type of cuisine: Cuban, Spanish, Italian Basic, Italian Exclusive, Other.
   d. Ranking: Top 20, Top 100, Other.
   e. Popularity: High (200+ opinions), Medium (100+ opinions), Low (Other number of opinions).

   Herein, for generating top n recommendations the proposed content-based group recommendation approach [28] intermediately calculates an individual score for each group member, and subsequently calculates an aggregated score for the group. Both scores will be, respectively, regarded as the outputs for the calculation of SHAP-based feature contribution and ranking at the individual and the group level.
2. *Feature contribution and ranking for each individual user*
   For accomplishing this goal, we identify feature importance as it provides a high-level summary of which variables the model relies on most to make predictions. In this case, we use CatBoost's feature importance method with the prediction values change type, which measures the average change in the model's output when a feature's value is varied, and for specific case of categorical features. This metric reflects the relative contribution of each feature across all samples in the dataset. While it does not show how a feature influences predictions, it is useful for discriminating the list of relevant features. It is also used for identifying redundant or irrelevant features, and serves as a valid starting point for further analysis.
   Figure 9 illustrates the feature importance for the individual recommendation score generated for each group member. Here it is clearly shown that *Type_of_cuisine* was the feature with the highest average impact in model output magnitude; having also *Ranking* and *Food_service_quality* a high importance. On the other side, *Prize* had an almost null impact.
   As a second stage of this step, we will go through performing SHAP dependence plots across features. Particularly, it plots the SHAP value (i.e., the marginal contribution of a feature to the prediction) against the actual value of the feature, revealing whether a feature's influence is positive or negative, and whether it is linear or non-linear. This visualization can also highlight interactions with other variables if a secondary color-coded feature is added. Unlike feature importance, which gives a global overview, SHAP dependence plots provide local interpretability by showing how specific input values translate into prediction changes across the dataset.
   Figure 10 illustrates the SHAP dependence plots the more relevant features for the recommendations (i.e. Type_of_cuisine versus Ranking and Type_of_cuisine

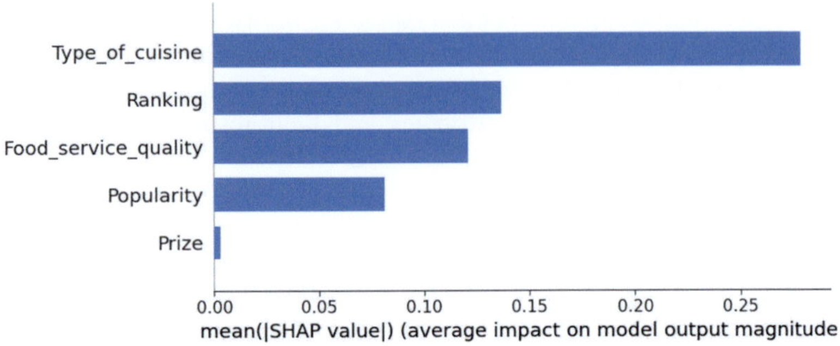

Fig. 9 Feature importance for the recommendation scores associated to the individual users in the groups

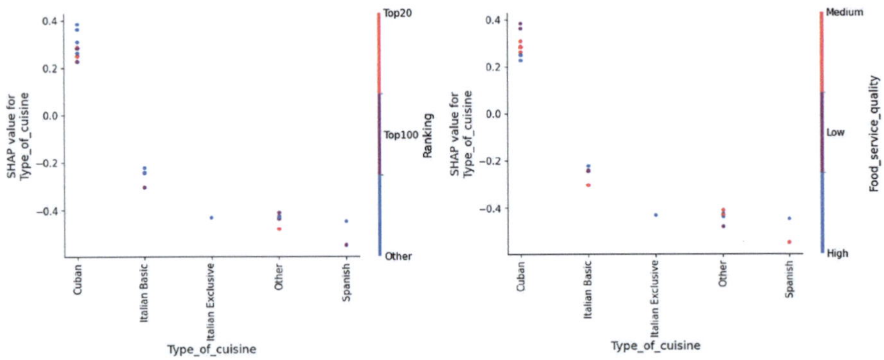

Fig. 10 SHAP dependence plots for the recommendation scores associated to the individual users in the groups

versus Food_service_quality). Herein, it can be appreciated that Cuban cuisine out of top 100 restaurants, interestingly has the highest values and benefiting best the recommendation scores. Concerning Food_service_quality, it is surprisingly to notice that the best expected scores are associated to low quality services, even though for the three possible values of the attributes positive SHAP values are obtained. On the other hand, we can identified that for the remaining types of cuisine, negative SHAP values are obtained, indicating that such attribute values globally contribute to a lower score.

3. *Feature ranking at the group level*
As previously discussed, one of the distinctive approach of the framework we are presenting across this contribution is the dual screening of explanations at the individual and group level. While in the previous step we explore feature ranking and SHAP dependence plots based on the individual recommendation scores, now we will be focused on performing the same graphics but regarding the group, aggregated scores.

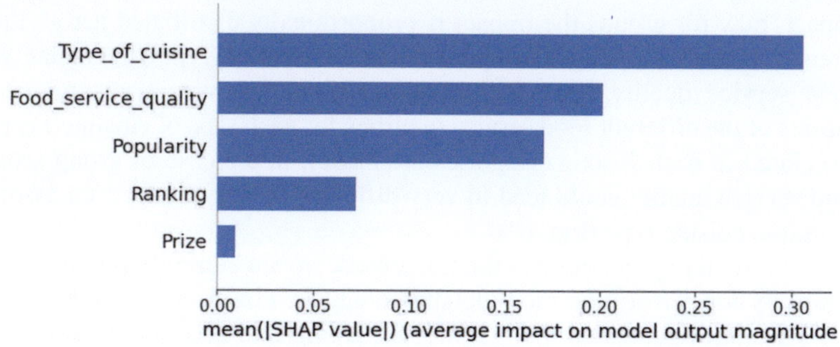

Fig. 11 Feature importance for the recommendation scores associated to the groups

Figures 11 and 12 illustrate the feature importance and the SHAP dependence plots, associated to the group scores. Here we have obtained a feature importance graph that is different in relation to the individual case. For this group scenario, type_of_cuisine was associated to an even larger value, while food_service_quality was the second feature with the higher importance values. Furthermore, Ranking receives less importance for the group case, and finally in a similar way to individuals, Prize had the lower importance value. We think that this issue highlights the complexity of the group recommendation approach used [28], that models the group as more than the simpler sums of its individuals. In contrast, for the case of the SHAP dependence plots, we have found more notable differences in relation to the individual case, mainly regarding the magnitude of the features' effect over the final score. Herein, the magnitude of the effect of the Cuban cuisine feature is lower in relation to the individual case. Also for Cuban cuisine, while for individuals the *Other* ranking had the higher

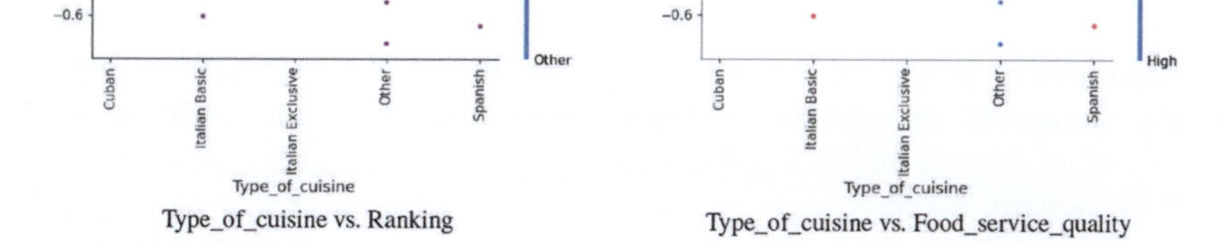

Fig. 12 SHAP dependence plots for the recommendation scores associated to the groups

impact, now for groups the impact is proportionally distributed across the different values of Ranking. Furthermore, in the case of Type_of_cuisine versus Food_service_quality, while for individuals it can be appreciated similar contributions of the different food service qualities for each type of cuisine (i.e. points associated to each X-axis categories are close), in the case of group score the food service quality could lead to very different contributions, even associated to a same cuisine type (Fig. 12).
It is important to point out that the framework we are currently proposing, gives flexibility for performing more detailed analysis. For example, both in Steps 2 and 3 we have generated the graphics by taking into account the whole set of generated scores. Possibly, taken into account only the scores of some specific users or groups, would lead to even more interesting findings.

4. *Generate the final explanations at individual and group level*
Finally, the previously identified findings would lead to some explanations concerning the recommendation process. Some of the text-based explanations for adding trustworthiness to the recommendation process could be:

   - Type of cuisine and food service quality are the attributes that mainly moderate the recommendation process.
   - Cuban cuisine usually leads to high recommendation scores.
   - Some attribute values as Type_of_cuisine = Other, contributes particularly negative to the recommendation scores at the group level.
   - At the individual score case, the contribution of different types of cuisine is not sensitive to the ranking and quality values.

Overall, the obtained results illustrate that our proposed two-stage framework is able to capture valuable knowledge both at the individual and group level, that can be used for building rich explanations.

Our next future works will be focused on exploring the score of most specific users or groups, aiming on identifying more distinctive explanations of the obtained results.

6 Conclusions

This contribution starts with the underexplored potential of SHAP values in explaining the outputs of both individual and group recommender systems, highlighting the limitations of current analyses in this area.

In this regard, the analysis of background and related works on post-hoc explanations for recommender systems demonstrated the presence of several established research directions, including the generation of post-hoc explanations using data mining techniques such as association rule mining or decision tree induction [46], and the use of generative text-based explanations [2]. However, despite the use of

SHAP by some researchers for this purpose, there is an important gap in the development of a comprehensive framework and methodological guidelines for conducting systematic research with this objective in mind. A PRISMA-based systematic review has been performed for supporting this statement.

In the current work, we have proposed a general framework for explaining individual and group recommendations under the umbrella of the SHAP framework. It is composed of four main stages: (1) Identify the item features and their associated output, (2) Calculate the feature contribution and ranking for each individual user, (3) Generate the feature ranking at the group level, and (4) Generate the final explanations at individual and group level. Furthermore, special attention has been given to the necessity of establishing appropriate evaluation frameworks for both offline analysis and user studies.

A case study has been developed, being focused on the use of the presented framework in a specific group recommendation scenario, showing how it would be able to add transparency to the individual and group recommendation scenario. Next future works will be focused on: (1) exploring the use of the presented procedure over specific users and groups instead of the whole available data, (2) evaluating the effect of the generated explanations across user studies, and (3) the benchmarking with other explainability techniques.

We hope that the current contribution will help to consolidate the research efforts to be made in the near future regarding the use of SHAP to support explanations for individual and group recommendations.

Acknowledgements Raciel Yera is supported by the European Union's Horizon Europe research and innovation program under the Marie Sklodowska-Curie Grant Agreement No.101106164. Bapi Dutta acknowledges the support of the Spanish Ministry of Science, Innovation and Universities, and the Spanish State Research Agency through the Ramón y Cajal Research grant (RYC2023-045020-I), Spain.

References

1. Adomavicius, G., Tuzhilin, A.: Toward the next generation of recommender systems: a survey of the state-of-the-art and possible extensions. IEEE Trans. Knowl. Data. Eng. **17**(6), 734–749 (2005)
2. Ariza-Casabona, A., Boratto, L., Salamó, M.: A comparative analysis of text-based explainable recommender systems. In: Proceedings of the 18th ACM Conference on Recommender Systems, pp. 105–115 (2024)
3. Ariza-Casabona, A., Salamó, M., Boratto, L., Fenu, G.: Towards self-explaining sequence-aware recommendation. In: Proceedings of the 17th ACM Conference on Recommender Systems, pp. 904–911 (2023)
4. Ataei, P., Litchfield, A.: The state of big data reference architectures: a systematic literature review. IEEE Access (2022)

5. Barile, F., Draws, T., Inel, O., Rieger, A., Najafian, S., Ebrahimi Fard, A., Hada, R., Tintarev, N.: Evaluating explainable social choice-based aggregation strategies for group recommendation. User Model. User-Adap. Inter. **34**(1), 1–58 (2024)
6. Chanson, A., Labroche, N., Verdeaux, W.: Towards local post-hoc recommender systems explanations. In: Proceedings of the 23rd International Workshop on Design, Optimization, Languages and Analytical Processing of Big Data (DOLAP) (2021)
7. Coba, L., Confalonieri, R., Zanker, M.: Recoxplainer: a library for development and offline evaluation of explainable recommender systems. IEEE Comput. Intell. Mag. **17**(1), 46–58 (2022)
8. Contreras, D., Salamó, M., Boratto, L.: Integrating collaboration and leadership in conversational group recommender systems. ACM Trans. Inf. Syst. (TOIS) **39**(4), 1–32 (2021)
9. Critical Appraisal Skills Programme (CASP): CASP qualitative studies checklist (2018). https://casp-uk.net/casp-tools-checklists/. Accessed: 28 Apr 2025
10. Dara, S., Chowdary, C.R., Kumar, C.: A survey on group recommender systems. J. Intell. Inf. Syst. **54**(2), 271–295 (2020)
11. Pessemier, T., Dooms, S., Martens, L.: Comparison of group recommendation algorithms. Multimedia Tools Appl. **72**, 2497–2541 (2014)
12. Díaz, S.B., Coussement, K., De Caigny, A.: From collaborative filtering to deep learning: advancing recommender systems with longitudinal data in the financial services industry. Eur. J. Oper. Res. (2025)
13. Ekstrand, M.D., Riedl, J.T., Konstan, J.A., et al.: Collaborative filtering recommender systems. Found. Trends® Hum. Comput. Inter. **4**(2), 81–173 (2011)
14. Felfernig, A., Boratto, L., Stettinger, M., Tkali, M.: Group Recommender Systems: An Introduction. Springer, Berlin, Incorporated (2018)
15. Gatt, A., Krahmer, E.: Survey of the state of the art in natural language generation: core tasks, applications and evaluation. J. Artif. Intell. Res. **61**, 65–170 (2018)
16. Guo, M., Yan, N., Cui, X., Hughes, S., Jadda, K.A.: Online product feature recommendations with interpretable machine learning (2021). arXiv preprint arXiv:2105.00867
17. Hancock, J.T., Khoshgoftaar, T.M.: Catboost for big data: an interdisciplinary review. J. Big Data **7**(1), 94 (2020)
18. Ismailoglu, F.: Aggregating user preferences in group recommender systems: a crowdsourcing approach. Decis. Supp. Syst. **152**, 113663 (2022)
19. Kitchenham, B.A., Pfleeger, S.L., Pickard, L.M., Jones, P.W., Hoaglin, D.C., El Emam, K., Rosenberg, J.: Preliminary guidelines for empirical research in software engineering. IEEE Trans. Softw. Eng. **28**(8), 721–734 (2002)
20. Koren, Y., Rendle, S., Bell, R.: Advances in collaborative filtering. In: Recommender Systems Handbook, pp. 91–142 (2021)
21. Le, T.T.H., Kim, H., Kang, H., Kim, H.: Classification and explanation for intrusion detection system based on ensemble trees and shap method. Sensors **22**(3), 1154 (2022)
22. Li, L., Zhang, Y., Chen, L.: Personalized transformer for explainable recommendation. In: Proceedings of the 59th Annual Meeting of the Association for Computational Linguistics and the 11th International Joint Conference on Natural Language Processing (Volume 1: Long Papers), pp. 4947–4957 (2021)
23. Li, L., Zhang, Y., Chen, L.: Prompt distillation for efficient llm-based recommendation. In: Proceedings of the 32nd ACM International Conference on Information and Knowledge Management, pp. 1348–1357 (2023)
24. Lu, J., Wu, D., Mao, M., Wang, W., Zhang, G.: Recommender system application developments: a survey. Decis. Supp. Syst. **74**, 12–32 (2015)
25. Nóbrega, C., Marinho, L.: Towards explaining recommendations through local surrogate models. In: Proceedings of the 34th ACM/SIGAPP Symposium on Applied Computing, pp. 1671–1678 (2019)
26. Page, M.J., McKenzie, J.E., Bossuyt, P.M., Boutron, I., Hoffmann, T.C., Mulrow, C.D., Shamseer, L., Tetzlaff, J.M., Akl, E.A., Brennan, S.E., et al.: The Prisma 2020 statement: an updated guideline for reporting systematic reviews. BMJ **372** (2021)

27. Peake, G., Wang, J.: Explanation mining: Post hoc interpretability of latent factor models for recommendation systems. In: Proceedings of the 24th ACM SIGKDD International Conference on Knowledge Discovery & Data Mining, pp. 2060–2069 (2018)
28. Pérez-Almaguer, Y., Carballo-Cruz, E., Caballero-Mota, Y., Yera, R.: Exploring content-based group recommendation for suggesting restaurants in havana city. J. Univ. Comput. Sci. **30**(1), 106–129 (2024)
29. Pérez-Almaguer, Y., Yera, R., Alzahrani, A.A., Martínez, L.: Content-based group recommender systems: a general taxonomy and further improvements. Expert Syst. Appl. **184**, 115444 (2021)
30. Pugoy, R.A., Kao, H.Y.: Unsupervised extractive summarization-based representations for accurate and explainable collaborative filtering. In: Proceedings of the 59th Annual Meeting of the Association for Computational Linguistics and the 11th International Joint Conference on Natural Language Processing (Volume 1: Long Papers), pp. 2981–2990 (2021)
31. Quijano-Sanchez, L., Sauer, C., Recio-Garcia, J.A., Diaz-Agudo, B.: Make it personal: a social explanation system applied to group recommendations. Expert Syst. Appl. **76**, 36–48 (2017)
32. Ribeiro, M.T., Singh, S., Guestrin, C.: "why should I trust you?" explaining the predictions of any classifier. In: Proceedings of the 22nd ACM SIGKDD International Conference on Knowledge Discovery and Data Mining, pp. 1135–1144 (2016)
33. Rong, Y., Leemann, T., Nguyen, T.T., Fiedler, L., Qian, P., Unhelkar, V., Seidel, T., Kasneci, G., Kasneci, E.: Towards human-centered explainable AI: a survey of user studies for model explanations. IEEE Trans. Patt. Anal. Mach. Intell. **46**(04), 2104–2122 (2024)
34. Shmaryahu, D., Shani, G., Shapira, B.: Post-hoc explanations for complex model recommendations using simple methods. In: IntRS@ RecSys, pp. 26–36 (2020)
35. Shrikumar, A., Greenside, P., Kundaje, A.: Learning important features through propagating activation differences. In: International Conference on Machine Learning, pp. 3145–3153. PMIR (2017)
36. Tintarev, N., Masthoff, J.: Evaluating the effectiveness of explanations for recommender systems: methodological issues and empirical studies on the impact of personalization. User Model. User-Adap. Inter. **22**, 399–439 (2012)
37. Tintarev, N., Masthoff, J.: Beyond explaining single item recommendations. In: Recommender Systems Handbook, pp. 711–756. Springer, Berlin (2022)
38. Tran, T.N.T., Atas, M., Felfernig, A., Le, V.M., Samer, R., Stettinger, M.: Towards social choice-based explanations in group recommender systems. In: Proceedings of the 27th ACM Conference on User Modeling, Adaptation and Personalization, pp. 13–21 (2019)
39. Wang, P., Cai, R., Wang, H.: Graph-based extractive explainer for recommendations. In: Proceedings of the ACM Web Conference 2022, pp. 2163–2171 (2022)
40. Waterschoot, C., Toledo, R.Y., Tintarev, N., Barile, F.: With friends like these, who needs explanations? Evaluating user understanding of group recommendations. In: Proceedings of the 33rd ACM Conference on User Modeling, Adaptation and Personalization (2025). https://doi.org/10.1145/3699682.3728345
41. Yalcin, E., Ismailoglu, F., Bilge, A.: An entropy empowered hybridized aggregation technique for group recommender systems. Expert Syst. Appl. **166**, 114111 (2021)
42. Yera, R., Alzahrani, A.A., Martínez, L.: Exploring post-hoc agnostic models for explainable cooking recipe recommendations. Knowl. Based Syst. **251**, 109216 (2022)
43. Yera, R., Alzahrani, A.A., Martínez, L.: A fuzzy content-based group recommender system with dynamic selection of the aggregation functions. Int. J. Approximate Reasoning **150**, 273–296 (2022)
44. Yera, R., Alzahrani, A.A., Martínez, L., Rodríguez, R.M.: A systematic review on food recommender systems for diabetic patients. Int. J. Environ. Res. Public Health **20**(5), 4248 (2023)
45. Yera, R., Carballo-Cruz, E., Maroto-Martos, J.: Group recommender systems for tourism: current state and future directions. Intell. Decis. Technol. (2025). https://doi.org/10.1177/18724981251335098

46. Yera, R., Martínez, L.: Lore4grouprs: explaining group recommendations supported by a local rule-based approach. In: International Conference on Intelligent Data Engineering and Automated Learning, pp. 300–312. Springer, Berlin (2024)
47. Yin, F., Fu, R., Feng, X., Xing, T., Ji, M.: An interpretable neural network TV program recommendation based on SHAP. Int. J. Mach. Learn. Cybern. **14**(10), 3561–3574 (2023)
48. Zhang, Y., Chen, X., et al.: Explainable recommendation: a survey and new perspectives. Found. Trends® Inf. Retrieval **14**(1), 1–101 (2020)
49. Zhong, J., Negre, E.: SHAP-enhanced counterfactual explanations for recommendations. In: Proceedings of the 37th ACM/SIGAPP Symposium on Applied Computing, pp. 1365–1372 (2022)

A *P*-based Multiple Criteria Ordinal Classification Method Where Classes Are Described by Subsets of "Non-central" Profiles

Eduardo Fernández, José Rui Figueira, Jorge Navarro, Efrain Solares, Anabella Hernández-Molina, and Víctor Molina-Morejón

Abstract A new multiple criteria ordinal classification method is proposed. Classes are described by two disjoint subsets of profiles, which are representative of their least and the most preferred zones. The paper argues in favor of this way to represent the classes. This is particularly important when the elicitation of either "central" or limiting profiles is a hard task for the Decision Maker (DM). This proposal allows us to work with subsets of profiles very close to the boundaries between adjacent classes. An asymmetric preference relation, P, is used to compare actions against those representative subsets composed by "non-central" profiles. Based on a relational system, (P, D), the method fulfills all the paradigmatic consistency properties of multiple criteria sorting methods, including symmetry through the transposition operation. The method is quite general since P is any asymmetric binary preference relation and D is a transitive relation subject to mild requirements. Its main

E. Fernández · E. Solares (✉) · V. Molina-Morejón
Faculty of Accounting and Administration, Universidad Autónoma de Coahuila, Saltillo, Mexico
e-mail: efrain.solares@uadec.edu.mx

E. Fernández
e-mail: eduardo.fernandez@uadec.edu.mx

V. Molina-Morejón
e-mail: ictormolina@uadec.edu.mx

J. R. Figueira
CEG-IST, Instituto Superior Técnico, Universidade de Lisboa, Lisbon, Portugal
e-mail: figueira@tecnico.ulisboa.pt

J. Navarro
Faculty of Informatics, Universidad Autónoma de Sinaloa, Culiacán, Mexico
e-mail: jnavarro@uas.edu.mx

A. Hernández-Molina
Universidad Autónoma de Sinaloa, Culiacán, Mexico
e-mail: anabella.hernandez@ciees.edu.mx

W. Pedrycz et al. (eds.), *Decision-Making in Computational Intelligence-Based Systems*, Studies in Systems, Decision and Control 628,
https://doi.org/10.1007/978-3-032-04056-5_2

practical importance consists of allowing a better characterization of classes, and consequently, more appropriate assignments.

Keywords Multiple criteria analysis · Ordinal classification · Asymmetric preference · Characteristic profiles

1 Introduction

Classification is the process of assigning objects to pre-existing classes. Ordinal classification differs from nominal classification because, in the former, objects are assigned to a set of ordered classes (*e.g.*, from worst to best). The main instance of ordinal classification is multiple criteria ordinal classification (multiple criteria sorting), where the objects to be classified are multiple criteria performance levels also called *actions* in the related literature. Among many examples of multiple criteria sorting problems, we can mention the assessment of (i) country risk, (ii) the probability of success of research projects, (iii) government performance, (iv) quality of financial stocks, and (v) quality of candidates to a certain position.

So-called Multicriteria Decision Aiding poses several methods that allow to address multiple criteria sorting; these methods can be grouped into four broad paradigms: value-function models (e.g., UTADIS, AHP-Sort), relational-based models (ELECTRE TRI-B, ELECTRE TRI-nB, ELECTRE TRI-nC, ELECTRE-Sort, PROMETHEE-TRI), rule-induction or decision-rule models (DRSA), and interactive or learning-oriented procedures [1, 2]. The present work sits in the outranking-based, ordinal-sorting paradigm. Nevertheless, it is important to emphasize that the method remains effective regardless of the specific form in which the P-relation is constructed. The only requirement for this relation is that it must be asymmetric, regardless of how the multiple criteria are aggregated. Moreover, the method is applicable even beyond the multiple criteria framework (see the example in Sect. 3).

Perhaps the most popular methods for multiple criteria ordinal classification belong to the ELECTRE TRI family. These methods use the outranking relation that was originally proposed in the context of the ELECTRE III method [3]. The ELECTRE TRI-B and ELECTRE TRI-nB methods make use of limiting actions (*profiles*) to characterize the frontier between adjacent classes, (also called *categories* in the related literature) [4–6]. These methods are composed by two decision rules. The pseudo-conjunctive rule uses an outranking relation S. The pseudo-disjunctive rule uses an asymmetric preference relation (the asymmetric part of S). Both procedures fulfil a set of consistency properties (Unicity, Homogeneity, Independence, Conformity, Monotonicity, Stability) firstly analyzed by [5]. Among others, methods that use limiting profiles have been proposed by [7–13]. However, the existence of the limiting profiles in several real-world problems has been questioned [14, 15]. In many applied settings (e.g., credit-risk scoring, research-project screening or staff selection) the boundaries between successive classes are intrinsically fuzzy, and experts seldom identify single "typical" alternatives that lie exactly on such boundaries.

Motivated by reducing the DM cognitive effort in setting limiting profiles, ELECTRE TRI-C and ELECTRE TRI-nC compare actions against characteristic actions (*profiles*) by using the outranking relation S [15, 16]. In ELECTRE TRI-C, the single characteristic profile is the most representative action of the related class. In ELECTRE TRI-nC, subsets of representative or informative profiles are used to characterize categories. The ELECTRE-based classification methods have been extended to different frameworks by [10, 17, 18].

Methods that use representative profiles to characterize the classes have been proposed, among others, by [10, 13, 14, 19, 20]. Requiring the decision maker to determine "central" or limiting profiles for every category is demanding. In practice that demand is hard to meet, since real-world classes often have fuzzy borders, and experts can rarely name perfectly representative cases. The new method, that will be detailed below, addresses this gap by letting each category be represented by two sets of non-central profiles, one from the class's least-preferred zone and one from its most-preferred zone. By providing reference profiles near the extremes, the decision maker's cognitive effort is reduced since this is more natural than defining a crisp midpoint. Assignments are made by comparing alternatives with those profile sets through an asymmetric preference relation P. Despite the simpler input, the procedure still satisfies the standard consistency properties (unicity, independence, conformity, monotonicity and stability). In short, the approach reduces elicitation effort, captures categories more faithfully, and delivers more reliable classifications when data are scarce or the decision maker's preferences are imprecise.

A wide generalization of methods that are similar to ELECTRE TRI was presented in [21]. That paper proposed new approaches based on comparing actions against either characteristic or limiting profiles, by using either general reflexive or asymmetric preference relations. Here, we are particularly interested in the approach that uses representative profiles.

Obviously, the capability of this kind of methods to suggest appropriate assignments depends on how appropriately the category is characterized by its representative profiles. It is interesting to note that little (if any) attention has been dedicated to analyzing this important issue. When the authors of [21] describe the method based on comparing actions with characteristic profiles through a reflexive preference relation, the characteristic profiles are defined only by two features: (i) being representative or informative of their category, and (ii) fulfilling certain separability conditions that are required to satisfy the set of consistency properties. There is no condition on the profiles belonging to the same category; the pairs of profiles can be indifferent, incomparable, or they can fulfill a strict preference relation. The same statement is valid for ELECTRE TRI-nC [16] and several similar methods [10, 17, 18]. Other methods do not even impose separability conditions (*e.g.*, [22]). A curious exception is the method that compare actions with characteristic profiles using an asymmetric preference, P, in [21]; in this approach, the pairs of profiles cannot fulfill a P-relation.

As an initial motivation of the importance of an appropriate distribution of profiles, consider that the DM is using the method by [21] to suggest assignments of actions to ordered categories in the sense of increasing preferences. In the method by [21], when a profile of a class, namely C_h, is indifferent[1] to an action x, C_h is a possible assignment for the action. It is like if x were "attracted" by the profile to be assigned to C_h. Let R_k and R_{k-1} be the subsets of representative profiles that describe classes C_k and C_{k-1}. Suppose that both subsets are composed of "non-central" profiles, which belong to the most preferred zone of their respective categories (operational definitions of "central" profiles, "the most preferred", and "the least preferred" zone of a category will be given in Sect. 4; in this introduction, we only provide intuitive ideas of these concepts). Describing each class with two compact sets of non-central profiles offers a cognitively lighter yet empirically richer representation.

Let x be an action that should be assigned to C_k but belongs to the least preferred zone of its class. If x is both indifferent to R_{k-1} and not indifferent to R_k, the action would be assigned to C_{k-1} by the method of [21]. This would be an inappropriate assignment due to an ill-characterization of the least preferred zone of C_k. Contrarily, if the subsets of profiles belong to the least preferred zone of the categories, then actions that belong to the most preferred zone of C_k would be assigned to C_{k+1}. This is again a consequence of an inappropriate characterization of the most preferred zone of the class.

Under the above "gravitational" analogy, having profiles in different preference zones is a necessary condition for a good characterization of the related class. When the S-based method in [21] is used, the profiles can be distributed in diverse preference zones, whenever they fulfil the separability conditions with the profiles of other classes. However, in the P-based method, additionally to the separability conditions, the pairs of profiles cannot fulfill a P-relation; otherwise, the conformity property (stating that each profile has to be classified into the category to which it belongs) is not satisfied by the method. But, if the pair of profiles should be indifferent or incomparable, to achieve a good characterization of the most and the least preferred zones of the categories would be very troublesome.

Being P stronger than S, the DM can be interested in modeling her preferences using P instead of S. Thus, a P-based ordinal classification method is required with capacity to represent categories through:

(a) Only non-central profiles, when the DM (i) is not able to define any central action; or (ii) explicitly prefers to focus on the most and least preferred zones to obtain a richer description of the category.
(b) A combination of central and non-central profiles, when the DM (i) can identify some central profiles but also wants to capture the preference extremes, or (ii) is uncertain whether the available profiles are central, non-central, or a mix of both.
(c) Some categories are described by only central profiles, while others by non-central actions (perhaps combined with some central ones).

---

[1] In [21] the indifference relation is defined regarding a relation $Q = \{S, P\}$. That is, "x is indifferent to y" $\Leftrightarrow$ *xQy and yQx*.

In this paper we propose a new multiple criteria ordinal classification method based on comparing actions against characteristic profiles through a general asymmetric preference relation. The main distinctive feature of this method regarding the *P*-based proposal in [21] consists of using two disjoint subsets of characteristic profiles in each category. Each subset is representative of the least, (respectively, most), preferred zone of the category. Profiles in the "best" zone can be preferred to the ones in the "worst" zone of the category, since any internal differences fall below the relevance thresholds unless finer distinctions are purposely introduced. The subsets can include profiles that are neither in the most nor in the least preferred zone. Thus, the DM, in collaboration with the decision analyst, can achieve a better characterization of classes than using the *P*-based method by [21].

This proposal keeps the generality of the approach in [21]. For this purpose, we work with a relational system (P, D) where P is any asymmetric preference relation, and D is a transitive relation on the decision set. However, the new method is more general than the one by [21] due to its wide flexibility to handle different types of profiles, thus allowing a better characterization of the different preference zones of each category. This flexibility is particularly important to facilitate the elicitation of appropriate profiles from either a large set of known assignment examples or a set of actions for which the DM is comfortable making assignments.

The new method fulfills the consistency properties first established by [5] for ELECTRE TRI-B, revisited by [16] for ELECTRE TRI-nC, and generalized by [21]. These properties have become paradigmatic for multiple criteria ordinal classification.

The paper structure follows: Sect. 2 outlines the method by [21], which is the main motivating precedent of the new approach. Section 3 presents a motivating example, illustrating the limitation of the previous approach. The new method is described in Sect. 4 (fundamental concepts, requirements, assignment rules, and consistency properties). The example of Sect. 3 is revisited by using the new method (Sect. 5). Finally, some concluding remarks are discussed in Sect. 6. To improve the paper readability, most of the mathematical proofs are placed in Appendix.

2 Background

Fernández et al. [21] defined a relational system (P, D) which fulfills the following conditions. Let A denote a set of actions.

- P and D are defined on $A \times A$.
- D is a transitive relation on A.
- P is an asymmetric preference relation.
- $\forall (y, z, w) \in A \times A \times A$:

  a. wPy and $yDz \Rightarrow wPz$.
  b. wDy and $yPz \Rightarrow wPz$.

Fernández et al. [21] does not restrict the set A, nor the nature of the P relation, thus keeping a great generality. When the elements of A are assessed through multiple criteria, if D is the Pareto dominance relation (assuming that higher criterion scores are preferred, $yDz \Leftrightarrow g_j(y) \geq g_j(z)$ for all evaluation criteria g_j, and $yDz \Leftrightarrow g_j(y) > g_j(z)$ for at least one g_j), the ordered pair (P,D) satisfies the above conditions.

Each category C_k is related to a set R_k of representative profiles, $r_{k,j}$, $j = 1, \ldots$, *card* (R_k), fulfilling:

a. For $h > k$, there is no pair (z, w) in $R_k \times R_h$ that fulfills zPw.
b. For $h > k$, for each y in R_k, there is a z in R_h, fulfilling zPy.
c. For $k > h$, for each y in R_k, there is a z in R_h fulfilling yPz.
d. For $k = 1,\ldots M$, there is no pair (z, w) in $R_k \times R_k$ that fulfills zPw.
e. For $h > k$, for each y in R_k, there is a z in R_h, fulfilling zDy.
f. For $k > h$, for each y in R_k, there is a z in R_h, fulfilling yDz.

Note that Condition iv. makes very difficult having profiles in the least and the most preferred zones of the category. If the profiles are "concentrated" in the most (respectively, least) preferred zone, the least (resp., most) preferred zone would not be appropriately characterized. Thus, the method should use rather "central profiles", whose concept will be operationally defined in Sect. 4.

The P-relation involving profile sets and actions is defined as

(a) xPR_k iff there is a y belonging to R_k, fulfilling xPy.
(b) R_kPx iff there is a y in R_k, fulfilling yPx.

Fernández et al. [21] proposed two assignment rules, which are symmetric through the transposition operation (reversing both the order of categories and the sense of preferences). The rationality of this feature has been defended by [7, 15, 23]. The descending rule finds the first subscript k for which xPR_k, assigning x into C_k or C_{k+1}. In the ascending rule, the first h such that R_hSx is identified, assigning x into C_{h-1} or C_h. The conjoint assignment rule suggests assigning x into a class in the range C_h–C_k (or vice versa).

As the ELECTRE TRI family and other related methods, the proposal by [21] fulfills a set of consistency properties (Unicity, Homogeneity, Independence, Conformity, Monotonicity, Stability) firstly analyzed by [5].

3 A Motivating Example

The following toy data set is deliberately minimal. It shows how an action x can fall into an "assignment grey zone" when categories are represented only by central or limiting profiles, and how the proposal removes that ambiguity. We first run the P-based method in [21] step-by-step to reveal the mis-classification; then, in Sect. 5, we introduce two non-central profile subsets for each class and repeat the procedure, obtaining an unambiguous assignment.

Table 1 Actions and profiles

Action	Category	g_1	g_2	g_3	g_4	g_5
$r_{2,1}$	C_2	5	5	5	4	4
x	?	3	3	4	3	3
b	(C_1, C_2)	3	3	3	2	2
$r_{1,1}$	C_1	2	2	1	1	1

Table 1 shows actions described by five assessment criteria (g_1, g_2, g_3, g_4, g_5) (whose values increase with preference). Each score is the assessment done by an individual decision maker belonging to a group of five people. Suppose that the possible assessments are integer numbers in the range [0, 5]. Each action is assigned to one of two ordered categories, (C_2 is the best class), by a "supra decision maker". x is preferred to y if and only if the majority of the group ranks x better than y.

Let us suppose that action b is a limiting profile between C_1 and C_2. Note that $r_{2,1}$ is in the most preferred zone of C_2. $r_{2,1}$ is preferred to action x; action x is preferred to $r_{1,1}$ and b. Being preferred to the limiting profile b, x should be assigned to C_2. We have $r_{2,1}Px$ and $xPr_{1,1}$. Hence, the descending and ascending rules in [21] suggest C_1 and C_2, respectively, as possible assignments for x. So, the method in [21] does not precisely identify the appropriate assignment for x. The least preferred zone of C_2 is not well characterized by $r_{2,1}$; thus, the ambiguity arises because the least-preferred zone of C_2 is not represented by any profile.

4 Comparing Actions and Profiles

In this section we detail a method to be used when the DM is willing to represent each class using some profiles that belong to the most preferred and the least preferred zone of their class. First, we give an operational definition of such concepts:

Let $C = \{C_1, \ldots, C_k, \ldots, C_M\}$ be the set of predefined and ordered classes (assuming increasing preference) ($M \geq 2$). Let us consider an action z belonging to C_k and two alternative propositions:

(A) z may be indifferent with an action belonging to C_{k+1}.
(B) z may be indifferent with an action belonging to C_{k-1}.

If the DM is more confident with Proposition A (respectively, B), we say that z belongs to the most (respectively, least) preferred zone of C_k. In both cases we say that z is a non-central action. If the DM has equal confidence in both propositions, or if she/he rejects both propositions, z is a central action of C_k.

4.1 Requirements

Let us assume that the relational system (*P*, *D*) is the same as in [21] (see Sect. 2). Suppose that each category is described by a subset R_k of reference profiles, which is formed by two disjoint subsets R_k^{Least} and R_k^{Most}. The set $R_k^{Least} = \{r_{k,j}^{Least}, j = 1, \ldots, n_k^{Least}\}$ contains non-central profiles belonging to the least preferred zone of C_k. The set $R_k^{Most} = \{R_{k,j}^{Most}, j = 1,\ldots, n_k^{Most}\}$ contains non-central profiles belonging to the most preferred zone of C_k.

To guarantee the fulfillment of the consistency properties presented by [5] (in the context of ELECTRE TRI-B), and reexamined by [15], the following requirements should be imposed to the set of reference profiles:

Condition 1

Any set *R* of reference profiles must fulfill:

1. For $k = 1, \ldots, M$, for each profile $z \in R_k^{Most}$, there is no $w \in R_k^{Most}$ such that zPw.
2. For $k = 1, \ldots, M$, for each profile $z \in R_k^{Least}$ there is no $w \in R_k^{Least}$ such that wPz.
3. For $k = 1, \ldots, M$, there is no pair $(z,w) \in {R_k}^{Least} \times {R_k}^{Most}$ such that zPw.
4. For $k = 2, \ldots, M$, for each profile $z \in R_k^{Most}$, there is $w \in R_{k-1}^{Most}$ such that zPw.
5. For $k = 1, \ldots, M - 1$, for each profile $z \in R_{k-1}^{Most}$, there is $w \in R_{k+1}^{Least}$ such that wPz.
6. For $k = 2, \ldots, M$, for each profile $z \in R_k^{Least}$, there is $w \in R_{k-1}^{Most}$ such that zPw.
7. For $k = 1, \ldots, M - 1$, for each profile $z \in R_k^{Least}$, there is $w \in R_{k+1}^{Least}$ such that wPz.
8. For $k = 1, \ldots, M - 1$, for each profile $z \in R_k^{Least}$, there is a profile $w \in R_{k+1}^{Least}$ such that wDz.
9. For $k = 2, \ldots, M$, for each profile $z \in R_k^{Most}$, there is a profile $w \in R_{k-1}^{Most}$, such that zDw.
10. For all ordered pairs (k, h) $(h > k)$, for $k = 1, \ldots, M$, there is no pair $(z, w) \in R_k \times R_h$ such that zPw.

Except for Condition 1, no other requirements are imposed to profiles in R_k^{Least} and R_k^{Most}. Some of them can belong to the "central preference zone" whereas others to the most (respectively, least) preferred zone of their category; or they can be very close to the limiting boundary between adjacent classes. Profiles in R_k^{Most} can be preferred to actions in R_k^{Least}, but the contrary is not permitted. Pairs of profiles in the same "preference zone" cannot belong to the *P*-relation.

4.2 Description of the Method

The way to compare actions against subsets of profiles is given by the following definition:

Definition 1 (*P-relation involving profile sets and actions*)

(a) xPR_k iff there is a profile $z \in R_k^{Most}$ fulfilling xPz.
(b) R_kPx iff there is a profile $w \in R_k^{Least}$ fulfilling wPx.

Definition 1 is fundamentally different from the P-relation defined by [21] (see Sect. 2). The relation given by Definition 1 is stronger than the other. Strictly speaking, the relation xPR_k involves comparing an action x against a set of reference profiles R_k. Formally, this notation represents the existence of at least one element $y \in R_k$ such that xPy. This simplified notation is adopted throughout the paper to improve readability. The defined relation explicitly involves the relation P, ensuring that indifferent profiles do not trigger the relation. Thus, the described condition inherently avoids preference ambiguities and preserves internal consistency.

The definition of the P-relation between an action and a set of profiles is intentionally minimal. It only requires the existence of one profile in the set to satisfy the preference condition. This design is consistent with the logic of outranking approaches, such as those in the ELECTRE family, where a single strong argument of preference is often sufficient. Furthermore, the structural requirements imposed in Condition 1 ensure that profile sets are internally coherent and ordered in a way that prevents preference cycles or contradictions. These requirements act as consistency protection, allowing the method to remain logically sound even when working with small or non-central profile sets.

The relation defined above fulfills the following proposition:

Proposition 1

(i) $xPR_k \Rightarrow xPR_h$ for all $h < k$.
(ii) $R_kPx \Rightarrow R_hPx$ for all $h > k$.

Proof

1.i. It is a direct consequence of Definition 1(a) and Condition 1–9 combined with the property xPy and $yDz \Rightarrow xPz$.
1.ii. It is a direct consequence of Definition 1(b) and Condition 1–8 combined with the property zDy and $yPx \Rightarrow zPx$.

Remark 1

- xPR_k suggests that x belongs to either the most preferred zone of C_k or a class better than C_k. R_kPx suggests that x belongs to either the least preferred zone of C_k or a class worse than this category. Hence, xPR_k combined with R_kPx is contradictory as a result of an inappropriate definition of P.
- $not(xPR_j)$ suggests assigning x to C_k $(k \leq j)$, while $not(R_jPx)$ suggests the assignment to C_k $(k \geq j)$. Then $not(xPR_j)$ combined with $not(R_jPx)$ suggests assigning x to C_j.

Considering the above remarks, we introduce the following definition of binary relations involving sets of reference profiles and actions.

Definition 2 (*Indifference, preference, and incomparability relations involving sets of reference profiles and actions*)

(a) *Indifference*: $xIR_j \Leftrightarrow not(xPR_j)$ and $not(R_jPx)$.
(b) *Strict Preference*: $xP_SR_j \Leftrightarrow xPR_j$ and $not(R_jPx)$; $R_jP_Sx \Leftrightarrow R_jPx$ and $not(xPR_j)$.
(c) *Incomparability*: $xIncR_j \Leftrightarrow xPR_j$ and R_jPx.

Remark 2

- Definitions 2(a) and 2(c) seem to be contradictory with the paper of [21]. Definitions 2(a) (respectively, Definition 2(c)) is related to the second (resp., first) point of Remark 1. The apparent discrepancy is a consequence of a different and stronger P-relation between actions and subsets of profiles.
- Note that P_S is asymmetric. On one hand, xP_SR_j suggests that x should be assigned to a category at least as good as C_j, and if it is assigned to C_j, x should belong to the most preferred zone of this category, can be close to the limiting boundary between C_j and C_{j+1}. On the other hand, R_jP_Sx indicates the assignment to a category at most as good as C_j; and if x is assigned to C_j, it should belong to the least preferred zone of C_j, can be close to the limiting boundary between C_{j-1} and C_j. So, xIR_j for $k' \le j \le k''$ supports assigning x into a class in the range $C_{k'}$–$C_{k''}$. $xIncR_j$ for $k' \le j \le k''$ combined with xP_sR_j for $j < k'$ and R_jP_sx for $j > k''$ suggests the range $C_{k'-1}$–$C_{k''+1'}$ as the possible assignments for x.

Similarly to [21], the new method is composed of two decision rules:

Definition 3 (*Descending assignment rule*)

Let P and R follow Definition 1 and Condition 1. Let R_0 be a fictitious "anti-ideal" set fulfilling xPR_0 for all x.

a. For $k = M, \ldots, 0$, find the first k fulfilling xPR_k.
b. Assign x to C_M when $k = M$.
c. Assign x to C_1 when $k = 0$.
d. Define C_k and C_{k+1} as appropriate classes to assign x when $0 < k < M$.

The dual rule is defined as follows to achieve symmetry through the transposition operation.

Definition 4 (*Ascending assignment rule*)

Le P and R follow Definition 1 and Condition 1. Let R_{M+1} be a fictitious "ideal" set fulfilling $R_{M+1}Px$ for all x.

a. For $k = 1, \ldots, M + 1$, find the first k for which R_kPx.
b. Assign x to C_1 when $k = 1$.
c. Assign x to C_M when $k = M + 1$.
d. Define C_k and C_{k-1} as appropriate classes to assign x when $1 < k < M + 1$.

It should be underlined that, unlike [21], the descending rule is based on the comparison with R_k^{Most}, whereas the ascending one on the comparison with R_k^{Least}. These rules are identical to those in [21], except for the definitions of xPR_k and R_kPx. Note also that the rules have symmetry related to the transposition operation. Given their symmetry regarding the transposition operation, the above rules should be used conjointly. We should derive a conjoint assignment rule supported by Remark 2.

Suppose that, using the descending rule, an action x is assigned to C_k or C_{k+1} and, using the ascending procedure, it is assigned to C_{h-1} or C_h. This includes as particular case when the descending rule assignment is C_1 and the ascending one is C_M. The following four cases can be described.

1. First: $h = k + 1$ (same recommendation by both rules); we have xP_sR_k and $R_{k+1}P_sx$. Hence, x may be assigned to C_k or C_h, but it is near the limiting boundary separating these categories.
2. Second: $h = k$ (C_k is the assignment shared by the two rules); we have $xIncR_k$, xP_sR_j for $j < k$ and R_jP_sx for $j > k$. Then any class in the range C_{k-1}–C_{h+1} may be chosen to assign x.
3. Third: $h > k + 1$; we have xIR_j for $k < j < h$; xP_sR_j for $j \leq k$ and R_jP_sx for $j \geq h$; according to Remark 2, any class in the range C_{k+1}–C_{h-1} may be chosen to assign x.
4. Fourth: $h < k$; $xIncR_j$ for $h \leq j \leq k$; xP_sR_j for $j < h$ and R_jP_sx for $j > k$. Any class in the range C_{h-1}–C_{k+1} may be chosen to assign x (see Remark 2).

Definition 5 (*Conjoint assignment rule*)

(a) The classes in the range C_1–C_M may be chosen as possible assignments for x when x is assigned to C_M by the descending rule and to C_1 by ascending rule.

(b) The classes in the range C_k–C_{k+1} may be chosen as possible assignments for x when that is the range recommended by both the descending and ascending rules.

(c) Assume that x is assigned to the range C_{a_low}–C_{a_up} (respectively, C_{d_low}–C_{d_up}) by the ascending (resp., descending) rule. Note that the case $a\_up = d\_up$ and $a\_low = d\_low$ corresponds to Point (b). Then, (i) if $a\_up = d\_low$, take the classes of range C_{a_low}–C_{d_up} as possible assignments for x; (ii) if $a\_up < d\_low$, take the classes of range C_{a_low}–C_{d_up} as possible assignments for x; (iii) if $a\_low \geq d\_up$, then take the classes in the range C_{d_up}–C_{a_low} as possible assignments for x.

Figure 1 shows the general procedure of the proposal.

The computational complexity of the proposed method is $O(n \cdot m \cdot M \cdot s)$, where n is the number of actions to classify, m is the number of criteria, M is the number of classes, and s is the maximum number of profiles per subset R_k^{Most} or R_k^{Least}. For each action, the method compares it with at most $2Ms$ profiles, and each comparison involves evaluating the preference relation P, which typically requires $O(m)$ operations. In practice, since M and s are small, the method is computationally efficient and scales linearly with the number of actions and criteria.

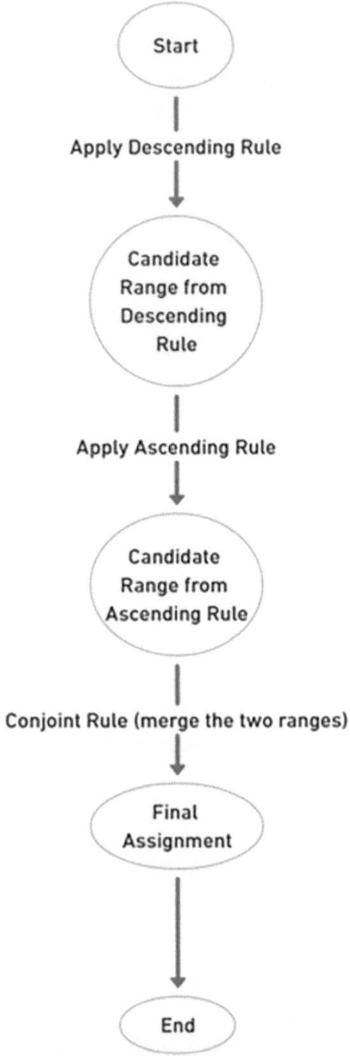

Fig. 1 Procedure of the proposal

4.3 Consistency Properties

In a general framework, the same properties were proved by [21] when an asymmetric preference relation is used to compare actions against either limiting or characteristic profiles. Here we present the analysis in the case of non-central profiles which are representative of the least and the most preferred zones of their category.

Proposition 2
The ascending and descending rules fulfill the following properties:

(i) *Uniqueness*: an action must be allocated to only one range of categories.
(ii) *Independence*: assigning an action to a class does not depend on the assignments of other actions.
(iii) *Homogeneity*: when the performance of two actions is exactly the same regarding the subsets of profiles, the actions have to be assigned to the same range of classes.
(iv) *Conformity*: each profile $z \in R_k$ is assigned to a range containing C_k.
(v) *Monotonicity*: when the descending (respectively, ascending) rule fulfills the following property: when x is assigned to the range C_d–C_{d+1} (resp. C_a–C_{a+1}) and yDx, then y is assigned to a range not lower than C_d–C_{d+1} (resp. C_a–C_{a+1}).

(See the proof in Appendix).

Remark 3 The conjoint assignment rule is also monotonic. That is, if yDx, then y is allocated to a range of categories not lower than the range to which x is assigned to. The proof follows from Monotonicity of Proposition 2 combined with Definition 5.

Proposition 3 (Conformity of the conjoint assignment rule)

If x belongs to R_j, then, using the conjoint assignment rule (Definition 5), x is allocated to C_j.
(See the proof in Appendix).

Remark 4 The conformity property is a theoretical support of the "attraction effect" made by the profiles on the actions that are "preferentially close" to them. Moreover, if x is an action rather central in C_k, it is likely that, according to Definition 1, $not(xPR_k)$ and $not(R_kPx) \Leftrightarrow xIR_k \Rightarrow x$ is assigned either to C_k or a range than contains C_k (Remark 2).

Definition 6 (*Merging*)

This modification in the set of categories combines two contiguous classes, C_k and C_{k+1}, into a new one, denoted C'_k, that will be characterized by two disjoint subsets of profiles $R'_k{}^{Least}$ and $R'_k{}^{Most}$. These subsets are formed by the profiles that described previously the merged classes, in the following way:
$R'^{Most}_k = R^{Most}_{k+1}$ and $R'^{Least}_k = R^{Least}_k$

Definition 7 (*Splitting*)

This operation on the set of categories consists in splitting a category, say C_k, in two new contiguous categories, denoted C'_k and C'_{k+1}. We assume that the new categories are described in the same way as the old class.

Proposition 4 (Analysis of stability)

The rules of Definitions 6 and 7 fulfill:

1. *When joining two contiguous classes* (C_k and C_{k+1}):

   a. If the joined classes do not belong to the range of classes to which x was assigned, then x is allocated to that range.
   b. Regarding the range to which x was assigned, if its lower class is merged, then its upper class is maintained but its lower class will be the new class. Similarly, if its upper class is merged, then its lower class is maintained but its upper class will be the new class.
   c. If the range to which x was assigned contains the joined classes, then x is allocated to a range that contains the new class.

2. *When splitting into two contiguous classes* (C_k *into* C'_k and C'_{k+1}):

   a. If the modified class does not belong to the range of classes to which x was assigned, then x is allocated to that range.
   b. If x has been allocated to a range containing the original class, then there are two options: (i) x is allocated to the range made up of the new classes, and (ii) x is allocated to the range comprising a new class and other class in the original range before the operation.

(See the proof in Appendix).

5 Revisiting the Example of Sect. 3

Let us consider the information shown in Table 2. Table 2 contains several of the actions previously given in Table 1.

$r_{2,2}$ (respectively, $r_{1,2}$) is a non-central profile in the most preferred zone of C_2 (resp. C_1). $r_{2,1}$ and $r_{1,1}$ are non-central profiles, representative of the least preferred zone of their category. Applying the descending rule *not*(xPR_2) and xPR_1; hence, x is assigned to C_1 or C_2. Applying the ascending rule, we have *not*(R_1Px) and *not*(R_2Px). This implies that, using the ascending rule, x is allocated to C_2. Since $h - 1 = k + 1 = 2$, using the conjoint assignment rule, x is appropriately allocated to C_2 (Definition 5).

Table 2 Actions, reference and limiting actions

Action	Category	g_1	g_2	g_3	g_4	g_5
$r_{2,2}$	C_2	5	5	5	4	4
$r_{2,1}$	C_2	3	4	3	4	2
X	?	3	3	4	3	3
B	(C_1, C_2)	3	3	3	2	2
$r_{1,2}$	C_1	3	3	2	2	2
$r_{1,1}$	C_1	2	2	1	1	1

6 Conclusions

This proposal is relevant to multiple criteria ordinal classification problems when the DM's preferences are aggregated in an asymmetric binary preference relation, and any of the following conditions are met: (i) the DM wants to enrich the description of the non-central preference zones of categories; (ii) the DM cannot set central profiles, or she is not able to distinguish between central and non-central profiles, because the related category is not well-defined in her mind. Point (ii) is usually relevant when the DM is an inaccessible person or an entity with ill-defined preferences (*e.g.*, a heterogenous group, or the public opinion).

Unlike the paper by [21] where an asymmetric preference relation is used to compare actions against rather central profiles, here the classes are described by two disjoint subsets, a "lower" (respectively, "upper") subset describing the least (resp., most) preferred zone of the category. Taken separately, profiles are not typical examples of their category, nor are they limiting actions in the boundaries between classes. But the set of profiles, taken as a whole, provides an appropriate characterization of its respective class. Some profiles can represent the central preference zone, but others should be representative of the most and least preferred zones of each category. As a limit case, the lower and upper subsets could be very close to the limiting boundaries that separate adjacent classes.

The method is based on a very general relational system (P, D), where P is any asymmetric preference relation and D is a transitive relation on the decision set. This general feature allows the approach to be used with diverse preference models, even beyond the multiple criteria framework. Under certain requirements on the set of profiles, the proposal fulfills a set of important consistency properties, which are an expression of rationality in relational-based multiple criteria sorting.

Combined with its theoretical interest, the practical importance of the new approach consists of allowing a better characterization of classes, and consequently, more appropriate assignments. Moreover, the flexibility of using non-central profiles may be helpful to choice appropriate profiles from either a set of known assignment examples, or actions for which the DM is comfortable making assignments, whenever the separability requirements are fulfilled.

Several avenues remain open for future research. First, the proposed method could be extended to handle imprecise or uncertain evaluations, for instance, by integrating fuzzy or interval-valued preferences into the P-relation. Second, while the current approach assumes the representative profile subsets are elicited by the decision maker, future work could explore automated or semi-automated learning of these subsets from assignment examples. Third, the method could be adapted to hierarchical or multi-level classification structures, allowing for greater expressiveness in complex decision contexts. Finally, it would be valuable to test the method in real-world applications, such as credit scoring, grant evaluation, or environmental risk assessment, to better understand its practical impact and guide further refinements.

Acknowledgements Eduardo Fernández thanks the Autonomous University of Coahuila. José Rui Figueira acknowledges the Portuguese national funds through the FCT—Foundation for Science and Technology, I.P., under the project UIDB/00097/2020. Jorge Navarro thanks the Autonomous University of Sinaloa. The work of Efrain Solares was supported by the National Board for Science and Technology under Project 321028, project supported by "CONACYT" in 2022-2023.

Appendix

Proof of Proposition 2

The proof of Uniqueness, Independence and Homogeneity are trivial.

Conformity

Let us prove the property for the descending rule:

Suppose that $x \in R_{j2}$. Then, there is no $w \in R_{j2}$ such that xPw (Condition 1.1.) $\Rightarrow$ $not(xPR_j)$.

If $j = 1$.

We know that xPR_0 (Definition 3) $\Rightarrow$ x is assigned to $C_1 = [C_j, C_j]$.

If $j > 1$.

From Condition 1.4. there is $w \in R_{j-1,2}$ such that $xPw \Rightarrow xPR_{j-1}$; also by Condition 1.10. we have $not(xPR_h)$ for all $h > j \Rightarrow not(xPR_h)$ for all $h \geq j$.

$\Rightarrow x$ is allocated to $[C_{j-1}, C_j]$.

Assume that $x \in R_{j1}$. Then there is no $w \in R_{j1}$ such that xPw (Condition 1.2.) $\Rightarrow$ $not(xPR_j)$.

If $j = 1$.

We know that xPR_0 (Definition 3) $\Rightarrow$ x is assigned to $C_1 = [C_j, C_j]$.

If $j > 1$.

From Condition 1.6., there is $w \in R_{j-1,2}$ such that $xPw \Rightarrow xPR_{j-1}$; also by Condition 1.10. we have $not(xPR_h)$ for all $h > j \Rightarrow not(xPR_h)$ for all $h \geq j$.

$\Rightarrow x$ is assigned to $[C_{j-1}, C_j]$.

Proof of Monotonicity

Suppose that x is assigned to $[C_d, C_{d+1}]$ by the descending rule.

$\Rightarrow xPR_d$ (Definition 3) $\Rightarrow$ there is $w \in R_{d2}$ such that xPw (Definition 1(a)). We have yDx and $xPw \Rightarrow yPw \Rightarrow yPR_d \Rightarrow yPR_h$ for all h < d (Proposition 1.i) $\Rightarrow x$ is assigned to $[C_k, C_{k+1}]$ with $k \geq d$.

Suppose that x is allocated to $[C_a, C_{a+1}]$ using the ascending rule.

$\Rightarrow R_{a+1}Px$ and $not(R_hPx)$ for all $h < a + 1$ (Definition 4).

Suppose R_kPy for some $k < a + 1 \Rightarrow$ there is $w \in R_k^{Least}$ such that wPy (Definition 1(b)). We have wPy and $yDx \Rightarrow wPx \Rightarrow R_kPx$ for some $k < a + 1$ this is a contradiction $\Rightarrow not(R_kPy)$ for all $k < a + 1 \Rightarrow x$ is assigned to $[C_k, C_{k+1}]$ with $k + 1 \geq a + 1$.

This completes the proof.

The proof for the ascending rule is justified by the equivalence via the transposition operation.

Proof of Proposition 3

Let us consider $x \in R_j$.

If $j = 1$.

by the descending (ascending) rule x is assigned to $[C_1,C_1]$ ($[C_1,C_2]$) $\Rightarrow x$ is assigned to $[C_1,C_1] = C_1 = C_j$ (Definition 5(c), Point iii).

If $j = M$.

by the descending (ascending) rule x is assigned to $[C_{M-1},C_M]$ ($[C_M,C_M]$) $\Rightarrow x$ is assigned to $[C_M,C_M] = C_M = C_j$ (Definition 5(c), Point iii)).

If $1 < j < M$.

by the descending (ascending) rule x is assigned to $[C_{j-1},C_j]$ ($[C_j,C_{j+1}]$) $\Rightarrow x$ is assigned to $[C_j,C_j] = C_j$ (Definition 5(c), Point iii)).

This completes the proof.

Proof of Proposition 4

1.a

Suppose that x is assigned to $[C'_i,C'_s]$ ($i \leq s$) before the merging.

Descending Rule Analysis

- If $i > k + 1$, we have $xPR_i \Rightarrow xPR_{k+1}$ (Proposition 1. i)) $\Rightarrow xPR'_k$ (Definition 6) $\Rightarrow xPR'_h$ for all $h \leq i - 1$ and $not(xPR'_l)$ for all $l > i - 1$ (Definition 3) $\Rightarrow x$ is assigned to $[C'_{i-1},C'_{s-1}]$ where $C'_{i-1} = C_i$ and $C'_{s-1} = C_s$.
- If $s < k$, we have $not(xPR_{k+1})$ (Definition 3) $\Rightarrow not(xPR'_k)$ (Definition 6) $\Rightarrow xPR'_h$ for all $h \leq i$ and $not(xPR'_l)$ for all $l > i$ (Definition 3) $\Rightarrow x$ is assigned to $[C'_i,C'_s]$ where $C'_i = C_i$ and $C'_s = C_s$.

Ascending Rule Analysis

- If $i > k + 1$then $s \geq k + 2$, we have R_hPx for all $h \geq s$ and $not(R_lPx)$ for all $l < s$ (Definition 4) $\Rightarrow not(R_kPx) \Rightarrow not(R'_kPx)$ (Definition 6) $\Rightarrow R'_hPx$ for all $h \geq s - 1$ and $not(R'_lPx)$ for all $l < s - 1 \Rightarrow x$ is assigned to $[C'_{i-1},C'_{s-1}]$ where $C'_{i-1} = C_i$ and $C'_{s-1} = C_s$.
- If $s < k$, we have R_hPx for all $h \geq s$ and $not(R_lPx)$ for all $l < s$ (Definition 4) $\Rightarrow R_kPx \Rightarrow R'_kPx$ (Definition 6) $\Rightarrow R'_hPx$ for all $h \geq s$ and $not(R'_lPx)$ for all $l < s$ $\Rightarrow x$ is allocated to $[C'_i,C'_s]$ where $C'_i = C_i$ and $C'_s = C_s$.

1.b.

Descending Rule Analysis

- If $i = k + 1 < M$ then xPR_{k+1} and $not(xPR_h)$ for all $h > k + 1$ (Definition 3) $\Rightarrow$ xPR'_k and $not(xPR'_h)$ for all $h > k$ (Definition 6) $\Rightarrow x$ is assigned to $[C'_k,C'_{k+1}]$.
- If $i = k + 1 = M$ then $s = M$ and xPR_M (Definition 3) $\Rightarrow xPR'_{M-1}$ (Definition 6) $\Rightarrow x$ is assigned to $C'_{M-1} = C'_k = [C'_k,C'_k]$.
- If $s = k > 1$ then xPR_{k-1} and $not(xPR_h)$ for all $h > k - 1$ (Definition 3) $\Rightarrow not(xPR_{k+1}) \Rightarrow xPR'_{k-1}$ and $not(xPR'_k)$ (Definition 6) $\Rightarrow x$ is assigned to $[C'_{k-1},C'_k]$.

- If $s = k = 1$ then $i = 1$ and $not(xPR_k)$ (Definition 3) $\Rightarrow$ $not(xPR'_k)$ (Definition 6) $\Rightarrow$ x is assigned to $C'_k = [C'_k, C'_k]$.

Ascending Rule Analysis

- If $i = k + 1 <$ M then $R_{k+2}Px$ and $not(R_hPx)$ for all $h < k + 2$ (Definition 4) $\Rightarrow$ $not(R_kPx) \Rightarrow R'_{k+1}Px$ and $not(R'_kPx)$ (Definition 6) $\Rightarrow x$ is assigned to $[C'_k, C'_{k+1}]$.
- If $i = k + 1 = M$ then $s = M$ and $not(R_MPx) = not(R_{k+1}Px)$ (Definition 4) $\Rightarrow$ $not(R'_kPx) = not(R'_{M-1}Px)$ (Definition 6) $\Rightarrow$ x is assigned to $C'_{M-1} = C'_k = [C'_k, C'_k]$.
- If $s = k > 1$ then R_kPx and $not(xPR_h)$ for all h < k (Definition 4) $\Rightarrow$ R'_kPx and $not(R'_hPx)$ for all $h < k$ (Definition 6) $\Rightarrow$ x is assigned to $[C'_{k-1}, C'_k]$.
- If $s = k = 1$ then $i = 1$ and R_kPx (Definition 4) $\Rightarrow$ R'_kPx (Definition 6) $\Rightarrow$ x is assigned to $C'_k = [C'_k, C'_k]$.

1.c.

Descending Rule Analysis

- If $i = k > 1$ and $s = k + 1$ then xPR_k and $not(xPR_h)$ for all h > k (Definition 3) $\Rightarrow$ xPR_{k-1} (Proposition 1.i)) and $not(xPR_{k+1}) \Rightarrow xPR'_{k-1}$ and $not(xPR'_k)$ (Definition 6) $\Rightarrow$ x is assigned to $[C'_{k-1}, C'_k]$.
- If $i = k = 1$ and $s = k + 1$ then $not(xPR_{k+1})$ (Definition 3) $\Rightarrow$ $not(xPR'_k)$ (Definition 6) $\Rightarrow$ x is assigned to $C'_1 = C'_k = [C'_k, C'_k]$.

Ascending Rule Analysis

- If $i = k$ and $s = k + 1 <$ M then $R_{k+1}Px$ and $not(R_hPx)$ for all h < k + 1 (Definition 4) $\Rightarrow$ $not(R_kPx)$ and $R_{k+2}Px$ (Proposition 1.ii)) $\Rightarrow$ $not(R'_kPx)$ and $R'_{k+1}Px$ (Definition 6) $\Rightarrow$ x is assigned to $[C'_k, C'_{k+1}]$.
- If $i = k$ and $s = k + 1 =$ M then $not(R_kPx)$ (Definition 4)$\Rightarrow$$not(R'_kPx)$ (Definition 6)$\Rightarrow$$not(R'_{M-1}Px)$ $\Rightarrow$ x is assigned to $C'_{M-1} = C'_k = [C'_k, C'_k]$.

2.a

Let x be assigned to $[C'_i, C'_s]$ $(i \leq s)$ before the splitting.

Descending Rule Analysis

- If $i > k$, we have $xPR_i \Rightarrow xPR_{k+1}$ (Proposition 1.i)) $\Rightarrow xPR'_{k+1}$ (Definition 7) $\Rightarrow$ xPR'_h for all $h \leq i - 1$ and $not(xPR'_l)$ for all $l > i - 1$ (Definition 3 and Proposition 1.i)) $\Rightarrow$ x is assigned to $[C'_{i-1}, C'_{s-1}]$ where $C'_{i-1} = C_i$ and $C'_{s-1} = C_s$.
- If $s < k$, we have $not(xPR_h)$ for all $h > i$ and xPR_l for all $l \leq i$ (Definition 3 and Proposition 1.i)). If $i = s$ then $i = s = 1 \Rightarrow not(xPR_1)$ (Definition 3) $\Rightarrow xPR_{k-1}$ $\Rightarrow not(xPR'_k)$ (Definition 7); if $i < s \Rightarrow not(xPR_s) \Rightarrow not(xPR_{k-1}) \Rightarrow not(xPR'_k)$ (Definition 7). We have xPR'_h for all $h \leq i$ and $not(xPR'_l)$ for all $l > i$ (Definition 3 and Proposition 1.i)) $\Rightarrow$ x is assigned to $[C'_i, C'_s]$ where $C'_i = C_i$ and $C'_s = C_s$ $\Rightarrow$ x is assigned to $[C'_i, C'_s]$ where $C'_i = C_i$ and $C'_s = C_s$.

Ascending Rule Analysis

- If $i > k$, we have R_hPx for all $h \geq s$ and $not(R_lPx)$ for all $l < s$ (Definition 4 and Proposition 1.ii). If $i = s$ then $i = s = M \Rightarrow not(R_MPx)$ (Definition 4) $\Rightarrow not(R'_{k+1}Px)$ (Definition 7); if $i < s \Rightarrow not(R_iPx) \Rightarrow not(R'_{k+1}Px)$ (Proposition 1.ii and Definition 7) $\Rightarrow R'_hPx$ for all $h \geq s + 1$ and $not(R'_lPx)$ for all $l < s + 1$ $\Rightarrow x$ is assigned to $[C'_{i+1},C'_{s+1}]$ where $C'_{i+1} = C_i$ and $C'_{s+1} = C_s$.
- If $s < k$, we have R_hPx for all h $\geq$ s and $not(R_lPx)$ for all $l < s$ (Definition 4 and Proposition 1 ii) $\Rightarrow R_{k-1}Px \Rightarrow R'_kPx$ (Definition 7) $\Rightarrow R'_hPx$ for all $h \geq s$ and $not(R'_lPx)$ for all $l < s \Rightarrow x$ is assigned to $[C'_i,C'_s]$ where $C'_i = C_i$ and $C'_s = C_s$.

2.b

Descending Rule Analysis

- If $i = k > 1$ and $s = k + 1$ then xPR_k and $not(xPR_h)$ for all $h > k$ (Definition 3) $\Rightarrow xPR_{k-1}$ (Proposition 1.i) and $not(xPR_{k+1}) \Rightarrow xPR'_{k-1}$ and $not(xPR'_{k+2})$ (Definition 7) $\Rightarrow x$ is assigned to $[C'_{k-1},C'_k]$ or $[C'_k,C'_{k+1}]$ or $[C'_{k+1},C'_{k+2}]$.
- If $i = k = 1$ and $s = k + 1$ then $not(xPR_{k+1})$ (Definition 3) $\Rightarrow not(xPR'_{k+2})$ (Definition 7) $\Rightarrow x$ is assigned to $C'_1 = [C'_k,C'_k]$ or $[C'_k,C'_{k+1}]$ or $[C'_{k+1},C'_{k+2}]$.
- If $i = k - 1$ and $s = k < M$ then xPR_{k-1} and $not(xPR_h)$ for all h > k − 1 (Definition 3) $\Rightarrow xPR_{k-1}$ and $not(xPR_{k+1}) \Rightarrow xPR'_{k-1}$ and $not(xPR'_{k+2})$ (Definition 7) $\Rightarrow x$ is assigned to $[C'_{k-1},C'_k]$ or $[C'_k,C'_{k+1}]$ or $[C'_{k+1},C'_{k+2}]$.
- If $i = k - 1$ and $s = k = M$ then xPR_{k-1} and $not(xPR_k)$ (Definition 3) $\Rightarrow xPR'_{k-1}$ (Definition 7) $\Rightarrow x$ is assigned to $[C'_{k-1},C'_k]$ or $[C'_k,C'_{k+1}]$ or $C'_{k+1} = [C'_{k-1},C'_{k+1}]$.
- If $k = M$ and $i = s = M$ then xPR_k (Definition 3) $\Rightarrow xPR_{k-1}$ (Proposition 1.i)) $\Rightarrow xPR'_{k-1} \Rightarrow x$ is assigned to $[C'_{k-1},C'_k]$ or $[C'_k,C'_{k+1}]$ or $C'_{k+1} = [C'_{k+1},C'_{k+1}]$.
- If $k = 1$ and $i = s = 1$ then $not(xPR_k)$ (Definition 3) $\Rightarrow not(xPR_{k+1}) \Rightarrow not(xPR'_{k+2})$ $\Rightarrow x$ is assigned to $C'_k = [C'_k,C'_k]$ or $[C'_k,C'_{k+1}]$ or $[C'_{k+1},C'_{k+2}]$.

Ascending Rule Analysis

- If $i = k > 1$ and $s = k + 1$ then $R_{k+1}Px$ and $not(R_hPx)$ for all $h < k + 1$ (Definition 4) $\Rightarrow R'_{k+2}Px$ and $not(R'_{k-1}Px)$ (Definition 7) $\Rightarrow x$ is assigned to $[C'_{k-1},C'_k]$ or $[C'_k,C'_{k+1}]$ or $[C'_{k+1},C'_{k+2}]$.
- If $i = k = 1$ and $s = k + 1$ then $R_{k+1}Px$ (Definition 4) $\Rightarrow R'_{k+2}Px$ (Definition 7) $\Rightarrow x$ is assigned to $C'_k = [C'_k,C'_k]$ or $[C'_k,C'_{k+1}]$ or $[C'_{k+1},C'_{k+2}]$.
- If $i = k - 1$ and $s = k < M$ then R_kPx and $not(xPR_h)$ for all $h < k$ (Definition 4) $\Rightarrow R_{k+1}Px$ (Proposition 1.ii)) $\Rightarrow R'_{k+2}Px$ and not $(R'_{k-1}Px)$ (Definition 7) $\Rightarrow x$ is assigned to $[C'_{k-1},C'_k]$ or $[C'_k,C'_{k+1}]$ or $[C'_{k+1},C'_{k+2}]$.
- If $i = k - 1$ and $s = k = M$ then $not(R_{k-1}Px)$ (Definition 4) $\Rightarrow not(R'_{k-1}Px)$ (Definition 7) $\Rightarrow x$ is assigned to $[C'_{k-1},C'_k]$ or $[C'_k,C'_{k+1}]$ or $C'_{k+1} = [C'_{k+1},C'_{k+1}]$.
- If $k = M$ and $i = s = M$ then $not(R_kPx)$ (Definition 4) $\Rightarrow not(R_{k-1}Px) \Rightarrow not(R'_{k-1}Px)$ (Definition 7) $\Rightarrow x$ is assigned to $[C'_{k-1},C'_k]$ or $[C'_k,C'_{k+1}]$ or $C'_{k+1} = [C'_{k+1},C'_{k+1}]$.
- If $k = 1$ and $i = s = 1$ then R_kPx (Definition 4) $\Rightarrow R_{k+1}Px$ (Proposition 1.ii)) $\Rightarrow R'_{k+2}Px$ (Definition 7) $\Rightarrow x$ is assigned to $C'_k = [C'_k,C'_k]$ or $[C'_k,C'_{k+1}]$ or $[C'_{k+1},C'_{k+2}]$.

The proof is complete.

References

1. Figueira, J., Mousseau, V., Roy, B.: ELECTRE methods. In: Figueira, S.G., Ehrgott, M. (eds.) International Series in Operations Research and Management Science, pp. 133–162. Springer Science+Business Media Inc., New York (2005)
2. Greco, S., Ehrgott, M., Figueira, J.: Multiple Criteria Decision Analysis: State of the Art Surveys. Springer, Berlin (2016)
3. Roy, B.: The outranking approach and the foundations of ELECTRE methods. Theory Decis. **31**, 49–73 (1991)
4. Yu, W.: Multi-criteria decision-making in the context of sorting: concepts, methods and applications (1992)
5. Roy, B., Bouyssou, D.: Multi-criteria decision support: methods and cases. Econ. Paris (1993)
6. Fernández, E., Figueira, J.R., Navarro, J., Roy, B.: ELECTRE TRI-nB: a new multiple criteria ordinal classification method. Eur. J. Oper. Res. **263**, 214–224 (2017)
7. Bouyssou, D., Marchant, T.: On the relations between ELECTRE TRI-B and ELECTRE TRI-C and on a new variant of ELECTRE TRI-B. Eur. J. Oper. Res. **242**, 201–211 (2015)
8. Perny, P.: Multicriteria filtering methods based onconcordance and non-discordance principles. Ann. Oper. Res. **80**, 137–165 (1998)
9. Bouyssou, D., Marchant, T., Pirlot, M.: A theoretical look at ELECTRE TRI-nB (2020)
10. Fernández, E., Figueira, J.R., Navarro, J.: Interval-based extensions of two outranking methods for multi-criteria ordinal classification. Omega (Westport) **95**, 102065 (2020)
11. Kadziński, M., Martyn, K., Cinelli, M., Słowiński, R., Corrente, S., Greco, S.: Preference disaggregation for multiple criteria sorting with partial monotonicity constraints: application to exposure management of nanomaterials. Int. J. Approximate Reasoning **117**, 60–80 (2020)
12. Ishizaka, A., Pearman, C., Nemery, P.: AHPSort: an AHP-based method for sorting problems. Int. J. Prod. Res. **50**, 4767–4784 (2012)
13. Nemery, P., Lamboray, C.: Flow ort: a flow-based sorting method with limiting or central profiles. TOP **16**, 90–113 (2008)
14. Köksalan, M., Özpeynirci, S.B.: An interactive sorting method for additive utility functions. Comput. Oper. Res. **36**, 2565–2572 (2009)
15. Almeida-Dias, J., Figueira, J.R., Roy, B.: Electre Tri-C: a multiple criteria sorting method based on characteristic reference actions. Eur. J. Oper. Res. **204**, 565–580 (2010)
16. Almeida-Dias, J., Figueira, J.R., Roy, B.: A multiple criteria sorting method where each category is characterized by several reference actions: the Electre Tri-nC method. Eur. J. Oper. Res. **217**, 567–579 (2012)
17. Corrente, S., Greco, S., Słowiński, R.: Multiple criteria hierarchy process for ELECTRE Tri methods. Eur. J. Oper. Res. **252**, 191–203 (2016)
18. Fernández, E., Navarro, J., Solares, E.: A hierarchical interval outranking approach with interacting criteria. Eur. J. Oper. Res. **298**, 293–307 (2022)
19. Buğdaci, A.G., Köksalan, M., Özpeynirci, S., Serin, Y.: An interactive probabilistic approach to multi-criteria sorting. IIE Trans. **45**, 1048–1058 (2013)
20. Ishizaka, A., Nemery, P.: Assigning machines to incomparable maintenance strategies with ELECTRE-SORT. Omega (Westport) **47**, 45–59 (2014)
21. Fernández, E., Figueira, J.R., Navarro, J., Solares, E.: A generalized approach to ordinal classification based on the comparison of actions with either limiting or characteristic profiles. Eur. J. Oper. Res. **305**, 1309–1322 (2023)
22. Fernandez, E., Navarro, J.: A new approach to multi-criteria sorting based on fuzzy outranking relations: The THESEUS method. Eur. J. Oper. Res. **213**, 405–413 (2011)
23. Roy, B.: Presentation and interpretation of the ELECTRE TRI method for assigning areas to risk categories (2002)

A Framework of Targeted Marketing Based on the Outranking Approach

Juan Carlos Leyva López and Omar Alejandro Reyna Gutiérrez

Abstract This chapter presents a framework for targeted marketing using the outranking approach inspired by multicriteria decision analysis (MCDA). The framework aims to identify and evaluate the best matches between customer groups and products, focusing on transaction data and customer-product profiles. The primary goal is to prioritize products or groups of products for specific customer groups or vice versa by applying fuzzy outranking relations to model customer preferences. The chapter outlines two marketing strategies: product-oriented, which ranks customers based on their preferences for a product, and customer-oriented, which ranks products based on the preferences of a customer or customer group. This methodology overcomes the limitations of traditional models by addressing the complexity of aggregating various criteria. Four sub-models are proposed to construct fuzzy outranking relations.

Keywords Targeted marketing · Customer preference · Products · Multicriteria decision analysis · Outranking approach · Web intelligence

1 Introduction

The evolution of the World Wide Web has transformed how companies conduct business. Companies can gather customer attributes through customer profiles, product features via product profiles, and transaction data that links customers and products from their sales and other product-related websites, enabling them to enter and compete in the global market. For instance, companies can intelligently utilize this transaction data to determine which products to select for targeted marketing.

J. C. Leyva López (✉) · O. A. Reyna Gutiérrez
Economic and Management Science Department, Universidad Autónoma de Occidente, Culiacán, Sinaloa, Mexico
e-mail: juan.leyva@uadeo.mx

O. A. Reyna Gutiérrez
e-mail: omar.reyna@uadeo.mx

W. Pedrycz et al. (eds.), *Decision-Making in Computational Intelligence-Based Systems*, Studies in Systems, Decision and Control 628,
https://doi.org/10.1007/978-3-032-04056-5_3

However, these selected products must capture the attention of as many customers as possible to enhance the likelihood of selling more than their competitors. This chapter addresses this type of product selection problem.

Analyzing competitive products is critically important for companies to thrive in the global market and has garnered significant interest from the community [1–3]. For example, the data science department can perform this analysis to identify customer groups most likely to purchase their products. In contrast, the marketing department can design targeted promotions, advertising campaigns, coupons, or similar initiatives to boost product sales. Typically, these promotional actions aim to increase product sales, thereby enhancing overall revenue. However, some products may be appealing to the same customers, and not all products contribute equally to attracting customers in the market. As a result, companies seek to identify only a subset of products that are most likely to interest the largest number of customers, allowing promotions and other costs to reach a broader audience.

Targeted marketing is an effective strategy that focuses on identifying specific groups of products related to particular customers or specific customer segments related to certain products while also developing suitable solutions for those targeted groups. It represents a significant application of data mining [4] and is a key area within web Intelligence [5]. Data mining techniques are often used to achieve the objectives of targeted marketing [5]. For instance, by utilizing a transaction database, we can uncover associations between products that are purchased together. Based on these associations, when a customer buys one set of products, we can suggest the associated products as targeted recommendations [6]. This use of product association is straightforward and effective. However, its effectiveness is limited because it often overlooks customer attributes and product features.

The market value functions model can be applied to the targeted marketing issue described above [7]. In this context, we assume that each object is defined by its values across a finite set of attributes. Typically, it is believed that the market values of objects can be calculated using a linear market value function. This market value function offers several advantages: first, it can rank individuals based on their market value rather than categorize them; and second, the market value function is interpretable. The potential usefulness and effectiveness of a market value function model largely depend on the estimation of individual utility functions and attribute weights, i.e., the coefficients of the linear market value function. This type of model is compensatory; this assumption is grounded in the idea that humans make computerlike calculations, which may be questionable, at least in targeted marketing situations wherein the customer is unwilling or does not have the time to consider all attribute levels [8].

There could also be other models that are not dependent on utilities; for example, outranking relations are non-compensatory models that capture customer's preferences in pairwise evaluations. An outranking relation is a method for determining customer preferences by forming preference relations through pairwise comparisons among potential alternatives for each criterion [9]. The outranking relation determines whether, for a criterion, one alternative, a, is "at least as preferred" as another, b. This type of preference can have a rich structure of relations and can be used to inform decisions according to customer preferences. These models have been successful because of their ease of use and realistic assumptions [10].

To develop a comprehensive and realistic targeted marketing model, it is essential to consider at least three types of information and knowledge. These include customer attributes represented by customer profiles, product features represented by product profiles, and transaction data that connect customers and products [11]. Together, these elements provide valuable and critical information for Customer Relationship Management (CRM) [12]. The primary aim of this work is to propose a framework for targeted marketing based on the outranking approach. The development of the proposed framework relies on studies of customers, products, customer groups, product groups, and the associations between customers and products.

A set of customers on one side and a set of products on the other serve as the subjects of studies in a marketing problem. We employ two targeted marketing strategies: product-oriented targeted marketing and customer-oriented targeted marketing [11]. The product-oriented strategy ranks a set of customers based on a fuzzy outranking relation, given a product or a group of products. Conversely, the customer-oriented strategy ranks a set of products based on a customer or a group of customers. These two strategies complement each other, sharing the common goal of maximizing profits through targeted marketing. Product-oriented marketing is also referred to as direct marketing, while customer-oriented targeted marketing is known as product recommendation [13].

We can identify four models for targeted marketing. These models represent relationships (i) between a group of products and a group of customers, (ii) between a group of customers and a single product, (iii) between a single customer and a group of products, and (iv) between a single product and a single customer, respectively [11]. These four models are exhaustive in the sense of the relation customer: {a single customer, a group of customers}–product: {a single product, a group of products}. In this relation, there are only four possible combinations. Each model can be used for both customer-oriented and product-oriented marketing, although some models may be more appropriate for one in particular. We assume that customers and products are represented by their profiles. The relationships between customers and products are implicitly indicated by transaction data. Customer preferences for products qualitatively determine the value of products to customers. These preferences can be calculated using a fuzzy outranking relation [14]. Model (ii) can be applied to interpret both direct marketing and recommender systems utilizing collaborative filtering. Model (iii) can be employed to interpret recommender systems using content-based filtering [15]. Theoretically, the challenge of targeted marketing can be framed as creating models that establish a fuzzy outranking relation and then deriving a ranking of customers or products from that relation.

The outranking relation is established based on customer preferences and two types of data: customer features, represented by profiles, and transaction data that links customers to products. In the proposed framework, we also consider customer similarity, grouping those with comparable preferences. Typically, customer profiles are applied for personalization, product profiles are used for stock management, and transaction data drives recommendations [14].

This chapter contributes to the literature by developing a novel targeted marketing framework that integrates multicriteria decision analysis (MCDA) with fuzzy outranking relations to enhance the precision of marketing strategies. It provides a robust methodology for identifying the most relevant products for specific customer groups and vice versa. This framework incorporates customer preferences, product features, and transaction data in a comprehensive and flexible approach.

The main contributions of this chapter are specifically as follows:

- **Integration of Fuzzy Outranking Relations in Targeted Marketing**: This chapter introduces fuzzy outranking relations in Sects. 3.4–3.9 and Sect. 4 to model customer preferences more realistically. By addressing the limitations of compensatory models that assume customer choices are linear and fully rational, these relations facilitate more nuanced, non-compensatory decision-making, which better reflects the complexities of consumer behavior.
- **Development of a Unified Framework**: This chapter outlines a multicriteria framework that supports both product-oriented and customer-oriented marketing strategies. Section 3.9 proposes four models representing different relationships between customers and products. The framework allows marketers to select the most appropriate strategy based on their target audience or product focus.
- **Clustering for Better Segmentation**: Sect. 3.7 introduces clustering based on partitions for customers and products, allowing marketers to group them into equivalence classes. This approach better addresses the needs of individual customers or products by focusing on group-level preferences, which is especially valuable when working with large datasets.
- **Real-World Applicability**: The proposed framework is based on real-world business needs, particularly for companies that may not have access to Big Data resources yet still want to implement advanced marketing strategies. It emphasizes a non-Big Data approach that is advantageous for small and medium-sized companies.

This chapter is divided into four more sections. Section 2 discusses related studies and motivations. Section 3 outlines the model formulation. Section 4 addresses the construction of outranking relations within the framework outlined in Sect. 3. Finally, the Sect. 5 offers conclusions and potential future work.

2 Related Studies and Motivations

Targeted marketing is not only an efficient way to maintain strong relationships with existing customers but also a crucial method for identifying potential customers [16]. Numerous data mining techniques have been employed to address the direct marketing problem using customer profiles as input [11]. These methods can be categorized into two main types: segmentation models and response models. Segmentation models involve dividing customers into clusters based on their similarities. Customers within segments are as homogeneous as possible within each segment and as heterogeneous as possible between segments. Clusters with the highest likelihood of responding are selected for promotion. Market segmentation is a widely used method for examining the diversity of purchasing behavior and customer profitability. RFM (Recency, Frequency, Monetary) is one of the most widely used methods for customer segmentation, particularly in targeted marketing.

RFM serves as a foundational tool for identifying key customer segments based on their transactional behavior. By calculating the RFM scores for each customer, we can classify them into distinct segments, allowing businesses to target their marketing efforts more effectively. This is especially important when trying to identify high-value customers who are most likely to engage with future marketing initiatives [17]. For example, [18] conducted market segmentation experiments using modified best-fit regression techniques, Expectation–Maximization (EM), and KMeans++ clustering algorithms. Subsequently, they were evaluated through cluster quality assessment. They introduced a new method for Customer Lifetime Value (CLTV) and proposed three innovative variations of the RFM model.

Ernawati et al. [19] proposed a framework to assist decision-makers in selecting a target market based on the RFM score model enhanced with spatial and temporal analysis. Hwang and Lee [20] carried out a study using website access data to identify prospects for new products. They used an extension of RFM models that include website-specific weights to build an RF model using frequency and recency information from customers' website access data and an RwF model by adding website weights to the frequency of access. A TextRank algorithm was used to analyze weights for each website based on the access frequency, thus defining the weights in the RwF model. Beheshtian-Ardakani et al. [21] propose a novel model for product bundling in e-commerce websites by using market segmentation variables and customer loyalty analysis. RFM model is employed to calculate customer loyalty.

RFM fits well within the larger context of targeted marketing, which aims to customize marketing strategies for specific customer segments based on various criteria. While RFM serves as a valuable starting point for segmentation, it does not consider the diversity of customer attributes. For example, RFM mainly relies on transactional data, which can be limiting as it fails to capture the demographic, psychographic, or behavioral characteristics of customers.

To address this, more integrated approaches that combine RFM with other data sources have been developed. Demographics (e.g., age, gender, income) and psychographics (e.g., lifestyle, values, and preferences) provide a deeper understanding of customer motivations that RFM alone cannot capture [22]. By incorporating these characteristics, businesses can create a more holistic view of their customers, allowing them to better align products with customer needs. Moreover, modern customer behavior analysis tools, such as sentiment analysis and engagement metrics, can complement RFM by offering a more nuanced understanding of customer preferences [23]. This approach enhances the ability to tailor marketing campaigns not only based on past purchasing behaviors but also on customer attitudes, feelings, and interactions with the brand.

While RFM is widely used and effective for segmentation, it has limitations in handling large-scale datasets. Traditional RFM models can struggle with high-dimensional customer data, particularly when working with thousands of features, as seen in some modern databases. The total volume of data may overcome the model, making it challenging to analyze all features effectively without sacrificing predictive power [24]. To address these issues, RFM combined with machine learning techniques has been proposed. For instance, clustering methods such as K-means or DBSCAN can be used to group customers based on patterns across multiple features. These techniques can uncover insights that RFM might overlook, especially when the data involves complex, non-linear relationships between features [17].

Another solution is dimensionality reduction, which uses techniques like Principal Component Analysis (PCA) or t-Distributed Stochastic Neighbor Embedding (t-SNE) to decrease the number of features in a dataset while maintaining its underlying patterns. By utilizing these methods, businesses can tackle the challenge of working with large datasets, enhancing the scalability and performance of their segmentation efforts [22].

Response models assess the probability of each individual's response and select the individuals with the highest probability of responding for promotion. These models include the Rough Set Model, Logit/Probit Model, Genetic Model, and Neural Network Model [25], as well as market value functions [26], among others. Utilizing response models, Gubela et al. [27] introduce uplift models designed to maximize incremental revenues. In [28] propose consolidating an improvement model of uplift models designed for multiple treatments and continuous outcomes, with a specific focus on causal models of machine learning for targeted campaigns. Baier and Stöcker [29] propose a profit uplift modeling approach that combines one-stage ordinary regression and random forests with two-stage Heckman sample selection and zero-inflated negative binomial regression for parameter estimation, aiming to identify the most profitable customers for direct marketing campaigns. Sarkar and De Bruyn [30] illustrate that long short-term memory (LSTM) neural networks can accurately predict customer behavior. Traditional customer response models rely significantly on feature engineering, and their effectiveness is contingent upon domain knowledge and analyst expertise in creating relevant predictors. However, as data complexity increases, traditional models become increasingly complex.

Recommender systems can be categorized into two broad categories: collaborative filtering and content-based filtering. According to [13, 31–33] collaborative filtering utilizes transaction data to identify customers' preferences for a set of products [34]. Assuming that similar types of customers tend to make comparable decisions and choose similar products, collaborative filtering recommends products to a specific customer based on the purchase histories of both the customer and like-minded individuals. Collaborative filtering techniques are classified into two types: memory-based methods and model-based methods [35]. The memory-based method leverages transaction data to identify customers with similar preferences. A key challenge lies in measuring the similarity between two customers. The two most commonly used techniques for this purpose are rating-based similarity and structural similarity [33]. Rating-based similarity is determined by customers' ratings of products, utilizing the cosine index, Pearson coefficient, and weighted Pearson coefficient to estimate the similarity between two customers [34].

The structural similarity is calculated based on transaction data without ratings. In this measure of structural similarity, the transaction data are transformed into a bipartite graph. Within the bipartite graph, methods such as Common Neighbors, Local Path Index, and Random Walk are utilized to assess the similarity between two customers [33]. These methods demonstrate that memory-based collaborative filtering recommends products based on the choices of a group of similar customers. Conversely, the model-based approach learns from transaction data to develop models, which it then uses for prediction. Models are typically created by employing data mining or machine learning algorithms, including association rules, Bayesian networks, and genetic algorithms, among others. Association rules extracted from transaction data reveal relationships between products. A Bayesian network offers practical recommendations through probability inference. Genetic algorithms generate recommendations based on clustering and user models [36]. Model-based collaborative filtering also relies on the decisions of similar customers to formulate recommendations.

Content-based filtering utilizes product profiles to identify items that a specific customer is likely to like. This method typically recommends products similar to those the customer has previously purchased. Like memory-based collaborative filtering methods, estimating the similarity between two products is a core challenge. Item-based methods utilize the Cosine index, Pearson coefficient, and Weighted Pearson coefficient to determine similarity from transaction data [35]. Other approaches apply information retrieval and information filtering results to assess similarity [13]. Content-based filtering methods suggest products based on a set of items with similar characteristics.

Many methods have been proposed to enhance the performance of recommender systems. One solution is to combine collaborative filtering and content-based filtering techniques by utilizing customer profiles, product profiles, and transaction data. Balabanovic and Shoham [37], Basilico and Hofmann [38], Vozalis and Margaritis [39] discussed various strategies for developing collaborative, content-based, and demographic filtering in recommender systems. Zhong et al. [4] proposed a framework for explanation-oriented data mining that utilizes customer features and product characteristics to identify explanations for the associations found in transaction data.

Most existing studies rely on classification methods. Precisely, they classify customers or products into different categories and then select those within these categories. In these methods, customers or products within the same category are considered to have the same market value. This approach can lead to an overestimation or underestimation of potential customers or products. An alternative solution is to create a ranked list of customers or products rather than using classification [13, 32, 40–42]. A market value function [7, 11, 15, 26] has been proposed to rank customers or products. This value indicates the degree of association between a product and a customer concerning their potential profit. Customers or products at the top of the ranked list are deemed to possess higher market values. A cutoff point is established on the ranked list based on various criteria, such as financial constraints. Customers or products above this cutoff point are selected as the target group.

Some researchers have proposed creating ranked lists of customers (or products) and then selecting the highest-ranked ones for promotion [43, 44]. Customer ranking can be achieved by assigning a specific score to each customer and then establishing a hierarchical order based on that score. A customer's score is a numerical value that reflects their performance as assessed by the organization based on one or more criteria [43]. Zunk and Koch [45] identify various variables to qualify customers.

The study of direct marketing systems and recommender systems has attracted increased attention. Zhong et al. [5] proposed a direct marketing framework based on a data mining process, which divides direct marketing systems into three components: data preparation and preprocessing, pattern discovery, and customer promotion. Herlocker et al. [46] introduced a collaborative filtering framework from an algorithmic perspective. Pazzani [47] presented a unified framework combining collaborative filtering with content-based filtering. Huang et al. [15] offered a conceptual framework for targeted marketing that integrates various approaches. Hernandez and Gaudioso [48] proposed a framework that incorporates methods for guiding customers with those for delivering valuable products. Huang and Zhong [49] developed a unified probabilistic inference model for targeted marketing that utilizes the relationship between individual products and specific customers. Bobadilla et al. [36] introduced a framework based on prediction quality, recommendation effectiveness, novelty, and trust. These studies primarily focus on constructing either direct marketing systems or recommender systems. However, there is no comprehensive model that unifies direct marketing and recommender systems. To bridge this gap, we propose a general framework grounded in customer preferences and outranking preference relations.

On the other hand, in recent years, deep learning methods, metaheuristics, and data science applications have also been utilized for customer selection. However, these methods are only effective for organizations that possess Big Data resources [43]. Unfortunately, small and medium enterprises (SMEs) often underutilize Big Data technologies for various reasons. Willetts et al. [50] identify several barriers to the adoption of Big Data by SMEs. The significance of SMEs in the global economy stems from their role in comprising the majority of companies worldwide and their vital contribution to job creation and economic development. They account for approximately 90% of businesses and more than 50% of the global workforce. Formal SMEs contribute up to 40% of a nation's GDP in emerging economies [51].

The targeted marketing problem can be formulated as construction models that construct a market value function and estimate the required parameters of the function. A set of customers, on the one hand, and a set of products, on the other, are the subjects of study in a marketing problem. Focusing on each of them, we can identify two distinct marketing strategies: a customer-oriented marketing strategy and a product-oriented marketing strategy, respectively [11]. To address this issue, the development of multicriteria models and methods for product-targeted marketing is proposed based on the similarity profile of customer preferences. This approach is designed for companies positioned in the market as small, medium, or large who wish to target marketing without reaching the Big Data environment. In the development of this research, it is assumed that methods for segmenting and grouping exist, as well as that similarity of preference can be used to define groups that share a preference for customers and/or products.

A non-big data approach will be beneficial for companies that have customers, products, or transactional data in the hundreds or thousands. MCDA techniques can help us develop a customer classification for companies that lack Big Data capabilities. We present an alternative solution to the multicriteria targeted marketing problem by extending the results of the MCDA methodology. The methods developed for MCDA can be used to identify customers that have potential market value. The targeted marketing problem can be formulated as a multicriteria ranking problem. Many MCDA results can be applied immediately. The exploitation of the fuzzy outranking relation enables us to produce a ranked list, which provides more flexibility in specific targeted marketing problems.

3 The Formulation of the Model

3.1 Preference Relations

Let $A = \{p_1, p_2, \ldots, p_m\}$ be the set of decision alternatives and assume that there exist defined criteria $g_k, k = 1, 2, \ldots, r$. Traditional preference modeling assumes the following two preference relations hold for two alternatives $(p_i, p_j) \in A \times A$:

$$\left.\begin{array}{ll} p_i P_k p_j \ \left(p_i \, \textit{is preferred to}\, p_j\right) & \Leftrightarrow g_k(p_i) > g_k(p_j) \\ p_i I_k p_j \ \left(p_i \, \textit{is indifferent to}\, p_j\right) & \Leftrightarrow g_k(p_i) = g_k(p_j) \end{array}\right. \text{for each } k = 1, 2, \ldots, r$$

It is often inadequate to consider that slight differences in evaluation should give rise to clear-cut distinctions. This is particularly true when different formulations of criteria are acceptable. Imprecision underlying criteria design can be modeled using the following discrimination thresholds [52]:

- An indifference threshold allows for two alternatives with close criterion values to be judged as equivalent. The indifference threshold draws the boundary between an indifference and a preference situation.
- A preference threshold is introduced when we want or need to be more precise when describing a preference situation. Therefore, it establishes the boundary between a situation of a strict preference and a hesitation between an indifference and a preference situation, namely a weak preference.

A criterion g_k, having indifference and preference thresholds, q_k and t_k, respectively ($t_k \geq q_k \geq 0$), is called a pseudo-criterion. A set of criteria should consider compliance with certain axioms as operational, exhaustive, monotonic, and non-redundant [53]. Comparing two alternatives p_i and p_j according to a pseudo-criterion

g_k leads to the following partial preference relations:

$$p_i I_k p_j \Leftrightarrow \left|g_k(p_i) - g_k(p_j)\right| \leq q_k$$
$$p_i Q_k p_j \Leftrightarrow q_k < g_k(p_i) - g_k(p_j) \leq t_k$$
$$p_i P_k p_j \Leftrightarrow g_k(p_i) - g_k(p_j) > t_k$$

where I_k, Q_k and P_k represent respectively indifference, weak preference and strict preference relations restricted to criterion g_k. These three relations could be grouped into a partial outranking relation $S_k = (I_k \cup Q_k \cup P_k), S_k \subseteq A \times A$ such that $p_i S_k p_j \Leftrightarrow g_k(p_i) - g_k(p_j) \geq -q_k$ which corresponds to the assertion p_i "is at least as good as" p_j with respect to the aspects covered by criterion g_k. Operatively the method tests whether the assertion $p_i S_k p_j$ is true or not for each pair $(p_i, p_j) \in A \times A$.

To model situations where a very low assessment of an alternative p_j with respect to p_i, according to some criterion g_k, cannot be compensated by a good assessment on one or several other criteria, we use a veto threshold v_k $(v_k \geq t_k)$ and define the following veto relation $V_k : p_i V_k p_j \Leftrightarrow g_k(p_i) - g_k(p_j) > v_k$. In this case, p_j cannot be considered as "at least as relevant as" p_i, whatever the assessments on other criteria.

The choice of thresholds intimately affects whether a particular binary relation holds. While the choice of appropriate threshold is not easy, in most realistic decision-making situations, there are good reasons for choosing non-zero values for t and q, where for consistency reasons $v_k > t_k \geq q_k \geq 0$.

To obtain a global preference model for the set of alternatives, we use outranking approaches [52]. These approaches are quite appropriate for the specificities of targeted marketing and are based on a partial compensatory logic. They consist of two phases: an aggregation phase and an exploitation phase. We will provide details about both phases hereafter.

3.2 Aggregation Phase

Outranking approaches take as input the partial preference outranking relations induced by the criterion family and aggregate them into one or more global preference outranking relation(s) S. These types of preference relations are relevant in targeted marketing since they (i) permit imprecision in product or customer evaluations, (ii) can handle criteria expressed on heterogeneous scales, (iii) use all the available information on product or customer performances, and (iv) do not necessarily require inter-criteria information, such as weights.

According to the global model of decision-maker (DM) preferences, $p_i S p_j$ is true if there are enough arguments to proclaim that "p_i is at least as good as p_j" or "p_i is not worse than p_j," and there are no strong motive to refute such an assertion. There are different types of outranking relations:

- Single outranking $(p_i S p_j)$: p_i is preferred or indifferent to p_j $(p_i P p_j \vee p_i I p_j)$.
- Double outranking ($p_i S p_j$ and $p_j S p_i$): p_i is indifferent to p_j $(p_i I p_j)$.

- Non-outranking $\neg(p_iSp_j)$: Both alternatives are incomparable "p_i no outranks"p_j.

As soon as two alternatives $p_i, p_j \in A$ are compared, this gives rise to one and only one of the following four situations:

- p_iSp_j and $\neg(p_jSp_i)$: p_i outranks p_j and p_j does not outrank p_i(preference).
- $\neg(p_iSp_j)$ and p_jSp_i: p_i does no outrank p_j and p_j outranks p_i(preference).
- p_iSp_j and p_jSp_i: p_i outranks p_j and p_j outranks p_i(indifference).
- $\neg(p_iSp_j)$ and $\neg(p_jSp_i)$: p_i does not outrank p_j and p_j does not outrank p_i (incomparability).

Each pair of p_i and p_j alternatives is then tested to check if assertion p_iSp_j is valid or not. In the ELECTRE III outranking method [54], the test to accept assertion p_iSp_j is implemented using two conditions:

- A concordance condition, which ensures that a majority of criteria, after considering their relative importance, are concordant with p_iSp_j (majority principle), and
- A non-discordance condition, which ensures that none of the discordant criteria strongly refutes p_iSp_j (respect of minorities principle).

The operational implementation of these two conditions is now discussed, assuming that all criteria are to be maximized.

We first consider the outranking relation defined for each of the r criteria, that is, $p_iS_kp_j$ means that "p_i is at least as good as p_j" with respect to the kth criterion, $k =$ 1,2,...,r.

The *kth* criterion is in concordance with assertion p_iSp_j if and only if $p_iS_kp_j$, that is, if $g_k(p_i) \geq g_k(p_j) - q_k$. Thus, even if $g_k(p_i)$ is less than $g_k(p_j)$ by an amount up to q_k, it does not contravene assertion $p_iS_kp_j$ and therefore is in concordance.

The *kth* criterion is in discordance with assertion p_iSp_j if and only if $p_jP_kp_i$, that is, if $g_k(p_j) \geq g_k(p_i) + t_k$. That is, if p_j is strictly preferred to p_i for criterion k, then it is clearly not in concordance with the assertion that p_iSp_j.

Each criterion g_k has an associated weight that symbolizes the relative importance of the criterion within the set of criteria [55]. The weights are denoted by w_k, $k = 1, 2, \ldots, r$, and are assumed to be positive and their sum is equal to 1, i.e., $w_k > 0$ for each g_k, and $\sum_{k=1}^{r} w_k = 1$.

With this information, it is now possible to define the indifference (I_k), strict preference (P_k), and weak preference (Q_k) relations for each criterion g_k. Let $C(p_iIp_j)$, $C(p_iPp_j)$, and $C(p_iQp_j)$ represent the set of criteria such that $p_iI_kp_j$, $p_iP_kp_j$, and $p_iQ_kp_j$, respectively. The three preference relations can be grouped into one partial outranking relation S_k comprising the three corresponding situations $S_k = P_k \cup Q_k \cup I_k$, where $p_iS_kp_j$ (p_i outranks p_j) means that "p_i is at least as good as p_j" on criterion g_k.

For the construction of an outranking relation, p_iSp_j, we use the ELECTRE III method which considers the principles of concordance and non-discordance, where the first represents the strength of the coalition of criteria g_k, being in favor of p_iSp_j and, the second requires that within the minority of criteria, which do not support the assertion p_iSp_j, none of them is strongly against the assertion. The principles are modeled with a comprehensive concordance index and per- elementary criterion discordance indices respectively. They are then combined to form a credibility index of the relation "p_i outranks p_j". We assume that all criterion g_k are increasing monotone with respect to the preference, i.e., the greater the evaluation, the better it is. These concepts are modeled through a *concordance index* $c(p_i, p_j)$ and a *criterion discordance index* $d_k(p_i, p_j)$ for each criterion g_k. $c(p_i, p_j)$ is defined as follows:

$$c\left(p_i, p_j\right) = \sum_{g_k \in C\left(p_i\{I,Q,P\}p_j\right)} w_k + \sum_{g_k \in C\left(p_jQp_i\right)} \varphi w_k \tag{1}$$

where

$$\varphi_k = \frac{t_k - \left[g_k(p_j) - g_k(p_i)\right]}{t_k - q_k} \in [0, 1], \sum_{k=1}^{r} w_k = 1.$$

Notice then that $c(p_i, p_j) \in [0, 1][0, 1]$ and $c(p_i, p_j) = 0$ if $g_k(p_i) + t_k \leq g_k(p_j)$, for all g_k (p_j is strictly preferred to p_i on all criteria), and $c(p_i, p_j) = 1$ if $g_k(p_i)+q_k \geq g_k(p_j)$, for all g_k (p_i outranks p_j on all criteria).

The *criterion discordance index* for each criterion g_k is defined as follows:

$$d_k\left(p_i, p_j\right) = \begin{cases} 1, & if\ g_k(p_j) - g_k(p_i) \geq v_k, \\ \frac{\left[g_k(p_j) - g_k(p_i)\right] - t_k}{v_k - t_k} & if\ t_k < g_k(p_j) - g_k(p_i) < v_k, \\ 0, & if\ g_k(p_j) - g_k(p_i) \leq t_k. \end{cases} \tag{2}$$

where q_k, t_k, and v_k are the indifference, preference, and veto thresholds, and w_k is the relative importance of the criterion g_k.

In short, for each criterion g_k, and for each pair of alternatives $\left(p_i, p_j\right) \in A \times A$, there are now a concordance and a discordance measure. The final step in the model building phase is to combine these two measures to produce a measure of the degree of outranking, that is, a *credibility index* $\sigma(p_i, p_j)(0 \leq \sigma(p_i, p_j) \leq 1)$ for each criterion g_k, which assesses the strength of the assertion that "p_i is at least as good as p_j, p_iSp_j." A manner of modeling it is with the following expression:

$$\sigma(p_i, p_j) = c(p_i, p_j) \prod_{g_k} T_k(p_i, p_j) \tag{3}$$

where

$$T_k(p_i, p_j) = \begin{cases} \frac{1 - d_k(p_i, p_j)}{1 - c(p_i, p_j)} & if\ d_k(p_i, p_j) > c(p_i, p_j), \\ 1 & otherwise. \end{cases}$$

From (3), we can note that if the strength of the concordance surpasses that of the discordance, then the concordance value should not be altered. Else, the validity of the assertion p_iSp_j is questioned and adjust $c(p_i, p_j)$ according to the previous model. In the case that a discordance index is equal to 1 for any $(p_i, p_j) \in A \times A$ and any criterion g_k, then we have no confidence that p_iSp_j; consequently, $\sigma(p_i, p_j) = 0$. In this way, ELECTRE III constructs a fuzzy outranking relation S_A^σ; where $\sigma(p_i, p_j)(0 \leq \sigma(p_i, p_j) \leq 1)$ indicates the degree of strength of the arguments favouring the crisp outranking p_iSp_j, for each $(p_i, p_j) \in A \times A$.

3.3 Exploiting an Outranking Relation to Derive a Ranking of a Set of Alternatives

This completes the construction of the outranking model. The second phase of the method is to exploit the model to derive a global ranking of the alternatives in a non-increasing order of preference from the fuzzy outranking relation S_A^σ. It is carried out with the net flow rule (*NFR*) [56, 57]. First, the so-called outranking flows are calculated. In this phase, the main idea is to analyze how an alternative p_i is preferred over all the other alternatives and, inversely, how the other alternatives are preferred over p_i. For that purpose, the leaving and entering flows are computed as follows:

$$\phi^+(p_i) = \sum_{\substack{j=1 \\ j \neq i}}^{m} \sigma(p_i, p_j), \quad \phi^-(p_i) = \sum_{\substack{j=1 \\ j \neq i}}^{m} \sigma(p_j, p_i)$$

The leaving flow is the sum of the values of the arcs that leave node p_i and, therefore, yields a measure of the "outranking character" of p_i. $\varphi^+(p_i)$ represents the "strength" of the alternative p_i with respect to all the other alternatives.

$\varphi^-(p_i)$ is a measure of the strength of all the other alternatives with respect to the alternative p_i and a measure of the weakness of the alternative p_i measuring the "outranked character" of p_i. The *net outranking flow* is defined as follows:

$$\varphi(p_i) = \varphi^+(p_i) - \varphi^-(p_i) = \sum_{\substack{j=1 \\ j \neq i}}^{m} \sigma(p_i, p_j) - \sum_{\substack{j=1 \\ j \neq i}}^{m} \sigma(p_j, p_i). \tag{4}$$

The best alternatives correspond to large positive values of $\varphi(p_i)$ and the worst ones to large negative values. The net preference flows thus induce a complete ranking of the products of A.

3.4 Customer Preferences of Products

In targeted marketing, we deal with a set of products, a set of customers, and relationships between them. Let C and P_r denote the sets of customers and products in an enterprise, respectively. Their Cartesian product $C \times P_r$ contains all (customer, product) pairs. We can state customer preferences of products as a preference relation on $C \times P_r$.

A customer preference relation on $C \times P_r$ can be defined for each of the r criteria in the following form:

- for two (customer, product) pairs (c_i, p_j), $(c_l, p_v) \in C \times P_r$, $(c_i, p_j)P_k(c_l, p_v)$ ((c_i, p_j) is preferred to (c_l, p_v))$\leftrightarrow$ customer c_i likes product p_j more than c_l likes p_v with respect to the *kth* criterion, $k = 1,2, \ldots,r$.
- $(c_i, p_j)I_k(c_l, p_v)$ ((c_i, p_j) is indifferent to (c_l, p_v))$\leftrightarrow$ customer c_i likes product p_j equal to c_l likes p_v with respect to the *kth* criterion, $k = 1,2, \ldots,r$.
- $(c_i, p_j)S_k(c_l, p_v)$ ((c_i, p_j)outranks (is preferred or indifferent to) (c_l, p_v)) $\leftrightarrow$ customer c_i likes product p_j more than or equal to c_l likes p_v with respect to the *kth* criterion, k = 1,2, …,r.
- $(c_i, p_j)S(c_l, p_v)$ means that, according to the global model of *DM* preferences, there are good reasons to consider that " (c_i, p_j) is at least as good as (c_l, p_v)" or " (c_i, p_j) is not worse than (c_l, p_v)." In the context of targeted marketing, the outranking relation S assesses the assertion that a (customer, product) pair is at least as good as another one.

3.5 Two Targeted Marketing Strategies

The above customer preference relation on $C \times P_r$ has limited application in real-word problems of targeted marketing, and a lot of times the comparison of two (customer, product) pairs (c_i, p_j), $(c_l, p_v) \in C \times P_r$ may no make too much sense. To remedy this problem, Adomavicius and Tuzhilin [11, 13] propose to perform projections on one of the variables, that is, we can project the preference relation onto C and P_r from the standing point of a product and a customer, respectively, fixing in this form either a customer or a product.

Let us consider the outranking relation S on $C \times P_r$. The projections of S onto C and P_r lead naturally to two targeted marketing strategies. For customer-oriented marketing, we focus on a particular customer c and project S onto P_r to obtain a preference outranking relation S_c on P_r. For product-oriented marketing, we focus on a product p and project S onto C to obtain a preference outranking relation S_p on C. The projections of the outranking relation S result in S_c and S_p, respectively.

For a customer $c \in C$, we can project S onto P_r as follows:

$p_i S_c p_j \leftrightarrow (c, p_i)S(c, p_j) \leftrightarrow$ for the customer c, product p_i outranks (is prefered or is indiferent) to product p_j.

According to S_c, we can compare products pairwise based on the customer's preferences. In so doing, we have a customer-oriented marketing strategy. The projected outranking relation S_c can be exploited and derive a ranking of products. A cut-off point of the ranked list may be chosen based on certain criteria. We can select the top products above the cutoff points as targeted products.

For a given product p, we project S onto C to obtain an outranking relation S_p on C as follows:

$c_i S_p c_j \leftrightarrow (c_i, p) S (c_j, p) \leftrightarrow$ customer c_i outranks (is preferred or indifferent to) customer c_j with respect to product p.

This leads to a product-oriented marketing strategy. The projected outranking relation S_p can be exploited and derive a ranking of customers in decreasing order of preferences. A cut-off point of the ranked list may be chosen based on certain criteria. A top-ranked portion of the list can be chosen as the targeted customers.

3.6 Examples of Problem Formulation

Before introducing the formulation for both the customer-oriented and product-oriented targeted marketing problems based on the outranking approach, we first outline the expected outcomes of these formulations.

- **Customer-Oriented Targeted Marketing Problem**:

This formulation will generate a ranking of products that are most aligned with a given customer's preferences. These preferences are based on a set of criteria such as price, quality, and past purchase history. Using fuzzy outranking relations, the framework will calculate the degree of preference for each product, considering the customer's attributes. The expected outcome is a ranking of products, tailored to meet the specific needs and desires of the customer, which can then be used to inform marketing strategies and product recommendations.

- **Product-Oriented Targeted Marketing Problem**:

In this case, the formulation will generate a ranking of customers who are most likely to purchase a particular product. By applying the fuzzy outranking relations, this framework will calculate how different customer groups prefer a product, generating an ordered list of the most promising customer segments for that product. This allows the marketing team to target the right customer segments for a given product effectively.

3.6.1 Formulation of the Product-Oriented Targeted Marketing Problem Based on the Outranking Approach

Let $A = \{c_1, c_2, ..., c_m\}$ represent a large-sized set of customers for a company for whom a particular product p is being promoted. Each customer is characterized by a set of attributes, denoted as $H = \{h_1, h_2, ..., h_r\}$. These attributes may include demographic, behavioral, or other features relevant to the marketing campaign. The set A, is partitioned into three pair-wise disjoint classes, namely, positive instances A_1, negative instances A_2, and undecided instances A_3, i.e., $A = A_1 \cup A_2 \cup A_3$ and $\forall r, q \in \{1, 2, 3\} : A_r \cap A_q = \emptyset, r \neq q$., where:

- A_1: Set of current customers who have previously purchased the product p.
- A_2: Set of customers who have refused to purchase the product p.
- A_3: Set of customers for whom no prior information about their purchase decision is available.

The product-oriented targeted marketing problem can be formally defined as follows:

1. **Criteria Definition**: Define a family of criteria denoted by $G = \{g_1, g_2, ..., g_r\}$, based on the attributes in set $H = \{h_1, h_2, \ldots, h_r\}$ and the set A where $g_j(c_i)$ represents the evaluation of the customer c_i by criterion g_j.
2. **Preference Aggregation**: The aggregation of preference similarity between customers on A_3 based on A_1(preference pattern-seeking between customers that already have bought the product) can be modeled through a fuzzy outranking, denoted as $S^{\sigma}_{A_3}$ $(\sigma(c_i, c_j))$, where this relation reflects the degree of strength of the arguments favoring the crisp outranking $c_i S_{A_3} c_j$, where $c_i S_{A_3} c_j$ means that the customer c_i likes the product p more or equal than customer c_j, or customer c_i outranks customer c_j with respect to product p.
3. **Partitioning and Classifications**: Construct $P_k(A) = \{C_1, C_2, \ldots, C_k\}$, a partition of A into k classes of customers, where, $P_k(A) = \bigcup_{i=1,2,\ldots,k} C_i$, $\forall r, q \in \{1, 2, 3\} : C_r \cap C_q = \emptyset, r \neq q$. The customers inside a class are indifferent to each other with respect to their preferences.
4. **Comparing Customer Groups**: Develop a comparison mechanism between customers and customer classes on the defined preference relations. These comparisons will allow us to rank customer classes in terms of their preferences for the product, where a higher rank implies a greater likelihood of purchasing the product.
5. **Outranking Relation of Customer Classes**: Determining the reflexive and anti-symmetric crisp outranking relation of classes of customers of A, $S^*_{P_k(A)}$ of $P_k(A)$ that is most compatible with the information contained in the fuzzy outranking relation S^{σ}_{A}. This relation serves to establish a partial order among the customer classes, where customers in higher-ranked classes are more likely to be interested in the product.
6. **Ranking and Selection**: Using the partial order of customer classes $O^*_{P_k(A)}$ derived from $S^*_{P_k(A)}$, establish a cut-off point of the partial order based on specific criteria (e.g., financial constraint, campaign budget). The top-ranked customer classes above this cut-off will constitute the target group for product promotion.

3.6.2 Formulation of the Customer-Oriented Targeted Marketing Problem Based on the Outranking Approach

Let $P_r = \{p_1, p_2, \ldots, p_m\}$ represent a large set of products offered by a company for a particular customer. These products are characterized by a set of attributes denoted as $H = \{h_1, h_2, \ldots, h_r\}$, where $h_j(p_i)$ represents the evaluation of the product p_i by attribute h_j. We assume that these products can be divided into three subsets P_r1, P_r2, and P_r3, which are pair-wise disjoint subsets, i.e., $P_r = \mathrm{P}_r1 \cup A_r2 \cup A_r3$ and $\forall i, j \in \{1, 2, 3\} : P_ri \cap P_qj = \emptyset, i \neq j$, where;

- P_r1: Set of products that have been purchased at least once by the customer c,
- P_r2: Set of products that the customer c has never purchased.
- P_r3: Set of products under consideration for recommendation to the customer c, based on available marketing strategies.

The customer-oriented targeted marketing problem can be formally formulated as follows:

1. **Criteria Definition**: Define a family of criteria $G = \{g_1, g_2, \ldots, g_r\}$, based on the set of attributes $H = \{h_1, h_2, \ldots, h_r\}$ and the set P_r where $g_j(p_i)$ represents the evaluation of the product p_i by criterion g_j. These criteria may include features such as price, quality, brand, or past purchase history.
2. **Preference Aggregation**: The aggregation of customer preferences can be modeled using a fuzzy outranking relation, denoted as $S^{\sigma}_{P_r3}$ $(\sigma(p_i, p_j))$, which must be interpreted as the degree of strength of the arguments favoring the crisp outranking $p_i S_{P_r3} p_j$, where $p_i S_{P_r3} p_j$ means that the product p_i is at least as preferred as product p_j to customer c or product p_i outranks product p_j to customer c.
3. **Ranking and Selection**: Using the partial order of products derived from $S^{\sigma}_{P_r3}$, establish a cut-off point of the partial order based on specific criteria. The top-ranked products above this cut-off will constitute the targeted products.

We want to find the k-most preferable products (k-MPP) from P_r3 for targeted marketing. We may model the k-MPP problem and propose an algorithmic framework for customer-oriented marketing. Our framework may utilize a multicriteria outranking approach to solve the k-MPP problem.

3.7 *Customer and Product Clustering*

Customer-oriented and product-oriented marketing strategies focus on specific customers and products, respectively. In practice, however, obtaining sufficient data to develop reliable models for a single customer or product can be a significant challenge [58]. Traditional methods like collaborative filtering and recommendation systems have been used to overcome this limitation by clustering customers and products into similar groups. This grouping allows for the consideration of group-level preferences, which helps in better addressing the needs of individual customers or products.

To handle this, we introduce a partition-based clustering method. Let *EC* be an equivalence relation on the set of customers *C*, and EP_r be an equivalence relation on the set of products P_r. These relations partition the customer and product sets into equivalence classes, resulting in the quotient customer space *C/EC* and the quotient product space P_r/EP_r. For any customer *c*, the equivalence class is denoted as $[c]$, and for any product *p*, the equivalence class is denoted as $[p]$. These partitions lead to the generation of six new types of outranking relations for both customer-oriented and product-oriented marketing strategies.

The six key new outranking relations are:

1. **Customer preference for product-group**: $[p_i]S_c[p_j] \leftrightarrow$ For a customer *c*, a product group $[p_i]$ outranks (is preferred or is indifferent) to another product group $[p_j]$.
2. **Customer-group preference for single product**: $p_iS_{[c]}p_j \leftrightarrow$ For a customer group $[c]$, a product p_i outranks (is preferred or is indifferent) to another product p_j.
3. **Customer-group preference for product-group**: $[p_i]S_{[c]}[p_j] \leftrightarrow$ For a customer group $[c]$, a product group $[p_i]$ outranks (is preferred or is indifferent) to another product group $[p_j]$.
4. **Product preference for customer-group**: $[c_i]S_p[c_j] \leftrightarrow$ For a product *p*, a customer group $[c_i]$ outranks (is preferred or is indifferent) to another customer group $[c_j]$.
5. **Product-group preference for customer**: $c_iS_{[p]}c_j \leftrightarrow$ For a product group $[p]$, a customer c_i outranks (is preferred or is indifferent) to another customer c_j.
6. **Product-group preference for customer-group**: $[c_i]S_{[p]}[c_j] \leftrightarrow$ For a product group $[p]$, a customer group $[c_i]$ outranks (is preferred or is indifferent) to another customer group $[c_j]$.

In fact, they correspond to the customer and product spaces $(P_r/EP_r, S_c)$, $\left(P_r, S_{[c]}\right)$, $\left(P_r/EP_r, S_{[c]}\right)$, $\left(C/EC, S_p\right)$, $\left(C, S_{[p]}\right)$, and $\left(C/EC, S_{[p]}\right)$, respectively (see Table 1).

This expansion of the outranking relations allows for more flexibility in targeted marketing. We can now rank customer groups by product preferences and product groups by customer preferences. Additionally, we can focus on individual customers/

Table 1 Customer preferences for targeted marketing

	Single product	Product group
Single customer	$S_C \subseteq P_r \times P_r$ $S_{P_r} \subseteq C \times C$	$1S_C \subseteq P_r/EP_r \times P_r/EP_r$ $5S_{[P_r]} \subseteq C \times C$
Customer group	$2S_{[C]} \subseteq P_r \times P_r$ $4S_{P_r} \subseteq C/EC \times C/EC$	$3S_{[C]} \subseteq P_r/EP_r \times P_r/EP_r$ $6S_{[P_r]} \subseteq C/EC \times C/EC$

products or their respective groups, providing a comprehensive view of preferences within the market.

The approach is particularly effective in marketing contexts where large datasets are involved, as it helps avoid the limitations associated with focusing only on single customer or product models. The clustering of customers and products enhances the precision of preference modeling, allowing for more targeted and effective marketing strategies.

3.8 Information Sources

Three types of information and knowledge related to customers and products are particularly important for targeted marketing: (i) customer features represented by customer profiles, (ii) product features represented by product profiles, and (iii) transaction data linking customers and products [11]. For our proposal, the customer profile may be either sociodemographic or based on the customer's preferences, which can be explicitly provided by the customer or inferred from implicit information derived from transaction data. The product profile typically relates to the inherent attributes of a product. Transaction data reflects customers' preferences regarding products, where these preferences are generally implied, as we can ascertain whether a customer has purchased a product from purchase records. While there are other types of transaction data, we will not focus on them in this work.

The information above can be represented in a matrix structure. The rows of the tables correspond to objects in the universe, while the columns correspond to their attributes, and each cell contains the value of an object relative to an attribute. For example, consider a restaurant database. An alternative method for processing the database is as follows: the customer profile table displays information about customers' age, gender, education, and more. The product profile table outlines characteristics of restaurants, such as title, genre, and additional details.

The transaction table displays customers' options for restaurants, distinguished by varying scales and values. In this example, these options are indicated by "0" or "1", where "0" means that the customer has not visited the restaurant, and "1" signifies that the customer has visited the restaurant. Let the symbols c_1,c_2,... represent customers and r_1, r_2,... represent restaurants. Figure 1 illustrates the relationship among the three tables. From Fig. 1, we can observe that the customer profile table and the product profile table provide information regarding customers and restaurants in the transaction table, respectively, while the transaction table links the customer profile table and the product profile table. Collectively, they present valuable insights into targeted marketing.

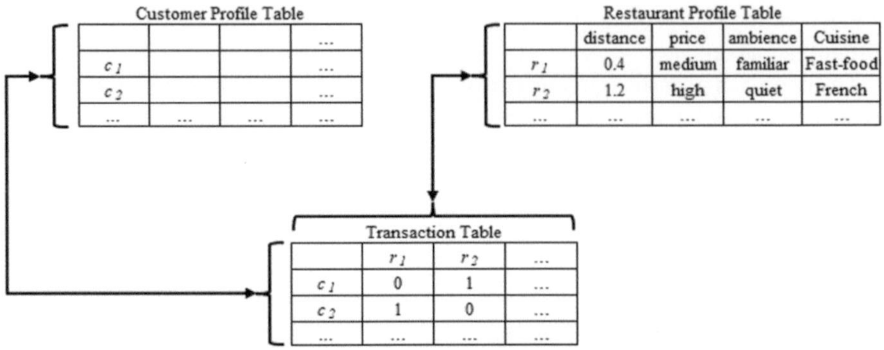

Fig. 1 An example of customer and product profiles and transaction tables for targeted marketing

3.9 Models for Constructing Fuzzy Outranking Relations

The types of information related to customers and products can be analyzed in various forms, and the results can be expressed using fuzzy outranking relations. These outranking relations calculate the degree of preference for marketing a specific product (or group of products) to a particular customer (or group of customers).

To study fuzzy outranking relations, we adapt the value function model proposed by [11], which is based on an information retrieval model comprising the following four sub-models, as shown in Fig. 2.

i. Model 1 describes the relationship between a given customer c and a given product p.
ii. Model 2 describes the relationship between a given customer c and a given group of products $[p]$.
iii. Model 3 describes the relationship between a given group of customers $[c]$ and a given product p.
iv. Model 4 describes the relationship between a given group of customers $[c]$ and a given group of products $[p]$.

Fig. 2 The targeted marketing models

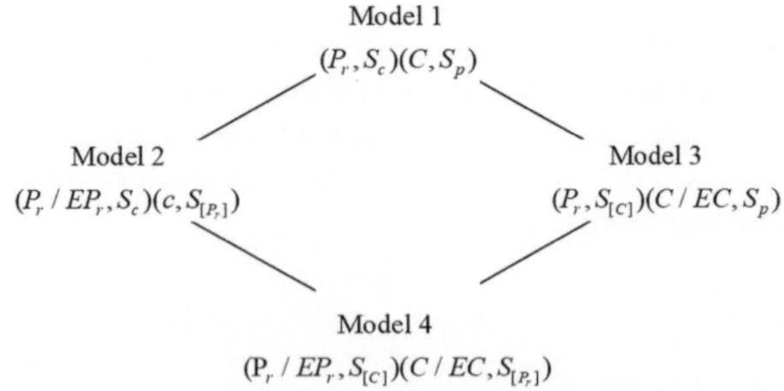

We can use Model 3 to construct an outranking relation $(P_r, S_{[C]})$ for a given group of customers, or we may group customers together and construct an outranking relation $(C/EC, S_p)$ for a given product. On the other hand, given a customer, we may group products together according to their properties and construct an outranking relation $(P_r/EP_r, S_c)$ by using a Model 2. Model 2 can also be used to construct an outranking relation $(C, S_{[P_r]})$ for different customers. Otherwise, we can group customers and products, respectively, and construct an outranking relation $(P_r/EP_r, S_{[C]})$ and $(C/EC, S_{[P_r]})$ by using Model 4. We can also construct an outranking relation (P_r, S_c) and (C, S_p) by using Model 1, which describes the relationship between a given customer c and a given product p.

Multicriteria decision analysis assumes that customers with similar preferences tend to make similar decisions and choose similar products. In this case, Model 3 can be used to construct an outranking relation $(P_r, S_{[C]})$. Our approach is based on product clusters under the assumption that customers tend to choose products that are similar to what they have bought in the past, and cluster products to obtain product clusters P_r/EP_r. In this case, Model 2 is used to construct an outranking relation $(P_r/EP_r, S_c)$. In direct marketing, one aims to single out likely buyers for a product who share the same or similar features for a product. Model 3 is used to construct an outranking relation $(C/EC, S_p)$. Additional outranking relations can be obtained in this form.

For models 2, 3, and 4, a fundamental issue is how to classify individual customers or products based on the available information. Customers can be categorized in various ways, depending on their attributes, the products they have purchased, and the characteristics of those products. One approach to clustering focuses on customer preferences regarding these attributes. Customers with similar preferences are placed in the same group. Alternatively, customers are grouped together if they tend to purchase the same or similar products. According to product properties, customers are organized into the same group when they buy products with equal or similar features. Similarly, products can be grouped based on their attributes, the customers who bought them, and the characteristics of those customers.

4 Construction of Outranking Relations and Deriving a Final Ranking

The outranking approach for the multicriteria ranking problem consists of two phases [52, 59]. The first phase involves constructing a decision maker's preferences aggregation model based on a set of alternatives evaluated against a set of criteria, as presented in Sects. 3.1 and 3.2. This model is represented through one or more outranking relations. The second phase focuses on utilizing the model to generate a global ranking of the alternatives in a non-increasing order of preference, as detailed in Sect. 3.3. In outranking relations, the alternatives may represent products or customers, depending on the strategy employed to address the targeted marketing problem.

We create three types of outranking relations based on the available information. One outranking relation utilizes product profiles and transaction data. Another outranking relation incorporates customer profiles along with transaction data, while a third outranking relation combines transaction data, customer profiles, and product profiles.

Customer profiles, product profiles, and transaction data are given by information tables [11]. An information table is a quadruple:

$$IT = (U, H, \{V_{h_k} : h_k \in H\}, \{I_{h_k} : h_k \in H\})$$

where U is a finite nonempty set of objects, H is a finite nonempty set of attributes, V_{h_k} is a nonempty set of values for $h_k \in H$, and $I_{h_k} : U \to V_{h_k}$ is an information function for $h_k \in H$. We have three types of tables for targeted marketing. If U is the product set P_r and H is the set of the properties of products that can be denoted by H_P, we call it a product profile table. If U is the customer set C and H is the set of the socioeconomic and demographic information of customers, which can be denoted by H_C, we call it a customer profile table. If U is the customer set C, H is the product set P_r, and I_{h_k} denotes the judgments of customers on products; we call it a transaction table. In a transaction table, each row represents a customer's judgments on various products, while each column reflects the judgments of different customers for the same product.

4.1 Using Product Information to Construct an Outranking Relation for a Given Customer

For customer-oriented targeted marketing based on the outranking approach, we focus on a particular customer c and a set of products P_r to obtain an outranking relation S_c. The outranking relation is established through pairwise comparisons of products based on the implicitly provided customer preferences. By doing this, we create a customer-oriented marketing strategy. The outranking relation S_c can be exploited to derive a ranking of products. A cut-off point of the ranked list may be chosen based on certain criteria. We can identify the top products above the cutoff points as our targeted products. In the following, we will explain this strategy in detail.

Here, we assume that there are many products that the sales department wants to recommend to a particular customer. The problem then becomes how to select products that will interest the customer, considering the implicitly preferential information about the customer in an multicriteria context so that the recommendations generated align with those preferences.

We can view the product profiles as the input for the outranking relation. For a specific customer c, the outranking relation of products can be defined using the concordance and non-discordance conditions, along with the indifference, preference, and veto thresholds and weights as parameters in the outranking relation model provided by the decision-maker.

The product set P_r can be decomposed into three subsets P_r1, P_r2, and P_r3. P_r1 is the set of products of the company that the customer c has purchased at least once, and there are records of their purchases (note that the set P_r1 can have repeated products), P_r2 represents the set of products of the company which the customer has never bought them and the company does not want to offer them, and finally, P_r3 is the set of products under study to be offered to the customer c. A discrete or continuous criterion function g_{h_k} can be deduced from the attribute h_k, for each $h_k \in H$, and it may be defined based on the historical distribution of attribute values. In the following, we present in detail the determination of these two types of criteria.

4.1.1 The Criterion g_{h_k} is Discrete

Suppose the set of attribute values V_{h_k} is discrete, and, let $r(v)$ be the subset of P_r1 defined by

$$r(v) = \{p \in P_r1 : I_{h_k}(p) = v, \quad v \in V_{h_k}\}.$$

Let $|r(v)|$ denote the cardinality of the set $r(v)$. We define a frequentist version of the function f_{h_k} as

$$f_{h_k}(v) = \frac{|r(v)|}{\sum_{v' \in V_{h_k}} |r(v')|} = \frac{|r(v)|}{m_2}$$

where m_2 is the number of products in P_r1.

In terms of products, we have an equivalent definition:

$$f_{h_k}\left(I_{h_k}(p)\right) = \frac{\left|r\left(I_{h_k}(p)\right)\right|}{m_2}$$

The function g_{h_k} is defined as

$$g_k = g_{h_k} = f_{h_k} \circ I_{h_k} : P \to [0, 1]$$

is a discrete criterion function deduced from the attribute h_k, for each $h_k \in H$ for which the set of attribute values V_{h_k} is discrete.

According to the functions f_{h_k}, an attribute value contributes more to the interest of a product if it appears in a greater number of products.

In terms of the multicriteria decision analysis, the set P_r3 represents the set of decision alternatives under study, and the set $G = \{g_1, g_2, \ldots, g_r\}$ represents a family of decision criteria.

Let $p_i, p_j \in P_r3$ be two products which are candidates to promote, and $g_k \in G$ is a criterion function, then, we can use the outranking approach of MCDA of Sect. 3 for comparing both products.

$$p_i P_k p_j (p_i \textit{ is preferred to } p_j) \leftrightarrow g_k(p_i) > g_k(p_j)$$

$$p_i I_k p_j (p_i \textit{ is indifferent to } p_j) \leftrightarrow g_k(p_i) = g_k(p_j)$$

4.1.2 The Criterion g_{h_k} is Continuous

Now, Suppose the set of attribute values V_{h_k} is continuous. The set V_{h_k} represents an scale associated with $I_{h_k} : U \to V_{h_k}$, the information (performance) function for $h_k \in H$.

Considering the previously defined attributes in the formulation of the model, we will give some examples of preference continuous criteria that can appear in customer-oriented marketing:

These continuous criteria are based on the types of attributes where the customer's preferences range on an interval of real values related to the attributes of the products that the customer has historically purchased:

Case 1. The customer expresses a weak preference for those products whose attribute value is between 0 and x_1, while he shows a strong preference for those products whose attribute value is between x_1 and x_2, and expresses a weak preference for those products that have a value of the attribute between x_2 and x_3. Any product whose attribute value is less than 0 and greater than x_3 is not attractive to the customer, where $0 < x_1 < x_2 < x_3$. The behavior of this function can be modeled as a fuzzy trapezoidal function, as shown graphically in Fig. 3, and algebraically by means of Eq. (5):

$$g_{h_k}(x) = \begin{cases} \frac{x}{x_1} & \textit{if } 0 \le x \le x_1 \\ 1 & \textit{if } x_1 \le x \le x_2 \\ \frac{x_3 - x}{x_3 - x_2} & \textit{if } x_2 \le x \le x_3 \\ 0 & \textit{if } x > x_3 \end{cases} \tag{5}$$

Other examples of customer preference modelization on a specific attribute based on historical purchased products are the following:

Case 2. The customer expresses a strong preference for those products whose attribute value is up to x_1, while he shows a weak preference for those products whose attribute

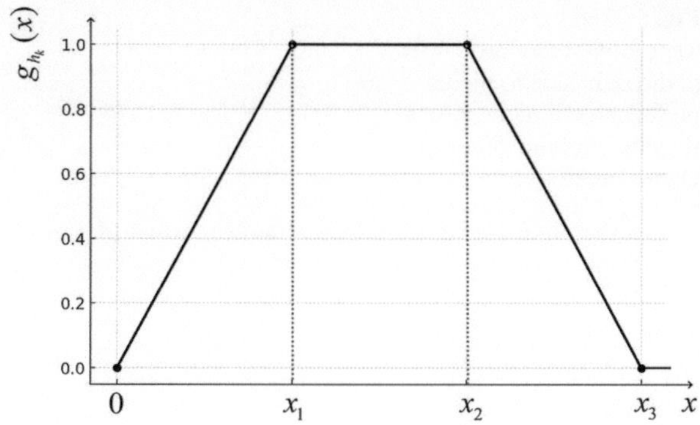

Fig. 3 Trapezoidal fuzzy function to model the preferences of a customer. *Source* Own elaboration

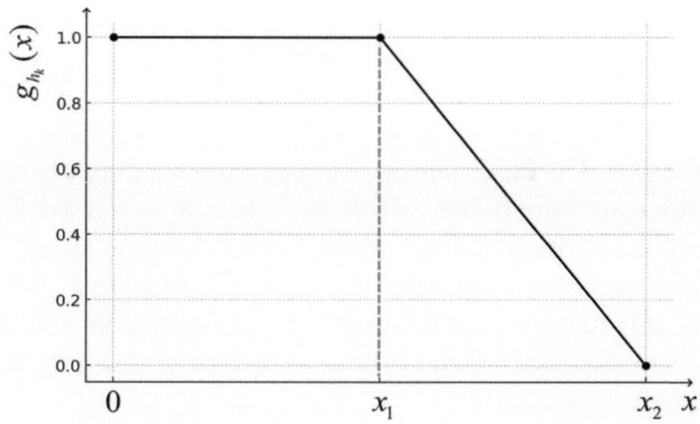

Fig. 4 The trapezoidal fuzzy function opened to the left side to model the preferences of a client on a specific attribute. *Source* Own elaboration

value is between x_1 and x_2. Any product whose attribute value is greater than x_2 is not attractive to the customer where $0 < x_1 < x_2$. The behavior of this function can be modeled as a fuzzy trapezoidal function opened to the left side, as shown graphically in Fig. 4, and algebraically by means of Eq. (6):

$$g_{h_k}(x) = \begin{cases} 1 & if \quad x \leq x_1 \\ \frac{x_2 - x}{x_2 - x_1} & if \; x_1 \leq x \leq x_2 \\ 0 & if \quad x > x_2 \end{cases} \tag{6}$$

Case 3. The customer expresses a weak preference for those products whose attribute value is from 0 to x_1, while he shows a strong preference for those products whose attribute value is equal or greater to x_1, where $0 < x_1$. The behavior of this function can be modeled as a fuzzy trapezoidal function opened to the right side, as shown graphically in Fig. 5, and algebraically by means of Eq. (7):

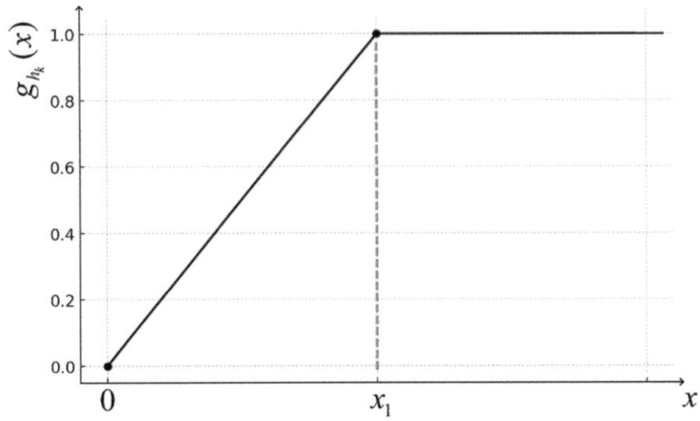

Fig. 5 The fuzzy trapezoidal function is open to the right side to model preferences of a client on a specific attribute. *Source* Own elaboration

$$g_{h_k}(x) = \begin{cases} \frac{x}{x_1} \; if \; 0 \leq x \leq x_1 \\ 1 \; \; if \; x \geq x_1 \end{cases} \tag{7}$$

Case 4. The customer does not express preference for the products whose attribute values is from 0 to x_1, while, he/she expresses a weak preference for those products whose attribute value is from x_1 to x_2, and he shows a strong preference for those products whose attribute value is at least equal to x_2, where $0 < x_1 < x_2$. The behavior of this function can be modeled as a fuzzy trapezoidal function opened to the right side, shifted from the origin, as shown graphically in Fig. 6, and algebraically by means of Eq. (8):

$$g_{h_k}(x) = \begin{cases} 0 & if \; 0 \leq x \leq x_1 \\ \frac{x-x_1}{x_2-x_1} & if \; x_1 \leq x \leq x_2 \\ 1 & if \; x \leq x_2 \end{cases} \tag{8}$$

Therefore, it can be defined that a product p_i is preferable to a product p_j or "at least as good as" if $g_{h_k}(p_i) \geq g_{h_k}(p_j)$, where $g_{h_k}(p_i)$ is the preference degree of the

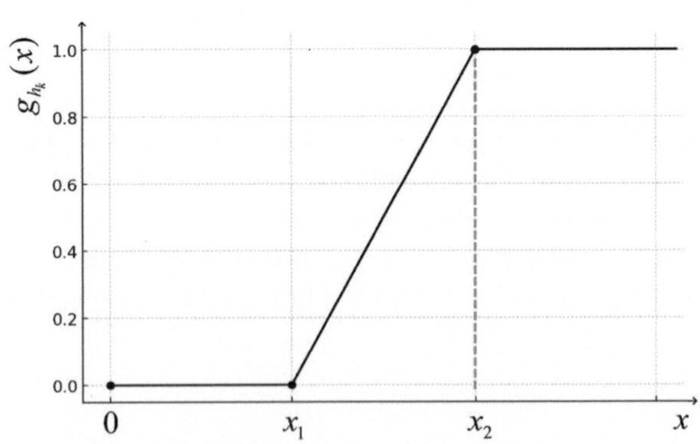

Fig. 6 Open fuzzy trapezoidal function to the right side according to the origin to model preferences of a client on a specific attribute. *Source* Own elaboration

attribute h_k within of a product p_i. The decision-maker is interested in modeling strong preferences of customer purchases on a specific attibute. Then, it should be desirable to offer to the customer products whose attribute value range is in the category of strong preference modelization according to his/her pattern of historical purchases of products in the company. All these criteria are to maximize.

In general, a trapezoidal fuzzy function can have various shapes depending on the application concerned.

4.1.3 The Outranking Model and De Associated Ranking

Suppose we have a large set of products $P_r = \{p_1, p_2, \ldots, p_m\}$ of a company that a marketer can offer to a particular customer. The products can be analyzed by a family of criteria $G = \{g_1, g_2, \ldots, g_r\}$, determined with the processes in Sect. 4.1.1 and 4.1.2, where $g_j(p_i)$ represents the evaluation of the product p_i by criterion g_j. The product set P_r can be decomposed into three subsets P_r1, P_r2, and P_r3. P_r1 is the set of products of the company that the customer c has purchased at least once, and there are records of their purchases, P_r2 represents the set of products of the company which the customer has never bought them, and finally, P_r3 is the set of products under study to be offered to the customer c.

In Algorithm 1, we present an algorithmic framework for customer-oriented marketing to find the k-most preferable products (k-MPP) from P_r3 for targeted marketing. The framework utilizes a multicriteria outranking approach to solve the k-MPP problem.

5 Conclusion

In this chapter, we propose a multicriteria framework for targeted marketing based on the outranking approach, where a set of criteria is introduced and used to conduct pairwise comparisons of products or customers according to a series of decision rules. The proposed method addresses the limitations of classical analytical targeted marketing models, which do not accommodate complex logic when aggregating various criteria.

In this framework, customer and product attributes, along with transaction data, are used to establish outranking relations related to four models for targeted marketing through preference relations. The four models correspond to the relationships among a group of customers and a group of products, between a group of customers and a specific product, between a specific customer and a group of products, and between a particular customer and a particular product. Customer-oriented marketing and product-oriented marketing strategies are applied across these four models. Customer-oriented marketing ranks products concerning a specific customer or group of customers, while product-oriented marketing ranks customers concerning

```
Input: P_r={p_1,p_2,...,p_m}:Set of products
       C_t={c_1,c_2,...,c_t}:Set of customers
       H_n={h_1,h_2,...,h_n}: Set of attributes
       V_{h_k}: Set of values for h_k ∈ H_n
       I_{h_k}: P_r → V_{h_k}: Information function for h_k ∈ H_n
       G_{P_r} : Product profile table
       T_{C_t×P_r} : C_t × P_r → □ : Transaction table
Output: The k-most preferable products (k-MPP)
        from P_r3 for targeted marketing.
P_r ⊇ P_r1 ∪ P_r2 ∪ P_r3 // Decomposition of P_r
G_n={g_1,g_2,...,g_n} // Determination of the set of criteria
[g_j(p_i)]_{|P_r3|×n} // Calculate the performance matrix g_j ∈ G_n, p_i ∈ P_r3
q⃗ ← q⃗_0, p⃗ ← p⃗_0, v⃗ ← v⃗_0, w⃗ ← w⃗_0 // Define the indiference,
  preference and veto threshold and weights of the criteria.
θ ← {1,2,...,n} // Set of indices of the criteria with veto.
k ← 1
while (k ≤ n) do:
   Compute I_h, P_h, Q_h and S_h
   i ← 1, j ← 1
   while(i ≤ |P_r3|) do:
      while(j ≤ |P_r3|)do:
         compute c(p_i I_k p_j), c(p_i P_k p_j), <c(p_i Q_k p_j)
         j ← j+1
      end while
      i ← i+1
   end while
   k ← k+1
end while
for each(k ∈ θ) do
   i ← 1, j ← 1
   while (i ≤ |P_r3|) do
      while (j ≤ |P_r3|) do
         compute d_k(p_i, p_j) // Discordance index
      end while
   end while
end for
i ← 1, j ← 1
while(i ≤ |P_r3|)do
   while (j ≤ |P_r3|)do
      compute c(p_i, p_j) // Concordance Index
      compute σ(p_i, p_j) // Credibility Index
   end while
end while
i ← 1
while(i ≤ |P_r3|) do
   Compute ℓ+(p_i) // The leaving flow
   Compute ℓ-(p_i) // The entering flow
   Compute ℓ(p_i) // The net flow
end while
Define a cut-off point "k" of the ranking
Return the k-MPP from P_r3
```

Algorithm 1 Algorithm framework for customer-oriented marketing a specific product or group of products. Within the proposed framework, collaborative filtering can be viewed as constructing the relationship between a group of customers and a product, content-based filtering as connecting a customer and a group of products, and direct marketing systems as linking a group of customers and a product.

Fuzzy outranking relations are utilized to implement the proposed framework, which can be leveraged to generate a ranking of products or customers. Based on specific criteria, a cut-off point can be established. We can identify the top products above these cut-off points as targeted products. We believe the primary value of this framework stems from the pairwise comparison of products or customers in fuzzy outranking relations, resulting from a partially compensatory balance between concordant and discordant arguments.

Future work will involve operationally extending our framework of targeted marketing to construct outranking relations related to models 2, 3, and 4, as well as conducting experiments to strengthen the results. More specifically, applying our approach within a customer-oriented marketing strategy and a product-oriented marketing strategy, considering groups of customers and products, would be an intriguing extension of our work. Furthermore, this work acknowledges that each criterion is neither predominant nor negligible. It will also be interesting to explore a specific outranking approach that does not consider the weight of the criteria when there is no evidence indicating that some criteria are more important than others.

References

1. Peng, Y., Wong, R.C., Wan, Q.: Finding top-k preferable products. IEEE Trans. Knowl. Data Eng. **24**, 1774–1788 (2012)
2. Lin, C.-Y., Koh, J.-L., Chen, A.L.P.: Determining (k)-most demanding products with maximum expected number of total customers. IEEE Trans. Knowl. Data Eng. **25**, 1732–1747 (2013)
3. Koh, J.-L., Lin, C.-Y., Chen, A.L.P.: Finding k most favorite products based on reverse top-t queries. VLDB J. **23**, 541–564 (2014)
4. Zhong, N., Liu, J., Yao, Y.Y.: Web Intelligence. Springer, Berlin, Germany (2003)
5. Zhong, N., Yao, Y.Y., Liu, C., Huang, J., Ou, C.: Data mining for targeted marketing. In: Intelligent Technologies for Information Analysis, pp. 109–131. Springer, Berlin, Germany (2004)
6. Han, J.W., Kamber, M.: Data Mining Concepts and Techniques. Higher Education Press and Morgan Kaufmann Publishers (2002)
7. Huang, J., Zhong, N., Liu, C., Yao, Y.Y.: Adaptive linear market value functions for targeted marketing. In: Proceedings of the 4th International Conference on Rough Sets and Current Trends in Computing (RSCTC 2004), pp. 743–751. Springer, Berlin, Germany (2004)
8. Figueroa-Perez, J.F., Leyva-Lopez, J.C., Santillan, L.C., Pérez Contreras, E.O., Sánchez, P.J.: The use of marketing decision support systems for new product design: a review. IJCIS **12**, 761 (2019). https://doi.org/10.2991/ijcis.d.190624.001
9. Perny, P., Roy, B.: The use of fuzzy outranking relations in preference modelling. Fuzzy Sets Syst. **49**, 33–53 (1992). https://doi.org/10.1016/0165-0114(92)90108-G
10. Greco, S., Ehrgott, M., Figueira, J.R. (eds.): Multiple Criteria Decision Analysis: State of the Art Surveys. Springer, New York (2016)

11. Huang, J., Zhong, N., Yao, Y.: A unified framework of targeted marketing using customer preferences. Comput. Intell. **30**, 451–472 (2014). https://doi.org/10.1111/coin.12003
12. Babu, G., Bhuvaneswari, T.: A data mining technique to find optimal customers for beneficial customer relationship management. J. Comput. Sci. **8**, 89–98 (2012). https://doi.org/10.3844/jcssp.2012.89.98
13. Adomavicius, G., Tuzhilin, A.: Toward the next generation of recommender systems: a survey of the state-of-the-art and possible extensions. IEEE Trans. Knowl. Data Eng. **17**, 734–749 (2005). https://doi.org/10.1109/TKDE.2005.99
14. Leyva López, J.C., Reyna Gutiérrez, O.A.: Finding Top-k preferable products for customer-oriented marketing based on the outranking approach: a case study on Mexican restaurants. J. Univ. Comput. Sci. **31**, 210–238 (2025). https://doi.org/10.3897/jucs.150597
15. Huang, J., Zhong, N., Yao, Y.Y., Liu, C.: A general framework of targeted marketing. In: Szczepaniak, P.S., Kacprzyk, J., Niewiadomski, A. (eds.) Advances in Web Intelligence, pp. 197–203. Springer, Berlin Heidelberg (2005)
16. Liu, L., Yang, Z., Benslimane, Y.: Conducting efficient and cost-effective targeted marketing using data mining techniques. In: 2013 Fourth Global Congress on Intelligent Systems, pp. 102–106. IEEE, USA (2013)
17. Zhao, H.-H., Luo, X.-C., Ma, R., Lu, X.: An extended regularized K-means clustering approach for high-dimensional customer segmentation with correlated variables. IEEE Access **9**, 48405–48412 (2021). https://doi.org/10.1109/ACCESS.2021.3067499
18. Yoseph, F., Ahamed Hassain Malim, N.H., Heikkilä, M., Brezulianu, A., Geman, O., Paskhal Rostam, N.A.: The impact of big data market segmentation using data mining and clustering techniques. JIFS **38**, 6159–6173 (2020). https://doi.org/10.3233/JIFS-179698
19. Ernawati, Kamal, B.S.S., Kasmin, F., Purwanugraha, H.A.: Geo-marketing promotional target selection using modified RFM with spatial and temporal analysis: a case study. J. Syst. Manage. Sci. **12**, 156–180 (2022). https://doi.org/10.33168/JSMS.2022.0309
20. Hwang, S., Lee, Y.: Identifying customer priority for new products in target marketing: using RFM model and TextRank. Innov. Market. **17**, 125–136 (2021). https://doi.org/10.21511/im.17(2).2021.12
21. Beheshtian-Ardakani, A., Fathian, M., Gholamian, M.: A novel model for product bundling and direct marketing in e-commerce based on market segmentation, 39–54 (2018). https://doi.org/10.5267/j.dsl.2017.4.005
22. Kaur, M., Sharma, A., Kishore, M., Shubhnath, S.S., Verma, H.: Customer segmentation with RFM analysis using ML models. In: 2023 3rd International Conference on Innovative Mechanisms for Industry Applications (ICIMIA), pp. 743–749. IEEE, USA (2023)
23. Ali, N., Shabn, O.S.: Customer lifetime value (CLV) insights for strategic marketing success and its impact on organizational financial performance. Cogent Bus. Manag. **11** (2024). https://doi.org/10.1080/23311975.2024.2361321
24. Theodorakopoulos, L., Theodoropoulou, A.: Leveraging big data analytics for understanding consumer behavior in digital marketing: a systematic review. Hum. Behav. Emerg. Technol. 3641502 (2024). https://doi.org/10.1155/2024/3641502
25. Potharst, R., Kaymak, U., Pijls, W.: Neural networks for targeted selection in direct marketing. In: Networks in Business: Techniques and Applications, pp. 89–110. Idea Group, London (2001)
26. Yao, Y.Y., Zhong, N., Huang, J., Ou, C., Liu, C.: Using market value functions for targeted marketing data mining. Int. J. Patt. Recog. Artif. Intell. **16**, 1117–1131 (2002)
27. Gubela, R.M., Lessmann, S., Jaroszewicz, S.: Response transformation and profit decomposition for revenue uplift modeling. Eur. J. Oper. Res. **283**, 647–661 (2020). https://doi.org/10.1016/j.ejor.2019.11.030
28. Gubela, R.M., Lessmann, S., Stöcker, B.: Multiple treatment modeling for target marketing campaigns: a large-scale benchmark study. Inf. Syst. Front. (2022). https://doi.org/10.1007/s10796-022-10283-4
29. Baier, D., Stöcker, B.: Profit uplift modeling for direct marketing campaigns: approaches and applications for online shops. J. Bus. Econ. **92**, 645–673 (2022). https://doi.org/10.1007/s11573-021-01068-3

30. Sarkar, M., De Bruyn, A.: LSTM response models for direct marketing analytics: replacing feature engineering with deep learning. J. Interact. Mark. **53**, 80–95 (2021). https://doi.org/10.1016/j.intmar.2020.07.002
31. Burke, R., Felfernig, A., Goker, M.H.: Recommender systems: an overview. Artif. Intell. Mag. **32**, 13–18 (2011)
32. Ricci, F., Rokach, L., Shapira, B., Kantor, P.B. (eds.): Recommender Systems Handbook. Springer, US, Boston, MA (2011)
33. Lu, L.Y., Medo, M., Yeung, C.H., Zhang, Y.C., Zhang, Z.K., Zhou, T.: Recommender systems. Phys. Rep. **519**, 1–49 (2012)
34. Breese, J., Heckerman, D., Kadie, C.: Empirical analysis of predictive algorithms for collaborative filtering. In: Proceedings of the 14th Conference on Uncertainty in Artificial Intelligence (UAI'98), pp. 43–52. Morgan Kaufmann, San Francisco (1998)
35. Sarwar, B., Karypis, G., Konstan, J., Riedl, J.: Item-based collaborative filtering recommendation algorithms. In: Proceedings of the 10th International World Wide Web Conference (WWW 2001), pp. 285–295. ACM, New York (2001)
36. Bobadilla, J., Hernando, A., Ortega, F., Bernal, J.: A framework for collaborative filtering recommender systems. Expert Syst. Appl. **38**, 14609–14623 (2011)
37. Balabanovic, M., Shoham, Y.: Fab: Content-based, collaborative recommendation. Commun. ACM **40**, 66–72 (1997)
38. Basilico, J., Hofmann, T.: Unifying collaborative and content-based filtering. In: Proceedings of the 21st International Conference on Machine Learning (ICML 2004). ACM Press, New York (2004)
39. Vozalis, M., Margaritis, G.: On the enhancement of collaborative filtering by demographic data. Web Intell. Agent Syst. Int. J. **4**, 117–138 (2006)
40. Cohen, W., Schapire, R.E., Singer, Y.: Learning to order things. Artif. Intell. Res. **10**, 243–270 (1999)
41. Sai, Y., Yao, Y.Y., Zhong, N.: Multiagent systems: A survey from a machine learning perspective. In: Proceedings of the 2001 IEEE International Conference on Data Mining (ICDM 2001), pp. 497–504. IEEE Press, New York (2001)
42. Jin, R., Si, L., Zhai, C., Callan, J.: Collaborative filtering with decoupled models for preferences and ratings. In: Proceedings of the 12th International Conference on Information and Knowledge Management. ACM, New York (2003)
43. Hadad, Y., Keren, B.: A decision-making support system module for customer segmentation and ranking. Expert Syst. (2022). https://doi.org/10.1111/exsy.13169
44. Hashemi, A., Zare Chahooki, M.A.: GroupRank: ranking online social groups based on user membership records. JAIDM **9** (2021). https://doi.org/10.22044/jadm.2020.8337.1973
45. Zunk, B.M., Koch, V.: Customer ranking model for project businesses: a case study from the automotive industry. Int. J. Eng. Bus. Manag. **6**, 10 (2014). https://doi.org/10.5772/58454
46. Herlocker, J.L., Konstan, J.A., Terveen, L.G., Riedl, J.T.: An algorithmic framework for performing collaborative filtering. In: Proceedings of the 22nd Annual International Conference on Research and Development in Information Retrieval (SIGIR 1999), pp. 230–237. ACM Press, New York (1999)
47. Pazzani, M.J.: A framework for collaborative, content-based and demographic filtering. Artif. Intell. Rev. **13**, 393–408 (1999)
48. Hernandez, F., Gaudioso, E.: Evaluation of recommender systems: a new approach. Expert Syst. Appl. **35**, 790–804 (2008)
49. Huang, J., Zhong, N.: A unified probabilistic inference model for targeted marketing. In: Communications and Discoveries from Multidisciplinary Data, pp. 171–186. Springer, Berlin, Germany (2008)
50. Willetts, M., Atkins, A.S., Stanier, C.: Barriers to SMEs adoption of big data analytics for competitive advantage. In: 2020 Fourth International Conference On Intelligent Computing in Data Sciences (ICDS), pp. 1–8. IEEE, Fez, Morocco (2020)
51. World Bank: SME finance: Development news, research, Data
52. Roy, B.: Multicriteria Methodology for Decision Aiding. Springer, US, Boston, MA (1996)

53. Bouyssou, D., Vincke, Ph.: Ranking alternatives on the basis of preference relations. J. Multicriteria Decis. Anal. **6**, 77–85 (1997)
54. Roy, B.: ELECTRE III: Un algorithme de classement fondé sur une représentation floue des préférences en présence de critères multiples. Cahiers du CERO **20**, 3–24 (1978)
55. Figueira, J.R., Roy, B.: Determining the weights of criteria in the electre type methods with a revised Simos' procedure. Eur. J. Oper. Res. **139**, 317–326 (2002)
56. Brans, J.P., Vincke, Ph.: A preference ranking organization method. Manag. Sci. **31**, 647–656 (1985)
57. Behzadian, M., Kazemzadeh, R.B., Albadvi, A., Aghdasi, M.: PROMETHEE: a comprehensive literature review on methodologies and applications. Eur. J. Oper. Res. **200**, 198–215 (2010). https://doi.org/10.1016/j.ejor.2009.01.021
58. Masthoff, J.: Group recommender systems: Combining individual models. In: Recommender Systems Handbook, pp. 677–704. Springer, Berlin, Germany (2011)
59. Govindan, K., Jepsen, M.B.: ELECTRE: A comprehensive literature review on methodologies and applications. Eur. J. Oper. Res. **250**, 1–29 (2016). https://doi.org/10.1016/j.ejor.2015.07.019

SINDy Meets Schelling. Transforming Agent-Based Model Spatial Outputs Into Dynamical Systems

Jorge Zazueta Gutiérrez

Abstract Agent-based models (ABMs) are standard tools for modeling social and physical phenomena from the ground up by building detailed simulations in which individual agents interact and study emergent behavior. However, since they are simulations, it is challenging to generalize or even quantitatively interpret the results. A frequent output of an ABM model is a grid with points of one or more classes that represent the agents' configuration over time. A canonical example is the Schelling segregation model, where agents of two types follow a simple relocation rule based on their tolerance to the proportion of different agents in a given location, resulting in a segregated configuration that is visually revealing but not quantitative. In this work, we propose assigning a quantitative measure of entropy, based on the spatial configuration of the steady state of the Schelling model, to a range of population values in the model using Topological Data Analysis (TDA) techniques. The resulting dataset of quantitative metrics related to the original configuration is analyzed via Sparse Identification of Nonlinear Dynamics (SINDy) methods to obtain a representation of the system dynamics in the form of an ordinary differential equation.

Keywords Sparse identification of nonlinear dynamics · Differential equations · ABM · TDA · Topological data analysis

1 Overview

In many scientific disciplines, tension exists between the detail-oriented formulation characteristic of Agent-Based Modeling (ABM) and the somewhat coarse view of Dynamical Systems represented as Differential Equations (DE). While the ABM framework allows us to construct fine-grained models based on an agent label and explore the system's potential emerging properties, the model's simulation-based

---

J. Zazueta Gutiérrez (✉)
Universidad Autónoma de San Luis Potosí, San Luis Potosí, Mexico
e-mail: jorge.zazueta@uaslp.mx

W. Pedrycz et al. (eds.), *Decision-Making in Computational Intelligence-Based Systems*, Studies in Systems, Decision and Control 628,
https://doi.org/10.1007/978-3-032-04056-5_4

nature can often be challenging to interpret and generalize. On the other hand, the necessary idealization to craft a parsimonious DE model risks missing behavioral details of the system under study. Nardini, Baker, Simpson, and Flores [1] explore equation learning methods for modeling the behavior of simulated biological systems, arguing against mean-field approaches. Nieman, Klus, and Schütte [2] use Koopman operator theory [3] to derive reduced models of agent-based models (ABMs). Elsenbroich and Polhill [4] review the various points of view of ABMs and prediction from methodological and epistemological positions, indirectly stressing the advantages of deriving more analyzable models.

The outcome of an ABM is typically a trajectory or an image; Schelling's segregation model is a canonical example [5], illustrating significant agent segregation due to mild individual preferences.

Several methods exist to identify the governing equations in the case of a trajectory [6–8]. This paper will focus on Sparse Identification of Nonlinear Dynamics (SINDy) [9, 10], a regression-based method to learn the governing differential equations (DEs) of a nonlinear dynamical system. This approach has been successfully applied to various physical problems, including the recovery of the Lorentz system from simulated data and multiple applications in fluid dynamics and engineering [9, 11, 12] We use Schelling's model as a case study to illustrate the value of applying SINDy in combination with topological data analysis techniques (TDA) [13] to identify governing equations for the ABM output.

When the ABM's output is an image, it is not immediately clear how to represent the result or its aspects quantitatively. In Schelling's model, a lattice populated by two types of agents evolves into different configurations depending on its size, density, and the individual agents' tolerance to diversity or homophily. While the structure of the final configuration is intuitively interpreted as segregation or integration, it is unclear how to translate that intuition into a governing equation.

A natural first thought is to measure Shannon's entropy [14] on the model's final state. However, given that the number of agents of each type remains fixed, entropy is invariant to the model's evolution. One way to capture the structure of a state is to represent its image as a point cloud and compute the entropy of the persistence barcode associated with its simplicial complex [15–19].

A *persistence barcode* is a visual representation used in TDA to summarize how topological features—such as connected components, holes, and voids—appear and disappear across different spatial or scale parameters in a dataset. Each bar in the barcode corresponds to a specific feature, with the bar's start and end points indicating the "birth" and "death" of that feature as the scale changes. These features are typically extracted from a mathematical structure known as a *simplicial complex*, which is a combinatorial object built from simple building blocks called *simplices*—points (0-simplices), line segments (1-simplices), triangles (2-simplices), and their higher-dimensional counterparts. A simplicial complex provides a discrete way to capture the shape or topology of a space. It serves as the foundation for computing persistent homology, from which persistence barcodes are derived. A detailed discussion can be found in [15].

Entropy dynamics can then be learned through symbolic machine-learning techniques such as SINDy.

A notable disadvantage of ABMs is that they are computationally expensive, making it challenging to study their behavior through exhaustive simulations. By identifying a formal system that faithfully represents the model's dynamics, the researcher is freed from the computational cost and can study the formal system either numerically or formally, with the added advantage of insights provided by the mathematical properties of the differential equation.

2 Methodology

Departing from a standard implementation of an ABM that produces a graphical output, we loop over a parameter of interest (i.e., population density) until convergence, collecting the resulting image for each parameter value and transforming it to a point cloud C_p. Notice that in the case of a n-dimensional output where there are m agent types, the resulting point cloud will be in $\mathbb{R}^{n\times(m-1)}$, where each agent type is represented on a plane that is orthogonal to all others. From C_p, we construct a simplicial complex S_p and encode its persistence homology as a barcode from which we calculate Shannon's entropy as follows:

$$S = -\sum_i \frac{l_i}{L} \log\left(\frac{l_i}{L}\right). \tag{1}$$

l_i is the persistence of the ith feature, and L is the sum of all persistencies.

Looping over the range of values of the parameter of interest, we obtain a list of entropy *observations* associated with the corresponding end state of the model run. This process is depicted in Fig. 1.

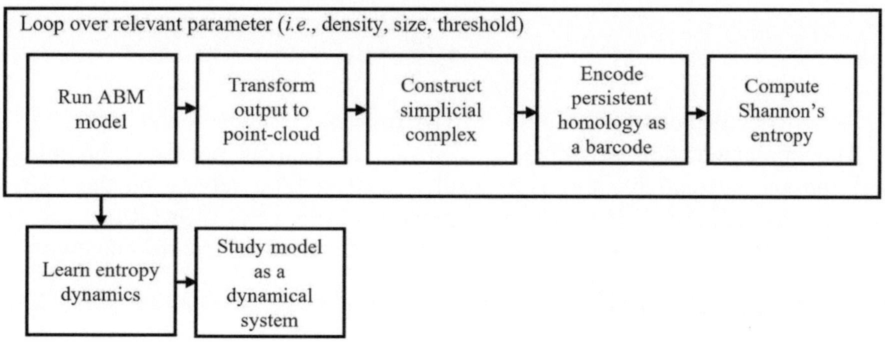

Fig. 1 Quantitative data generation and nonlinear system identification

To identify the governing equation for these observations, we rely on the SINDy framework.

Consider a dynamical system represented by

$$\dot{\mathbf{x}}(t) = f(\mathbf{x}(t)), \quad \mathbf{x}(t) \in \mathbb{R}^n \tag{2}$$

SINDy approximates the dynamics of (2) as

$$f(\mathbf{x}) \approx \sum_{k=1}^{l} \theta_k(\mathbf{x})\xi_k. \tag{3}$$

where $\theta(\mathbf{x}) = (\theta_1(\mathbf{x}), \theta_2(\mathbf{x}), \ldots, \theta_l(\mathbf{x}))$ is a library of predetermined basis functions and the coefficients ξ_i are chosen to be mostly zero by using standard regularization methods such as lasso of sequentially thresholded least squares (STLSQ). The use of regularization stems from the somewhat philosophical assumption that a good model should be parsimonious.

The sparse regression problem approximates.

$$p\dot{\mathbf{X}} = \Theta((X))\Xi, \tag{4}$$

where

$$\mathbf{X} = \begin{pmatrix} x_1(t_1) & x_2(t_1) & \cdots & x_n(t_1) \\ x_1(t_2) & x_2(t_2) & \cdots & x_n(t_2) \\ \vdots & \vdots & \ddots & \vdots \\ x_1(t_m) & x_2(t_m) & \cdots & x_n(t_m) \end{pmatrix}, \tag{5}$$

$$\Theta(\mathbf{X}) = (\theta_1(\mathbf{X}), \theta_2(\mathbf{X}, \ldots, \theta_l(\mathbf{X})), \tag{6}$$

and

$$\Xi = (\xi_1, \xi_2, \ldots, \xi_l)^T \tag{7}$$

A more detailed description of the SINDy framework can be found in [9].

The implementation of Schelling's model, as well as persistent homology and entropy calculations, was performed in Python [20], relying on the *Ripser* package [21, 22]. The SINDy discovery process relied on *PySINDy* [10].

The resulting model can then be analyzed as a standard differential equation.

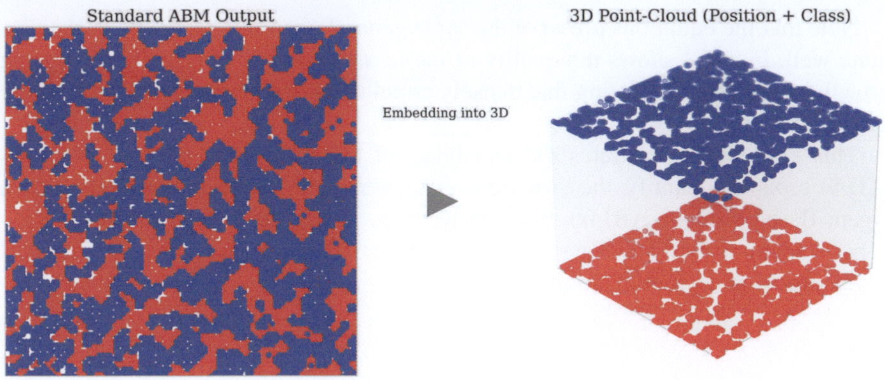

Fig. 2 Representing a grid with two agent types as a set of orthogonal planes in $\mathbb{R}^3$

Table 1 Schellling ODE model performance metrics for different homophily values

Homophily	Model	R^2	RMSE
h = 0.1	$\frac{dS}{d\delta} = 4.625 + 5.278S - 1.956S^2$	0.993874	0.004260
h = 0.3	$\frac{dS}{d\delta} = 4.605 + 5.961S - 2.245S^2$	0.993428	0.005716
h = 0.5	$\frac{dS}{d\delta} = 4.663 + 6.804S - 2.823S^2$	0.993609	0.006359

3 From Schelling to ODE

Schelling's model simulates agents of two types occupying a 2D grid. Agents prefer to have a minimum fraction of similar neighbors defined by the homophily threshold. Unhappy agents move to random empty spots until they are all happy.[1]

Agents check happiness based on a local Moore neighborhood (i.e., A Moore neighborhood consists of the eight surrounding cells (horizontal, vertical, and diagonal) around a central cell in a 2D grid.) Unhappy agents relocate randomly to empty cells.

The model's output is a grid with two color-coded agent types. To construct a simplicial complex, we represented the output as a point cloud consisting of two orthogonal planes in $\mathbb{R}^3$ where each agent type lies on its plane type (see Fig. 2).

We used SINDy to identify a differential equation that governs the dynamics of the entropy of the simplicial complex associated with the grid density. To test the robustness of our model, we ran three simulations with different levels of homophily and fitted them. Table 1 lists the resulting ODEs, along with their performance metrics.

[1] In our simulation, we set a convergence criteria of 1% of unhappy agents to avoid excessive computation time.

Note that the equations preserve the same general quadratic form and fit the data quite well. Figure 3 shows the quality of the fit and exhibits marginally decreasing growth in entropy, suggesting that densely populated areas segregate less than more sparse ones.

This case study illustrates the feasibility of using Topological Data Analysis (TDA) tools to quantify the structural characteristics of the output from certain Agent-Based Models (ABMs). It allows us to model aspects of their emerging complex behavior as dynamical systems, which, in turn, becomes a means of theoretical development and generalization.

4 Further Examples

To illustrate the versatility of the modeling framework, we will present two other ABM models and their SINDy representation. The Deffuant-Weisbuch model of opinion dynamics [23] and a simple voter model [24].

4.1 The Deffuant–Weisbuch Model of Opinion Dynamics

The Deffuant–Weisbuch model is a stochastic, agent-based model used to study the evolution of continuous opinions in a population. Each agent i holds an opinion vector $\mathbf{x}_i \in \mathbb{R}^d$, representing their stance on one or more issues. At each time step,

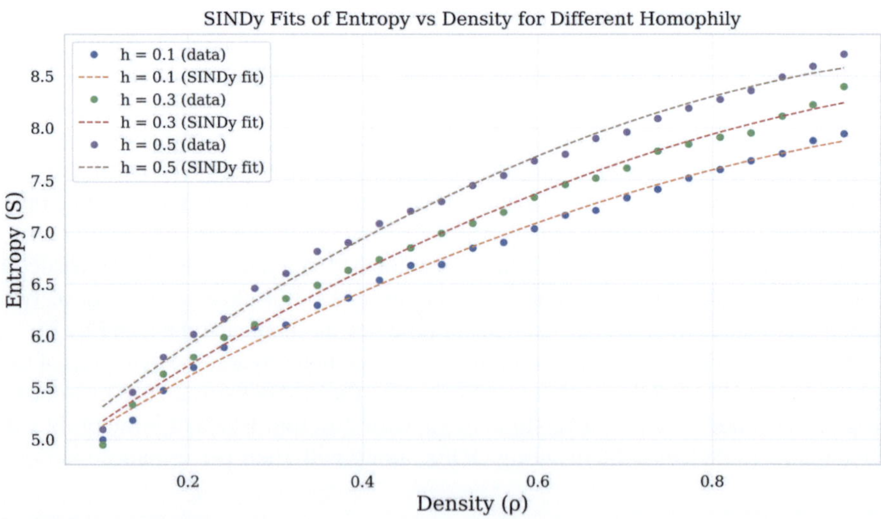

Fig. 3 Model predictions versus Schelling's ABM simulated data

a pair of agents (i, j) is selected at random. They interact only if their opinions are sufficiently similar, measured by a confidence bound $\varepsilon > 0$. Specifically, interaction occurs if the Euclidean distance between their opinions satisfies $\|\mathbf{x}_i - \mathbf{x}_j\| < \varepsilon$.

If the condition is met, both agents adjust their opinions toward each other according to a convergence parameter $\mu \in (0, 0.5]$. The update rule is given by:

$$\mathbf{x}_i \leftarrow \mathbf{x}_i + \mu(\mathbf{x}_j - \mathbf{x}_i), \quad \mathbf{x}_j \leftarrow \mathbf{x}_j + \mu(\mathbf{x}_i - \mathbf{x}_j).$$

This symmetric update ensures that opinions become more similar after each interaction. Over time, the model may lead to consensus (where all agents share the same opinion), polarization (with a few distinct clusters), or fragmentation (resulting in many small clusters), depending on the value of ε. Smaller values of ε tend to inhibit consensus by limiting interactions, while larger values promote agreement between populations.

To quantify the diversity of opinions in a population of N agents, one common approach is to compute the *Shannon entropy* of the opinion distribution. This requires discretizing the continuous opinion space into a finite number of bins or cells. Suppose the opinion space is partitioned into K bins, and let n_k denote the number of agents whose opinions fall into bin k. The empirical probability of bin k is then:

$$p_k = \frac{n_k}{\sum_{j=1}^{K} n_j} = \frac{n_k}{N}.$$

The (base-2) Shannon entropy is defined as:

$$H = -\sum_{k=1}^{K} p_k \log_2 p_k,$$

where the convention $0 \log 0 = 0$ is used for empty bins. Entropy is maximized when opinions are uniformly distributed across all bins, and minimized (equal to zero) when all agents fall into a single bin, indicating consensus. Therefore, entropy provides a quantitative measure of opinion fragmentation or diversity in the system.

Table 2 and Fig. 4 show the identified models and the quality of the continuous representation for several confidence bounds.

Table 2 Deffuant–Weisbuch ODE model performance metrics for different confidence bounds

Confidence	Model	R^2	RMSE
$\varepsilon = 0.1$	$\frac{dS}{dt} = 4.625 + 5.278S - 1.956S^2$	0.9862	0.0000
$\varepsilon = 0.3$	$\frac{dS}{dt} = 0.042 + -0.028S + 0.005S^2$	0.9812	0.0001
$\varepsilon = 0.5$	$\frac{dS}{dt} = 0.004 + -0.006S + 0.001S^2$	0.9865	0.0050

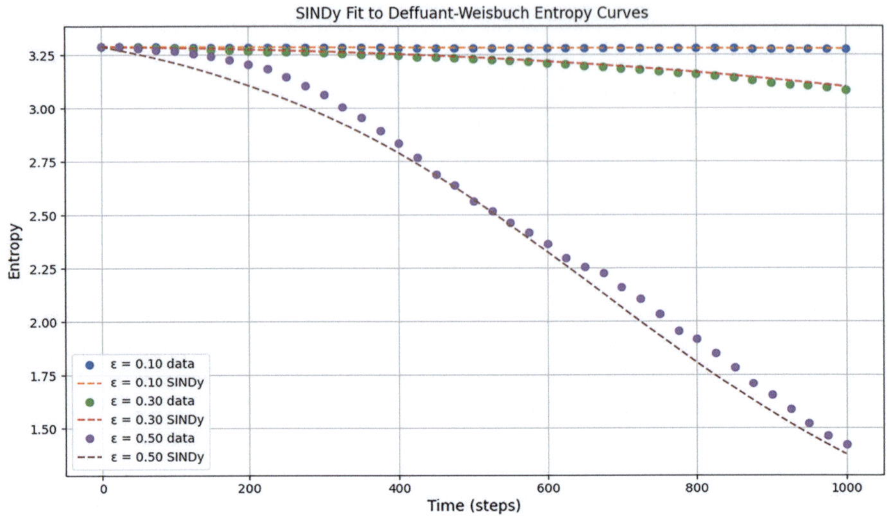

Fig. 4 Model predictions versus Deffuant-Weisbuch ABM simulated data

4.2 Continuous Voter Model

The *continuous voter model* is an extension of the classical binary voter model, where each agent holds a continuous opinion $x_i(t) \in [0, 1]$ rather than a binary choice. This framework enables more nuanced modeling of opinion dynamics in a population, where agents gradually adjust their opinions in response to interactions with others. At each discrete time step t, an agent i interacts with a randomly chosen neighbor j and updates its opinion by partially adopting the neighbor's opinion. The update rule is given by:

$$x_i(t+1) = x_i(t) + \mu \left(x_j(t) - x_i(t) \right),$$

where $\mu \in (0, 1)$ is the convergence parameter that controls how strongly agent i is influenced by agent j. This rule ensures that the agent's opinion moves closer to the neighbor's opinion, without completely adopting it unless $\mu = 1$.

Unlike bounded confidence models such as the Deffuant-Weisbuch model, the continuous voter model permits all agents to interact regardless of opinion distance. If the interaction network is connected, the system will converge to a consensus where all opinions are equal. This final consensus value is the mean of the initial opinions.

In this case, the model's output is a time series and can be modeled directly. Table 3 shows the models identified by SINDy, along with their performance metrics.

In this case, the identified model has a linear derivative with negative slope for all the tested values of μ, implying exponential decay. The model fit is remarkable (see Fig. 5).

Table 3 Voter ODE model performance metrics for different receptiveness values

Receptiveness	Model	R^2	RMSE
$\mu = 0.01$	$\frac{dv}{dt} = -0.020v$	1.000000	0.000001
$\mu = 0.02$	$\frac{dv}{dt} = -0.040v$	1.000000	0.000004
$\mu = 0.3$	$\frac{dv}{dt} = -0.673v$	0.999033	0.000145
$\mu = 0.4$	$\frac{dv}{dt} = -0.892v$	0.995893	0.000196

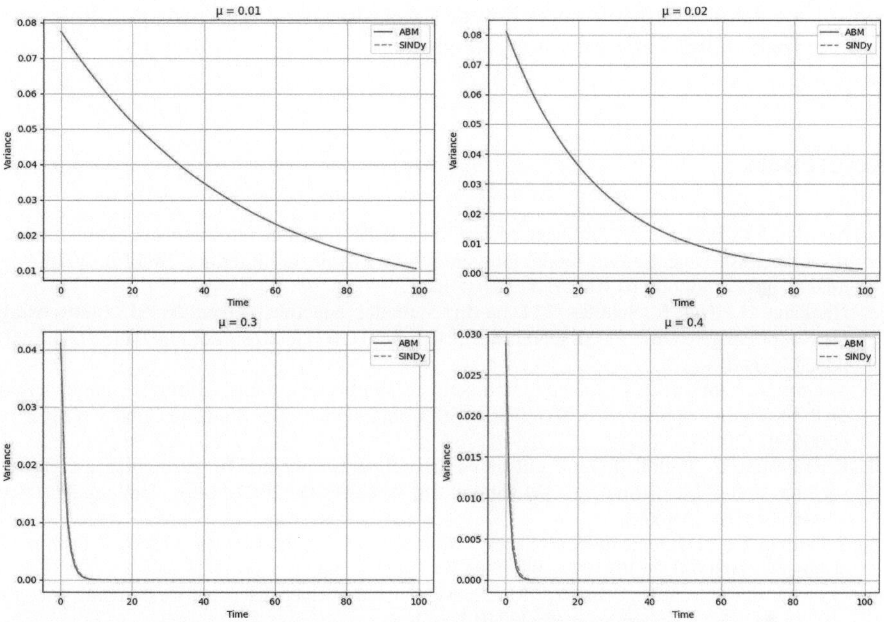

Fig. 5 Model predictions versus continuous voter model simulated data

4.3 Concluding Remarks

The ability of ABMs to capture subtle details of a social process by simulating action at the agent level is hampered by their computational cost and a notable difficulty in making predictions or being studied formally. By representing the dynamics generated by the model with a differential equation, the researcher can perform both numerical and symbolic analysis at a much lower computational cost and with greater interpretability, as well as the potential for generalization and identification of properties that are not easily spotted through simulation alone.

The SIDNy framework provides a means to do that, although not always in the most intuitive fashion. In this chapter, we encountered models that output a simple time series that can be modeled directly. In addition, we demonstrated techniques for addressing more complex problems, such as transforming a visual representation of segregation into an entropy curve using topological data analysis.

The case studies presented in this chapter generated a dynamic described by a single ordinary differential equation. However, more complex ABM models require a system of ODEs to describe the interaction between different features of the output (e.g., various topological properties in a visual representation or the several factors playing a role in a contagion model). ABM simulations are frequently spatial and require a partial differential equations model. Further refinements might incorporate control mechanisms to inform policy and decision-making. There is a long and exciting road ahead. Let's walk it together.

References

1. Nardini, J.T., Baker, R.E., Simpson, M.J., Flores, K.B.: Learning differential equation models from stochastic agent-based model simulations. J. Royal Soc. Interface **18**(176), 20200987 (2021). https://doi.org/10.1098/rsif.2020.0987
2. Niemann, H., Klus, S., Schütte, C.: Data-driven model reduction of agent-based systems using the Koopman generator. PLOS ONE **16**(5), e0250970 (2021). https://doi.org/10.1371/journal.pone.0250970
3. Mauroy, A., Goncalves, J.: Linear identification of nonlinear systems: a lifting technique based on the Koopman operator. In: 2016 IEEE 55th Conference on Decision and Control (CDC), p. 65006505 (2016)
4. Elsenbroich, C., Polhill, J.G.: Agent-based modelling as a method for prediction in complex social systems. Int. J. Soc. Res. Methodol. **26**(2), 133–142 (2023). https://doi.org/10.1080/13645579.2023.2152007
5. Schelling, T.C.: Dynamic models of segregation. J. Math. Sociol. **1**(2), 143–186 (1971). https://doi.org/10.1080/0022250X.1971.9989794
6. North, J.S., Wikle, C.K., Schliep, E.M.: A review of data-driven discovery for dynamic systems (2022). ArXiv. https://arxiv.org/abs/2210.10663
7. Wang, W., Lai, Y., Grebogi, C.: Data based identification and prediction of nonlinear and complex dynamical systems. Phys. Rep. **644**, 1–76 (2016). https://doi.org/10.1016/j.physrep.2016.06.004
8. Velázquez-Sánchez, R.D., Escobedo-Alva, J.O., Peña-García, R., Tapia-Herrera, R., Meda-Campaña, J.A.: Identification of high-order nonlinear coupled systems using a data-driven approach. Appl.Sci. **14**(9):3864 (2024). https://doi.org/10.3390/app14093864
9. Brunton, S.L., Proctor, J.L., Kutz, J.N.: Discovering governing equations from data by sparse identification of nonlinear dynamical systems. Proc. Natl. Acad. Sci. **113**(15), 3932–3937 (2016). https://doi.org/10.1073/pnas.1517384113
10. De Silva, B.M., Champion, K., Quade, M., Loiseau, J., Kutz, J.N., Brunton, S.L.: PySINDy: A Python package for the sparse identification of nonlinear dynamics from data (2020). ArXiv. https://arxiv.org/abs/2004.08424
11. Wei, B.: Sparse dynamical system identification with simultaneous structural parameters and initial condition estimation. Chaos Solitons Fractals **165**, Part 2, 112866, ISSN 0960-0779 (2022). https://doi.org/10.1016/j.chaos.2022.112866
12. Joseph, B., Kathleen, C., Nathan Kutz, J., Brunton Steven, L.: Discovering governing equations from partial measurements with deep delay autoencoders. Proc. R Soc. A.47920230422. https://doi.org/10.1098/rspa.2023.0422
13. Frédéric, C.al, Bertrand, M.: An introduction to topological data analysis: fundamental and practical aspects for data scientists. Front. Artif. Intell. (2021). https://doi.org/10.3389/frai.2021.667963

14. Shannon, C.E.: A mathematical theory of communication. Bell Syst. Tech. J. **27**(3), 379–423 (1948). https://doi.org/10.1002/j.1538-7305.1948.tb01338.x
15. Chazal, F., Michel, B.: An introduction to topological data analysis: fundamental and practical aspects for data scientists (2017). ArXiv. https://arxiv.org/abs/1710.04019
16. Hensel, F., Moor, M., Rieck, B.: A survey of topological machine learning methods. Front. Artif. Intell. **4**, 681108 (2021). https://doi.org/10.3389/frai.2021.681108
17. Leykam, D., Angelakis, D.G.: Topological data analysis and machine learning. Adv. Phys.: X **8**(1) (2023). https://doi.org/10.1080/23746149.2023.2202331
18. Murugan, J., Robertson, D.: An introduction to topological data analysis for physicists: from LGM to FRBs (2019). ArXiv. https://arxiv.org/abs/1904.11044
19. Wasserman, L.: Topological data analysis. Annu. Rev. Stat. Appl. **5**(1), 501–532 (2018)
20. Van Rossum, G., Drake, F.L.: Python 3 Reference Manual. CreateSpace, Scotts Valley, CA (2009)
21. Christopher, T., Nathaniel, S., Rann B.O.: Ripser.py: a lean persistent homology library for Python. J. Open Source Softw. (2018). https://doi.org/10.21105/joss.00925
22. Bauer, U.: Ripser: efficient computation of Vietoris-Rips persistence barcodes. J. Appl. Comput. Topol. **5**(3), 391–423 (2021). https://doi.org/10.1007/s41468-021-00071-5
23. Deffuant, G., Neau, D., Amblard, F., Weisbuch, G.: Mixing beliefs among interacting agents. Adv. Complex Syst. 87–98 (2000). https://doi.org/10.1142/S0219525900000078
24. Olfati-Saber, R., Murray, R.M.: Consensus problems in networks of agents with switching topology and time-delays. IEEE Trans. Autom. Control. **49**(9), 1520–1533 (2004). https://doi.org/10.1109/TAC.2004.834113

Optimizing Electric Vehicle Routes: A Hybrid Approach of Genetic Algorithm with Q-Learning

J. C. Hernández-Marín, Laura Cruz Reyes, Patricia Ruiz, Bernabé Dorronsoro, Claudia Gomez-Santillan, Nelson Rangel-Valdez, and Maria Lucila Morales-Rodriguez

Abstract The rapid expansion of electric mobility introduces significant challenges in optimizing vehicle routes, particularly under constraints related to energy consumption, limited battery capacity, and the availability of charging infrastructure. The Electric Vehicle Traveling Salesman Problem (EV-TSP) extends the classical Traveling Salesman Problem (TSP) by incorporating these constraints, thereby increasing its computational complexity. This chapter proposes a hybrid approach that combines Genetic Algorithms (GA) with Q-Learning, a reinforcement learning technique, to improve electric vehicle route optimization. Four GA variants are analyzed: three using fixed crossover operators (Order Crossover, Cycle Crossover, and Position Crossover), and an adaptive variant (GA-QL) that dynamically selects the most effective operator through Q-Learning. The methodology utilizes benchmark instances derived from the TSP-Lib dataset, which have been modified to

J. C. Hernández-Marín (✉) · L. Cruz Reyes · C. Gomez-Santillan · N. Rangel-Valdez · M. L. Morales-Rodriguez
Department of Computer Science, National Technological Institute of Mexico, Ciudad Madero Institute of Technology, Ciudad Madero, Mexico
e-mail: juancarloshdzm726@gmail.com

L. Cruz Reyes
e-mail: laura.cr@cdmadero.tecnm.mx

C. Gomez-Santillan
e-mail: claudia.gs@cdmadero.tecnm.mx

N. Rangel-Valdez
e-mail: nelson.rv@cdmadero.tecnm.mx

M. L. Morales-Rodriguez
e-mail: lucila.mr@cdmadero.tecnm.mx

P. Ruiz · B. Dorronsoro
School of Engineering, University of Cádiz, Cádiz, Spain
e-mail: patricia.ruiz@uca.es

B. Dorronsoro
e-mail: bernabe.dorronsoro@uca.es

W. Pedrycz et al. (eds.), *Decision-Making in Computational Intelligence-Based Systems*, Studies in Systems, Decision and Control 628,
https://doi.org/10.1007/978-3-032-04056-5_5

reflect the constraints of electric vehicles. A comprehensive experimental evaluation is conducted, including statistical validation via the Bonferroni-Dunn post-hoc test. Results demonstrate that GA-QL achieves superior performance in terms of route efficiency, energy consumption, and solution stability. Its dynamic operator selection accelerates convergence and enhances adaptability to real-world factors such as traffic variations and stochastic energy usage. In addition to its theoretical contributions, this work outlines practical applications in urban logistics and smart mobility. The proposed framework supports energy-efficient routing and scalability, with future directions including multi-objective optimization, real-time adaptability, and deep reinforcement learning strategies.

Keywords EV-TSP · GA · Q-learning · Machine learning · ML

1 Introduction

Electric mobility has emerged as a key solution to address the environmental and energy challenges associated with urban transportation. However, the widespread adoption of electric vehicles (EVs) faces significant obstacles, particularly in route optimization considering range limitations and charging infrastructure availability. In this context, the Electric Vehicle Traveling Salesman Problem (EV-TSP) has become a priority research area, as it combines the complexity of the classical route optimization problem with the specific constraints of EVs, such as battery capacity and charging station locations.

In addition to these challenges, charging infrastructure in many cities is still underdeveloped, limiting EV autonomy and increasing driver anxiety about running out of energy. This highlights the importance of developing optimization algorithms that not only minimize travel distance but also efficiently manage available energy and strategically plan charging stops. The lack of strategically placed charging stations and the variability in energy consumption due to factors such as traffic, topography, and weather conditions make the EV-TSP a particularly complex and challenging problem.

Artificial intelligence (AI) and machine learning (ML) are playing a crucial role in transforming logistics and transportation. Genetic Algorithms (GAs) and Reinforcement Learning (RL) have proven to be powerful tools for solving complex and dynamic optimization problems. GAs, inspired by the principles of biological evolution, efficiently explore the solution space using selection, crossover, and mutation operators. On the other hand, RL, specifically the Q-Learning algorithm, provides a robust framework for sequential decision-making, adapting to changing environments and maximizing long-term rewards.

This chapter is structured into several interconnected sections that together provide a comprehensive view of optimizing routes for electric vehicles. Initially, the theoretical framework presents the fundamentals of genetic algorithms and reinforcement learning, emphasizing key aspects such as solution encoding, individual selection, crossover and mutation operators, and the implementation of the Q-Learning algorithm. Subsequently, the methodology employed to address the EV-TSP is detailed by describing the implementation of various genetic algorithm variants, the experimental design, and specific parameters like population size, crossover and mutation rates, and Q-Learning intervals. Based on the experiments conducted, the results are analyzed, with the findings indicating that the variant integrating Q-Learning achieved the best solution quality, a conclusion supported by rigorous statistical analysis. Finally, the discussion extends to the implications of these results for optimizing urban logistics, exploring future research directions, and practical applications that could lead to reduced operating costs, improved energy efficiency, and the promotion of sustainable mobility. Through this chapter, the goal is not only to present a robust theoretical framework but also to inspire future research that combines classical optimization techniques with advanced artificial intelligence methods. The integration of these tools has the potential to transform electric mobility, contributing to a more sustainable and efficient future.

2 Theoretical Framework

Developing effective solutions for route optimization in the context of electric mobility requires a solid understanding of theoretical foundations and available methodological approaches in the fields of artificial intelligence and combinatorial optimization. This section presents a detailed analysis of two key areas: Genetic Algorithms (GA) and Reinforcement Learning (RL), along with their application to the EV-TSP problem.

By understanding the underlying principles of GA and RL, as well as their application in the specific context of the EV-TSP, we establish the theoretical foundation necessary for the development and implementation of efficient and sustainable route optimization strategies for electric vehicles.

2.1 Genetic Algorithms

Genetic Algorithms (GA) are a class of optimization algorithms inspired by the biological processes of evolution and natural selection. Their application extends across a wide range of optimization problems in various disciplines, from engineering to computer science and computational biology [1–3].

At the core of Genetic Algorithms lies the concept of evolution. Like nature, a GA operates with an initial population of potential solutions to a given problem. These solutions, called individuals, are typically represented by a string of symbols or numbers, known as a chromosome or genotype [2, 4, 5].

The quality of each individual is determined using an evaluation function, known as the fitness function, which quantifies how good that solution is with respect to the problem being addressed. One of the main characteristics of Genetic Algorithms is the encoding of solutions.

This process aims to represent problem solutions as sequences of genes, also known as chromosomes. This representation can be binary, integer, real-valued, or even more complex structures like trees.

Another key feature of these algorithms is the selection of individuals. This process aims to choose the fittest individuals for reproduction, where individuals with higher fitness have a greater probability of being selected.

Crossover and Mutation Operators are additional key characteristics of GAs. The crossover operator facilitates the exchange of genetic information between two individuals, simulating the recombination process in biological reproduction. Meanwhile, mutation introduces random changes in the genes of individuals, adding genetic diversity to the population and preventing premature convergence to local optima.

The final characteristic is fitness evaluation. This process determines the quality of a solution using the predefined fitness function, which assigns a numerical value to each individual based on its effectiveness in solving the problem.

The combination of these components enables Genetic Algorithms to efficiently explore the solution space, dynamically adapting to different types of problems and environmental conditions. Their ability to find optimal or near-optimal solutions in complex problems makes them a powerful tool in the modern scientist's or engineer's toolkit.

2.2 Reinforcement Learning (RL)

Reinforcement Learning (RL) is a machine learning paradigm focused on sequential decision-making in dynamic environments. Unlike supervised learning, where labeled input–output pairs are provided, and unsupervised learning, where the algorithm must find patterns in data by itself, reinforcement learning involves an agent interacting with an environment, making decisions, and receiving feedback in the form of rewards or penalties.

A classic example of RL is its application in robotics, where a robot learns to navigate an unknown environment through exploration and exploitation of actions that maximize cumulative rewards. In the context of EV-TSP, RL can be used to dynamically adapt routes based on changes in traffic, charging station availability, or energy demand. For example, in [6], RL was used to optimize routes in real-time dynamic urban environments, demonstrating its ability to adapt to changing conditions.

In reinforcement learning, an agent learns through experience, aiming to maximize a cumulative reward signal over time. This requires balancing exploration of new actions to discover potential rewards and exploitation of known actions that have yielded good results in the past. The agent must find a balance between exploration and exploitation to learn effectively [3, 7, 8].

2.3 Q-Learning Algorithm

One of the most well-known algorithms in reinforcement learning is Q-Learning. This algorithm is designed to learn an optimal policy, a strategy that maximizes long-term cumulative rewards.

In Q-Learning, the agent maintains a Q-value table representing the quality of each state-action pair. These Q-values are updated iteratively using the Bellman equation, which incorporates immediate rewards and estimated future values. Through exploration of the action space and updating Q-values, the agent gradually converges to an optimal policy that maximizes cumulative rewards [6].

The success of Q-Learning lies in its ability to balance exploration and exploitation of actions. By assigning Q-values to possible actions in each state, the agent can make informed decisions based on its current knowledge of the environment and its goal of maximizing long-term rewards [6].

Reinforcement Learning and the Q-Learning algorithm offer a powerful framework for addressing sequential decision-making problems in dynamic environments, ranging from games to robotics and resource management. Their ability to learn optimal policies through interaction with the environment makes them essential tools in the field of machine learning.

2.4 Exploration Versus Exploitation in Reinforcement Learning (RL)

One of the main challenges in reinforcement learning (RL) is balancing exploration and exploitation. Exploration involves testing new actions to discover potentially better outcomes, while exploitation consists of selecting actions that, based on current knowledge, offer the highest expected reward. This trade-off is crucial in dynamic environments where conditions may change over time, such as optimizing routes for electric vehicles.

In the context of Q-Learning, this balance is typically managed using an ε-greedy strategy. This approach introduces a probability ε of selecting a random action rather than following the best-known option according to Q-values. As learning progresses, it usually decreases gradually, favoring the exploitation of the best decisions identified so far.

When combined with genetic algorithms (GA), reinforcement learning can significantly enhance the optimization process by dynamically adapting crossover operator selection. Traditional GA implementations apply crossover and mutation operators statically, without considering their effectiveness in real-time. However, by integrating Q-Learning, the algorithm can evaluate the performance of different operators and adjust its strategy accordingly based on solution quality.

For example, in the hybrid approach proposed in this chapter, the Q-Learning agent assigns reward values to crossover operators based on their past performance. This allows the algorithm to favor operators with higher intensification capacity (exploitation) or higher diversification capacity (exploration), depending on the scenario. As a result, the algorithm improves its ability to avoid stagnation in local optima and maximizes search efficiency.

Integrating Q-Learning into GA introduces an adaptive mechanism that responds to problem evolution over time, preventing the algorithm from relying on a single crossover operator. This approach not only enhances solution quality but also reduces the number of iterations needed to converge on an optimal result.

2.5 Limitations of Traditional Methods

Traditional optimization methods, such as linear programming and heuristic algorithms, often struggle when applied to complex problems like EV-TSP. These limitations include the inability to handle large search spaces, a lack of adaptability to environmental changes, and premature convergence to suboptimal solutions.

Hybrid approaches, such as those combining GA with RL, overcome these limitations by leveraging the strengths of both techniques. GA provides efficient exploration of the solution space, while RL dynamically adapts the search strategy based on problem conditions. For example, in [3], it was demonstrated that integrating RL into metaheuristic algorithms significantly improves solution quality in complex optimization problems.

2.6 Route Optimization and Electric Vehicle Autonomy

In the context of route optimization and electric vehicle autonomy, a complex challenge arises in balancing the efficiency of service delivery while maximizing energy use. This problem resembles the classic Traveling Salesman Problem (TSP) but incorporates constraints related to electric vehicle range and the availability of charging stations [9].

One of the main factors influencing the autonomy of electric vehicles is energy consumption variability, which depends on multiple factors such as terrain topology, traffic conditions, and acceleration and braking patterns. Accurately modeling these factors within a route optimization framework is crucial for obtaining realistic and applicable solutions in urban and logistics environments. Various strategies have been proposed in the literature, ranging from heuristic and metaheuristic approaches to machine learning models that predict energy consumption and adjust routes in real time [10].

Another key challenge in optimizing electric vehicle routes is the efficient use and scheduling of charging stations. Unlike internal combustion vehicles that can refuel in minutes, electric vehicle charging takes significantly longer, imposing additional constraints. Some studies have addressed this by incorporating mathematical programming models and combinatorial optimization techniques to minimize wait times and balance the charging infrastructure load [11].

Additionally, the integration of multi-objective optimization techniques has gained importance in this field, allowing for the balance of criteria such as minimizing travel time, reducing energy consumption, and maximizing battery lifespan. Evolutionary algorithms and reinforcement-based strategies have proven to be promising tools for finding optimal solutions in large-scale problems. These approaches not only improve the operational efficiency of electric fleets but also contribute to urban transportation sustainability by reducing emissions and the environmental impact of energy consumption [10].

2.7 Electric Vehicle Traveling Salesman Problem

The Electric Vehicle Traveling Salesman Problem (EV-TSP) involves determining the shortest route for an electric vehicle to visit a set of locations and return to its starting point, while considering range limitations and the availability of charging stations along the way. Unlike the traditional TSP, where vehicles are assumed to have unlimited fuel, the EV-TSP requires careful management of battery capacity and strategic charging station placement [9].

To address the EV-TSP, various optimization strategies have been developed. One effective approach involves metaheuristic algorithms, such as Genetic Algorithms and Simulated Annealing, which efficiently explore the solution space to identify an optimal route that satisfies both range and time constraints. These algorithms are specifically adapted for electric vehicles by incorporating energy consumption and recharging dynamics at charging stations [10].

Another widely used strategy formulates the problem, such as an Integer Linear Programming (ILP) or Mixed-Integer Linear Programming (MILP) model, optimizing the route while accounting for range constraints and available charging infrastructure. These models offer a precise problem representation and can efficiently handle large-scale instances [12–14].

The Electric Vehicle Traveling Salesman Problem (EV-TSP) is a complex optimization problem that aims to determine the most efficient route for an electric vehicle (EV) to visit a given set of locations and return to its starting point. Unlike the traditional Traveling Salesman Problem (TSP), where vehicles are assumed to have unlimited fuel, the EV-TSP introduces additional constraints related to battery capacity, energy consumption, and charging station availability [9]. These factors make the problem significantly more challenging, requiring strategies that balance route optimization, energy efficiency, and charging logistics.

2.8 Challenges in Solving the EV-TSP

One of the primary challenges in solving the EV-TSP is battery autonomy management. Unlike conventional fuel-powered vehicles, electric vehicles have a limited driving range, and their energy consumption is affected by factors such as terrain conditions, vehicle load, and driving behavior. This necessitates the integration of intelligent charging strategies to ensure that vehicles can complete their routes without depleting their batteries, while also minimizing detours for recharging.

Additionally, charging station placement plays a crucial role in the feasibility of solutions. Unlike gas stations, which are widely available, charging infrastructure is still developing, and stations may have limited capacity, different charging speeds, and varying energy costs. As a result, solving the EV-TSP requires accounting for charging constraints, waiting times, and the impact of recharging on overall route efficiency.

Another key challenge is computational complexity. The EV-TSP belongs to the class of NP-hard problems, meaning that finding an optimal solution becomes computationally expensive as the number of locations increases. This has led to the development of advanced optimization techniques to efficiently explore the solution space and obtain high-quality solutions within a reasonable timeframe.

3 Related Work

This subsection presents solutions reported in the literature to address the Electric Vehicle Traveling Salesman Problem (EV-TSP), aiming to improve the efficiency and sustainability of urban transportation in the context of increasing demand for delivery and distribution services in urban areas. The following review highlights the most relevant approaches and identifies current trends in electromobility optimization.

3.1 Electric Vehicle Traveling Salesman Problem (EV-TSP)

In a recent study, Ahsini et al. [9] studied route optimization for electric delivery vehicles, incorporating topographic and traffic data into urban graphs to represent more realistic scenarios. They adapted the Steiner Traveling Salesman Problem (STSP) for electric vehicles, allowing certain nodes to be skipped, distinguishing it from the traditional Traveling Salesman Problem. This approach is particularly relevant in urban contexts, where the availability of charging infrastructure and variations in energy consumption directly impact the planning of efficient routes.

In contrast, Cuch'y et al. [10] proposed an innovative approach to solving daily mobility problems by preprocessing route planning under energy constraints and organizing activities with spatial and temporal limitations. Their methodology integrates real-time data sources, allowing dynamic adaptation to changing conditions such as traffic congestion or fluctuations in transportation demand. Their proposal combines an optimal algorithm with suboptimal acceleration heuristics, significantly reducing computation times without compromising solution quality.

Unlike models based solely on distance or fixed consumption rates, Doppstadt et al. [12] introduced a neural network approach to estimate energy consumption between nodes, motivated by the rise of online delivery services. Accurate estimation of energy consumption is crucial for maximizing electric vehicle autonomy and reducing the need for unexpected recharging. Their STSP-based approach improves estimation accuracy by incorporating driving patterns and by accounting for variations in energy consumption across different urban and rural environments.

Extending beyond energy estimation, Liao et al. [13] addressed the challenge of integrating charging strategies into route planning for electric vehicles. They developed algorithms, including a Greedy Heuristic, to compute the shortest path between locations while respecting order constraints in node visits. The inclusion of charging management strategies significantly improves the operational viability of electric fleets by optimizing total travel time without compromising battery availability.

Building on previous efforts to improve routing efficiency, Roberti and Wen [14] proposed a mixed-integer linear programming (MILP) formulation capable of solving instances with up to 20 customers in short computation times. In addition, they developed a heuristic algorithm that combines Variable Neighborhood Search with Dynamic Programming to provide efficient solutions for electric vehicle routing problems. These approaches have proven effective in reducing operational costs and improving electric fleet performance, particularly in scenarios with dynamically changing demand.

Pushing the boundaries of traditional heuristic design, Wu et al. [15] introduced a deep reinforcement learning framework that employs self-attention mechanisms within neural networks to learn improvement heuristics for routing problems. This innovative approach stands out for its strong generalization capability, enabling the model to adapt to various types of route optimization problems without requiring extensive manual tuning. It not only outperforms state-of-the-art models in both the TSP and CVRP but also scales effectively to different problem sizes and real-world datasets, opening new avenues for optimizing logistics systems based on electric fleets.

Finally, Shi et al. [16] presented an innovative strategy for the dynamic allocation of electric vehicles in routing environments with dynamic changes (DEVD). The memory-based MACO approach utilizes advanced pheromone relocation and local search techniques, demonstrating significant improvements in both simulations and real-world data. The adaptability of this approach makes it a key tool for optimization in highly uncertain scenarios, where demand and charging infrastructure can change unexpectedly.

3.2 *Current Trends in Electromobility Optimization*

This section examines recent trends in electromobility optimization, focusing on electric vehicle routing, energy management, and the location of charging infrastructures. Various studies have addressed these topics by applying techniques such as genetic algorithms, heuristics, and machine learning methods to improve route planning, optimize charging station placement, and reduce operational costs. Tables 1 and 2 summarize, respectively, the constraints and objectives considered in these studies, providing an overview of the diversity and complexity of the addressed problems.

3.2.1 Constraints in Electromobility Optimization

As shown in Table 1, electromobility problems incorporate multiple constraints derived from the complexity of real-world transportation scenarios. These constraints include energy consumption cost (ECC), time windows (TW), activity duration (AD), charging constraints (CC, SCS, CTR), battery capacity (BC), and constraints related to vehicle assignment and availability (VA, VAV), among others. Such restrictions help accurately model critical operational factors, including the required information (such as date/time data or specific routes) and the type of problem being solved (routing, location, or scheduling). Incorporating these elements into the model is essential to realistically simulate urban operational conditions.

Table 1 Constraints in optimization problems derived from electromobility

References	Solved problem (SP)	Required information (RI)	Constraints (C)
[9]	Routing (R)	DT, T	ECC
[10]	Routing (R)	C	TW, AD, CC
[12]	Routing (R)	DT, C	VN, AT, MRD
[13]	Routing (R)	TR, E	BC
[14]	Routing (R)	TW, E	TW, BC, SCS, CTR
[16]	Routing (R)	DT, RV	BC, VA, VAV
[15]	Routing (R)	C, TW	CV, VC, MR
[17]	Routing, Scheduling (R, S)	CI, E	PT, VT, LC, CI, CS, VN, BC
[18]	Routing (R)	E, CI	N/A
[19]	Routing (R)	C, TR	LC, BC
[20]	Location (L)	DT, E	VC, FA
[21]	Location (L)	DT, C	SCS, VC, BC
[22]	Location (L)	E, C	SPL, UCP, UA
This proposal	Routing, Scheduling, Location (R, S, L)	DT, RV, TR, C, CI	BC, TW, MRD, SCS, VAV, VC

Table 2 Objectives in optimization problems derived from electromobility

Reference	Objectives (O)	Dynamic problem (DP)	Stochastic uncertainty (SU)	Real case study (RCS)	Problem size (PS)
[9]	MEC	N	N	Y	M
[10]	MTT, MCT	N	N	N	M
[12]	MOC	N	N	Y	M
[13]	MTT	N	N	N	S
[14]	MBT	N	N	N	S
[16]	MWT, MCT	Y	Y	N	L
[15]	MDR	Y	N	N	L
[18]	IPP, ECP, ERP, SCP	N	N	Y	M
[19]	MDR	N	N	N	M
[20]	MFC	N	N	N	M
[21]	MDR	N	N	Y	M
[22]	MLF, MRL	N	N	N	M
This Proposal	MDR, MTT, MCT, MFC	Y	Y	Y	L, M, S

3.2.2 Optimization Objectives

Table 2 summarizes the objectives pursued in the optimization of electromobility-related problems, highlighting goals such as the minimization of energy consumption (MEC), travel time (MTT), charging times (MCT), and operating costs (MOC), as well as more specific objectives such as predicting energy consumption, recharge planning, and service continuity (IPP, ECP, ERP, SCP). Some studies have tackled dynamic problems and stochastic uncertainty, with these models being applied in real-case studies. This underscores the practical applicability of these optimization models across different problem scales, whether small, medium, or large.

3.2.3 Problem Size and Practical Application

Another fundamental aspect to consider is the problem size, classified as small (S), medium (M), or large (L). The studies vary in the scale of the cases analyzed, which directly influences the complexity and scalability of the proposed solutions. Integrating constraints and objectives based on problem size allows for a precise evaluation of algorithm performance in scenarios ranging from controlled simulations to real-world urban applications.

3.3 Intelligent Optimization in Charging Networks and Energy Management

The rise of electric mobility has driven advancements in route optimization and charging network management. The availability and strategic location of charging stations remain key challenges, especially in cities where infrastructure is still developing. In this context, optimization systems based on AI and reinforcement learning are emerging as key solutions.

A current trend is the use of hybrid optimization and prediction models, where algorithms such as Deep Q-Learning and deep neural networks (DNN) are combined to forecast energy demand and dynamically adjust charging plans. This enables a more efficient distribution of resources, reducing waiting times at charging stations and minimizing operational costs.

For example, recent studies have demonstrated that integrating real-time data (traffic conditions, battery charge status, energy consumption) into optimization models significantly improves route efficiency. Companies like Tesla and NIO have implemented AI-based strategies to manage fleet charging and optimize vehicle autonomy.

Electromobility and Multi-Objective Route Optimization.

Another significant advancement in electromobility is the incorporation of multi-objective optimization into routing problems, simultaneously considering factors such as:

(1) Minimizing travel time and energy costs.
(2) Maximizing battery lifespan (avoiding unnecessary charge cycles).
(3) Reducing emissions and congestion in urban areas.

The combination of metaheuristics with reinforcement learning strategies has proven effective in planning dynamic routes, adapting to real-time changes such as traffic conditions and charging station availability. This has led to the development of autonomous route planning systems, reducing human intervention and improving logistics efficiency in electromobility.

3.4 Towards a Future of Autonomous Mobility and Sustainable Energy

The impact of electric vehicle route optimization extends beyond conventional transportation. Today, applications are being explored in autonomous electric vehicles and microgrid energy management, where vehicles can act as distributed energy storage (V2G, Vehicle-to-Grid). Route optimization not only enhances operational efficiency but also contributes to more sustainable models of urban mobility and energy management.

3.4.1 Contribution of This Proposal

The analysis of the reviewed studies reveals several key trends in electric vehicle routes, optimization and energy management. One prominent aspect is the frequency with which constraints such as Battery Capacity (BC) are addressed, as seen in works [13, 14, 16]. This highlights the importance of energy management in designing efficient solutions, where limited autonomy and consumption planning are crucial.

Regarding objectives, the Minimization of Energy Consumption (MEC) is one of the most recurrent, reflecting a clear focus on reducing energy impact. Studies like [9] optimize MEC using techniques such as Genetic Algorithms and Mixed-Integer Programming, adjusting routes based on energy demands and the availability of charging stations.

The Minimization of Total Travel Time (MTT), present in works such as [10, 13], demonstrates a continuous interest in improving operational efficiency and reducing service times. These optimizations are achieved through heuristic techniques such as Tabu Search or metaheuristics like Ant Colony Algorithms.

Handling dynamic problems and stochastic uncertainty is another distinctive factor in studies like [16], which allow greater flexibility in adapting to changing situations, such as traffic conditions or fluctuating energy demand. However, many studies still do not consider these aspects, representing an opportunity for future research.

Finally, although some studies such as [16, 18] have used real-world data to validate their models, the majority still rely on artificially generated instances. This reliance limits the direct applicability of their solutions to real-world scenarios.

3.4.2 Innovations of This Proposal

In contrast to the reviewed studies, our proposal combines Genetic Algorithms with Reinforcement Learning (Q-Learning) to dynamically select crossover operators, automatically adapting to problem conditions. Unlike traditional approaches that use fixed operators, this method enhances the exploration of the solution space and improves the overall search efficiency, especially in dynamic scenarios such as energy demand variability or traffic conditions.

During experimentation, the incorporation of reinforcement learning significantly improved the Genetic Algorithm's search capability. For example, in tests with instances ranging from 50 to 100 nodes, the machine learning-enhanced algorithm reduced computation time by 20% and improved solution quality by 15% compared to the algorithm without learning. These results demonstrate that the proposed approach is not only more efficient but also more adaptable to large-scale problems and dynamic conditions.

Additionally, our proposal integrates traffic data and detailed stochastic uncertainty management, enabling more robust and adaptable route planning. This makes it a more versatile and applicable solution for real-world scenarios, where flexibility and adaptability are critical.

In summary, this proposal not only addresses a wide range of constraints and objectives but also introduces innovative approaches that enhance energy efficiency and adaptability in dynamic environments, representing a significant contribution to the field of electromobility.

4 Materials and Methods

To address the EV-TSP problem, a methodology was developed that includes the implementation and comparison of various Genetic Algorithm (GA) variants, as well as a hybrid version integrating Q-Learning. This section details the characteristics of the EV-TSP problem, the parameters used, and the experimental procedure, ensuring the study's replicability.

4.1 *Problem Definition*

The Electric Vehicle Traveling Salesman Problem (EV-TSP) is a variant of the classic Traveling Salesman Problem (TSP) that incorporates specific limitations of electric vehicles (EVs), such as battery autonomy and charging station locations. This problem seeks to find the most efficient route for an electric vehicle that must visit a set of locations and return to the starting point, minimizing total energy consumption while considering battery capacity constraints and charge/discharge cycles.

4.2 *Model Formulation*

To tackle the EV-TSP, a mathematical model is proposed, including decision variables, an objective function, and a set of constraints that ensure the feasibility of the solution. The model is described in detail below.

4.2.1 Sets and Parameters

Before presenting the mathematical model, it is important to define the sets and parameters used in the problem formulation:

(1) Sets

V:: Set of nodes (including the depot), $V = \{0, 1, \ldots, n-1\}$.
A: Set of possible arcs, $A = \{(i, j) : i, j \in V, i \neq j\}$.

(2) Parameters

Battery:

B: Maximum initial battery capacity (total energy in units).
α: Degradation rate per cycle ($0 < \alpha \leq 1$).
$q_0 = 0$: Initial accumulated cycles (new battery).

Consumption and Recharge:

c: Energy consumption per unit of distance.
d_{ij}: Distance between nodes i and j.
r_j: Energy that can be recharged at node j(0 if j has no charging station).

Trips

n: Total number of nodes (including the depot).

4.2.2 Model Formulation

To address the EV-TSP, a mathematical model is proposed that includes decision variables, an objective function, and a set of constraints to ensure feasibility. The detailed formulation is presented below.

Decision Variables

- $x_{ij} \in \{0, 1\}$: Binary variable, where $x_{ij} = 1$ if the vehicle travels from node i to node j, and 0 otherwise.
- $b_i \geq 0$: Remaining battery energy upon arrival at node i.
- $q_j \geq 0$: Accumulated cycles upon arrival at node j.

Objective Function

Minimize total energy consumption over two complete cycles of the route:

$$\min \sum_{i=0}^{n-1} \sum_{j=0}^{n-1} c \cdot d_{ij} \cdot x_{ij}$$

where:

c is the energy consumption factor per unit of distance.
d_{ij} is the distance between nodes i and j.
x_{ij} is the binary variable indicating if the vehicle travels from i to j.

The objective function aims to minimize total energy consumption over two full iterations of the route, optimizing the electric vehicle's energy efficiency.

Constraints

i. Unique Visit to Each Node (Entry and Exit)

$$\sum_{j=0}^{n-1} x_{ij} = 1, \quad \forall i \in V$$

$$\sum_{i=0}^{n-1} x_{ij} = 1, \quad \forall j \in V$$

ii. Subtour Elimination (Subtour Constraint)

$$u_i - u_j + n \cdot x_{ij} \leq n - 1, \quad \forall i, j \in V,\ i \neq j$$

iii. Initial Battery Capacity

$$b_0 \leq B$$

iv. Battery Level Update When Traveling from i to j

$$b_j = b_i - c \cdot d_{ij} + r_j, \quad \forall (i, j) \in A$$

v. Dynamic Residual Battery Capacity

$$b_j \leq B' \quad \forall j \in V$$

where:

$$B' = B \cdot \left(1 - \alpha \cdot q_j\right)$$

vi. Accumulated Cycles Calculation (q_j)

$$q_j = q_0 + \sum_{i=0}^{n-1} \frac{\left(|b_i - b_j| + r_j\right)}{B} \cdot x_{ij}, \quad \forall j \in V$$

vii. Non-Negativity of Battery

$$b_j \geq 0, \quad \forall j \in V$$

viii. Binary Restriction

$$x_{ij} \in \{0, 1\}, \quad \forall (i, j) \in A$$

To evaluate the effectiveness of the proposed model and validate the performance of the implemented algorithms, it was necessary to generate benchmark problems that included specific characteristics of electromobility. Traditional TSP instances available in the TSP-Lib library do not account for autonomy or energy consumption constraints, which limit their applicability in the context of electric vehicles. Therefore, modifications were made to the original instances, adapting them to the EV-TSP problem. This process included the generation of new attributes and the structuring of the data in a suitable format for the experiments.

4.3 *Generation of Benchmark Problems*

To evaluate the effectiveness of the proposed model and validate the performance of the implemented algorithms, it was necessary to generate benchmark problems that included specific characteristics of electromobility. Traditional TSP instances available in the TSP-Lib library do not account for autonomy or energy consumption constraints, which limit their applicability in the context of electric vehicles. Therefore, modifications were made to the original instances, adapting them to the EV-TSP problem. This process included the generation of new attributes and the structuring of the data in a suitable format for the experiments.

4.3.1 Modification of TSP-Lib Instances

The base instances were obtained from TSP-Lib and adjusted to incorporate additional information relevant to the optimization of electric vehicle routes. In particular, the graph structure was modified to include not only the distance between nodes but also the energy consumption associated with each arc.

To maintain fidelity to the real-world problem, energy costs were defined as a function of distance and of a random factor within a predefined range. This allowed for modeling variations in energy consumption due to factors such as terrain slope, traffic, and vehicle efficiency.

4.3.2 Data Structuring

To facilitate the integration of the modified instances into the model, a structured data format was designed using CSV files. Each row in the file represents a connection between nodes and consists of the following attributes:

(1) Start Node: Represents the starting point of the connection.
(2) End Node: Represents the destination point of the connection.

(3) Distance: Value taken from the original TSP-Lib instance.
(4) Energy Consumption: Attribute randomly generated within a defined range, considering the aforementioned factors.

Each CSV file includes a header with additional information, such as the instance name, the number of nodes, and the properties of the graph.

4.3.3 Generation of Attributes

The distance attribute was taken directly from the original TSP-Lib instance, ensuring consistency with the base problem. The energy consumption attribute was randomly generated within a predefined range, ensuring a realistic distribution of values. Table 3 exemplifies the structure of one of the modified instances:

The generation of benchmark problems tailored to the EV-TSP context was a critical step in ensuring the relevance and applicability of the proposed model. By modifying traditional TSP instances to include energy consumption constraints and structuring the data in a systematic format, the experiments were able to accurately reflect the challenges of optimizing routes for electric vehicles. This approach not only enhances the realism of the problem but also provides a solid foundation for evaluating the performance of the algorithms in scenarios that closely mimic real-world conditions. The results obtained from these adapted instances will serve as a valuable benchmark for future research in the field of electromobility and route optimization, offering insights into the integration of energy efficiency and operational constraints in transportation planning.

4.4 Genetic Algorithms (GA)

To solve the EV-TSP problem, four genetic algorithms were implemented to compare their performance. These algorithms were designed to explore different crossover strategies while maintaining consistent approaches to initial population generation, selection, and mutation. The characteristics of these implementations are summarized

Table 3 Example of a modified instance for EV-TSP

NAME: Burma14_Modified			
Dimension: 14			
Start node	End node	Distance	Energy consumption
1	2	123.45	7.891
1	3	234.56	10.12
2	3	98.76	6.34
…	…	…	…

Table 4 Operator descriptions

Operators	GA-OCO	GA-CCO	GA-PSO	GA-QL
Initial population	Random	Random	Random	Random
Selection	Ranking	Ranking	Ranking	Ranking
Crossover	Order cross over	Cycle cross over	Position cross over	Q-learning
Mutation	1-Flip	1-Flip	1-Flip	1-Flip

Note The GA-QL column represents the research proposal

in Table 4, which provides a detailed overview of the operators used in each algorithm.

4.4.1 Operator Descriptions

Initial Population

The initial population for all four algorithms was generated randomly to ensure a diverse set of solutions. This randomness is crucial for exploring a wide solution space and avoiding premature convergence to suboptimal solutions. By starting with a varied population, the algorithms increase the likelihood of discovering high-quality solutions over successive generations.

Selection

The selection process was based on a ranking mechanism, where the top 20% of individuals from the population were chosen for crossover. This approach prioritizes individuals with better fitness scores, ensuring that the most promising solutions are carried forward. Ranking selection helps maintain a balance between exploitation (refining good solutions) and exploration (discovering new solutions).

Crossover

Crossover is a critical operator in genetic algorithms, as it combines genetic material from two parents to produce offspring. The four algorithms implemented different crossover strategies to evaluate their effectiveness in solving the EV-TSP problem:

- Order Cross Over (OCO): This operator selects a contiguous segment from one parent and inserts it directly into the child. The remaining positions are filled with elements from the second parent, preserving their order while avoiding duplicates. OCO is particularly effective for problems where the sequence of nodes is crucial, as it maintains the integrity of the parent solutions [23].

- Cycle Cross Over (CCO): CCO identifies cycles between the two parents and alternates copying elements from each parent to the child. This method ensures that the child inherits genetic material from both parents in a structured manner, promoting diversity while preserving the feasibility of the solution [24].
- Position Cross Over (PSO): In PSO, specific positions are copied from one parent to the child, and the remaining positions are filled with elements from the second parent. This approach allows for a more flexible combination of genetic material, making it suitable for problems where certain positions in the solution have a higher impact on fitness [25].
- Q-Learning: Unlike the other crossover operators, Q-Learning dynamically selects the most effective crossover strategy in each iteration based on a reward and penalty function. This adaptive approach allows the algorithm to respond to problem conditions in real time, optimizing its performance as it evolves. The integration of Q-Learning, which is the primary proposal presented in this chapter, introduces an intelligent layer to the crossover process, making it particularly effective for addressing complex and dynamic problems like the EV-TSP.

Mutation

All four algorithms employed the 1-Flip mutation technique, which involves randomly selecting two positions within a solution and swapping their values. This operator introduces diversity into the population, preventing the algorithm from stagnating at local optima. By making small, random changes to the solutions, mutation ensures that the search process continues to explore new areas of the solution space.

4.4.2 The Proposed GA-QL Algorithm

The proposed algorithm is divided into two fundamental processes: the adaptive selection of operators using Q-Learning (QL), and the execution of the evolutionary process of the Genetic Algorithm (GA) using the sequence of selected operators. This approach aims to combine exploration and exploitation of the search space more efficiently by prioritizing crossover operators with historically better performance.

Operator Selection Through Q-Learning

The first process involves training a Q-Learning agent to identify an optimal sequence of crossover operators. The training begins with a randomly generated population. From this population, a ranking-based selection is performed, and new populations are generated by independently applying each of the available crossover operators.

Each generated population is evaluated, and the average fitness per operator is calculated. These values represent the reward assigned to each crossover operator based on its performance when selected as an action by the Q-Learning agent. The results are recorded in a Q-table, which stores the average utility of each operator according to its performance. The operator with the highest reward is selected, and its generated population becomes the basis for the next iteration of the training process.

This procedure is repeated until the complete sequence of operators is selected, with the stopping condition being the predefined number of operators to select. This process is described in Algorithm 1, referred to as the Adaptive Genetic Algorithm with Unified Operator Selection. As a result, an ordered list of operators is obtained, which constitutes the policy generated by the Q-Learning algorithm and will later be used during the evolutionary process.

Algorithm 1. Adaptive Genetic Algorithm with Unified Operator Selection

---

Input
- N: Population size
- O: Set of available crossover operators {C1, C2,..., Cm}
- K: Number of operators to be selected
- f: Objective function

Output
- S: Ordered sequence of selected operators

Process

```
Initialize population P with N randomly generated individuals.
Initialize sequence S ← []
While length(S) < K do:
    Select subset P′ ⊆ P using ranking selection
    For each operator C_i ∈ O do:
        Apply C_i to P′ to generate population P_i
        Evaluate P_i using f
        Compute average fitness q_i of P_i
    End for
    Record the values {q_1, q_2, ..., q_m} in a new row of the Q-table
    Select the operator C_best with the highest q_i
    Add C_best to S
    P ← population generated by C_best
End While
Return S
```

Execution of the Genetic Algorithm with Operator Sequence

Once the operator sequence has been determined through the previous process, the classic genetic algorithm is executed, integrating the resulting sequence. Each operator in the sequence is applied for a predefined number of n generations, using the population resulting from the previous operator as a base.

During each block of n generations, corresponding to a specific operator in the sequence, a ranking selection is performed on the current population, after which the crossover operator defined in the previously obtained sequence is applied. This crossover generates a new population, which is then evaluated using the problem's objective function. Finally, the population is updated according to the criteria established in the genetic algorithm, allowing for a progressive and adaptive evolution throughout the different execution blocks.

This approach enables the dynamic exploitation of the specific strengths of each crossover operator, based on the information learned during the training phase. The process is described in Algorithm 2.

Algorithm 2. Genetic Algorithm with Operator Sequence and Repeated Execution

Input
  S : Sequence of selected operators $C_1, C_2, \ldots C_k$
  n : Number of generations per operator
  N : Population size
  f : Objective function
Output
  P : Final optimized population
Process
  Initialize population P with N randomly generated individuals
  For each operator C_i in S do:
    For $i = 1$ to n do:
      Select subset $P' \subseteq P$ using ranking selection
      Apply C_i to P′ to generate new population P_i
      Evaluate P_i using f
      Update $P \leftarrow P_i$
    End For
  End For
  Return P

The integration of reinforcement learning, specifically Q-Learning, into the dynamic selection of crossover operators within a genetic algorithm is expected to improve solution quality and accelerate convergence toward higher-quality optima. Adaptability in the choice of operators could lead to more efficient evolutionary trajectories, particularly in complex or non-stationary environments. This synergy between evolutionary methods and reinforcement learning is projected as a robust alternative to static or heuristic approaches, especially in highly dynamic or complex problem scenarios.

4.4.3 Discussion

The implementation of these four genetic algorithms highlights the importance of crossover strategies in solving the EV-TSP problem. While all algorithms share common features like random initial population generation, ranking-based selection, and 1-Flip mutation, their performance is heavily influenced by the choice of crossover operator. The inclusion of Q-Learning as an adaptive crossover mechanism represents a significant innovation, as it allows the algorithm to learn and adapt to the problem's specific characteristics over time.

By comparing the performance of these algorithms, we can gain valuable insights into the strengths and weaknesses of each crossover strategy. This analysis not only helps identify the most effective approach for the EV-TSP but also provides a foundation for further research into hybrid and adaptive genetic algorithms for complex optimization problems.

4.4.4 Experimental Design

The objective of the experimental design is to identify which implemented algorithm yields the best results in terms of solution quality. The experiments were conducted with the following characteristics:

- Number of experiments: 31 per algorithm.
- Iterations per experiment: 100.
- Population size: 100 individuals.
- Crossover rate: 30% (applied only to the top 20% of individuals).
- Mutation rate: 5%.
- Q-Learning intervals: 10 iterations to select the crossover operator.
- Statistical Analysis
- To evaluate the results, the following indicators were used:
- Min: Average minimum distance.
- Max: Average maximum distance.
- Mean: Average distance.
- Standard deviation: Variation in the average distance.
- Additionally, the following statistical tests were performed:
- Kolmogorov–Smirnov Test: To verify data normality.
- Friedman Test: To compare algorithm performance.
- Post-hoc Bonferroni-Dunn Test: To identify significant differences between algorithms.
- Methodological Contribution
- The proposed methodology introduces two main contributions:
- Integration of Q-Learning: Allows for the dynamic selection of the most effective crossover operator, improving solution quality over time.
- Rigorous Statistical Analysis: Includes non-parametric tests to validate results and ensure the robustness of conclusions.

5 Results

This section presents the results of the experimentation and statistical analysis conducted to compare the performance of the implemented algorithms. The genetic algorithm incorporating reinforcement learning (GA-QL) proved to be the most effective, as detailed below.

Table 5 Experimentation results (Min, Max, Average, Standard deviation)

Algorithm	Min	Max	Average	Standard deviation
Position	16,586.007	20,719.422	18,744.938	1067.516
Cycle	16,594.125	20,295.275	18,706.650	839.752
Order	23,932.558	27,076.003	25,709.776	700.192
QL	16,081.589	20,074.877	17,985.270	912.254

5.1 Experimentation Results

Table 5 shows the experimental results, including minimum, maximum, average values, and standard deviation for each algorithm.

As observed in Table 5, the GA-QL algorithm achieved the best results in terms of minimum, maximum, and average distance, with a competitive standard deviation. This indicates that GA-QL is not only more efficient but also more consistent in its performance.

5.2 Operator Selection Analysis (Q-Learning)

The GA-QL algorithm uses Q-Learning to automatically choose the most effective crossover operator at each iteration. Figure 1 presents the histogram of decisions made by Q-Learning during experimentation.

For the next phase of statistical analysis, the Kolmogorov–Smirnov normalization test was conducted on the solution performance obtained by the algorithms. The results indicated that the test to be performed was non-parametric, as the solutions were not normally distributed. The results are shown in Table 6.

The next phase of statistical analysis involved applying the non-parametric Friedman test, which was based on the following null hypothesis:

(a) "The mean performance of the implemented algorithms is equal, with a significance level of 0.05."

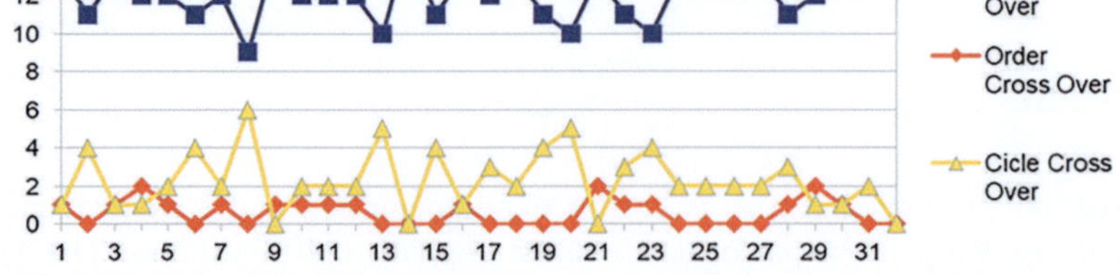

Fig. 1 Histogram of Q-learning decisions

Table 6 Normalization test—Kolmogorov–Smirnov

Algorithm	Normalized sample
Position	NO
Cycle	NO
Order	NO
QL	NO

Table 7 Bonferroni-Dunn post-hoc test

Comparison	Statistic	*P*-value	Null hypothesis
QL versus order	7.84279	0.00000	Rejected
QL versus cycle	2.51744	0.03546	Rejected
QL versus position	2.42061	0.04648	Rejected

(b) The test resulted in a *P*-value of 0.05, rejecting the null hypothesis. Therefore, it was concluded that there is a significant difference between the evaluated algorithms.

In the final phase, the Bonferroni-Dunn post-hoc test was conducted using GA-QL as the reference algorithm, comparing it pairwise against the other algorithms. This test was based on the null hypothesis from the previous test, which was rejected in all comparisons, as shown in Table 7.

As shown in Fig. 1, the Position Cross Over operator was selected most frequently (in 60% of iterations), suggesting that it is the most effective for optimizing routes in the EV-TSP problem. This operator helps maintain the structure of the routes while introducing diversity, which contributes to improving GA-QL's performance. Q-Learning enabled the dynamic adaptation of operator selection, resulting in a significant improvement in solution quality.

5.3 Statistical Tests

To validate the results, the following statistical tests were conducted:

I. Kolmogorov–Smirnov Test

- Used to verify data normality. The results in Table 6 indicate that the data do not follow a normal distribution pattern.
- Justification: This justifies using non-parametric tests to compare algorithm performance.

II. Friedman Test

- Used to compare algorithm performance. The null hypothesis was: "The mean performance of the implemented algorithms is equal, with a significance level of 0.05."

- Result: A P-value of 0.0 led to rejecting the null hypothesis, concluding that there is a significant difference between the algorithms. This indicates that at least one algorithm performs significantly better, with GA-QL standing out.

III. Bonferroni-Dunn Post-hoc Test:

- Used to identify significant differences between algorithms, using GA-QL as the reference. The results in Table 4 confirm that GA-QL significantly outperforms the other algorithms in all comparisons.

These results reinforce the superiority of GA-QL, demonstrating that its performance is statistically better than that of the other algorithms.

5.4 Conclusions

Experimental results and statistical analysis confirm that the GA-QL algorithm is the most effective in solving the EV-TSP problem, significantly outperforming the other variants in terms of performance. GA-QL's ability to dynamically select the crossover operator through Q-Learning was crucial to this success, as it enabled the algorithm to adapt to problem conditions and improve the quality of the solutions obtained. This efficiency in operator selection underscores how integrating machine learning techniques into genetic algorithms can yield substantial benefits in complex optimization problems.

Furthermore, the findings suggest that GA-QL holds great potential for real-world urban logistics scenarios, where it could be employed to optimize delivery routes, reduce energy consumption in electric vehicle fleets, and improve charging station planning. These improvements would likely lead to lower operational costs and a reduced environmental impact. The adaptability and scalability of GA-QL were evident, as it can adjust to various problem conditions, opening the possibility of applying it to larger instances or to problems with additional constraints, such as time windows or load limitations.

Moreover, certain limitations were identified in the study, such as the fact that experiments were conducted on moderately sized instances and that the computational time required might pose challenges for real-time applications. These limitations, however, pave the way for future research directions, including the integration of other machine learning techniques, such as deep learning, and the evaluation of GA-QL under more dynamic conditions with larger datasets.

Overall, these results not only confirm the effectiveness of GA-QL but also highlight its practical relevance in the transportation and logistics sectors, offering a promising path toward more efficient and sustainable solutions in this field.

Acknowledgements This publication is part of eMob project (PID2022-137858OB-I00) funded by MCIN/AEI/10.13039/501100011033/FEDER, UE.

JCHM gratefully acknowledges the CONAHCYT grant for doctoral studies (Grant 002206). The authors are also thankful for the support obtained from Tecnológico Nacional de México through the Instituto Tecnológico de Ciudad Madero, for the use of the Laboratorio Nacional de Tecnologías de Información (LaNTI), and for the support provided through the TecNM project “Design of an intelligent virtual agent for decision support with impact on the navigation of electric vehicles.”

AI Use Statement Some portions of this manuscript were enhanced with the assistance of generative AI tools, under the careful supervision of the authors. The authors carefully reviewed, edited and approved all AI-generated content. The authors take full responsibility for the accuracy and integrity of the entire chapter.

References

1. Deb, K., Pratap, A., Agarwal, S., Meyarivan, T.: A fast and elitist multiobjective genetic algorithm: NSGA-II. IEEE Trans. Evol. Comput. **6**, 182–197 (2002). https://doi.org/10.1109/4235.996017
2. Peralta-Abarca, J.: Metaheurísticas. Inventio **14**, 25–32 (2018). https://doi.org/10.30973/inventio/2018.14.34/3
3. Zambrano-Gutierrez, D., Cruz-Duarte, J., Avina-Cervantes, J., Ortiz-Bayliss, J., Yanez-Borjas, J., Amaya, I.: Automatic design of metaheuristics for practical engineering applications. IEEE Access **11**, 7262–7276 (2023). https://doi.org/10.1109/ACCESS.2023.3236836
4. Ali, I., Sallam, K., Moustafa, N., Chakraborty, R., Ryan, M., Choo, K.: An automated task scheduling model using non-dominated sorting genetic algorithm II for fog-cloud systems. IEEE Trans. Cloud Comput. **10**, 2294–2308 (2022). https://doi.org/10.1109/TCC.2020.3032386
5. Catalao, J., Mariano, S., Mendes, V., Ferreira, L.: Scheduling of head-sensitive cascaded hydro systems: A nonlinear approach. IEEE Trans. Power Syst. **24**, 337–346 (2009). https://doi.org/10.1109/TPWRS.2008.2005708
6. Talbi, E.: Machine learning into metaheuristics: A survey and taxonomy. ACM Comput. Surv. **54**, 1–32 (2021)
7. Ai, G., Zuo, X., Chen, G., Wu, B.: Deep reinforcement learning based dynamic optimization of bus timetable. Appl. Soft Comput. **131**, 109752 (2022). https://doi.org/10.1016/j.asoc.2022.109752
8. Bengio, Y., Lodi, A., Prouvost, A.: Machine learning for combinatorial optimization: a methodological tour d'horizon. Eur. J. Oper. Res. **290**, 405–421 (2021). https://doi.org/10.1016/j.ejor.2020.07.063
9. Ahsini, Y., D'ıaz-Masa, P., Ingl'es, B., Rubio, A., Mart'ınez, A., Magraner, A., Conejero, J.: The electric vehicle traveling salesman problem on digital elevation models for traffic-aware urban logistics. Algorithms **16**, 402 (2023). https://doi.org/10.3390/a16090402.
10. Cuch'y, M., Stolba, M., Jakob, M.: Integrated route, charging and activity planning for whole day mobility with electric vehicles. In: Van Den Herik, J., Rocha, A. (eds.) Agents and Artificial Intelligence, pp. 274–289. Springer, Cham (2019). https://doi.org/10.1007/978-3-030-05453-3_13.
11. Schiffer, M., Walther, G.: Strategic planning of electric logistics fleet networks: A robust location-routing approach. Omega **80**, 31–42 (2018). https://doi.org/10.1016/j.omega.2017.09.004
12. Doppstadt, C., Koberstein, A., Vigo, D.: The hybrid electric vehicle–traveling salesman problem. Eur. J. Oper. Res. **253**, 825–842 (2016). https://doi.org/10.1016/j.ejor.2016.03.006
13. Liao, C., Lu, S., Shen, Z.: The electric vehicle touring problem. Transp. Res. Part B: Methodol. **86**, 163–180 (2016). https://doi.org/10.1016/j.trb.2016.02.002

14. Roberti, R., Wen, M.: The electric traveling salesman problem with time windows. Transp. Res. Part E Logist. Transp. Rev. **89**, 32–52 (2016). https://doi.org/10.1016/j.tre.2016.01.010
15. Wu, Y., Tian, Y., Zuo, X., Liu, M., Zhang, Q.: Learning improvement heuristics for solving routing problems via self-attention and reinforcement learning. Neurocomputing **453**, 260–274 (2021). https://doi.org/10.1016/j.neucom.2021.03.053
16. Shi, L., Zhan, Z.-H., Liang, D., Zhang, J.: Memory-based ant colony system approach for multi-source data associated dynamic electric vehicle dispatch optimization. IEEE Trans. Intell. Transport. Syst. **23**, 17491–17505 (2022). https://doi.org/10.1109/TITS.2022.3150471
17. Ghandriz, T., Jacobson, B., Murgovski, N., Nilsson, P., Laine, L.: Real-time predictive energy management of hybrid electric heavy vehicles by sequential programming. IEEE Trans. Veh. Technol. **70**, 4113–4128 (2021). https://doi.org/10.1109/TVT.2021.3069414
18. Fiori, C., Marzano, V.: Modelling energy consumption of electric freight vehicles in urban pickup/delivery operations: analysis and estimation on a real-world dataset. Transp. Res. Part D: Transp. Environ. **65**, 658–673 (2018). https://doi.org/10.1016/j.trd.2018.09.020
19. Jia, Y.-H., Mei, Y., Zhang, M.: A Bilevel ant colony optimization algorithm for capacitated electric vehicle routing problem. IEEE Trans. Cybern. **52**, 10855–10868 (2022). https://doi.org/10.1109/TCYB.2021.3069942
20. Campaña, M., Inga, E., Cárdenas, J.: Optimal sizing of electric vehicle charging stations considering urban traffic flow for smart cities. Energies **14**, 4933 (2021). https://doi.org/10.3390/en14164933
21. Yang, Q., Zhang, J., Shi, G., Hu, J., Wu, Y.: Maneuver decision of UAV in short-range air combat based on deep reinforcement learning. IEEE Access. **8**, 363–378 (2020). https://doi.org/10.1109/ACCESS.2019.2961426
22. Kang, N., Feinberg, F.M., Papalambros, P.Y.: Autonomous electric vehicle sharing system design. J. Mech. Des. **139**, 011402 (2017). https://doi.org/10.1115/1.4034471
23. Cicirello, V.A.: Non-wrapping order crossover: an order preserving crossover operator that respects absolute position. In: Proceedings of the 8th Annual Conference on Genetic and Evolutionary Computation, pp. 1125–1132. ACM, Seattle Washington USA (2006). https://doi.org/10.1145/1143997.1144177
24. Hussain, A., Muhammad, Y.S., Nauman Sajid, M., Hussain, I., Mohamd Shoukry, A., Gani, S.: Genetic algorithm for traveling salesman problem with modified cycle crossover operator. Comput. Intell. Neurosci. **2017**, 1–7 (2017). https://doi.org/10.1155/2017/7430125
25. Umbarkar, A.J., Sheth, P.D.: Crossover operators in genetic algorithms: a review. IJSC **06**, 1083–1092. https://doi.org/10.21917/ijsc.2015.0150. Walchand College of Engineering, Government College of Engineering, Karad

Optimized Switched Reluctance Generator Operation in Wind Energy Applications

Zeineb Touati, Rui Esteves Araújo, and Adel Khedher

Abstract Switched reluctance generators (SRG) are one of the machines with huge potential in wind power generation due to their reliability and robust design. However, the inherent characteristics of SRGs lead to significant challenges in achieving high efficiency and low output current and torque ripple simultaneously. The performance of SRGs is hindered by conflicting requirements. To address these issues, this chapter presents an optimization control strategy aimed at improving the static performance of SRGs. The chapter discusses the application of the Particle Swarm Optimization (PSO) technique to optimize the commutation angles, specifically the turn-on (θ_{on}) and turn-off (θ_{off}) angles, for an 8/6 SRG. The proposed strategy consists of two main steps. First, a Maximum Power Point Tracking (MPPT) algorithm is implemented to maximize power output at varying rotor speeds, combined with a direct power control method to regulate the power generated by the SRG. Second, a multi-objective function is developed to evaluate the SRG performance, considering key factors such as power output, output current ripple, and torque ripple. The simulation results indicate that implementing optimized turn-on and turn-off angles leads to a reduction in torque ripple from -1.78 Nm using the conventional technique to -0.66 Nm with the proposed method, corresponding to an impressive 63% decrease. Furthermore, the optimization strategy effectively maximizes the efficiency of the system employing an MPPT approach, ensuring optimal energy conversion under varying operating conditions. Future research directions include experimental validation of the proposed control system on real hardware to assess its practical feasibility and performance under real-world operating conditions.

Z. Touati (✉) · A. Khedher
LATIS-Laboratory of Advanced Technology and Intelligent Systems, National Engineering School of Sousse, University of Sousse, Sousse, Tunisia
e-mail: zeineb.touati@eniso.u-sousse.tn

A. Khedher
e-mail: adel_kheder@yahoo.fr

R. E. Araújo
INESC TEC and Faculty of Engineering, University of Porto, Porto, Portugal
e-mail: raraujo@fe.up.pt

W. Pedrycz et al. (eds.), *Decision-Making in Computational Intelligence-Based Systems*, Studies in Systems, Decision and Control 628,
https://doi.org/10.1007/978-3-032-04056-5_6

Keywords Switched reluctance generator · Modeling · Finite element analysis · Particle swarm optimization · Power control · Renewable energy sources

1 Introduction

Compared with other electrical machines, industrial interest in Switched Reluctance Machines (SRMs) has grown, mainly due to their absence of permanent magnets. SRMs are also known for their simple structure, rotor without windings, excellent fault tolerance, and strong performance [1–4]. These advantages make SRMs an attractive solution for various applications, such as water pumping, turbocharger [5], electric propulsion [6] and gas turbine [7].

In certain applications, such as hybrid electric vehicles and flywheel energy storage, the SRM can function both as a motor and a generator [8, 9]. Recent reviews on the various applications of SRMs are available in [10, 11].

In recent years, the Switched Reluctance Generator (SRG) has gained popularity in distributed generation applications [12–17]. In these systems, a turbine is connected to the SRG, which supplies power to a power electronic converter with a DC output. In standalone applications, the DC output of the SRG can either be used to power a DC load directly or be inverted to supply an AC load. In grid-connected systems, the DC output of the SRG can either be fed into a DC grid or converted to AC and injected into an AC grid. Despite these advantages, there are several challenges in the design, control, and operation of SRGs [10, 11, 18]. Researchers have focused on addressing DC-bus voltage control and reducing voltage ripple. Significant efforts have also been made to optimize output power control and improve overall performance [12–17, 19].

The open-loop instability of the DC-bus voltage in self-excited SRGs has been explored in [14]. DC-bus voltage control strategies for SRGs are discussed in [12]. The voltage ripple is notably high due to the inherent phase commutation. A novel pulse train strategy aimed at reducing DC-bus voltage ripple is presented in [20]. In the case of DC grid-connected SRGs, [21] introduces an additional interfacing converter to regulate the output voltage of the SRG.

Many researchers have focused, either directly or indirectly, on maximizing the efficiency of SRGs [22–29]. Given the widespread application of SRGs, this focus can be expanded to include both efficiency and reliability. The current through the capacitor is significantly affected by the switching angles, which in turn influence the capacitor's loss, temperature rise, and lifespan. Maximizing efficiency requires estimating machine losses, including both core and copper losses. This evaluation needs to be conducted over a wide range of possible operating points for the SRG system. The phase RMS current directly determines copper loss and significantly influences core loss. Therefore, the maximization of efficiency is often addressed indirectly through the minimization of phase RMS. This phase RMS current is easily evaluated through simulation runs during the optimization of switching angles. The optimized switching angles are then applied in closed-loop power control. Experimental studies in [30] highlighted a strong correlation between reduced output current ripple and increased efficiency. A reduction in output current ripple or DC-bus capacitor current leads to lower capacitor losses and heating, which in turn enhances the capacitor's lifespan and reliability, thus improving the overall reliability of the converter. Voltage ripple, a significant challenge in SRG control, can also be mitigated by reducing capacitor current.

Through computational experimental design, an approach is used to determine the optimum firing angles and DC link voltage that guarantee the best system performance at each rotor speed, as reported in [31].

In multi-objective optimization [23, 24, 32, 33], selecting the objectives for a given application and assigning appropriate weights to them can be challenging. In [23], a multi-objective function is constructed as a result of SRG performance evaluation, which takes into account system efficiency, output voltage ripple and power converter loss. In [34] the generated output power, torque smoothness and current smoothness defined above are selected as optimization targets to be improved. Two techniques for determining optimum switching angles for an SRG operating in single-pulse mode are presented in [35]. The first technique obtains the pair of switching angles to provide the desired output power with minimum rms phase current. The second technique obtains the pair of switching angles to provide the desired output power with minimum DC bus RMS current. An optimized fuzzy logic-based controller for turn-on and turn-off angles is proposed in [36], taking into account output voltage ripple and the deviation from the reference voltage. In [37], an SRG is employed as a battery charger, incorporating a power converter and a smart search control (SSC) strategy to reduce current ripple at maximum output power. The SSC method dynamically adjusts the excitation, turn-on, and turn-off angles for each SRG phase during operation. In [38], differential evolution strategy is introduced to find optimal firing angles and DC bus voltage level under multiple operating conditions Optimum ignition angles are determined in [39] to reduce torque ripple and minimize acoustic noise. In [40], the optimum turn-on and turn-off angles for maximum output power are controlled, and the relationship between excitation current and conduction time is examined.

However, the efficiency of SRG is significantly impacted by high torque, which lead to increased mechanical vibrations, acoustic noise, and efficiency losses. To address these challenges, numerous multi-objective optimization approaches have been proposed, each introducing different control strategies and optimization techniques. Despite these variations, the primary objective remains the same: enhancing the overall performance of SRGs by minimizing ripple and maximizing efficiency. The key multi-objective design and optimization approaches for SRMs are summarized and compared in Table 1, considering the specific modeling and optimization methods, the performance indices targeted for optimization, and the integration of current profile and torque ripple. The SRGs are widely used in wind energy applications due to their robust structure and fault tolerance.

To consider the effects of nonlinear dynamic characteristics of the machine and improve the comprehensive performance of wind application connected to an SRG, a direct power control strategy based on optimization of SRG drive system is proposed in this chapter. This strategy is presented to achieve maximum electrical power with

Table 1 Comparison of various multi-objective design and optimization of SRG

Reference	Optimization method	Voltage ripple	Torque ripple	Efficiency	Integration of current profile
[17]	Quasi-Newton method and the Genetic Algorithm	Yes	Yes	Yes	No
[23]	Simulated Annealing Algorithm	Yes	Yes	Yes	No
[24]	Particle Swarm Optimization	No	Yes	Yes	Yes
[29, 41]	Non-dominated Sorting Genetic Algorithm	No	Yes	Yes	No
[38]	Differential Evolution	No	No	Yes	No
[42]	Differential Evolution	No	Yes	Yes	Yes
[43]	Grey Wolf Optimization	No	Yes	Yes	No

minimum current and torque ripple. The method of optimization is an online control strategy that determines optimal turn-on and turn-off to achieve maximum electrical power. Simulation results show the effectiveness and applicability of the proposed power control with optimization. This chapter is organized as follows: Firstly, the structure and operation principle of the switched reluctance machine are established in Sect. 2. Then, the modeling of wind turbine with power maximization is proposed and performed in Sect. 3. An optimization method for SRG is suggested in Sect. 4. The Methodology are established in Sect. 5. Finally, the Simulation results and discussion are analyzed and compared in Sect. 6.

2 Structure and Operation Principle of the Switched Reluctance Machine

2.1 Basic Structure

The SR machine is a double-salient machine, featuring salient poles on both the stator and the rotor. Its stator contains field coils, similar to those in a DC motor, while the rotor is free of coils or magnets and is made of ferromagnetic material with regularly spaced salient poles.

The SRG is an electromechanical energy converter designed to transform mechanical energy into electrical energy. For the SRG to operate as a generator, it requires excitation during the phase of decreasing inductance, accompanied by the application of mechanical torque to the machine shaft. The magnetization of the phase windings, in conjunction with the mechanical input on the machine's shaft, induces an electromotive force (emf) that accelerates the rate of current increase within the system [19].

Figure 1 shows overview of SRG system. The drive system under consideration consists of a power converter and a closed-loop control system. This configuration is necessary because the SRG is unstable in the open loop. The converter depicted in Fig. 1 represents a single phase of the SRG, while the drive system for multi-phase SRGs is described subsequently. The SRG can either directly power a load, or transfer energy to the grid through an additional power electronic converter.

2.2 *Working Principle*

The operation of an SRG relies on three key stages: excitation, freewheeling, and generation. During the excitation stage, a voltage is applied to a specific phase of the SRG, causing current to flow through its windings, where the inductance is decreasing. This process magnetizes the phase and transfers energy from the voltage bus to

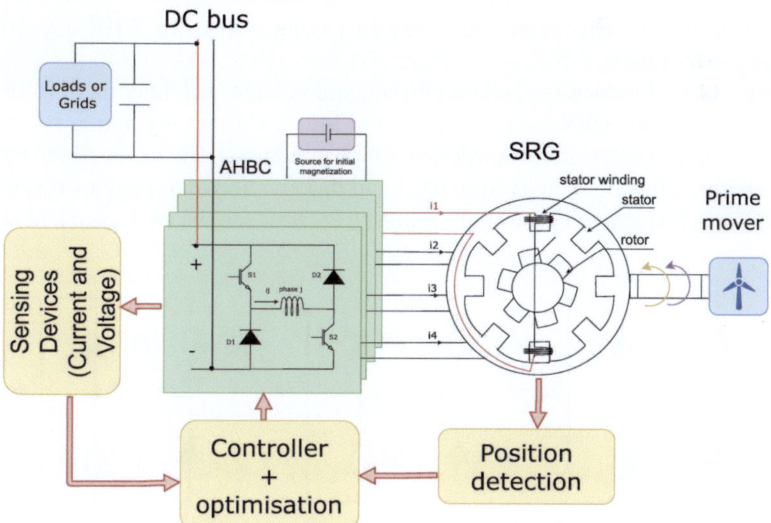

Fig. 1 An overview of SRG system

the magnetic field. In the freewheeling stage, the current is maintained within the winding without transferring energy to the load, allowing the magnetic field to store energy. Finally, in the generation stage, the stored magnetic energy, combined with the energy from mechanical-to-electrical conversion, is delivered to the load. These stages are cyclically repeated for each phase, ensuring continuous energy conversion. The SRG drive converter must not only facilitate these operational stages but also provide flexibility, efficiency, and stability. Among the various converter topologies available for driving SRGs, the Asymmetric Half-Bridge Converter (AHBC), is the most widely used due to its robustness and ability to manage energy regeneration and freewheeling as required [12]. The AHB converter makes the SRG self-excited but requires an initial external source, such as a battery, to charge the capacitor and start operation. Once the capacitor is charged, it supplies excitation to the SRG phases when the external source is removed. Additionally, the capacitor stabilizes the voltage delivered to the load. Figures 2 illustrate the AHB converter's operation stages.

During the excitation stage (Fig. 2a), the two switches of the active phase are turned on, allowing current to flow from the capacitor to magnetize the phase. After the excitation interval, the switches are turned off, and the diodes conduct the generated energy to the load while recharging the capacitor (Fig. 2b). In the freewheeling stage (Fig. 2c), the diodes allow energy to flow to the load and stabilize the system. This process repeats cyclically for all phases of the converter.

The AHB converter offers several advantages, including individual phase current control and inherent protection against DC bus short circuits due to its design, which connects switches in series with the machine windings [11]. However, this configuration requires two semiconductor switches per phase, making it less cost-effective compared to other topologies that use fewer switches. Although these alternative topologies, such as those described in [11, 44], reduce costs, they come with limitations in terms of performance and flexibility compared to the AHB converter, as noted in the literature.

The profiles of inductance, phase current, and voltage as a function of rotor position across the three regions are illustrated in Fig. 3. Current regulation is achieved through the use of a hysteresis controller, which operates within a predefined hysteresis band. The selection of activation (θ_{on}) and deactivation (θ_{off}) angles is crucial to the SRG's performance. These angles must be set between the unaligned and aligned

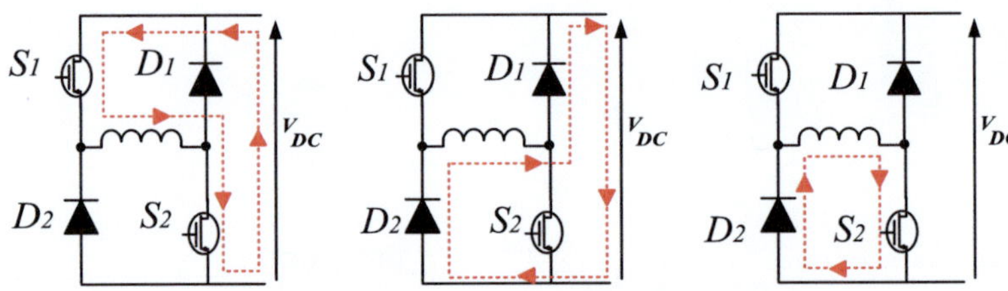

Fig. 2 Representation of the current direction during the operating states: **a** excitation stage, **b** generation stage, and **c** freewheeling stage

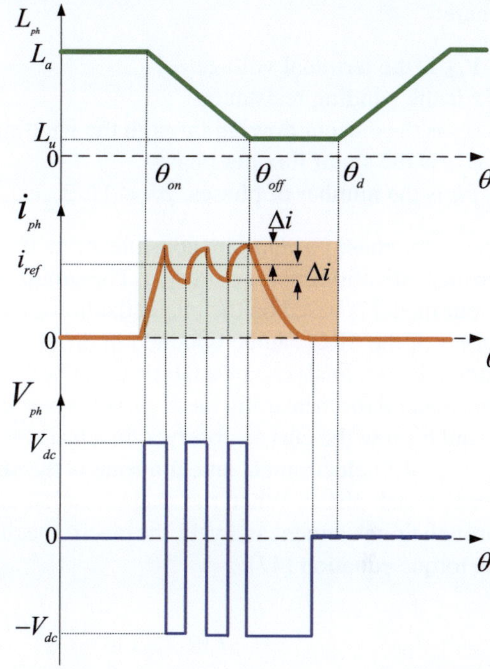

Fig. 3 Profile of inductance, phase current and phase voltage as a function of rotor position

rotor positions. Accurately determining these angles is essential for achieving optimal current regulation and ensuring the smooth operation of the generator across its entire operational range.

2.3 *Modeling of SRG*

To simplify the SRG model, the following assumptions are considered: (1) the mutual effects between different phases are neglected; (2) all phase parameters are assumed to be identical; (3) the winding resistance is considered constant and unaffected by temperature changes; (4) induced currents in the magnetic circuit and core losses are ignored; and (5) the converter switches are assumed to be ideal. These assumptions facilitate a more straightforward and efficient modeling and analysis of the SRG system. The electrical circuit of an SRG with n phases consists of a resistor in series with an inductance, and its behavior is nonlinear for each phase of the machine. The voltage equation of SRG for the ph-phase winding can be described as follows [45, 46]:

$$V_{ph} = ri_{ph} + \frac{d\phi_{ph}(\theta, i_{ph})}{dt} \tag{1}$$

where:

- V_{ph} is the terminal voltage,
- r is the winding resistance
- i_{ph} is the current flowing through the windings,
- ϕ_{ph} is the stator flux per phase.
- ph is the number of phases, $ph \in \{1, 2, 3, \ldots n\}$.

The stator phase currents are non-linear functions $i(\phi, \theta)$ that can be calculated from the magnetization curves $\phi(\theta, i_{ph})$. The simulation developed in MATLAB-Simulink of our model is based on the magnetization curve of the machine. The magnetization curves of the SRG can be obtained in different ways, such as through experiments, finite element analysis, or analytical methods using available machine parameters. One method for measuring these curves is based on finite element analysis. Figures 4, 5, and 6 show the curves observed in a test performed on an experimental 8/6 SRG.

The SRG electromechanical torque is the sum of each individual torques developed at each phase. When the magnetization curves are obtained experimentally or through finite element analysis, the electromechanical torque can be calculated using the torque equation [47]:

$$T(\theta, i_{ph}) = \frac{1}{2} {i_{ph}}^2 \frac{dL_{ph}(\theta, i_{ph})}{d\theta} \tag{2}$$

The global electromagnetic torque T_e is described by:

$$T_e(\theta, i_{ph}) = \sum_{k=1}^{n} T_k(\theta, i_{ph}) \tag{3}$$

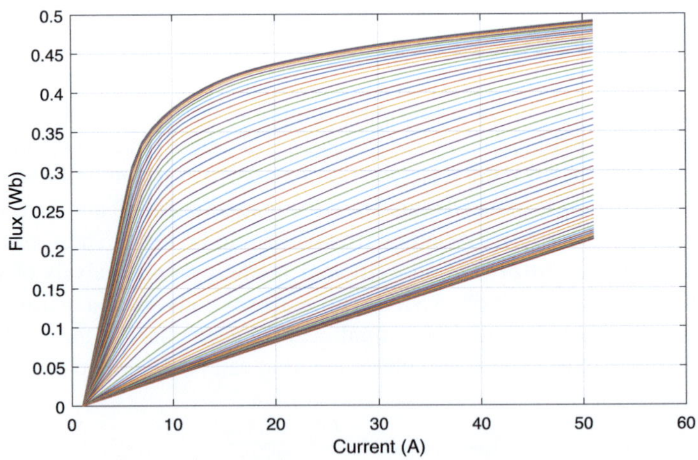

Fig. 4 Flux linkage characteristics as a function of current at different rotor position

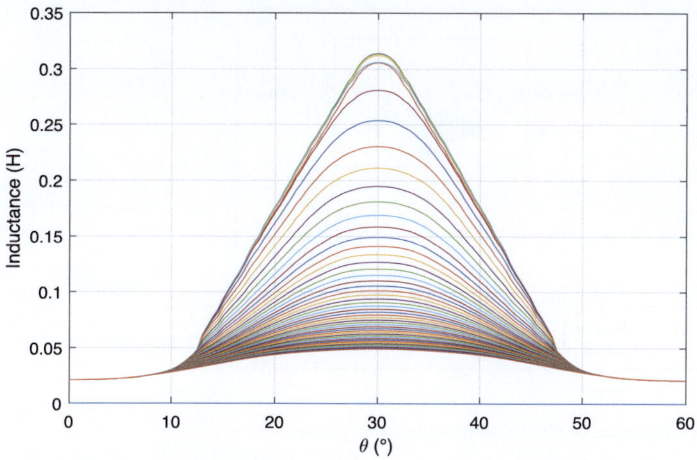

Fig. 5 Inductance characteristics as a function of rotor position at different current from 0.2 to 10 A

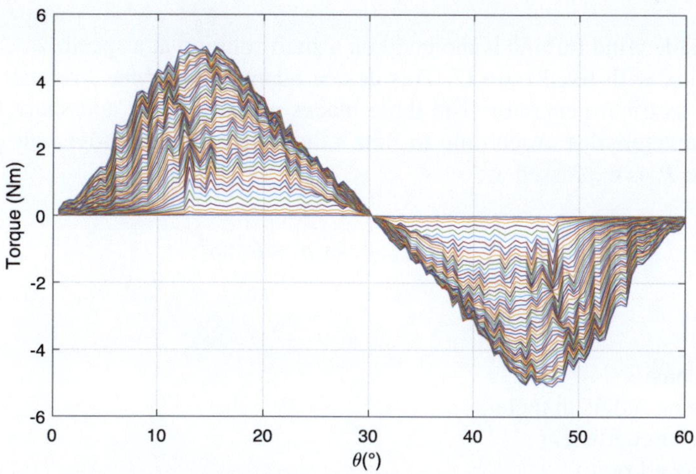

Fig. 6 Torque characteristics as a function of rotor position at different current from 0.2 to 10 A

The electrical circuit of an SRG phase is modeled as illustrated in Fig. 7.

We can observe from Fig. 6 two fundamental operating parts of the machine. The positive part corresponds to the operation of the SRM in motor mode, while the negative part corresponds to the operation in generator mode.

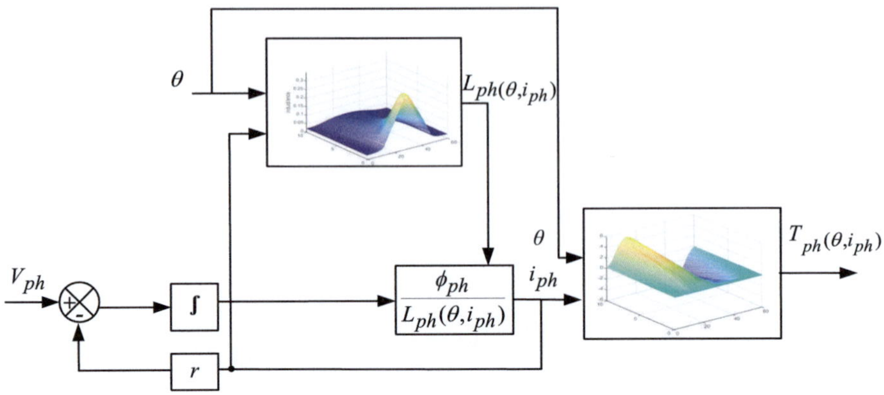

Fig. 7 Block diagram of electrical model of one phase

3 Modeling of Wind Turbine with Power Maximization

A three-blade wind turbine is mounted on a shaft rotating at a speed Ω_t, connected to a gearbox with fixed ratio G. This device adapts the turbine's rotational speed with the electrical generator. The three blades are adjustable and share the same design characteristics. According to Betz's theory [48], the aerodynamic power of the turbine P_t is expressed as:

$$P_t = \frac{1}{2}.C_p(\beta, \lambda).\rho.\pi.R^2.{v_t}^3 \tag{4}$$

where:

- ρ: Air density
- R: Turbine radius in meters
- C_p: Power coefficient
- λ: Tip speed ratio
- β: Blade pitch angle in degrees
- v_t: Wind speed

Using the above relation, the aerodynamic torque of the turbine is described by the following equation:

$$T_t = \frac{P_t}{\Omega_t} = \frac{P_t \cdot R}{\lambda \cdot v_t} \tag{5}$$

The tip speed ratio λ is defined as:

$$\lambda = \frac{\Omega_t \cdot R}{v_t} \tag{6}$$

By substituting Eq. 6 into Eq. 5, the wind turbine torque equation becomes:

$$T_t = \frac{1}{2 \cdot \lambda} \cdot C_p(\beta, \lambda) \cdot \rho \cdot \pi \cdot R^3 \cdot v_t^2 \quad (7)$$

The expression for C_p used in this analysis is formulated in Eq. 8 [49]:

$$C_p(\beta, \lambda) = c_1 \left(c_2 \left(\frac{1}{\lambda + 0.08\beta} - \frac{0.035}{\beta^3 + 1} \right) - c_3\beta - c_4 \right) e^{-c_5\left(\frac{1}{\lambda+0.08\beta} - \frac{0.035}{\beta^3+1} \right)} + c_6\lambda \quad (8)$$

with $c_1 = 0.5176$, $c_2 = 116$, $c_3 = 0.4$, $c_4 = 5$, $c_5 = 21$, and $c_6 = 0.0068$. By applying the fundamental law of dynamics, we considered no effect of torsion of the shaft, the mechanical speed is expressed by the following equation:

$$J \frac{d\Omega_g}{dt} = T_M - T_e - F \cdot \Omega_g \quad (9)$$

with:

$$\begin{cases} \Omega_g = G.\Omega_t \\ T_M = \frac{T_t}{G} \end{cases} \quad (10)$$

where:

- Ω_g: Rotational speed of the SRG,
- T_M: Aerodynamic torque,
- G: Gear ratio,
- T_e: Electromagnetic torque of the SRG,
- F: Friction coefficient,
- J: total inertia of the turbine with SRG.

The block diagram shown in Fig. 8 illustrates the complete kinematic model of the turbine.

The pitch angle β is used to control the desired maximum power by adjusting the orientation of the wind turbine blades, despite fluctuations in wind speed and direction. For a zero pitch angle, the power coefficient C_p reaches its maximum.

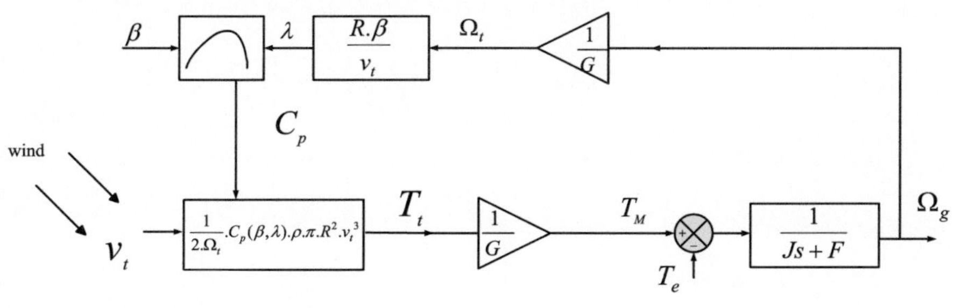

Fig. 8 Block diagram of the turbine model

Under this condition, we observe the variation of C_p as a function of the tip speed ratio λ for different values of β, as shown in Fig. 9.

A Maximum Power Point Tracking (MPPT) control strategy is applied to maintain the power coefficient C_p at its optimal level, ensuring maximum capture power even under variations in wind speed, which is the main source of disturbance in the wind turbine conversion chain. These variations result in fluctuations in energy production. The control strategy involves adjusting the torque on the turbine shaft to maintain it at the desired reference speed. To achieve this goal, a speed regulation system is implemented. The functional diagram of a turbine with MPPT is shown in Fig. 10. The optimal aerodynamic power of the wind turbine is expressed as:

$$P_t^{opt} = \frac{1}{2}.C_p{}^{\max} \cdot \rho \cdot \pi \cdot R^3 \cdot v_t^2 \tag{11}$$

To control the mechanical speed, a linear control law is applied, expressed as follows:

$$T_{e-ref} = R_\Omega(s) \cdot (\Omega_{g-ref} - \Omega_g) \tag{12}$$

where $R_\Omega(s)$ is the speed controller, and Ω_g is the mechanical reference speed. The transfer function relating speed and torque in Laplace is given by:

$$\frac{\Omega_g}{T_M - T_{e-ref}} = \frac{1}{Js + F} \tag{13}$$

Figure 11 illustrates the operational diagram for speed control.

Based on the principle of linear regulation, a PI controller is adopted to eliminate the error between the controlled variable and the reference variable. Consequently, the closed-loop transfer function is defined as:

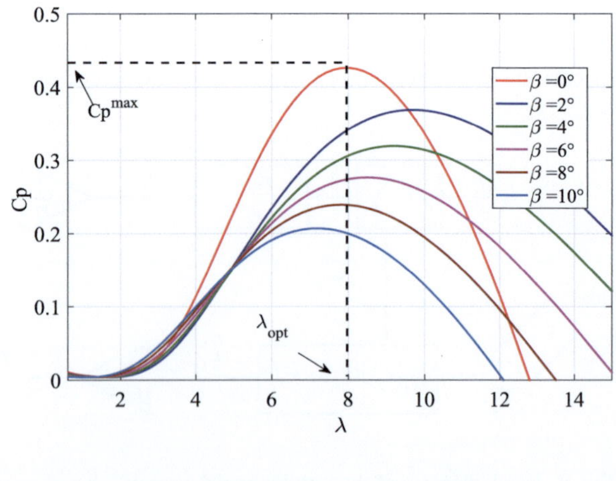

Fig. 9 Variation of the power coefficient as a function of β and λ

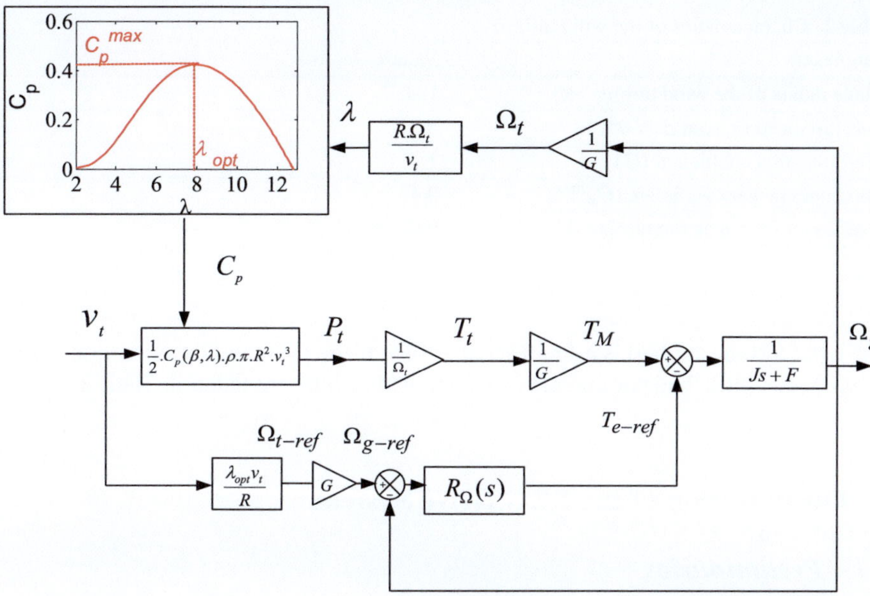

Fig. 10 Block diagram of the turbine model with MPPT

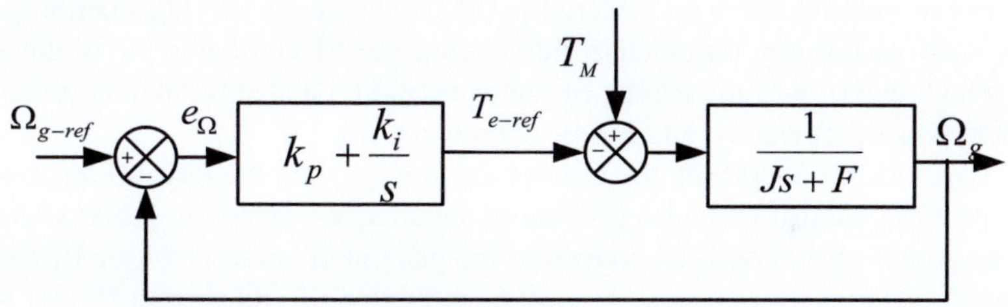

Fig. 11 Functional diagram of the PI controller

$$F_{BF}(s) = \frac{k_p s + k_i}{s^2 + \frac{k_p + F}{J} s + \frac{k_i}{J}} \tag{14}$$

The parameters K_p and K_i of the wind speed controller are determined using the pole placement method:

$$\begin{cases} \frac{F + K_p}{J} = 2\omega_n \xi \\ \frac{K_i}{J} = {\omega_n}^2 \end{cases} \tag{15}$$

From Eq. 15, the parameters K_p and K_i of the speed controller are:

$$\begin{cases} K_p = 2J\omega_n \xi - F \\ K_i = J{\omega_n}^2 \end{cases} \tag{16}$$

Table 2 Characteristics of the wind turbine

Parameters	Values
Blade radius of the wind turbine (R)	$1\ m$
Specific speed (λ_{opt} at $\beta = 0°$)	8
Multiplication coefficient (G)	1
Maximum power coefficient ($C_p{}^{\max}$)	0.42
Moment of inertia on the turbine side (J)	0.01 kg m^2

The wind turbine studied is a small wind turbine blade designed for operation at low wind speeds [50]. The parameters of the studied turbine are listed in Table 2.

4 Optimization of the SRG Drive System

4.1 Preliminaries

The optimization of the SRG drive system is important for improving its efficiency, torque ripple minimization, and overall performance under varying operating conditions. One of the key parameters influencing the SRG's behavior is the firing angle, which determines the timing of phase excitation and significantly affects the generated power, efficiency, and torque characteristics.

In a typical SRG operation, the turn-on angle (θ_{off}) and the turn-off angle (θ_{off}) are the primary control variables governing the energy conversion process. Adjusting these angles allows optimal control of the generated voltage, torque ripple suppression, reduced acoustic noise, as well as reduction of DC current ripple, which improves battery charging performance. However, determining the optimal firing angles involves a compromise between multiple objectives, such as maximizing efficiency while maintaining stable power output.

This section proposes a multi-objective optimization framework to fine-tune the firing angle parameters by considering the trade-offs between efficiency, torque ripple, and power quality. The selected optimization algorithm will be applied to determine the best set of firing angles for different operating conditions.

4.2 Formulation of the Objective Function

The power generated by an SRG is a result of electromechanical energy conversion, where electrical power is derived from the variation of inductance with rotor position. The average output power can be expressed in terms of phase voltage and current as:

$$P_{in} = average\left(\sum_{ph=A,B,C,D} (V_{ph} i_{ph})\right) \tag{17}$$

However, in an SRG, due to the switching nature of excitation and the nonlinear variation of inductance with rotor position, deriving a closed-form analytical expression for power is challenging. Instead, an approximate representation can be obtained under specific assumptions.

Considering some simplifications and neglecting losses, the average output power of the SRG can be expressed as follows [22]:

$$P_{SRG} = \frac{N_s\, N_r\, V_{dc}}{\Omega_g}\left(\int_{\theta_{on}}^{\theta_{off}} \frac{(\theta - \theta_{on})}{L(\theta)}\, d\theta + \int_{\theta_{off}}^{\theta} \frac{(\theta_{off} - \theta - \theta_{on})}{L(\theta)}\, d\theta\right) \tag{18}$$

where;

- θ_{on} and θ_{off} are the firing angles
- Nr is the number of rotor pole
- N_s is the number of stator pole
- V_{dc} is the DC link voltage
- θ is the angular position
- Ω_g is the mechanical speed of wind turbine

To analyze the impact of the firing angle on SRG behavior, which defines the timing of phase excitation, Fig. 12 presents typical examples of SRG control operation and compares three different cases. In Case 1, when $\theta_{on} < \theta_a$, the conducting period is extended, resulting in a high peak flux linkage and consequently increased iron losses. Despite the relatively low peak phase current, the prolonged conduction period causes higher stator copper losses. In Case 2, when $\theta_{on} > \theta_a$, the conducting period is significantly shortened as the turn-on angle shifts closer to the unaligned position. This reduces iron losses but increases the peak current, thereby raising copper losses. Finally, in Case 3, maximum efficiency is achieved by selecting an optimal conducting period, which minimizes total electrical losses. Where θ_a is the aligned position of the stator and rotor, and $\theta_d = \frac{2\pi}{N_r}$ represents the angle between successive rotor poles.

According to Eq. 18, the output power is directly influenced by the firing angles, the DC link voltage, and the operating speed. Different combinations of these driver variables can produce the same output power. However, these variables have a significant impact on electric and magnetic losses, as well as torque ripple. Therefore, it is crucial to determine the optimal activation parameters to ensure satisfactory SRG performance across a wide speed range. In the proposed system, the DC link voltage is maintained constant, allowing the generated power to be controlled through various combinations of firing angles, hysteresis current, and speed.

Fig. 12 The effect of firing angles on SRG performance

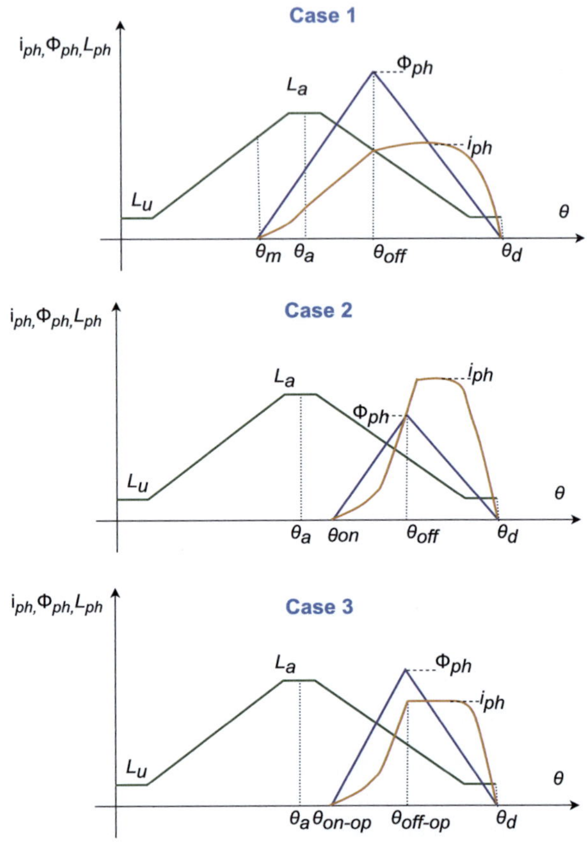

The primary optimization goals are to achieve higher energy generation efficiency, reduce torque ripple for smoother operation, and minimize current and voltage ripple to enhance system reliability and extend equipment lifespan. To optimize these objectives, three key optimization indices are considered: power generation, torque, and current ripples. The current and supply of the DC bus are:

$$\begin{cases} i_{dc} = i_{srg} - i_r \\ V_{dc} = \frac{1}{C} \int i_{dc} dt \end{cases} \tag{19}$$

where i_{srg} is the total currents of SRG, and i_r is the load current. To evaluate the power generation of the SRG system under lossless conditions, the output power, P, is selected as the objective and is defined as:

$$P = V_{dc} i_r \tag{20}$$

The mechanical input power provided by the wind turbine is obtained by:

$$P_M = T_M \Omega_g \tag{21}$$

The system efficiency can be obtained by solving:

$$\eta = \frac{P}{P_M} \tag{22}$$

Torque ripple can cause significant damage to the generator and shorten its lifetime. It is expressed as the difference between the maximum and minimum instantaneous torque, relative to the average torque, as follows:

$$T_{ripple} = \frac{T_{\max} - T_{\min}}{T_{ave}} \tag{23}$$

where $T_{\max}$ and $T_{\min}$ represent the instantaneous maximum and minimum torque values, respectively, and T_{ave} is the average torque. The level of torque ripple suppression can be quantified by the torque smoothness factor, defined as:

$$T_{SF} = \frac{1}{T_{ripple}} \tag{24}$$

However, large current fluctuations are highly detrimental to the battery, as they can directly shorten its lifespan. Therefore, it is crucial to reduce the current fluctuation, I_{ripple}. In this chapter, the current smoothness, I_{SF}, is considered as a metric to assess and minimize current fluctuations, and is expressed as:

$$I_{SF} = \frac{1}{I_{ripple}} = \frac{I_{ave}}{I_{\max} - I_{\min}} \tag{25}$$

where $I_{\max}$ and $I_{\min}$ represent the instantaneous maximum and minimum DC link current values, respectively, and I_{ave} is the average DC link current. The output generated power, torque smoothness, and current smoothness, as defined above, are selected as optimization objectives to enhance energy generation efficiency, minimize torque ripple and improve the power quality of DC bus power supply. The values of these objectives should be maximized to achieve optimal performance.

4.3 Particle Swarm Optimization (PSO)

Kennedy and Eberhart proposed Particle Swarm Optimization (PSO) in 1995, drawing inspiration from the social interactions observed in bird flocks [51]. This technique is highly effective for solving complex optimization problems. The process begins with a swarm of particles, representing a population of potential solutions. Each particle continuously adjusts its velocity and position as it searches the solution space. Through these iterative updates, the swarm collectively works toward identifying the optimal value of the objective function.

PSO is a robust stochastic optimization technique that relies on iterative velocity and position updates. At each iteration, each particle updates its position based on its own previous best position (P_{Best}) and the global best position (G_{best}) found by the swarm. Each particle retains memory of the best position it has explored in the search space, which is used to calculate a new velocity (v_{k+1}). This velocity is determined based on the particle's current velocity (v_k), the distance to its personal best position (P_{Best}), and the distance to the global best position (G_{best}).

The velocity and position are updated using the following equations:

$$V_i^{(k+1)} = \varpi V_i^{(k)} + C_1 rand_1 (P_{Best} - x_i^k) + C_2 rand_2 (g_{Best} - x_i^k) \tag{26}$$

$$x_i^{k+1} = x_i^k + V_i^{(k+1)} \tag{27}$$

where:

- $V_i^{(k)}$: Current velocity of the particle
- x_i^k: Current position of the particle
- P_{Best}: Particle's personal best position
- G_{Best}: Global best position found by the swarm
- ϖ: Inertia weight
- C_1, C_2: Acceleration coefficients
- $rand_1, rand_2$: Random values in the range [0, 1]

In this study, the inertia weight (w) is not fixed but varies dynamically according to the following expression [52]:

$$\varpi = \varpi_{\max} + \frac{(\varpi_{\max} - \varpi_{\min})}{k_{\max}} k \tag{28}$$

where:

- $\varpi_{\max}$: Maximum inertia weight

Fig. 13 PSO flowchart for the control system

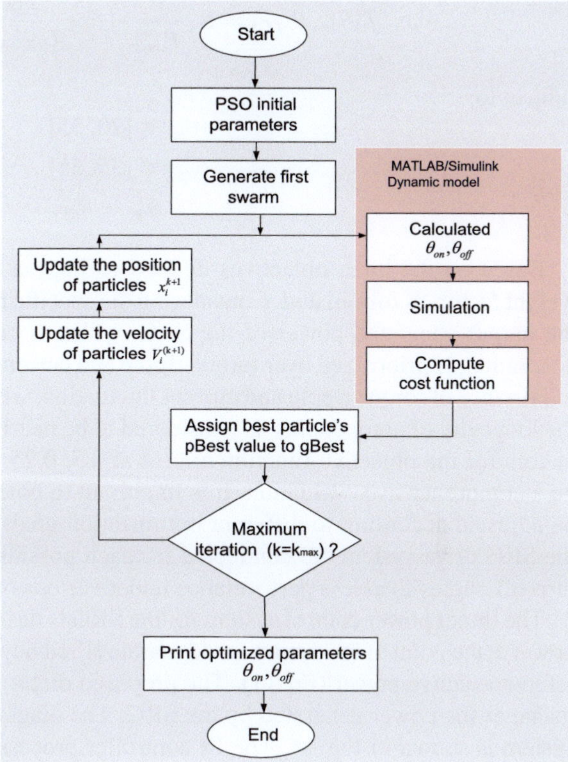

- $\varpi_{\min}$: Minimum inertia weight
- k: Current iteration number
- $k_{\max}$: Maximum number of iterations

In this study, the coefficients in Eq. 28 are set as constants: $C_1 = C_2 = 2$. The maximum number of iterations is set to 100, and the inertia weight values are defined as: $\varpi_{\min} = 0.4$ and $\varpi_{\max} = 0.9$. The PSO program has been developed using MATLAB due to its compatibility with control systems in Simulink, making it a suitable choice for this application. The suggested method for coordinating the PSO with the control system is represented by the following flowchart (Fig. 13).

5 Proposed Control Strategy

To improve the overall performance of wind turbines under varying operating conditions, an optimization strategy is proposed in this chapter as follows:

$$\max . f(\theta_{on}, \theta_{off}) = \omega_1 \frac{P}{P_{\max}} + \omega_2 \frac{T_{SF}}{T_{SF-\max}} + \omega_2 \frac{I_{SF}}{I_{SF-\max}} \tag{29}$$

subject to:

$$\theta_{on} \in [20, 35]$$
$$\theta_{off} \in [35, 55]$$
$$\theta_{on} < \theta_{off}$$

Based on the three objectives defined earlier, an objective function with three weight factors is formulated. Considering the specific requirements of wind turbines, the output generated power of the SRG, which is critical for maximizing energy generation, is prioritized over torque ripple and current fluctuation. Additionally, the importance of torque ripple and current fluctuation, which affect system stability and the longevity of components, is considered to be nearly equal. Therefore, the weight factors for the objective function are set at 0.5, 0.25, and 0.25, respectively, based on a comprehensive evaluation. It is important to note that these weight factors can be adjusted according to different optimization goals. Furthermore, simulations of the SRG drive system are conducted for each possible combination of turn-on and turn-off angles to assess performance under various operating conditions.

The direct power control system for the SRG is designed to regulate the generated power at the point of maximum aerodynamic efficiency, P_{ref}, where P_{ref} represents the reference active power (Eq. 11). The proposed direct power control system directly manages the power generated by the SRG. The diagram of the direct power control system is shown in Fig. 14. The PI controller processes the error between P_{ref} and the generated power P, and adjusts the reference current accordingly. The angle

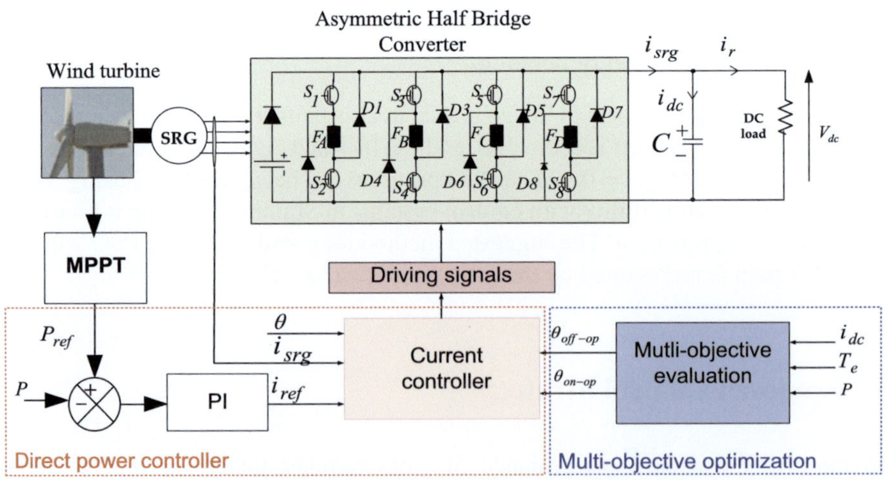

Fig. 14 Structure of multi-objective controller of the SRG proposed controller

Table 3 Parameters of the SRG

Parameters	Symbol	Values
Nominal output power	P_{rated}	500 W
Nomber of stator poles	N_s	8
Nomber of rotor poles	N_r	6
Nominal speed	Ω_{rated}	1500 rpm
Inductance (aligned position)	L_a	0.14 H
Inductance (unaligned position)	L_u	0.021 H
Viscous friction	F	0.05 Nms
Resistance of phase winding	r	5 Ω
Maximum current	i_{max}	10 A

optimization controller is developed based on the outcomes of the optimization process. By integrating the switching system with the angle optimization controller, an SRG optimization control system is established, as illustrated in Fig. 14. The optimal switching angle is determined in real-time using the output power, current of DC bus, and SRG torque. Additionally, the switching system is capable of switching between different control modes depending on the rotor speed.
The reference current is given by:

$$i_{ref} = K_p e + K_i \int edt \tag{30}$$

where $e = P_{\text{ref}} - P$, K_p and K_i are the proportional and integral gain.

6 Simulation Results and Discussion

In this chapter, the MATLAB/Simulink simulation results of an 8/6 pole, 4-phase SRG connected to a wind turbine are presented. The parameters of the SRG are summarized in Table 3, and the test bench with these dimensions has been validated in [12, 14, 16]. The proportional and integral gains, K_p and K_i, are set to 1 and 0.01, respectively. The simulations analyze the current, voltage, power, and switching signal waveforms for all SRG phases in both optimized and unoptimized systems.

Firstly, the performance results of the wind turbine operating with Maximum Power Point Tracking (MPPT) are illustrated in Fig. 15. These results provide a comprehensive analysis of the system's behavior under varying conditions. Figure 15a depicts the wind speed variations applied to the turbine, highlighting the dynamic nature of the input conditions. Figure 15b presents the tip-speed ratio (λ)), which plays a critical role in evaluating the efficiency of energy conversion. Figure 15c

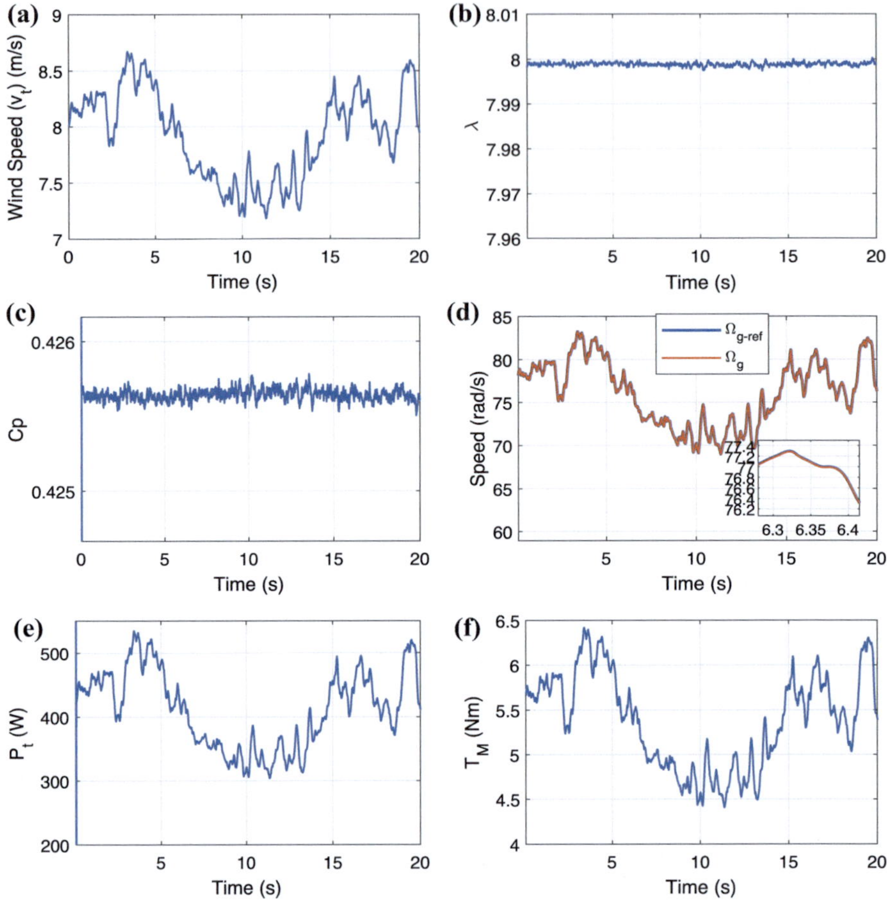

Fig. 15 **a** Wind speed (m/s), **b** tip-speed ratio (λ), **c** power coefficient (Cp), **d** rotor speed (rad/s), **e** turbine power (W), and **f** turbine torque (Nm)

shows the power coefficient (Cp), indicating how effectively the wind turbine extracts energy from the wind. Figure 15d compares the reference speed and the desired speed of the system, providing insights into the control strategy's accuracy. Figure 15e illustrates the generated power, demonstrating the system's overall energy output, while Fig. 15f highlights the wind torque, which reflects the mechanical interaction between the wind and the turbine. These detailed results collectively showcase the effectiveness of the MPPT algorithm and the system's operational performance.

The wind profile shown in Fig. 15a is applied as the system input, and the simulation is conducted over a duration of 20 s. Subsequently, the variations in the output parameters are analyzed for both systems, with the results presented in Figs. 16 and 17, respectively. This analysis provides a clear comparison of the system's performance under the given wind conditions.

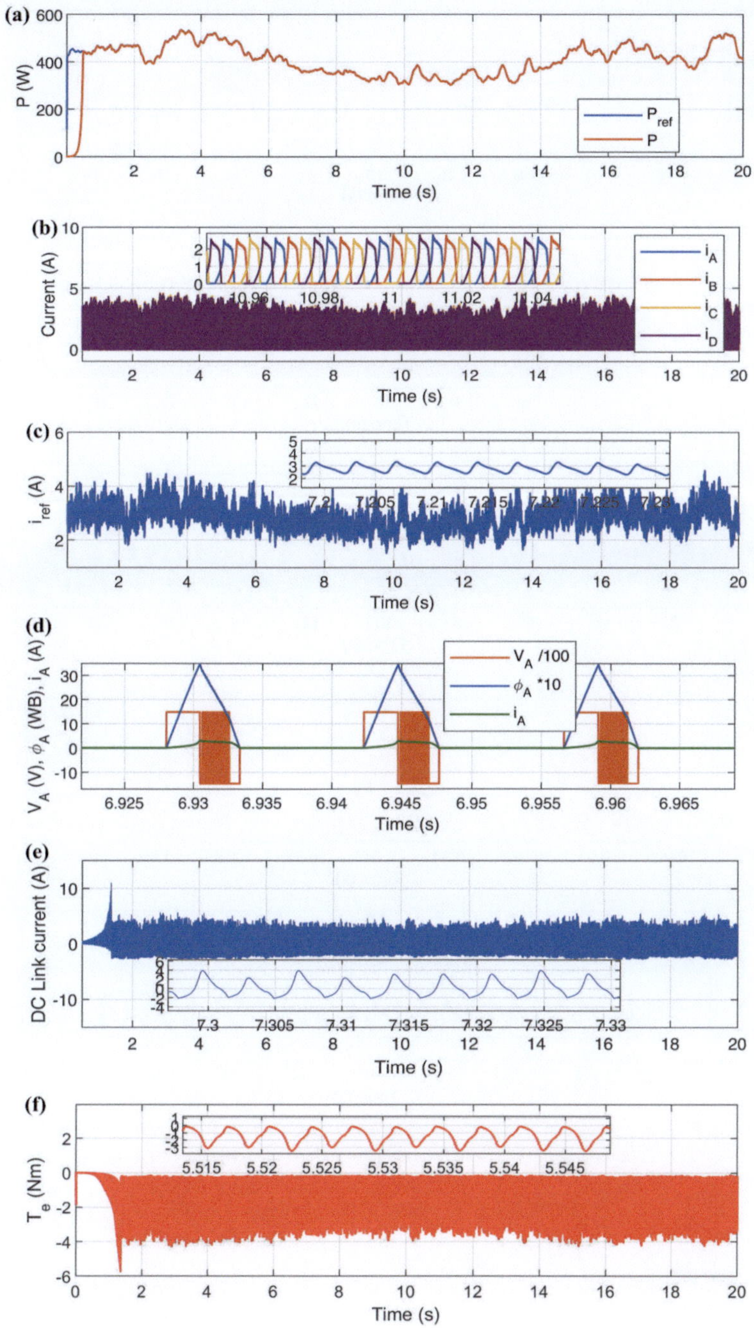

Fig. 16 The simulation results obtained from the unoptimized system: **a** SRG output power, **b** reference current (i_{ref}), **c** SRG phase current, **d** voltage, flux, and current of phase A, **e** DC link current, and **f** SRG torque (T_e)

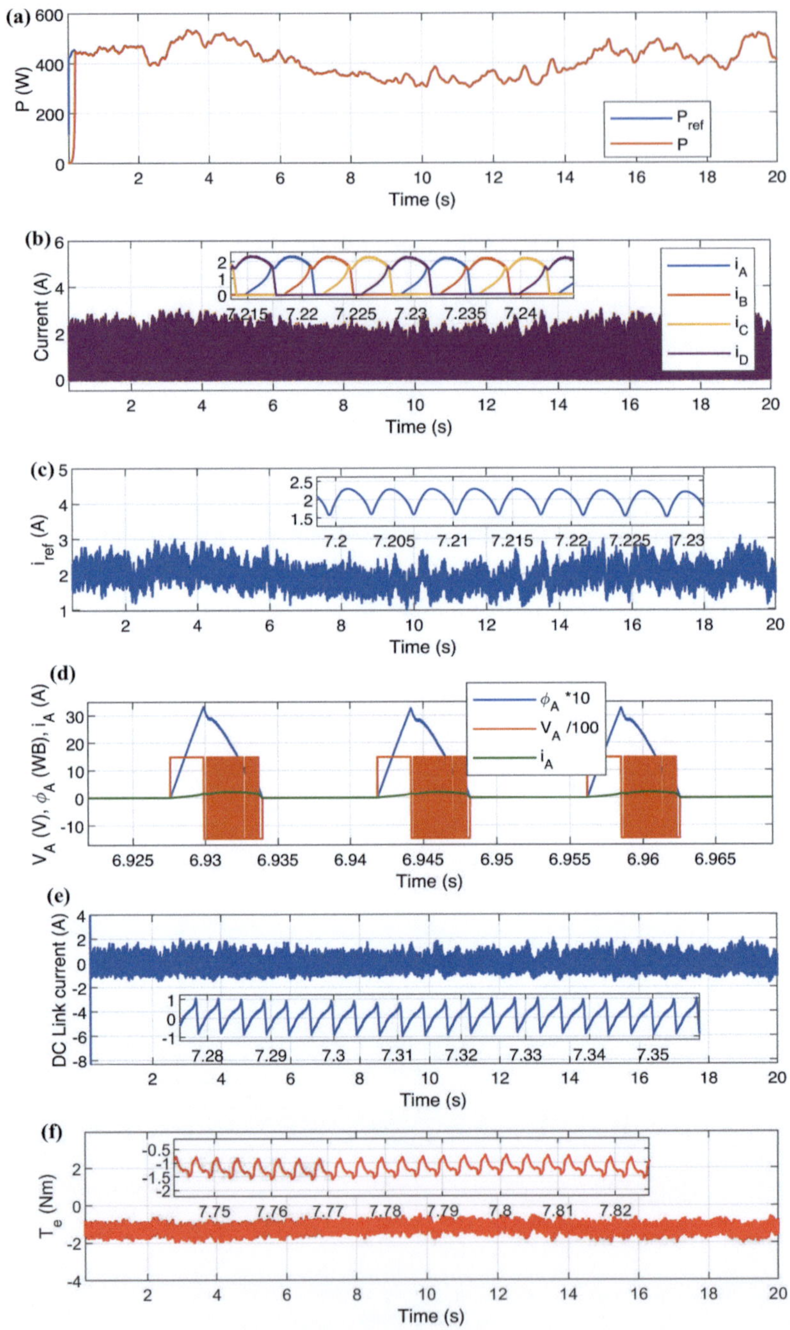

Fig. 17 Simulation results based on the optimized system: **a** SRG output power (P) with reference, **b** reference current (i_{ref}), **c** SRG phase current, **d** voltage, flux, and current of phase A, **e** DC link current, and **f** electromagnetic torque of SRG (T_e)

Table 4 Optimized firing angles for different wind speeds

Wind speed (m/s)	Rotor speed (rad/s)	θ_{on-opt}	θ_{on-opt}	η (%)
5	48	29.03	45.16	22
7	67	24.56	45.36	32.5
8	76.4	25.09	45.65	50

Figure 17 presents the simulation results of the optimized system during a switching period at an average wind speed of 8 m/s. In Fig. 17a, the power profile of the energy delivered to the load is shown, demonstrating that the output successfully aligns with the reference value. Figure 17b illustrates the waveform of the reference current, where ripples are observed at the phase transition points. The actual phase current waveforms are depicted in Fig. 17c, showing that the currents closely follow the reference. However, high-frequency fluctuations are present due to the hysteresis current control. The flux, current, and voltage profiles of the phase windings are analyzed in Fig. 17d, revealing that the maximum flux value reaches 0.35 WB near the zero-aligned position at θ_{on}. The flux subsequently drops to zero as the phase current decreases to zero.

Figure 17e illustrates the DC generation bus current. Due to the switching angles, two-phase currents are generated simultaneously, resulting in overlapping currents. The negative values in the curve indicate that the current flows from the excitation bus through the excitation diode. Figure 17f presents the curve of the electromagnetic torque T_e for the SRG with optimized control. The results demonstrate a significant reduction in torque ripple compared to the unoptimized control, leading to smoother and more stable torque output. As observed from the results, the optimized system exhibits significantly reduced torque ripple and current ripple in the SRG compared to the unoptimized system. Based on the above investigations, Table 4 summarizes the performance of the two operating modes, focusing on key parameters such as generated power, torque ripple, and DC link current across various wind speeds.

7 Conclusion

This book chapter presents a comprehensive proposal for the direct power control of a Switched Reluctance Generator (SRG) with optimization of power regulation. The chapter includes detailed modeling of both the SRG and the wind turbine, along with the validation of the Maximum Power Point Tracking (MPPT) strategy. The direct power control effectively regulates the generator's power through current loops, ensuring it closely tracks the maximum power point of the turbine.

To improve the overall performance of the SRG, an optimization control strategy is introduced. The optimization of the turn-on and turn-off angles is carried out to achieve the maximum output power range while minimizing torque ripple and DC current fluctuations. The proposed controller demonstrates satisfactory performance with minimal complexity, making it practical for implementation. Simulation results validate the effectiveness of the optimized power controller under variable-speed operating conditions of the generator. Consequently, the direct power control strategy, combined with optimization, proves to be a highly efficient and robust approach for managing the power output of an SRG in wind turbine applications. The proposed approach provides interesting results, but there are certain limitations that need to be highlighted. The assumption of ideal switches neglects real factors such as switching losses, conduction losses, and parasitic effects, which can affect the overall accuracy and efficiency of the system. Future work will focus on incorporating these non-idealities into the model to enhance its predictive capability. Moreover, efforts will be directed toward implementing nonlinear control strategies for the SRG to further enhance its dynamic performance. Additionally, experimental validation through field tests on a 500W turbine will be conducted to further assess the system's real-world performance. We also plan to explore advanced multi-objective optimization algorithms, such as NSGA-II for our application

References

1. Mapa, S., Bhuvaneswari, G.: Suitable choice of capacitor for a self-excited switched reluctance generator driven by a roof-top wind turbine at varying wind speeds. Trans. Indian Natl. Acad. Eng. 1–9 (2021). https://doi.org/10.1007/s41403-021-00265-4
2. Sedky, M.M., Abdel-Azim, W.E., Abdel-Khalik, A.S., Massoud, A.M.: Integrated switched reluctance starter/generator for aerospace applications: particle swarm optimization for constant current and constant voltage control designs. Appl. Sci. **12**(15), 7583 (2022). https://doi.org/10.3390/app12157583
3. Makhad, M., Zazi, K., Zazi, M., Loulijat, A.: Adaptive super-twisting terminal sliding mode control and lvrt capability for switched reluctance generator based wind energy conversion system. Int. J. Electrical Power Energy Syst. **141**, 108142 (2022). https://doi.org/10.1016/j.ijepes.2022.108142
4. Touati, Z., Araújo, R.E., Mahmoud, I., Khedher, A.: Analysis of skewing effects on radial force for different topologies of switched reluctance machines: 6/4 srm, 8/6 srm, and 12/8 srm. U Porto J. Eng. **9**(1), 55–71 (2023). https://doi.org/10.24840/2183-6493--009-001-001134
5. El-Shahat, A., Hunter, A., Rahman, M., Wu, Y.: Ultra-high speed switched reluctance motor-generator for turbocharger applications. Energy Proc. **162**, 359–368 (2019). https://doi.org/10.1016/j.egypro.2019.04.037
6. Howey, B., Bilgin, B., Emadi, A.: Design of an external-rotor direct drive e-bike switched reluctance motor. IEEE Trans. Veh. Technol. **69**(3), 2552–2562 (2020). https://doi.org/10.1109/tvt.2020.2965943
7. Pham, D., Klein-Hessling, A., De Doncker, R.W.: Control of a dc-dc converter as an active filter in combination with switched reluctance generators for gas turbine applications. In: 2019 AIAA/IEEE electric aircraft technologies symposium (EATS). IEEE, pp 1–17. https://doi.org/10.2514/6.2019-4471
8. Ho, C.Y., Wang, J.C., Hu, K.W., Liaw, C.M.: Development and operation control of a switched-reluctance motor driven flywheel. IEEE Trans. Power Electronics **34**(1), 526–537 (2018). https://doi.org/10.1109/tpel.2018.2814790

9. Cheng, H., Wang, L., Xu, L., Ge, X., Yang, S.: An integrated electrified powertrain topology with SRG and SRM for plug-in hybrid electrical vehicle. IEEE Trans. Ind. Electronics **67**(10), 8231–8241 (2019). https://doi.org/10.1109/tie.2019.2947854
10. Li, S., Zhang, S., Habetler, T.G., Harley, R.G.: Modeling, design optimization, and applications of switched reluctance machines—a review. IEEE Trans. Ind. Appl. **55**(3), 2660–2681 (2019). https://doi.org/10.1109/TIA.2019.2897965
11. Touati, Z., Pereira, M., Araújo, R.E., Khedher, A.: Integration of switched reluctance generator in a wind energy conversion system: an overview of the state of the art and challenges. Energies **15**(13), 4743 (2022). https://doi.org/10.3390/en15134743
12. Touati, Z., Pereira, M., Araújo, R.E., Khedher, A.: Comparative study of discrete pi and pr controller implemented in SRG for wind energy application: Theory and experimentation. Electronics **11**(8), 1285 (2022). https://doi.org/10.3390/electronics11081285
13. Catata, E.O.H., Neto, P.J.D.S., De Paula, M.V., Silveira, J.P.C., Barros, T.A.D.S., Ruppert Filho, E.: In-loop adaptive filters to improve the power quality of switched reluctance generator in WECS. IEEE Access **10**, 2941–2951 (2021). https://doi.org/10.1109/access.2021.3136203
14. Touati, Z., Mahmoud, I., Araújo, R.E., Khedher, A.: Fuzzy super-twisting sliding mode controller for switched reluctance wind power generator in low-voltage DC microgrid applications. Energies **17**(6), 1416 (2024). https://doi.org/10.3390/en17061416
15. Ahmad, S.S., Narayanan, G.: Modeling of single-pulse operated switched reluctance generator and its verification. IEEE Trans. Ind. Appl. **56**(5), 4966–4976 (2020). https://doi.org/10.1109/TIA.2020.3005586
16. Touati, Z., Pereira, M., Araújo, R.E., Khedher, A.: Improvement of steady state performance of voltage control in switched reluctance generator: Experimental validation. Machines **10**(2), 103 (2022). https://doi.org/10.3390/machines10020103
17. Araujo, W.R., Reis, M.R., Wainer, G.A., Calixto, W.P.: Efficiency enhancement of switched reluctance generator employing optimized control associated with tracking technique. Energies **14**(24), 8388 (2021). https://doi.org/10.3390/en14248388
18. Scalcon, F.P., Fang, G., Volpato Filho, C.J., Gründling, H.A., Vieira, R.P., Nahid-Mobarakeh, B.: A review on switched reluctance generators in wind power applications: fundamentals, control and future trends. IEEE Access **10**, 69412–69427 (2022). https://doi.org/10.1109/access.2022.3187048
19. Barros, T.A.S., Ruppert Filho, E.: Direct power control for switched reluctance generator in wind energy. IEEE Latin Am. Trans. **13**(1), 123–128 (2015). https://doi.org/10.5772/54162
20. Zan, X., Ni, K., Zhang, W., Jiang, Z., Cui, M., Yu, D., Zeng, R.: A new control strategy for SR generation system based on modified PT control. IEEE Access **7**, 179720–179733 (2019). https://doi.org/10.1109/access.2019.2959088
21. Wang, Q., Chen, H., Cheng, H., Yan, S., Abbas, S.: An active boost power converter for improving the performance of switched reluctance generators in DC generating systems. IEEE Trans. Power Electronics **35**(5), 4741–4755 (2019). https://doi.org/10.1109/tpel.2019.2944084
22. Barros, T.A.d.S., Neto, P.J.d.S., Filho, P.S.N., Moreira, A.B., Filho, E.R.: An approach for switched reluctance generator in a wind generation system with a wide range of operation speed. IEEE Trans. Power Electronics **32**(11), 8277–8292 (2017). https://doi.org/10.1109/tpel.2017.2697822
23. Wang, L., Liu, C., Jiang, Z., Xiao, W., Ren, S., Ding, J., Wang, Q.: Multi-objective control strategy for switched reluctance generators in small-scale wind power generations. Sustainability **15**(7), 6329 (2023). https://doi.org/10.3390/su15076329
24. Hajiabadi, H., Farshad, M., Shamsinejad, M.: Multi-objective optimization and online control of switched reluctance generator for wind power application. Int J Ind Electronics Control Optimization **4**(1), 33–45 (2021). https://doi.org/10.22111/ieco.2020.32687.1234
25. Parla, G., Yıldırım, M., Özdemir, M.: Maximum power control and optimization of switched reluctance generators for wind turbines. Dicle Üniversitesi Mühendislik Fakültesi Mühendislik Dergisi **14**(1), 55–64 (2023). https://doi.org/10.24012/dumf.1232026
26. Fabbina, C., Rahiman, R.A., Prabha, M., Janaki, N., Shadhik, M., Penthala, T.: High switched reluctance generator for pso optimized WECS. In: 2023 International Conference on Circuit

Power and Computing Technologies (ICCPCT), pp. 1793–1799. IEEE (2023). https://doi.org/10.1109/iccpct58313.2023.10245481
27. Zan, X., Lin, H., Xu, G., Zhao, T., Gong, Y.: Control strategy of parallel systems with efficiency optimisation in switched reluctance generators. Power Electronics Drives **6**(1), 61–74 (2021). https://doi.org/10.2478/pead-2021-0003
28. Afifi, M., Rezk, H., Ibrahim, M., El-Nemr, M.: Multi-objective optimization of switched reluctance machine design using jaya algorithm (mo-jaya). Mathematics **9**(10), 1107 (2021). https://doi.org/10.3390/math9101107
29. Abunike, E.C., Okoro, O.I., Davidson, I.E.: Finite element design and multi-objective optimization of four pole reluctance motor based on nsga-ii intelligent algorithm. In: 2021 IEEE AFRICON, pp. 1–6. IEEE (2021). https://doi.org/10.1109/africon51333.2021.9570964
30. Narla, S., Sozer, Y., Husain, I.: Switched reluctance generator controls for optimal power generation and battery charging. In: 2011 IEEE Energy Conversion Congress and Exposition, pp. 3575–3581. IEEE (2011). https://doi.org/10.1109/ecce.2011.6064253
31. dos Santos Neto, P.J., dos Santos Barros, T.A., de Paula, M.V., de Souza, R.R., Ruppert Filho, E.: Design of computational experiment for performance optimization of a switched reluctance generator in wind systems. IEEE Trans. Energy Conversion **33**(1), 406–419 (2017). https://doi.org/10.1109/tec.2017.2755590
32. Chaple, M.D., Deotale, N.T., Deotale, N.T.: Torque ripple minimization in an 8/6 switched reluctance motor drive: optimization using the azoa method. Austr. J. Electrical Electronics Eng. **22**(1), 27–44 (2024). https://doi.org/10.1080/1448837x.2024.2344876
33. Shahakar, D.A., Warkad, S.B.: An optimal angle control strategy for switched reluctance motor drive in EV applications. J. Control Decis. **11**(2), 271–282 (2022). https://doi.org/10.1080/23307706.2022.2150329
34. Zhu, Y., Wu, H., Zhang, J.: Regenerative braking control strategy for electric vehicles based on optimization of switched reluctance generator drive system. IEEE Access **8**, 76671–76682 (2020). https://doi.org/10.1109/access.2020.2990349
35. Verma, A., Ahmad, S.S., Narayanan, G.: Optimal control of single-pulse-operated switched reluctance generator to minimize rms phase and RMS DC-bus current. IEEE Trans. Ind. Appl. 1–13 (2023). https://doi.org/10.1109/tia.2023.3327977
36. Fan, B., Liu, Y., Li, Y., Zhou, Z.: Optimized control of SRG based on fuzzy logic by turn-on and turn-off angle. In: 2016 IEEE International Conference on Aircraft Utility Systems (AUS), pp. 82–86. IEEE (2016). https://doi.org/10.1109/aus.2016.7748025
37. Roshandel, E., Namazi, M.M., Rashidi, A., Saghaian-Nejad, S.M., Ahn, J.: SSC strategy for SRG to achieve maximum power with minimum current ripple in battery charging. IET Electric Power Appl. **11**(7), 1205–1213 (2017). https://doi.org/10.1049/iet-epa.2016.0770
38. Yahia, H., Liouane, N., Dhifaoui, R.: Differential evolution method-based output power optimisation of switched reluctance generator for wind turbine applications. IET Renew. Power Generation **8**(7), 795–806 (2014). https://doi.org/10.1049/iet-rpg.2013.0179
39. Zaharia, M., Laczko, A.A., Pop, A., Radulescu, M., Gillon, F.: Optimal commutation angles of a switched reluctance motor/generator. In: 2015 Tenth International Conference on Ecological Vehicles and Renewable Energies (EVER), pp. 1–8. IEEE (2015). https://doi.org/10.1109/ever.2015.7113015
40. Sun, N., Choi, D., Li, J., Cho, Y.: The angle control of switched reluctance generator for maximum output power. In: 2012 Sixth International Conference on Electromagnetic Field Problems and Applications, pp. 1–4. IEEE (2012). https://doi.org/10.1109/icef.2012.6310409
41. Li, J., Li, Y., Wang, Y.: Fuzzy inference NSGA-iii algorithm-based multi-objective optimization for switched reluctance generator. IEEE Trans. Energy Conversion **36**(4), 3578–3581 (2021). https://doi.org/10.1109/tec.2021.3099961
42. Yueying, Z., Chuantian, Y., Chengwen, Z.: Multi-objective optimization of switched reluctance generator for electric vehicles. In: 2018 21st International Conference on Electrical Machines and Systems (ICEMS), pp. 1903–1907. IEEE (2018). https://doi.org/10.23919/icems.2018.8549386

43. Shadhik, M., Senthil Kumaran, M.: Meta heuristic optimization algorithm based rectifier control in switched reluctance generator. IETE J. Res. **70**(4), 3941–3963 (2023). https://doi.org/10.1080/03772063.2023.2204848
44. Narayanan, V., Singh, B.: Seamless operation of an SRG based WEQS and a bidirectional converter interfaced BES microgrid under polluted grid conditions. IEEE Trans. Ind. Appl. **59**(6), 7744–7756 (2023). https://doi.org/10.1109/tia.2023.3306741
45. Touati, Z., Mahmoud, I., Khedher, A.: Hysteresis current control of switched reluctance generator. In: 2020 11th International Renewable Energy Congress (IREC), pp. 1–8. IEEE (2020). https://doi.org/10.1109/irec48820.2020.9310391
46. Touati, Z., Mahmoud, I., Khedher, A.: Nonlinear modelling of switched reluctance machine. In: 2019 International Conference on Signal, Control and Communication (SCC), pp. 256–261. IEEE (2019). https://doi.org/10.1109/scc47175.2019.9116143
47. Touati, Z., Mahmoud, I., Khedher, A.: Modelling and electromagnetic characteristics for two mode functions of switched reluctance machine. In: 2019 International Conference on Signal, Control and Communication (SCC), pp. 114–120. IEEE (2019). https://doi.org/10.1109/scc47175.2019.9116181
48. Ragheb, M., Ragheb, A.M.: Wind turbines theory-the Betz equation and optimal rotor tip speed ratio. In: Fundamental and advanced topics in wind power, p. 19 (2011). https://doi.org/10.5772/21398
49. Reyes, V., Rodriguez, J.J., Carranza, O., Ortega, R.: Review of mathematical models of both the power coefficient and the torque coefficient in wind turbines. In: 2015 IEEE 24th International Symposium on Industrial Electronics (ISIE), pp. 1458–1463. IEEE (2015). https://doi.org/10.1109/isie.2015.7281688
50. Chaudhary, M.K., Roy, A.: Design & optimization of a small wind turbine blade for operation at low wind speed. World J. Eng. **12**(1), 83–94 (2015). https://doi.org/10.1260/1708-5284.12.1.83/full/html
51. Kennedy, J., Eberhart, R.: Particle swarm optimization. In: Proceedings of ICNN'95-International Conference on Neural Networks, vol. 4, pp. 1942–1948. IEEE (1995). https://ieeexplore.ieee.org/document/488968
52. Labiod, C., Srairi, K., Mahdad, B., Dib, A., Benchouia, M.T., Benbouzid, M.E.H.: Optimum performances for non-linear finite elements model of 8/6 switched reluctance motor based on intelligent routing algorithms. Adv. Electrical Electronic Eng. **15**(1) (2017). https://doi.org/10.15598/aeee.v15i1.1906

Impact of Diversity on the Optimization of Solutions by Adaptive Hybridization in a Coevolutionary Algorithm

Lorena Rosas, Claudia Gómez-Santillán, Nelson Rangel-Valdez, Lucila Morales, Héctor Fraire, and Laura Cruz-Reyes

Abstract Dynamic multi-objective problems represent a challenge in fields such as labor, economics, and industry, due to the changing nature of the objectives or constraints over time. To address these challenges, various strategies have been proposed to improve the performance of algorithms under dynamic conditions. However, to our knowledge, no strategy has been proposed combining two specific approaches to adapting to change: one based on memory, and another based on introducing diversity. This paper presents the hybridization of these memory and diversity introduction strategies within a dynamic cooperative coevolutionary algorithm based on MOEA/D (D-CC-MOEA/D). The memory-based strategy preserves populations from previous generations to function as an external population (EP) during environmental changes, while the diversity-based strategy introduces variability through selective mutations. D-CC-MOEA/D solves FDA dynamic problems, where FDA may refer to *Farina, Deb & Amato*. Currently this algorithm version of algorithm solves FDA1 and FDA3 problems, using MOEA-D as the solver, random selection operators, SBX crossover, and polynomial mutation in each species. The results so far show that the hybrid strategy allows the algorithm to adapt to changes in the

L. Rosas (✉) · C. Gómez-Santillán · N. Rangel-Valdez · L. Morales · H. Fraire · L. Cruz-Reyes
Division of Postgraduate Studies and Research, National Technological Institute of Mexico/ Technological Institute of Madero City, Ciudad Madero, Mexico
e-mail: lorenarocio52@gmail.com

C. Gómez-Santillán
e-mail: claudia.gs@cdmadero.tecnm.mx

N. Rangel-Valdez
e-mail: nelson.rv@cdmadero.tecnm.mx

L. Morales
e-mail: lucila.mr@cdmadero.tecnm.mx

H. Fraire
e-mail: hector.fh@cdmadero.tecnm.mx

L. Cruz-Reyes
e-mail: laura.cr@cdmadero.tecnm.mx

W. Pedrycz et al. (eds.), *Decision-Making in Computational Intelligence-Based Systems*, Studies in Systems, Decision and Control 628,
https://doi.org/10.1007/978-3-032-04056-5_7

problem effectively, maintaining adequate diversity among them. To identify effective adaptation, we used the hypervolume of each of the five fronts for FDA1 and FDA3, finding that, on average, diversity is maintained in both problems between 74 and 95%.

Keywords Dynamic multi-objective optimization · Strategy for adaptation to change · Cooperative coevolution

1 Introduction

Multi-objective optimization is fundamental in various fields such as engineering, economics, and industry. However, implementing this approach in dynamic environments, where objectives, variables, or restrictions are in constant flux, presents significant challenges. These include: (1) the management of multiple objectives which may conflict with each other since optimizing one may harm another objective; (2) adapting to changes over time is also difficult since objectives, variables, or restrictions may evolve, requiring an agile and effective response from the algorithm; and (3) the complexity of exploring the search space, that increases when variations generate new regions of interest, making it difficult to locate optimal solutions. This necessitates the development of robust strategies to guide the search process.

To address these challenges, several strategies have been proposed to improve the performance of algorithms under dynamic conditions [1–3]. However, to our knowledge, regarding challenge 2, no hybrid change adaptation strategy has been proposed, where one is memory-based, and the other focuses on introducing diversity within a coevolutionary algorithm.

This work presents the integration of the strategies of diversity and memory within a dynamic cooperative coevolutionary algorithm based on MOEA/D (D-CC-MOEA/D). The memory-focused strategy preserves populations from previous generations [4] to function as an external population (EP) during environmental changes, while the diversity-based strategy introduces variability through selective mutations [1]. This hybridization capitalizes on the strengths of both memory-based solutions and the ability of diversification to explore new regions of the search space.

The article is organized as follows: Sect. 2 presents the background which includes a summary of the literature on dynamic multi-objective optimization and its mathematical definition. Subsequently, details are given on some strategies for adapting to change, FDA problems, and the performance metrics used to evaluate the quality of solutions in dynamic environments, together with a definition of the Pareto front. Section 3, details the solution methodology employed in this study, including the hybrid strategy of adaptation to change. Section 4 presents the results obtained by comparing the proposed algorithm with the hybrid strategy and the D-MOEA/D algorithm [5]. Finally, the conclusions are presented in Sect. 5.

2 Background

2.1 Dynamic Multi-Objective Optimization

Dynamic multi-objective optimization problems (DMOPs) are characterized by temporal variations in their fundamental components (objective functions, decision variables, and/or constraints) [6]. Unlike static problems, DMOPs require algorithms to not only find optimal solutions but also rapidly adapt to discrete or continuous environmental changes [5].

Currently coevolutionary algorithms (CA) have emerged as an effective strategy to address the unique challenges of DMOPs. Their design based on specialized subpopulations enables:

- **Rapid adaptation**: Through shared memory mechanisms, CAs preserve valid solutions during environmental shifts, avoiding complete reinitialization [7].
- **Exploration–exploitation balance**: Species coevolution (e.g., one per objective) maintains diversity and prevents premature convergence [8]—critical in dynamic environments where diversity loss is a risk [9].
- **Scalability**: By decomposing problems into subcomponents, CAs reduce computational complexity in high-dimensional problems [10], a key DMOP challenge [11].

In hyperparameter optimization for machine learning models in non-stationary environments (as in [12]), CAs allow subpopulations specialized in accuracy and computational efficiency to cooperatively adapt models to data shifts.

In dynamic industrial processes (e.g., chemical production optimization [13]), CCEAs outperform monolithic approaches by:

1. Distributing computational load among subpopulations monitoring different objectives (quality, cost, energy).
2. Reusing historical solutions from shared archives when operational conditions recur.

Integration with recurrent neural networks (RNNs) [14] enhances these advantages: RNNs predict future changes while CAs proactively adjust subpopulations. For instance, in resource planning, one subpopulation may optimize for current conditions while another prepares for forecasted scenarios.

2.1.1 Mathematical Definition

Formally, DMOP incorporates an explicit time dependence, where the elements of the problem can be modified continuously or in discrete intervals. The mathematical definition of a DMOP is expressed in the following equation [15] (Eq. 1):

$$F(x, t) = min\big[f_1(x, t), \ldots, f_M(x, t)\big]^T \quad (1)$$

where $x = [x_1, x_2, \ldots, x_N]^T \in \Omega$, and x consist of N decision variables, Ω is the search space, $f(x, t)$ consists of M objective functions, $f_i(x, t) \in R^M$, $i = 1, \ldots, M$, R^M denotes the objective space. As mentioned above, in DMOPs the objective functions are held constant for a given period of time before undergoing a change. This feature allows optimization algorithms to adapt and find optimal solutions within a specific time interval, before the problem conditions change again. The mathematical equation $t = \left(\frac{1}{n_t}\right) * \left\lfloor \frac{\tau}{\tau_T} \right\rfloor$ acts as a measurement mechanism that identifies the periods of stability and transition within the optimization problem, where n_t denotes the severity of the change, τ_T indicates the frequency of change and τ functions as the iteration counter [16].

Since DMOPs involve frequent changes in search space and solutions, specialized algorithms that can quickly adapt to these transformations are essential.

However, in order for an algorithm to effectively adapt to these changes, it is necessary to incorporate change adaptation strategies that allow it to adjust quickly to new conditions.

2.2 *Strategy for Adaptation to Change*

Optimization algorithms face unique challenges in dynamic environments, such as adapting to changes in objective functions, constraints, or search spaces. To address these challenges, it is essential to incorporate change adaptation strategies into the design of algorithms, as they ensure performance is maintained under varying conditions. These strategies are essential components for dynamic optimization. They help in the detection of environmental changes and enable rapid reconfiguration of solutions, thus ensuring optimality [17].

A review of the literature reveals several approaches to achieving effective adaptation. According to Wu [18], the most commonly used strategies can be classified into three main categories:

- **Diversity introduction** [1]: In literature, the two most common strategies for introducing diversity in dynamic optimization are (1) immediately increasing the population when a change is detected, either by incorporating randomly generated solutions or by applying hypermutation to existing solutions. Deb et al. [1] integrated both strategies into the non-dominated sorting genetic algorithm II (NSGA-II) [19], resulting in two dynamic versions of NSGA-II.

    DNSGA-II-A: This strategy consists of replacing a small part of the population (20%) with new feasible solutions that are randomly generated. Since these solutions are unrelated to the existing population, they introduce diversity into the search process, which facilitates the exploration of unknown regions of the search space. Thus, the algorithm quickly adapts to new conditions and avoids stagnation at local optima.

DNSGA-II-B: In this strategy, instead of replacing individuals directly, mutation operators are applied to a part of the population. Specifically, 20% of the existing solutions are randomly selected and subjected to a high-probability, low-distribution mutation. In this way, the new solutions generated maintain a relationship with the original population, preserving some of the information obtained in previous environments, while introducing diversity to facilitate adaptation to change.

DNSGA-II-A is suitable for scenarios with drastic environmental changes that cause a significant reduction in population diversity, as the solutions it proposes contribute significantly to increasing it. DNSGA-II-B, on the other hand, is more appropriate for scenarios with little environmental change, as the solutions it incorporates generate little variation in the population [20].

Brambila et al. [21] proposes a dynamic decomposition-based multi-objective evolutionary algorithm (D-MOEA/D) to solve dynamic multi-objective optimization problems. The problem is divided into subproblems, each associated with a vector of weights, which are optimized in parallel by exchanging information with their neighbors. To detect changes in the environment, a subset of solutions (detectors) is randomly selected. If their objective values vary significantly, an adaptation strategy is activated that introduces diversity by randomly replacing the solutions, ensuring a more efficient exploration of the search space and a better response to dynamic changes.

Li et al [22] proposes an adaptive strategy for dynamic selection of genetic operators (crossover and mutation) within a multi-objective decomposition-based evolutionary algorithm (MOEA/D). The strategy is based on the multi-arm bandit (MAB) approach, which balances exploration, by testing less used operators, and exploitation, by prioritizing those with better historical performance. This strategy improves the ability of MOEA/D to maintain diversity and convergence in dynamic problems, as it automatically adjusts operator selection based on the performance obtained at each stage of the optimization process. In dynamic environments, MAB readjusts the selection probabilities when it detects changes, such as the shift of the Pareto front, allowing the algorithm to quickly recover diversity and adapt efficiently to new conditions.

- **Predictive strategies** [2]: Another widely studied approach is the use of predictive models to anticipate the behavior of the Pareto front or set of optimal solutions in a new environment. These strategies use historical information from previous environments to generate initial solutions that are closer to the expected optimum after a change. This method is particularly useful for problems where changes follow predictable patterns or have some regularity [2].

Biao [23] employed a linear predictive model to reinitialize a portion of the population in the following environment, using the simplified model formula: $x_{t+1}^k = x_t^k + C_t^k - C_{t-1}^k + Gaussian(0, d)$ where k = 1, 2,..., n is the index of the decision variables and n represents the dimension of the decision vector. C_t^k and C_{t-1}^k are the kth decision variables of the population center points at the tth and

(t-1)th time steps, respectively. x_t^k and x_{t+1}^k represent the kth decision variables of an individual at the tth time step and predicted individual at the $(t + 1)$th time step, respectively. *Gaussian* $(0, d)$ denotes Gaussian perturbation with a mean value 0 and standard deviation d. The standard deviation d is defined as: $d = \frac{\|C_t - C_{t-1}\|}{n}$ where $\|C_t - C_{t-1}\|$ is Euclidean distance between centers C_t and C_{t-1}.

For their part, Chen et al. [24] proposed a hybrid population prediction strategy that combines fuzzy inference with one-step prediction, aiming to closely follow the movement of the Pareto optimal set and its corresponding front (PS/ PF). This strategy is integrated into an optimization algorithm based on the teaching-learning approach. Muruganantham et al [25] proposed a population prediction strategy based on the Kalman filter technique. This technique allows the search for new Pareto optimal solutions to be guided and a large number of high-quality initial individuals to be generated. The algorithm then finds the optimal solution at each time point using a differential evolution approach based on decomposition.

Zhou et al. [26] proposed a population prediction method designed to predict the entire population rather than single points. This algorithm uses central points to predict the next central point, while previous varieties are used to estimate the next variety. The optimal population at each time point is determined by a decomposition-based differential evolution algorithm.

Tao et al. [27] proposes a new predictive strategy to address dynamic multi-objective optimization problems (DMOPs) by combining Fuzzy C-Means (FCM) and Support Vector Machines (SVM). The strategy is based on two main pillars: first, it uses FCM to group historical non-dominated solutions into clusters, identifying patterns in the data that represent optimal regions of the search space. Second, it employs SVM to predict future changes in the environment, such as variations in objective functions or constraints. When a change is detected, the algorithm reinitializes the population with solutions from the clusters that are expected to be optimal in the new environment, allowing for a smoother transition and reducing convergence time. In addition, the approach integrates evolutionary optimization techniques to maintain population diversity and avoid premature convergence, ensuring effective exploration of the search space.

- **Memory-based strategies** [28]: Memory-based strategies store relevant solutions from previous environments and reuse them when an environmental change is detected. This approach is especially useful in situations where changes follow a periodic or repetitive pattern, as it allows leveraging solutions that have proven effective in similar contexts. Instead of generating new solutions from scratch each time a change occurs, algorithms can draw on a memory pool containing previously successful solutions, allowing them to accelerate convergence to high-quality solutions. This approach not only improves the efficiency of the process, but also helps maintain solution quality in dynamic environments, where changes are frequent and can reduce the diversity of the solution population.

Goh and Tan [7] implemented an approach within a competitive-cooperative algorithm in which they randomly selected solutions from the existing population, giving priority to those representing extreme values in the objectives just before an environmental change occurred. These solutions were stored in a memory buffer of limited size. When this buffer reached its maximum capacity, the oldest solutions were removed to make room for new solutions. Importantly, the solutions stored in this buffer were not directly reintroduced into the population of the current environment. Instead, they were used to update the non-dominated set file to mitigate the risk of previous solutions disrupting or adversely affecting the optimization process. This approach reflects the cooperative nature of the algorithm, as historical solutions are used to guide and improve the overall search, while competition is manifested in the selection of extreme solutions representing different regions of the Pareto front.

Koo [29] proposed to store the non-dominated sets obtained from previous environments in a memory pool. Given environmental change, a specific number γ of past solutions, sampled uniformly from the memory pool, is selected. This is done by recursively removing the most frequent solution in the decision space from a copy of the pool until exactly γ solutions remain. These γ solutions, together with others generated by normal mutation, make up a significant part of the population in the new environments.

Liu et al. [30] uses a cooperative coevolutionary approach based on multiple particle swarms (PSO) to solve DMOP. Each problem objective is assigned to an independent swarm, which evolves cooperatively using an information exchange strategy. During evolution, each swarm uses a modified velocity equation to update the particles, which improves the exploration of the search space. In addition, a boundary constraint technique is applied to handle solutions outside the feasible space and ensure convergence to valid regions. It incorporates a dynamic memory-based strategy that reintroduces optimal solutions from previous environments (stored in a history file) to partially reinitialize the swarms.

Sahmoud et al. [31] propose the MNSGA-II algorithm, which is an extension of the well-known NSGA-II (Non-dominated Sorting Genetic Algorithm II) algorithm that is widely used to solve multi-objective optimization problems in static environments. This work uses an explicit memory-based strategy that stores non-dominated solutions from previous generations. These stored solutions are reused to reinitialize part of the population when a change in the environment is detected, allowing the algorithm to adapt more quickly to changes. MNSGA-II uses two populations: a search population and a main population. The search population is randomly reinitialized when a change occurs, while the main population is updated from memory. The memory update process selects non-dominated solutions based on Euclidean distance in the decision space, ensuring that distributed solutions are stored in different regions of the search space.

Some work combines multiple strategies to improve adaptability in dynamic environments. These strategies integrate exploration, exploitation, memory and prediction techniques to optimize the performance of multi-objective evolutionary algorithms. The goal of these combinations is to mitigate diversity loss, accelerate convergence and improve responsiveness to changes in the optimization environment.

Liang et al. [2] presented a proposed hybrid approach to enhance adaptivity in DMOPs by integrating two fundamental strategies: memory and prediction. The memory strategy makes use of an external archive to store Pareto-optimal solutions identified in previous environments, facilitating their retrieval and adjustment to changes, which accelerates convergence to the new Pareto front. The prediction strategy uses regression or extrapolation models to anticipate future changes in the environment, facilitating proactive adaptation of the algorithm before they occur. This hybrid approach is integrated within a multi-objective evolutionary optimization (MOEA) framework.

While these approaches offer significant advantages, their implementation is not without challenges. A key issue is the trade-off between fast response and stability. While fast adaptation is crucial to track dynamic changes, overly aggressive adjustments can generate instability and affect the quality of solutions [17]. Furthermore, accurate change detection remains a challenge, especially in scenarios where modifications are gradual or subtle.

To address these challenges and evaluate the effectiveness of proposed approaches, specialized benchmarks that simulate dynamic conditions have been developed. A prominent example is the FDA problems [6], which provide a standardized framework for testing and comparing the performance of algorithms in dynamic environments. These benchmarks allow analysis of how algorithms manage adaptation and maintain the quality of optimal solutions over time.

2.3 FDA Problems

The FDA problems are a set of problems designed to address the complexity of dynamic environments. Based on the two- and three-objective ZDT [32] and DTLZ [33] problems, they incorporate an explicit time variable in their mathematical functions. Initially proposed by Farina, Deb and Amato [6], these problems simulate temporal changes in the objective functions, constraints or decision variables. This allows analyzing the ability of algorithms to adapt to changing environments and provides a structured framework for investigating their response to predictable modifications in the solution spaces.

The FDA problems are designed to model DMOPs by means of mathematical functions with parameters that evolve with time (t). Their main characteristics are as follows [6]:

- **Modelling scenarios with temporal changes**: they incorporate a time variable that modifies the objective functions and/or constraints, mimicking dynamically evolving environments.
- **Evaluation of the adaptability of multi-objective algorithms**: they allow analysis of how algorithms adjust their solutions to changes in the Pareto front or in the search space.
- **Analysis of behaviors under non-convexity conditions**: they include non-convex functions, evaluating the ability of algorithms to handle complex and discontinuous surfaces.

- **Exploration of responses to discontinuities in objective functions**: some instances introduce abrupt changes in the objective functions, evaluating the robustness of the algorithms to significant perturbations.

2.3.1 Structure of FDA1 Problem

FDA problems combine static functions with dynamic terms [6]. For example, version FDA1 is defined as (Eq. 2–10):

1. Definition of objectives f_1 and f_2:

$$Minimize f_1(x_I) = x_1 \tag{2}$$

where x_I is a random value in the range [0,1]

$$Minimize f_2(x) = g * h \tag{3}$$

where g and h are a function that depend on (4):

$$X_2 = (x_2, \ldots, x_n) \in [-1, 1] \tag{4}$$

2. Constrains on decision variables:

   n = Number of decision variables

$$x_1 \in [0, 1] \tag{5}$$

$$x_{n \geq 2} \in [-1, 1] \tag{6}$$

3. H function:

   The function h depends on f_1 and g. It is calculated as follows:

$$h(f_1, g) = 1 - \sqrt{\frac{f_1}{g}} \tag{7}$$

4. The function g is defined as:

$$g(X_{II}) = 1 + \sum_{x_i \in X_{II}} (x_i - G(t))^2 \tag{8}$$

   $G(t)$ is calculated as:

$$G(t) = \sin\sin(0.5\pi t) \tag{9}$$

where t is the time variable:

$$t = \frac{1}{n_t}\left\lfloor \frac{\tau}{\tau_T} \right\rfloor \tag{10}$$

The value of n_t indicates the severity of changes, while t represents the current generation and τ_T? the change frequency, that is, how often a change will occur. The time variable t acts as a counter that determines whether a change occurs in the objective functions. For example, if t changes from 0 to 1 or from 1 to 2, this indicates an alteration in the problem's objectives during the execution of the algorithm, which in turn depends on the value assigned to τ_T. For example, if τ_T? is set to 5, then t will change every 5 generations.

2.3.2 Structure of FDA3 Problem

The FDA3 problem is defined as (Eq. 11–19):

1. Definition of Objective f_1 and f_2

$$Minimize f_1(x_I) = \sum_{x_i \in X_I} x_i^{F(t)} \tag{11}$$

where x_I is a random value in the range [0,1]

$$Minimize f_2(x) = g * h \tag{12}$$

where g and h, are functions that depend on

$$X_{II} = (x_2, \ldots, x_n) \in [-1, 1] \tag{13}$$

2. The decision variable constraints specify that X_1? (Eq. 14) must take values in the range [0, 1], while the decision variables from X_2 onward can take values between - 1 and 1

$$X_1 = x_1 \in [0, 1] \tag{14}$$

$$X_2 = x_{n \geq 2} \in [-1, 1] \tag{15}$$

3. h function

The function h depends on f_1 and g. It is calculated as follows:

$$h(f_1, g) = 1 - \sqrt{\frac{f_1}{g}} \tag{16}$$

4. g function

The function g is defined as:

$$g(X_{II}) = 1 + G(t) + \sum_{x_i \in X_{II}} (x_i - G(t))^2 \tag{17}$$

$G(t)$ is calculated as:

$$G(t) = |\sin(0.5\pi t)| \tag{18}$$

where t is the time variable:

$$F(t) = 10^{2sin(0.5\pi t)}, t = \frac{1}{n_t}\left\lfloor \frac{\tau}{\tau_T} \right\rfloor \tag{19}$$

Farina et al. [6]. designed five dynamic variants (FDA1 to FDA5) to simulate different evolutionary scenarios in multi-objective optimization problems and evaluate specific capabilities of the algorithms:

- FDA1: Simulates a linearly moving Pareto front in the objective space, which tests the tracking capability to follow the trajectory of the moving optimum.
- FDA2: Generates a transition of the Pareto front from convex to non-convex, which tests the morphological adaptability of the solutions.
- FDA3: Introduces abrupt or gradual changes in the space of decision variables and evaluates the robustness to perturbations in the problem parameters.
- FDA4: Incorporates time-varying dynamic constraints, examining the management of feasible solutions under changing conditions.
- FDA5: Combines simultaneous alterations in the objective functions and constraints and replicates complex hybrid scenarios.

These variants allow the evaluation of three key competencies of the algorithms:

- Accurate tracking: Ability to dynamically track the optimum without lags.
- Sustained diversity: Maintenance of a uniform and representative distribution of solutions on the front end.
- Fast convergence: Efficient adaptation to changes before the next transition, minimizing performance loss.

To effectively evaluate the performance of algorithms in dynamic optimization problems, specific metrics are required to assess the quality of the solutions generated. These metrics play a crucial role in the analysis of the ability of the algorithm to adapt, to maintain its effectiveness, and to respond efficiently to temporal variations in the optimization environment.

2.4 Performance Metrics

To evaluate the effectiveness of algorithms in dynamic optimization problems, it is important to use quantitative metrics to analyze aspects such as convergence, diversity, adaptability, and stability of solutions over time. These metrics not only validate theoretical performance but also provide an objective comparison between strategies. As Van Veldhuizen [34] points out in his analysis of multi-objective evolutionary algorithms, metrics should reflect an algorithm's ability to balance optimality, diversity, and adaptation. Among these metrics are the hypervolume and the mean hypervolume.

- **Hypervolume (HV)** [16]: It is a metric that measures both the diversity and convergence of non-dominated solutions along the Pareto optimal front (POF). We define the POF^* as an approximation to the Pareto front obtained using a specific algorithm and a reference point in the objective space, denoted z^*. The hypervolume indicator measures the space covered by the set of solutions Q of the POF^* in the objective space, which corresponds to the non-dominated volume of all hypercubes formed between the reference point z^* and each solution in POF^*. The formula for calculating the hypervolume is given by:

$$HV = \cup_{i=1}^{Q} \{vec_i \in POF^*\} \tag{20}$$

where vec_i is a non-dominated vector of POF^*, and vol_i represents the volume of the hypercube formed by the reference point and the vector vec_i (Eq. 20). A high HV indicates that the algorithm achieves solutions close to the Pareto true front (POF) and with good diversity [32]. It is worth mentioning that, in dynamic problems, HV must be calculated at each instant t, since the POF can move or change shape [6].

- ***Mean hypervolume (MHV)*** [35]: In dynamic problems, the hypervolume, used to compute the volume of the region dominated by the set of obtained solutions, denoted as POF*, is extended by averaging its value over multiple time instants T, giving rise to the *MHV* (Eq. 21). Proposed in the context of dynamic optimization by Jiang and Yang [36], the *MHV* allows evaluating the continuous adaptability of the algorithm by providing a more accurate assessment of the quality and adaptability of the solutions over time.

$$MHV = \frac{1}{T} \sum_{i=1}^{T} HV_t(POF_t^*) \tag{21}$$

A high MHV with low variance among HV_t values indicates that the algorithm adapts efficiently, maintaining both convergence and diversity of solutions. In contrast, high MHV fluctuation may signal instability to perturbations, such as abrupt changes in the decision space [37].

2.5 *Pareto Front*

The Pareto Optimal Front (POF^*) is the set of objective vectors corresponding to the Pareto optimal set (POS). This is defined as follows [38]:

$$POF^* = \{f = (f_1(x^*), f_2(x^*), \ldots, f_m(x^*)) | x^* \epsilon POS^*\} \quad (22)$$

where each objective function $f_k(x^*)$ is evaluated at the optimal point x^*. This front represents the solutions that cannot be improved in any objective, without worsening at least one of the other objectives. In other words, an objective vector belongs to the Pareto optimal front if there is no other vector that dominates it, other words, that is better in all aspects considered [38].

The ideal POF* should encompass the full spectrum of possible solutions, from those that optimize a specific objective to those that achieve a balance between all objectives. Its form may vary from convex to non-convex depending on the nature of the problem. In particular, in problems with dynamic constraints, the POF* may undergo changes in its structure over time, reflecting the evolution of environmental conditions [6].

3 Solution Methodology

This work presents a hybrid strategy that combines memory and diversity strategy to adapt to changes in dynamic multi-objective optimization problems. This strategy has been integrated into a cooperative coevolutionary algorithm called CO-MOEA/D [39], wherein a D-MOEA/D algorithm is employed as the problem solver within each species, thereby enhancing the algorithm's ability to address dynamic environments. The overall architecture of the proposed algorithm (Fig. 1) is composed of several key phases, which are described below.

1. **Initialization Phase**

The algorithm starts with an initialization phase where the essential parameters for the execution are configured:

- **Weight vector (*w*)**: a vector of weights is generated, which is used to decompose the multi-objective problem into several single-objective subproblems.

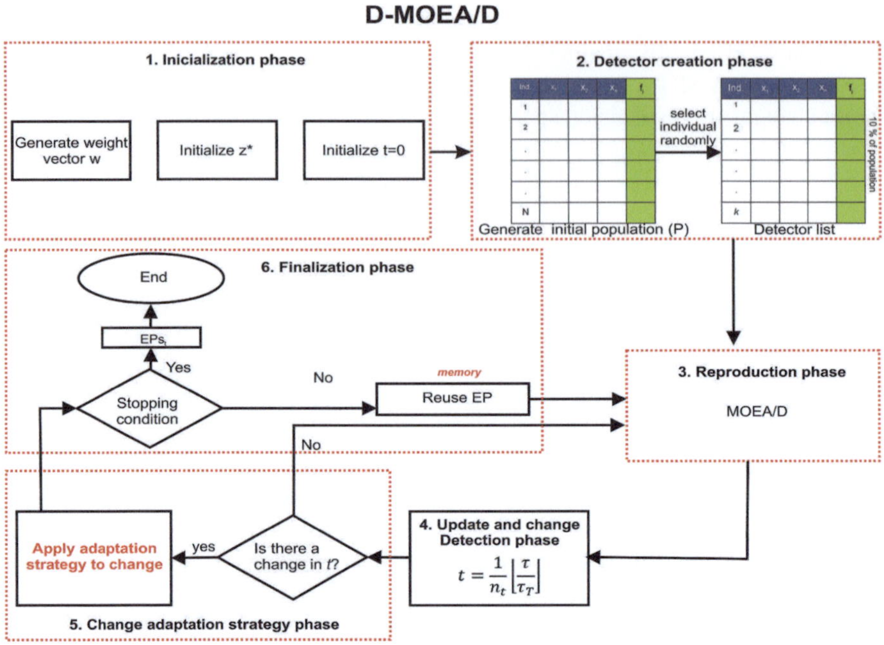

Fig. 1 Architecture of D-MOEA/D as a dynamic solver within CO-MOEA/D

- **Vector** z^*: The vector z^* is initialized, which stores the best values found for each objective function (minimum values in minimization problems and maximum values in maximization problems).
- **Counter** t: The counter t is initialized, which introduces dynamism to the problem by marking changes in the objectives over time.

From this configuration, an initial population (P) is generated, composed of individuals represented by a set of decision variables ($x_1, x_2, \ldots, x_n$) and their respective values of objective functions ($f_1, \ldots, f_j$).

2. **Detector Creation Phase**

Once the initial population has been generated, a list of detectors corresponding to 10% of the population is created on a random basis. These detectors are essential for identifying changes in the objectives during the evaluation process. Their main function is to monitor the dynamic environment and detect when modifications occur in the objective functions, which activates the strategy of adaptation to change.

3. **Reproduction Phase**

After initialization and detector creation, the algorithm enters the reproduction phase, which involves a series of iterative processes to generate new solutions:

- Selection: Individuals are selected from the current population to be parents in the offspring generation.

- Crossover: The characteristics of two selected individuals are combined to produce new offspring.
- Mutation: Small random modifications are made to the offspring to introduce diversity into the population.
- Repair and improvement: Unfeasible solutions are corrected, and existing solutions are improved to ensure their quality.

During this phase, three key components are updated:

- z^*: It is updated with the best values found for each objective function.
- *B*: It refers to the neighborhood matrix that defines the relationships between the subproblems.
- *EP*: It is the external population file that stores the non-dominated solutions found so far.
- **Update and Change Detection Phase**

The variable t, which introduces dynamicity into the problem, is updated according to the formula described above in Eq. 10 [6].

$$t = \frac{1}{n_t}\left\lfloor \frac{\tau}{\tau_T} \right\rfloor \tag{23}$$

During execution, z^* and t, are constantly updated.

5. **Change Adaptation Strategy Phase**

The algorithm checks for changes in t. If a change is detected, the adaptation strategy is activated, which combines memory and diversity strategy to adjust the population to the new environment. The diversity strategy involves applying a perturbation to the population, which consists of generating new random solutions that replace 10% of the current population. The memory strategy involves reusing previous optimal solutions stored in *EP* to accelerate convergence in the new environment.

If no change is detected, the algorithm continues its normal execution until the stopping condition, defined by the maximum number of iterations, is evaluated. If this condition is not met, the external *EP* file is reused as the initial population for the next iteration.

6. **Finalization Phase**

The algorithm will terminate when the predetermined stopping criterion is met, generating the EPs_t, which represent the set of external populations obtained at each change of the problem. These populations contain the non-dominated solutions found at each change of the problem.

The Multi-Objective Evolutionary Algorithm based on Decomposition (MOEA/D) decomposes a multi-objective optimization problem into a set of scalar subproblems, which are solved simultaneously. As illustrated in Fig. 2, MOEA/D begins with an initial population and iteratively evolves solutions by selecting neighbors, performing reproduction, evaluating offspring, and updating both the reference point

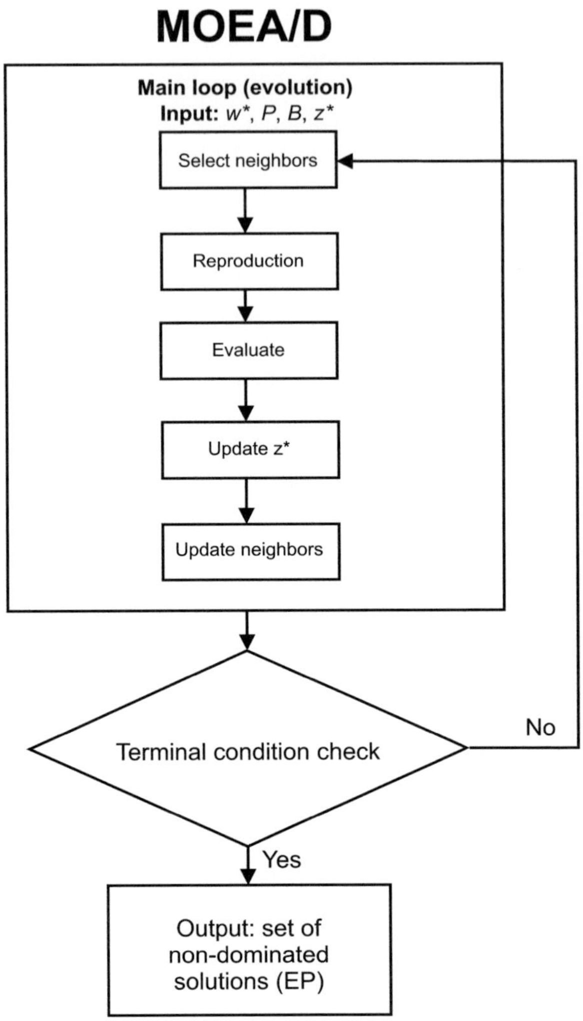

Fig. 2 Architecture of MOEA/D

z^* and the neighborhood solutions. This process continues until a predefined terminal condition is met, at which point the set of non-dominated solutions (EP) is returned.

3.1 Strategy for Adaptation to Change

The change adaptation strategy (Fig. 3) implements a mutation-based diversity introduction strategy, following the approach proposed by Farina in [6]. This process is divided into several steps, which are described below:

1. **Retrieval of the Detector List**

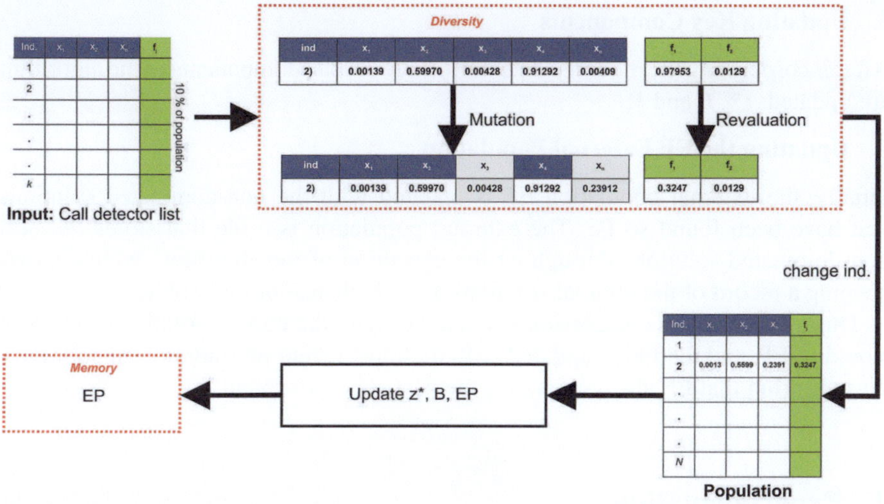

Fig. 3 Hybrid strategy for change adaptation

The process starts with the retrieval of the Detector List, which was previously generated in the initialization phase (Fig. 1). These detectors represent a portion of the population (10% in this case) and are essential for monitoring changes in the dynamic environment. Their main function is to identify when modifications occur in the objective functions, which activates the adaptation strategy.

2. **Application of the Mutation Operator**

Once the list of detectors has been retrieved, a mutation operator is applied to these individuals. The mutation alters the decision variables of the detectors, introducing random changes in their values. In this case, a 20% probability of a polynomial mutation was used, which allows the decision variables to be slightly modified. This ensures that the new solutions generated are different, but not too far from the original solutions.

3. **Re-Evaluation of the Detectors**

After the mutation, the modified detectors are re-evaluated to update their values in the objective functions. At each iteration, the new values are compared to the previous values to detect possible discrepancies. If a change is identified, it is determined that the problem has experienced a variation, which allows the algorithm to adjust its response to the dynamic environment.

4. **Incorporation of the Modified Detectors into the Population.**

After the re-evaluation, the modified detectors are incorporated into the *P* population by replacing 10% of the current population. This step ensures that the population is updated with solutions that reflect changes in the environment, maintain diversity and avoid premature convergence.

5. **Updating Key Components**

After incorporating the modified detectors, three crucial components of the algorithm are updated: z^*, B and P.

6. **Updating the EP External Population**

Finally, the external population EP is updated with the non-dominated solutions that have been found so far. The external population is a file that stores the best non-dominated solutions throughout the execution of the algorithm, which allows keeping a record of the optimal solutions at each change of the problem.

During this step, a comparison is made between the current solutions and those stored in EP, and the file is updated with the most recent non-dominated solutions, thus ensuring that EP always contains the best solutions found.

4 Experimentation

The experimental process was divided into two stages. The first corresponds to the validation of the proposed algorithm, with the objective of confirming its ability to approximate the shape of the Pareto fronts, as reported in the state of the art [6]. The second stage consists of a comparative analysis of the performance of the proposed algorithm, D-CC-MOEA/D, against D-MOEA/D [19, 21], using the metrics of hypervolume and mean hypervolume.

The comparison with D-MOEA/D is justified by its widespread use in solving dynamic problems, where it employs diversity as its adaptation strategy. In contrast, our proposed algorithm, D-CC-MOEA/D, incorporates a coevolutionary approach in which each species integrates the D-MOEA/D algorithm while also hybridizing two adaptation strategies: diversity and memory. Therefore, the objective of this comparison is to analyze whether this hybridization improves the diversity of the obtained Pareto fronts after each change in the problem, compared to D-MOEA/D, which relies solely on diversity as an adaptation strategy.

4.1 Experimental Design

The settings utilized during the experimentation are detailed in Table 1, which describes the algorithm setup, and in Table 2, which presents the problem configuration.

The algorithm configuration (Table 1) was designed to ensure a fair comparison between the proposed algorithm (D-CC-MOEA/D) and the reference algorithm (D-MOEA/D [21]). The key parameters are described below:

Table 1 Algorithms setup

Algorithm	D-MOEA/D [21, 19]	D-CC-MOEA/D
Population size	300	300
Iterations	100	100
Neighborhood size	15	15
Number of species	–	2
Experiments	30	30
Selection	Random	Random
Crossover	Differential evolution [20]	SBX [40]
Crossover probability	1.0	1.0
Mutation	Polynomial	Polynomial
Mutation probability	0.2	0.2

Table 2 Problem setting

Problem	Number of objectives	Number of variables	Periodicity of the change (τ_T)	Severity of the change n_T
FDA1	2	20	20	100
FDA3	2	10	100	10

- **Population Size**: Both algorithms used a population of 300 individuals, which allows adequate exploration of the search space without incurring excessive computational cost.
- **Number of Iterations**: This was set at 100 iterations for both algorithms, which provides a balance between execution time and quality of solutions.
- **Neighborhood Size**: It was set at 15 individuals for both algorithms, which defines the number of neighbors considered in the collaboration between subproblems.
- **Number of Species**: Only the D-CC-MOEA/D algorithm uses this parameter, with a value of 2 species, allowing for greater diversity in the population.
- **Number of Experiments**: 30 independent runs were performed for each algorithm, allowing for statistically significant results.
- **Selection:** Both algorithms used random selection of individuals for breeding.
- **Crossover**: D-MOEA/D employs Differential Evolution [22], a technique recognized for its efficacy in continuous optimization problems. D-CC-MOEA/D uses SBX (Simulated Binary Crossover), a prevalent crossover operator in evolutionary algorithms that emulate crossover behavior in nature.
- **Crossover Probability**: It was set to 1.0 for both algorithms, which means that the crossover operator is always applied to the selected individuals.
- **Mutation**: Both algorithms used polynomial mutation, which introduces small variations in the solutions to maintain diversity.
- **Mutation Probability**: It was set at 0.2, which means that 20% of the individuals undergo mutation in each iteration.

To evaluate the performance of the algorithms, two benchmark problems were selected from the literature: FDA1 and FDA3. These problems represent dynamic scenarios with different characteristics, allowing them to evaluate the adaptability of the algorithms to changes in the objectives. The key parameters are described below:

- **Number of objectives**: both problems have two objectives, which allows for clear visualization of the Pareto fronts.
- **Number of variables**
  - FDA1 has 20 variables, making it a higher-dimensional problem.
  - FDA3 has 10 variables, which makes it less complex in terms of search space.
- **Period of Change** (τ_T)
  - FDA1: Change occurs every 20 iterations, which simulates a dynamic environment with frequent changes.
  - FDA3: Change occurs every 100 iterations, which simulates an environment with less frequent but more severe changes.
- **Severity of Change** (n_T)
  - FDA1: Severity of change is 100, indicating that the changes are significant in terms of magnitude.
  - FDA3: Severity of change is 10, indicating less drastic but still relevant changes.

In the D-CC-MOEA/D algorithm, subpopulations are evolved sequentially. Therefore, the overall computational complexity remains in $O\left(\mathrm{pm}^2\mathrm{nr}\right)$, where p is the number of subpopulations, m the number of objectives, n the number of decision variables, and r the number of iterations.

4.2 Validation

Figure 4 presents the reference solution for the FDA1 problem, characterized by its convex Pareto front with a continuous and uniform distribution of points. Figure 5 displays the solutions obtained with D-CC-MOEA/D, where the color-coded results (with each hue representing a different time window) demonstrate:

1. Close alignment with the reference front's convex shape
2. Nearly identical point distribution density
3. Temporal progression visible through sequential color bands

The multi-colored solutions in Fig. 5 show only minor dispersion variations while successfully:

- Matching the spatial characteristics of the reference front (Fig. 4)
- Maintaining solution quality across all time periods
- Preserving geometric fidelity to the convex topology

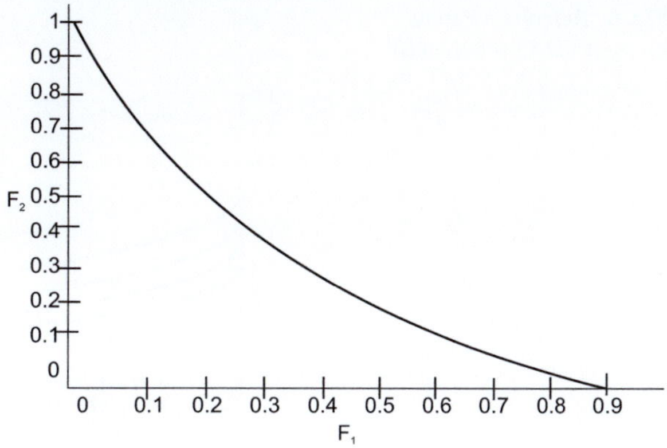

Fig. 4 Reference Pareto front of FDA1 problem [6]

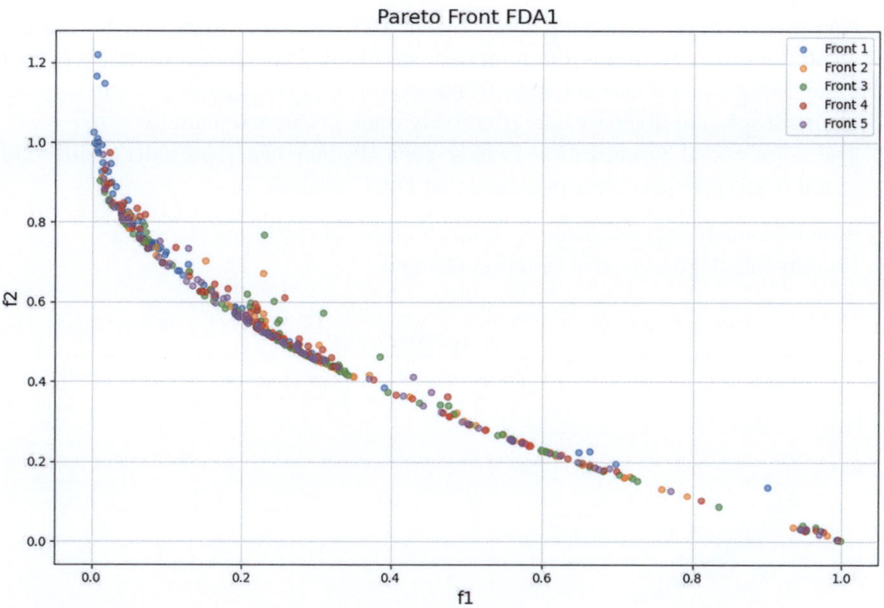

Fig. 5 Approximated Pareto Front of FDA1 by D-CC-MOEA/D

This chromatic representation confirms D-CC-MOEA/D's ability to:

- Accurately track the Pareto front's evolution
- Respond dynamically to changes
- Produce solutions comparable to state-of-the-art results [6].

The FDA3 problem features a time-varying convex Pareto front, as illustrated in Fig. 6. The figure displays a sequence of dynamic fronts represented by smooth curves that progressively shift, demonstrating the front's evolution in response to changing problem conditions.

Fig. 6 Reference Pareto Front of FDA3 problem [6]

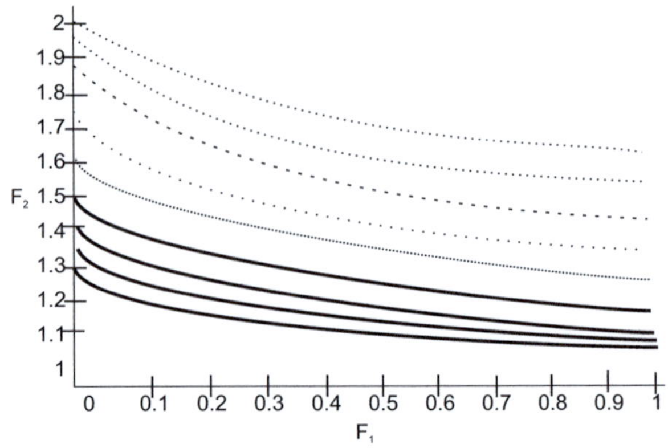

Analysis of Fig. 7 (showing D-CC-MOEA/D results) reveals that the generated solutions closely follow the reference structure. The solutions exhibit proper distribution across successive Pareto fronts, confirming the algorithm's capacity to: maintain solution diversity and effectively track problem dynamics.

The color-coded visualization (where each distinct hue represents a different temporal front) provides clear evidence that D-CC-MOEA/D:

- Simultaneously approximates optimal solutions at each time step, and
- Rapidly adapts to objective function changes.

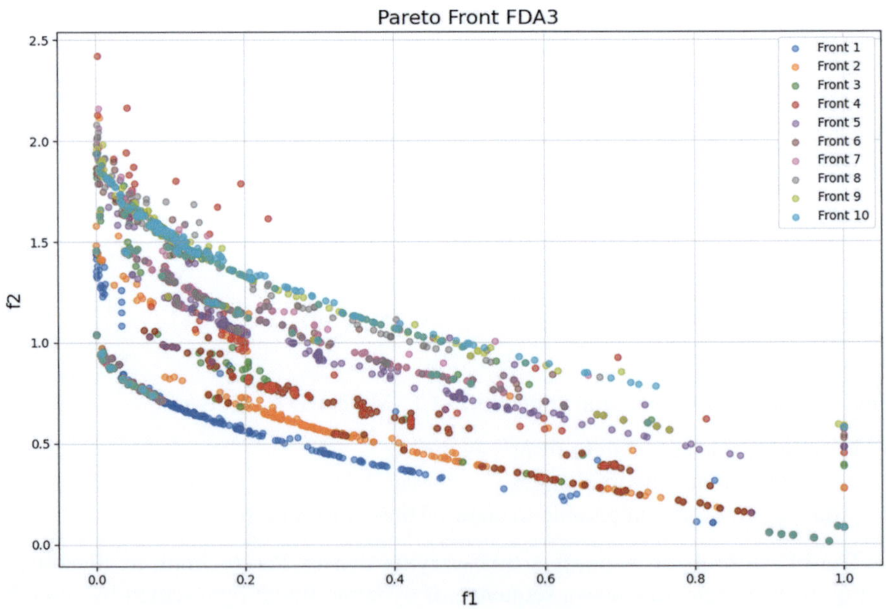

Fig. 7 Approximated Pareto front of FDA3 by D-CC-MOEA/D

- This multicolor representation allows for immediate visual differentiation between solution sets corresponding to different environmental conditions.

4.3 Comparison

This experiment evaluates the diversity of EP solutions in the fronts generated by D-CC-MOEA/D, comparing them with those obtained by D-MOEA/D. The objective is to assess whether the implementation of a hybrid diversity and memory strategy helps maintain diversity in EP solutions across different fronts in dynamic optimization problems.

In Fig. 8, it can be observed that the D-CC-MOEA/D algorithm consistently maintains higher MHV values compared to D-MOEA/D across the 30 runs. While D-CC-MOEA/D exhibits stable and elevated behavior, D-MOEA/D shows notable variations and lower values, indicating less stability and a reduced ability to preserve solution diversity. These results demonstrate that the hybrid diversity and memory strategy implemented in D-CC-MOEA/D favors better maintenance of diversity compared to the random insertion-based strategy used in D-MOEA/D.

Furthermore, the results obtained in Table 3 through the Wilcoxon test, applied to each of the fronts generated after every change in the problem, reinforce these graphical observations. The reported p-value (1.7344E-06) is significantly lower than the standard significance threshold of 0.05, allowing the rejection of the null hypothesis (H_0) and confirming the existence of statistically significant differences between the fronts generated by both algorithms. Thus, it is confirmed that the proposed hybrid strategy positively impacts the maintenance of solution diversity in the FDA1 problem.

- **Null hypothesis (H0)**: The medians of the differences between the two group samples are equal.
- Wilcoxon test (significance level of 0.05).

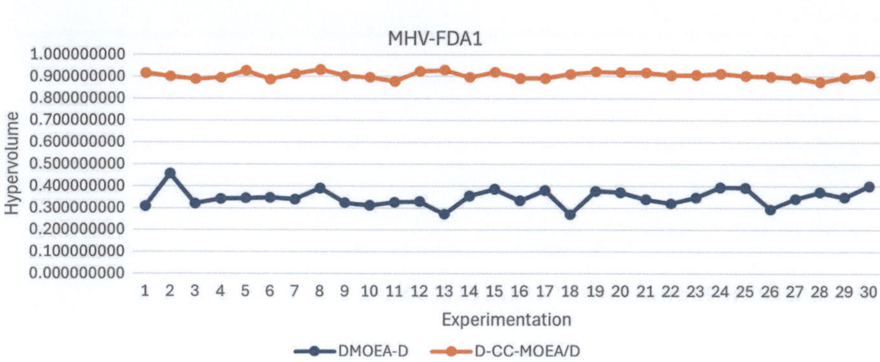

Fig. 8 MHV FDA1 problem

Table 3 Wilcoxon test FDA1

Front	FDA1	
	p-value	Result
F1	1.7344E-06	H0
F2	1.7344E-06	H0
F3	1.7344E-06	H0
F4	1.7344E-06	H0
F5	1.7344E-06	H0
MHV	1.7344E-06	H0

In Fig. 9, it can be observed that D-CC-MOEA/D outperforms D-MOEA/D in terms of MHV. Although both algorithms exhibit more stable behavior compared to the FDA1 problem, the MHV achieved by D-CC-MOEA/D consistently remains higher throughout all runs. This behavior demonstrates that the implemented hybrid strategy promotes better maintenance of diversity in more stable dynamic problems such as FDA3.

Additionally, the results presented in Table 4, obtained through the Wilcoxon test, show a *p*-value of 1.7344E-06, which is also below the significance threshold of 0.05. This result allows for the rejection of the null hypothesis (H_0) and confirms the existence of statistically significant differences between the fronts generated by D-MOEA/D and D-CC-MOEA/D in the FDA3 problem. Taken together, these findings confirm that the proposed hybrid strategy effectively contributes to maintaining diversity in FDA3 problems.

- **Null hypothesis (H0)**: The medians of the differences between the two group samples are equal.
- Wilcoxon test (significance level of 0.05).

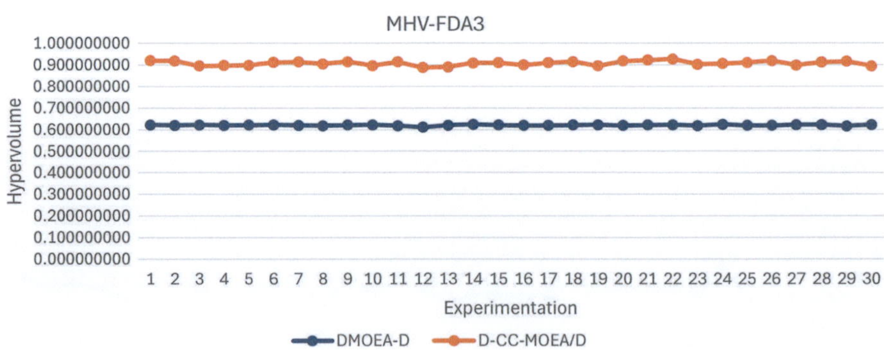

Fig. 9 MHV FDA3 problem

Table 4 Wilcoxon test FDA3

Front	FDA3	
	p- value	Result
F1	1.7344E-06	H0
F2	1.7344E-06	H0
F3	1.7344E-06	H0
F4	1.7344E-06	H0
F5	1.7344E-06	H0
MHV	1.7344E-06	H0

5 Conclusions

This work presents a hybrid strategy for adapting to change that integrates memory and diversity strategies in the CO-MOEA/D algorithm [39], giving rise to its dynamic version, D-CC-MOEA/D. The main objective was to evaluate the effectiveness of this approach in the solution of DMOPs, specifically in problems FDA1 and FDA3, its performance was compared with the D-MOEA/D algorithm [21]. The experimental results showed that:

1. D-CC-MOEA/D graphically demonstrated its ability to replicate the shape of the fronts obtained in the test instances, showing similarity with the POF reported in the state of the art [6], as well as adapting effectively to dynamic changes.
2. The results obtained, evaluated using the MHV, show that the hybrid diversity and memory strategy proposed in D-CC-MOEA/D allows for more effective preservation of solution diversity over time.
3. The incorporation of the hybrid strategy allowed D-CC-MOEA/D to significantly outperform the D-MOEA/D algorithm [21]. As shown in Tables 3 and 4, the *p*-values obtained through the Wilcoxon test were well below the significance threshold (0.05), confirming the presence of statistically significant differences between the fronts generated by both algorithms.

As a future line of research, it is proposed to extend the algorithm by incorporating interactive preference mechanisms into D-CC-MOEA/D, which would allow the user to guide the search for solutions based on their evolving interests and priorities.

AI Use Statement: Some portions of this manuscript were enhanced with the assistance of generative AI tools. The authors carefully reviewed, edited and approved all AI-generated content. The authors take full responsibility for the accuracy and integrity of the entire chapter.

Acknowledgments LRRS gratefully acknowledges the CONAHCyT grant for doctoral studies (Grant 960719). The authors are also thankful for the support obtained from: Tecnológico Nacional de México through the Instituto Tecnológico de Ciudad Madero, for the use of the Laboratorio Nacional de Tecnologías de Información (LaNTI), and for the support received through the TecNM project 'Design of an intelligent virtual agent for decision support with impact on the navigation of electric vehicles'.

References

1. Deb, K., Rao N.U.B., Karthik, S.: Dynamic multi-objective optimization and decision-making using modified NSGA-II: A case study on hydro-thermal power scheduling. In: Obayashi, S., Deb, K., Poloni, C., Hiroyasu, T., Murata, T. (eds.), Evolutionary Multi-Criterion Optimization, pp. 803–817. Springer (2007)
2. Liang, Z., Zheng, S., Zhu, Z., Yang, S.: Hybrid of memory and prediction strategies for dynamic multiobjective optimization. Inf. Sci. (2019). https://doi.org/10.1016/j.ins.2019.01.066
3. Chen, R., Li, K., Yao, X.: Dynamic multiobjectives optimization with a changing number of objectives. IEEE Trans. Evol. Comput. **22**(1), 157–171 (2017)
4. Muruganantham, A.: Dynamic multiobjective optimization using evolutionary algo-rithms (2017)
5. Coello, C., Lamont, G., Veldhuizen, D.: MOEA Local Search and Coevolution (2007). https://doi.org/10.1007/978-0-387-36797-2_3
6. Farina, M., Deb, K., Amato, P.: Dynamic multiobjective optimization problems: Test cases, approximations, and applications. IEEE Trans. Evol. Comput. **8**(5) (2004)
7. Goh, C.K., Tan, K.C.: A competitive-cooperative coevolutionary paradigm for dynamic multiobjective optimization. IEEE Trans. Evol. Comput. **13**(1), 103–127 (2009)
8. Liu, Y., Yen, G.G., Gong, D.: A multi-population cooperative coevolutionary algorithm for multi-objective optimization. IEEE Trans. Cybern. **50**(8), 3448–3461 (2019)
9. Zhou, A., Jin, Y., Zhang, Q., Sendhoff, B., Tsang, E.: Prediction-Based Population Re-initialization for Evolutionary Dynamic Multi-objective Optimization, pp. 832–846 (2006). https://doi.org/10.1007/978-3-540-70928-2_62.
10. Yang, S., Li, M., Liu, X., Zheng, J.: A grid-based evolutionary algorithm for many-objective optimization. IEEE Trans. Evol. Comput. **17**(5), 721–736 (2013)
11. Helbig, M., Engelbrecht, A.P.: Benchmarks for dynamic multi-objective optimisation algorithms. ACM Comput. Surv. (CSUR) **46**(3), 1–39 (2014)
12. Luo, S., Zhang, L., Fan, Y.: Dynamic multi-objective scheduling for flex-ible job shop by deep reinforcement learning. Comput. Ind. Eng. **159**, 107489, ISSN 360-8352 (2021). https://doi.org/10.1016/j.cie.2021.107489
13. Tian, H., Zhao, C., Xie, J., Li, K.: Dynamic operation optimization of complex industries based on a data-driven strategy. Processes **12**, 189 (2024). https://doi.org/10.3390/pr12010189
14. Hu, Y., Ou, J., Suganthan, P., Pedrycz, W., Wang, R., Zheng, J., Zou, J., Song, Y.: Dynamic multi-objective optimization algorithm guided by recurrent neural network. IEEE Transactions on Evolutionary Computation, pp 1–1 (2024). https://doi.org/10.1109/TEVC.2024.3419892.
15. Zhang, K., Shen, C., Liu, X., Yen, G.: Multiobjective evolution strategy for dynamic multiobjective optimization. IEEE Trans. Evol. Comput. **24**(5), 974–988 (2020)
16. Martínez Peñaloza, M.G.: A bio-inspired algorithm to solve dynamic multi-objective optimization problems. Ph.D. dissertation, Universidad Veracruzana, Centro de Investigación de Inteligencia Artificial, Región Xalapa (2018)
17. Macias, T.: Optimización de problemas dinámicos en presencia de un gran número de objetivos y preferencias del decisor. Ph.D. dissertation, Tijuana, B.C (2021)
18. Wu, F., Wang, W., Chen, J., Wang, Z.: A dynamic multi-objective optimization method based on classification strategies. Sci. Rep. **13**(1), 15221 (2023)
19. Deb, K., Pratap, A., Agarwal, S., Meyarivan, T.: Un algoritmo genético multiobjetivo rápido y elitista: NSGA-II. Trans IEEE. Evolución. Computador. **6**(2), 182–197 (2002)
20. Jiang, S., Zou, J., Yang, S., Yao, X.: Evolutionary dynamic multi-objective optimisation: A survey. ACM Comput. Surv. **55**(4), 47 (2022). https://doi.org/10.1145/3524495
21. Brambila-Hernández, J.A., et al.: Experimental evaluation of adaptive operators se-lection methods for the dynamic multiobjective evolutionary algorithm based on de-composition (DMOEA/D). In: Castillo, O., Melin, V. (eds.), Hybrid Intelligent Systems Based on Extensions of Fuzzy Logic, Neural Networks and Metaheuristics. Springer (2023)

22. Li, K., Fialho, Á., Kwong, S., Zhang, Q.: Adaptive operator selection with bandits for a multi-objective evolutionary algorithm based on decomposition. IEEE Trans. Evol. Comput. **18**(1), 114–130 (2014). https://doi.org/10.1109/TEVC.2013.2239648
23. Xu, B., Zhang, Y., Gong, D., Guo, Y., Rong, M.: Environment sensitivity-based cooperative co-evolutionary algorithms for dynamic multi-objective optimization. IEEE/ACM Trans. Comput. Biol. Bioinform. **15**(6), 1877–1890 (2017)
24. Chen, D., Zou, F., Lu, R., Wang, X.: A hybrid fuzzy inference prediction strategy for dynamic multi-objective optimization. Swarm Evolution. Comput. **43**, 147–165 (2018)
25. Muruganantham, A., Tan, K.C., Vadakkepat, P.: Evolutionary dynamic multiobjective optimization via kalman filter prediction. IEEE Trans. Cybern. **46**(12), 2862–2873 (2015)
26. Zhou, A., Jin, Y., Zhang, Q.: A population prediction strategy for evolutionary dynamic multiobjective optimization. IEEE Trans. Cybern. **44**(1), 40–53 (2013)
27. Zhang, T., Tao, Q., Yu, L., Yi, H., Chen, J.: A new prediction strategy for dynamic multi-objective optimization using hybrid Fuzzy C-means and sup-port vector machine. Neurocomputing **621**, 2025, 129291, ISSN 0925-2312, https://doi.org/10.1016/j.neucom.2024.129291
28. Masegosa, A., Pelta, D.: Problemas de optimización dinámicos: Enfoques y per-spectivas. Investig. Oper. **38**, 111–121 (2017)
29. Koo, W.T., Goh, C.K., Tan, K.C.: A predictive gradient strategy for multiobjective evolutionary algorithms in a fast-changing environment. Memetic Comput. 2(2), 87–110 (2010)
30. Liu, R., Li, J., Fan, J., Mu, C., Jiao, L.: A coevolutionary technique based on multi-swarm particle swarm optimization for dynamic multi-objective optimization. Eur. J. Oper. Res. **261**(3), 1028–1051, ISSN 0377-2217 (2017). https://doi.org/10.1016/j.ejor.2017.03.048.
31. Sahmoud, S., Topcuoglu, H.: A Memory-Based NSGA-II Algorithm for Dynamic Multi-objective Optimization Problems, pp. 296–310 (2016). https://doi.org/10.1007/978-3-319-31153-1_20
32. Zitzler, E., Deb, K., Thiele, L.: Comparison of multiobjective evolutionary algo-rithms: Empirical results. Evol. Comput. **8**(2), 173–195 (2000)
33. Deb, K., Thiele, L., Laumanns, M., Zitzler, E.: Scalable multi-objective optimization test problems. In: Proceedings Congress on Evolutionary Computation (CEC'02), pp. 825–830 (2002)
34. Van Veldhuizen, D.A.: Multiobjective evolutionary algorithms: classifications, analyses, and new innovations. Ph.D. Dissertation, Air Force Institute of Technology, Wright Patterson AFB (1999)
35. Xu, B., Zhang, Y., Gong, D., Guo, Y., Rong, M.: Environment sensitivity-based cooperative co-evolutionary algorithms for dynamic multi-objective optimization. In: IEEE/ACM Transactions On Computational Biology and Bioinformatics, vol. 15, no. 6, pp. 1877–1890, Nov–Dec (2018)
36. Jiang, S., Yang, S.: A steady-state and generational evolutionary algorithm for dynamic multiobjective optimization. IEEE Trans. Evol. Comput. **21**(1), 65–82 (2017)
37. Helbig, M., Engelbrecht, A.: Benchmarks for dynamic multi-objective optimisation. In: IEEE Symposium Computational Intelligence in Dynamic and Uncertain Enviroments (CIDUE), pp. 84–91 (2013)
38. Zitzler, E., Laumanns, M., Thiele, L.: SPEA2: Improving the Strength Pareto Evolutionary Algorithm (2001)
39. Rosas, L et al.: Experimental analysis of a cooperative coevolutionary algorithm with parameter tuning for multi-objective problem optimization with uncertainty. Computación y Sistemas **28**(3) (2024)
40. Deb, K., Sindhya, K., Okabe, T.: Self-adaptive simulated binary crossover for real-parameter optimization, pp 1187–1194 (2007). https://doi.org/10.1145/1276958.1277190

Soft Computing Models

Advanced Switched Reluctance Motor Control Methodologies for Electric Drive Applications

Zeineb Touati, Rui Esteves Araújo, and Adel Khedher

Abstract Switched Reluctance Motors (SRMs) are becoming increasingly popular for various applications, including automotive applications. However, challenges such as torque ripple and vibration persist, limiting their performance. This chapter investigates the application of intelligent control strategies, particularly fuzzy logic, to mitigate these issues. Fuzzy logic modeling does not require an accurate mathematical model which is very difficult to obtain from a SRM because of its inherit nonlinearities. In this work a Fuzzy Logic Controller (FLC) applied to the speed control of an SRM, highlighting the advantages of FL over traditional methods in terms of flexibility and performance. A comparison is made between the FLC, a Sliding Mode Control (SMC), and a Proportional Integral (PI) controller. Simulation results using MATLAB/Simulink show that the FLC substantially reduces torque ripple, offering better overall performance in terms of smoothness and robustness under varying operational conditions. The findings demonstrate that FLC offers a more effective solution than conventional approaches for SRM applications.

Keywords Switched reluctance motor · Modeling · Finite element analysis · Sliding mode control · Fuzzy logic controller · Electrical drives

Z. Touati (✉) · A. Khedher
LATIS-Laboratory of Advanced Technology and Intelligent Systems, National Engineering School of Sousse, University of Sousse, Sousse, Tunisia
e-mail: zeineb.touati@eniso.u-sousse.tn

A. Khedher
e-mail: adel_kheder@yahoo.fr

R. E. Araújo
INESC TEC and Faculty of Engineering, University of Porto, Porto, Portugal
e-mail: raraujo@fe.up.pt

W. Pedrycz et al. (eds.), *Decision-Making in Computational Intelligence-Based Systems*, Studies in Systems, Decision and Control 628,
https://doi.org/10.1007/978-3-032-04056-5_8

1 Introduction

Switched Reluctance Motors (SRMs) are gaining significant attention in modern electrical drive systems due to their unique design and operational characteristics. Unlike traditional motors, SRMs feature a simple and rugged structure, with no permanent magnets or windings on the rotor, making them highly cost-effective and robust. SRMs are being considered in a variety of applications requiring high performance, such as servomotors, electric vehicle, starter-generators, etc. SRMs have numerous intrinsic advantages, such as robust structure, maintenance-free operation, high torque-inertia ratio, robustness and reliable fault tolerance, high efficiency over a wide speed range, ability to operate in harsh environments [1–4]. Despite their numerous advantages, SRMs are not without their challenges. One of the primary drawbacks is the presence of significant torque ripple, which can lead to increased vibrations and noise during operation [5]. This characteristic can limit their application in environments where smooth operation is crucial. Additionally, the nonlinear magnetic characteristics of SRMs, including saturation and mutual coupling effects, make their control more complex and demand advanced algorithms to achieve optimal performance.

In recent years, significant advancements have been made in the design and development of various SRMs. These advancements include innovative structural designs [6–9] and the application of adaptive control techniques [10–12]. Currently, proportional, integral, and/or derivative (PI, PD, PID) controllers have demonstrated reliable control performance across numerous industrial applications [13–17]. These controllers, based on precise mathematical models and the operator's expertise, are capable of effectively compensating for system variations through appropriate tuning of their key parameters. However, their performance can degrade significantly when operating conditions change or system parameters deviate, highlighting their sensitivity to parameter variations and external disturbances.

Intelligent control techniques, such as Fuzzy Logic Control (FLC), genetic algorithms or neural network control, can achieve better performance when the exact analytical model of the controlled system is uncertain or difficult to characterize. These intelligent control methods aim to replicate and learn from the expertise of human operators to achieve optimal performance for the controlled system [18]. Among these, FLC stands out as a powerful tool that translates linguistic control rules into a practical operational framework. The efficacy of FLC is particularly effective for systems that are challenging to mathematically model due to their complexity, nonlinearity, or imprecise behavior. As a result, FLC has gained widespread adoption across numerous engineering disciplines, thanks to its adaptability and efficiency [19–21]. Research demonstrates that fuzzy control not only reduces hardware requirements and costs but also outperforms traditional controllers, in many scenarios [22–24]. The architecture of FLC closely mirrors the intuitive decision-making process of human experts. This feature makes it an appealing choice for integrating heuristic rules derived from human experience directly into the control system [25].

By leveraging this intuitive approach, FLC provides a robust and flexible solution for addressing the challenges posed by complex and nonlinear systems.

Fuzzy control theory has gained significant attention, leading to the development of various types of fuzzy controllers specifically designed for SRMs. These controllers have been introduced to further enhance the performance of SRM drives, addressing challenges such as nonlinearity, torque ripple, and efficiency under varying operating conditions. In [26], a fuzzy logic speed controller for SRM drives is proposed. This fuzzy controller takes the speed error and change of error as the input signals to generate an equivalent control term. It can generate smooth torque and improve the SRM system performance. An adaptive FLC for the speed control of SRM drives is proposed in [27]. The key approach in improving the performance of an FLC involves tuning fuzzy rules, membership functions, and scaling gains for both input and output variables. These adjustments significantly enhance both transient and steady-state system responses. Similarly, [28] introduces a fuzzy-logic-based PI-like controller for SRM speed control. Unlike conventional PI controllers, this approach does not rely on fixed proportional and integral gains. Instead, the control signal is directly derived from the knowledge base and fuzzy inference, allowing for a more flexible and adaptive control mechanism.

This chapter presents a comprehensive modeling and analysis of the drive system with detailed results of a FLC-based speed control scheme for a four-phase, 8/6 SRM. The study examines the current, torque, and speed responses across all operating conditions, including parameter variations. Key performance metrics, such as steady-state speed error, torque pulsations, and speed oscillations, are presented. Additionally, the performance indices of the SRM drive system using the FLC are compared with those achieved using a conventional controller. This chapter is organized as follows. Section 2 introduces the SRM dynamic modelling and the finite element analysis. The control structure of the proposed controllers are illustrated in detail in Sect. 3. Simulation results and comparisons are given in Sect. 4 to show the effectiveness of the developed controllers and to verify the performance improvement. Section 5 concludes this chapter.

2 Modeling of Switched Reluctance Motors

This section provides a comprehensive overview of different modeling approaches used for SRMs. Dynamic modeling is the starting point, describing the fundamental equations governing the behavior of SRMs. Next, Finite Element Analysis (FEA) is presented as a powerful tool for obtaining accurate electromagnetic characteristics. Finally, linearization of the SRM model is discussed to facilitate control system design and stability analysis.

Table 1 Key assumptions in SRM modeling

Assumption	Details
Neglect mutual effects between phases	No coupling between phases is considered
Identical phase parameters	All phases have the same properties
Constant winding resistance	Resistance is unaffected by temperature.
Ideal converter switches	No conduction or switching losses are assumed
No core losses	Eddy currents and hysteresis losses are ignored

2.1 Dynamic Modeling

To simplify the SRM model, we consider the following simplifying assumptions in Table 1. These simplifying assumptions facilitate a more straightforward modeling and analysis of SRMs.

The phase voltage of the SRM is given by [29]:

$$v_j = r \,.\, i_j + \frac{d\phi(\theta, i)}{dt} \tag{1}$$

with: j = 1,2,3,4.
where:

- v_j: The voltage of phase j,
- i_j: The current of phase j in the SRM,
- r: Stator resistance,
- θ: Rotor position,
- $\phi(\theta, i)$: Phase flux linkage as a function of rotor position and stator current.

The flux of each phase is determined by the rotor position and the current flowing through the corresponding phase. It is expressed as follows:

$$\frac{d\phi(\theta, i)}{dt} = \frac{\partial\phi(\theta, i)}{\partial\theta} \cdot \frac{\partial\theta}{\partial t} + \frac{\partial\phi(\theta, i)}{\partial i} \cdot \frac{\partial i}{\partial t} \tag{2}$$

where

$$\frac{\partial\theta}{\partial t} = \Omega \tag{3}$$

where Ω is the rotational speed.
In this case, Eq. 2 can be written as:

$$\frac{d\phi(\theta, i)}{dt} = \frac{\partial\phi(\theta, i)}{\partial i} \cdot \frac{\partial i}{\partial t} + \frac{\partial\phi(\theta, i)}{\partial\theta} . \Omega \tag{4}$$

The j-phase flux is defined also by the following expression :

$$\phi(\theta, i) = L(\theta, i)\ .\ i_j \tag{5}$$

where L represents the inductance of a phase j of an SRM, which depends on two parameters: position θ and current i.
The voltage Eq. 1 can then be reformulated as follows:

$$\nu_j = ri_j + L(\theta, i)\frac{di_j}{dt} + i_j\left(\frac{\partial L(\theta, i)}{\partial \theta}.\frac{\partial \theta}{\partial t} + \frac{\partial L(\theta, i)}{\partial i}.\frac{\partial i}{\partial t}\right) \tag{6}$$

When neglecting the term accounting for the saturation effect in Eq. 6, the voltage equation simplifies to:

$$\nu_j = ri_j + L(\theta)\frac{di_j}{dt} + i_j\ \Omega\ \frac{dL(\theta)}{d\theta} \tag{7}$$

The last term of Eq. 7 represents the electromotive force (e.m.f) generated by the fluctuations of the magnetic field. The e.m.f depends on the current and is expressed by the following formula:

$$e = i_j\ \Omega\ \frac{dL_j(\theta)}{d\theta} \tag{8}$$

Figure 1 shows the equivalent circuit of each phase of the SRM.

The electromagnetic torque for each phase is expressed by:

$$T_j = \frac{1}{2}{i_j}^2\frac{dL(\theta)}{d\theta} \tag{9}$$

We can observe that the direction of the torque does not vary with the direction of the current. It primarily depends on the slope of the inductance profile $\frac{dL(\theta)}{d\theta}$. The total electromagnetic torque T_e produced by the SRM represents the sum of the torques produced by each phase n, where n is the total number of phases. It is described by:

$$T_e = \sum_{j=1}^{n} T_j \tag{10}$$

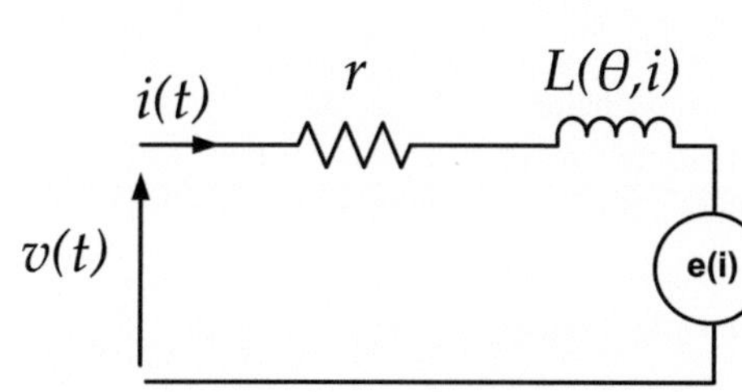

Fig. 1 Equivalent electrical circuit of the SRM

The fundamental equation of dynamics applied to the rotor of the machine is expressed by [29]:

$$J\frac{d\Omega}{dt} = T_e - T_L - f\Omega \tag{11}$$

where:

- T_L : Load torque,
- T_e: Total electromagnetic torque produced by the machine,
- J: Inertia of the rotating parts,
- f : Viscous friction coefficient.

2.2 Finite Element Analysis

Finite element modeling is used to determine the characteristics of the SRM. The meshing of the study domain is a crucial step that depends on the quality of the resolution. For the numerical modeling of the SRM, we use MagNet software for mesh distribution. The key parameters of the base design of the 8/6 SRM are summarized in Table 2. The materials used for the various components of our machine include several key elements. First, the air gap is typically considered empty, meaning it is not filled with any specific material. This assumption simplifies the modeling process. For the coils, copper is commonly used due to its excellent electrical conductivity, making it a preferred material in many electrical applications. As for the stator and rotor, magnetic materials are typically employed, and for our model, we use N020. Additionally, the number of turns in the coils is 75. Figure 2 shows the mesh and

Table 2 Parameters of the SRM

Parameters	Symbol	Values
Nominal output power	P_{rated}	500 W
Nomber of stator poles	N_s	8
Nomber of rotor poles	N_r	6
Nominal speed	Ω_{rated}	1500 rpm
Inductance (aligned position)	L_a	0.14 H
Inductance (unaligned position)	L_u	0.021 H
Viscous friction	f	0.05 Nms
Moment of inertia	J	0.01 kg·m^2
Resistance of phase winding	r	5 Ω
Maximum current	i_{max}	10 A
Maximum torque	$T_{e,max}$	6 Nm

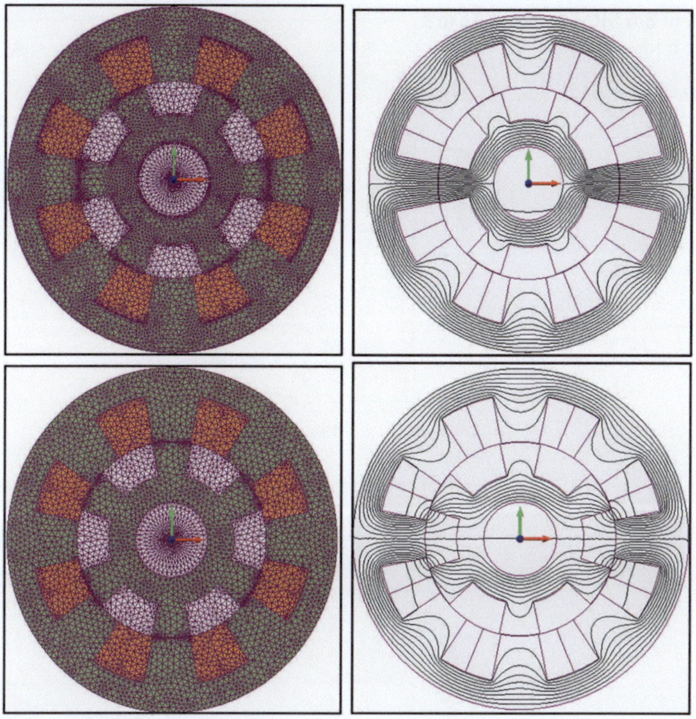

Fig. 2 8/6 SRM: field lines for the two extreme positions and meshing where the red vector represents the x-axis and the green vector represents the y-axis

flux lines of the 8/6 SRM for both conjunction and opposition positions. It is necessary for the mesh size near the airgap to be finer than that of the adjacent regions to ensure higher accuracy of the results. Figure 3a shows the inductance networks of 8/6 SRM as a function of current and position. According to the results, we observe that there are two factors affecting the inductance change: position and current. For low current values, there is no saturation effect. The saturation effect occurs when currents reach higher levels, primarily observed at the conjunction position where the magnetic field density is lower. Moreover, Fig. 3b illustrates the electromagnetic torque network as a function of position and current.

2.3 *Linearization of the SRM Model*

The set of equations that describe the SRM contains nonlinearities related to the product variables that appear in the voltage equations (Eq. 7) and in the electromagnetic torque expression (Eq. 9). A prevalent approach in the analysis and design of controllers involves linearizing the model [30, 31]. The main idea is to algebraically

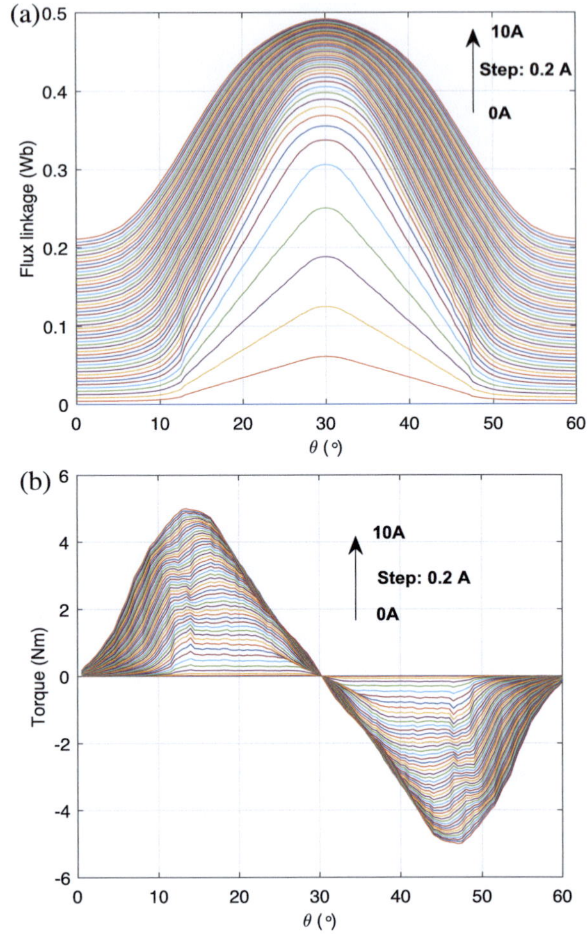

Fig. 3 FEA of 8/6 SRM, **a** flux linkage, **b** torque

transform the dynamics of nonlinear systems into linear ones. On the other hand, the dynamics of nonlinear systems retain all of their properties, as linearization simply transforms this dynamics from a complex form into a simpler one by transforming certain coordinates.

As mentioned previously, the SRM model is highly nonlinear, which complicates the design and synthesis of linear controllers. To ensure the robustness of SRM control, we linearize their model. This approach guides us toward the application of linear control techniques [32, 33].

When the system encounters a small variation around the operating point [32, 33] we have:

$$\begin{cases} i = i_0 + \delta i \\ v = v_0 + \delta v \\ \Omega = \Omega_0 + \delta\Omega \\ T_L = T_{L0} + \delta T_L \end{cases} \tag{12}$$

where the index 0 is used to define the value of the variable at the steady-state operating point, and the index δ represents the incremental deviations of the variable in the vicinity of this point. By substituting Eq. 12 into the set of electrical and mechanical Eqs. 7 and 11 of the SRM, they are reformulated as follows:

$$(v_0 + \delta v) = r(i_0 + \delta i) + L(\theta)\frac{d(i_0 + \delta i)}{dt} + (i_0 + \delta i)\frac{\partial L(\theta)}{\partial\theta}.(\Omega_0 + \delta\Omega) \tag{13}$$

$$J\frac{d(\Omega_0 + \delta\Omega)}{dt} = \frac{1}{2}(i_0 + \delta i)^2\frac{\partial L(\theta, i)}{\partial\theta} - f(\Omega_0 + \delta\Omega) - (T_{L0} + \delta T_L) \tag{14}$$

The linearization of the system consists of eliminating the zero-order terms with respect to the set of perturbation variables. Equations 13 and 14 then become:

$$\delta v = r.\delta i + L(\theta)\frac{d\delta i}{dt} + \delta i\frac{\partial L(\theta)}{\partial\theta}.\Omega_0 + i_0\frac{\partial L(\theta)}{\partial\theta}.\delta\Omega \tag{15}$$

$$J\frac{d\,\delta\Omega}{dt} = i_0\,\delta i\frac{\partial L(\theta, i)}{\partial\theta} - f\delta\Omega - \delta T_L \tag{16}$$

By reformulating Eqs. 15 and 16 as follows:

$$\frac{d\delta i}{dt} = -\frac{r + \frac{\partial L(\theta)}{\partial\theta}.\Omega_0}{L(\theta)}i_0 - \frac{i_0\,\frac{\partial L(\theta)}{\partial\theta}}{L(\theta)}\delta\Omega + \frac{1}{L(\theta)}\delta v \tag{17}$$

$$\frac{d\,\delta\Omega}{dt} = \frac{i_0\,\frac{\partial L(\theta,i)}{\partial\theta}}{J}\delta i - \frac{f}{J}\delta\Omega - \frac{1}{J}\delta T_L \tag{18}$$

By applying the Laplace transform, the last two equations become as follows:

$$s.\delta i = \frac{1}{T_{eq}}\delta i - \frac{i_0\frac{\partial L(\theta)}{\partial\theta}}{L(\theta)}\delta\Omega + \frac{1}{L(\theta)}\delta v \tag{19}$$

$$s.\delta\Omega = \frac{i_0\,\frac{\partial L(\theta,i)}{\partial\theta}}{J}\delta i - \frac{1}{T_m}\delta\Omega - \frac{1}{J}\delta T_L \tag{20}$$

with:

$$\begin{cases} R_{eq} = r + \frac{\partial L(\theta)}{\partial\theta}.\Omega_0 \\ T_{eq} = \frac{1}{R_{eq}} \\ T_m = \frac{J}{f} \end{cases} \tag{21}$$

From Eq. 20, we have:

$$\delta\Omega = \frac{K}{1 + s.T_m}\delta i \tag{22}$$

where

$$K = \frac{i_0 \frac{\partial L(\theta,i)}{\partial\theta} T_m}{J} \tag{23}$$

Then, by substituting the Eq. 22 into the Eq. 19, we obtain :

$$\frac{s^2 + \left(\frac{1}{T_{eq}} + \frac{1}{T_m}\right)s + \left(\frac{(i_0 \frac{\partial L(\theta,i)}{\partial\theta})^2}{J.L(\theta)} + \frac{1}{T_{eq}T_m}\right)}{\frac{1}{T_m}(1 + s.Tm)}\delta i = \frac{1}{L(\theta)}\delta v \tag{24}$$

The transfer function relating the current and voltage of the SRM can be described as follows:

$$\frac{\delta i}{\delta v} = k_2 \frac{(1 + s.Tm)}{s^2 + a.s + b} \tag{25}$$

with:

$$\begin{cases} k_2 = \frac{1}{T_m L(\theta)} \\ a = \left(\frac{1}{T_{eq}} + \frac{1}{T_m}\right) \\ b = \left(\frac{(i_0 \frac{\partial L(\theta,i)}{\partial\theta})^2}{J.L(\theta)} + \frac{1}{T_{eq}T_m}\right) \end{cases} \tag{26}$$

3 Speed Control

Figure 4 shows the speed control structure. It includes a speed controller that compares the rotational speed reference with the measured speed of the SRM. For this control strategy, several speed controllers are proposed to improve the transmission system's efficiency, focusing on reducing torque ripple, enhancing speed tracking, and minimizing response time. Due to the machine's switched nature, a static converter is necessary for SRMs. The most commonly used topology for SRMs is the Asymmetric Half-Bridge Converter (AHBC) [15]. In each phase, two diodes and two switches can be controlled by this structure. The direct torque control strategy of the SRM is based on the use of hysteresis comparators [58]. These comparators compare the reference torque with the estimated torque of the machine to determine the torque control error.

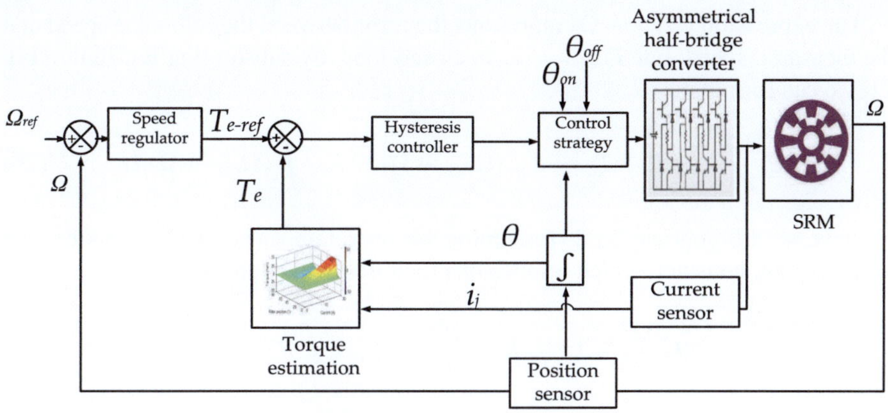

Fig. 4 Speed control strategy of the SRM

3.1 Classical PI Control

In this section, we focus on the design of a speed controller using the classical PI control strategy. The equation of the PI controller is expressed as follows:

$$T_{e-ref} = K_{pv}E(t) + K_{iv} \int E(t)\, d\tau \tag{27}$$

where $E(t)$ represents the error between the speed reference and the measured speed. The constants K_{pv} and K_{iv} correspond to the proportional and integral gains, respectively. The functional diagram of the closed-loop PI speed controller is shown in Fig. 5.

The control signal $u_v(t)$ in Eq. 27 corresponds to the reference torque of the machine T_{e-ref}, which can be formulated as follows:

$$T_{e-ref} = K_{pv}(\Omega_{ref} - \Omega) + K_{iv} \int (\Omega_{ref} - \Omega) d\tau \tag{28}$$

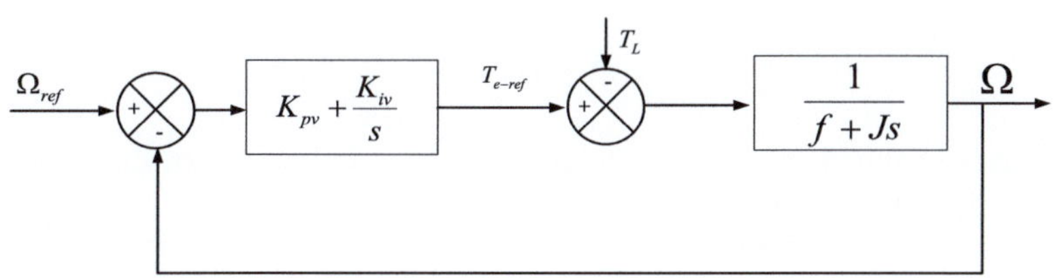

Fig. 5 Functional diagram of the closed-loop PI speed controller

The expression $(\Omega_{ref} - \Omega)$ represents the error between the reference speed and the measured speed. For $T_{e-ref} = T_e$ in closed-loop, by substituting Eq. 28 into Eq. 11, we can formulate:

$$J\frac{d\Omega}{dt} + f\Omega + T_L = K_{pv}(\Omega_{ref} - \Omega) + K_{iv}\int(\Omega_{ref} - \Omega)d\tau \tag{29}$$

Next, we differentiate Eq. 29 (assuming that the reference speed Ω_{ref} and the load torque T_L are constant, which implies that their derivatives are zero):

$$J\frac{d\Omega^2}{dt} + (f + K_{pv})\frac{d\Omega}{dt} + K_{iv}\Omega = K_{iv}\Omega_{ref} \tag{30}$$

Equation 30 can be reformulated as follows:

$$\frac{d\Omega^2}{dt} = -\frac{(f + K_{pv})}{J}\frac{d\Omega}{dt} - \frac{K_{iv}}{J}\Omega + \frac{K_{iv}\Omega_{ref}}{J} \tag{31}$$

A representation of the closed-loop state of Eq. 31 is described as follows:

$$\begin{bmatrix} \frac{d\Omega}{dt} \\ \frac{d\Omega^2}{dt} \end{bmatrix} = \begin{bmatrix} 0 & 1 \\ -\frac{K_{iv}}{J} & -\frac{f+K_{pv}}{J} \end{bmatrix}\begin{bmatrix} \Omega \\ \frac{d\Omega}{dt} \end{bmatrix} + \begin{bmatrix} 0 \\ \frac{K_{iv}}{J} \end{bmatrix}\Omega_{ref} \tag{32}$$

Equation 32 can be expressed alternatively as follows:

$$\begin{cases} \dot{x} = Ax + Bu \\ y = Cu \end{cases} \tag{33}$$

with $\dot{x} = \frac{d^2\Omega}{dt}$, $x = \frac{d\Omega}{dt}$, $u = \Omega_{ref}$,et $y = \Omega$.

and

$$A = \begin{bmatrix} 0 & 1 \\ -\frac{K_{iv}}{J} & -\frac{f+K_{pv}}{J} \end{bmatrix}$$

$$B = \begin{bmatrix} 0 \\ \frac{K_{iv}}{J} \end{bmatrix}$$

$$C = [1 \quad 0]$$

The transfer function $G(s)$ that relates the input and output of the closed-loop system is expressed as follows:

$$G(s) = \frac{\frac{K_{iv}}{J}}{s^2 + \left(\frac{f+K_{pv}}{J}\right)s + \frac{K_{iv}}{J}} \tag{34}$$

Where s is the Laplace transformation.
The closed-loop transfer function is represented by the denominator of a standard second-order system.

$$s^2 + \left(\frac{f + K_{pv}}{J}\right)s + \frac{K_{iv}}{J} = s^2 + 2\omega_n \xi s + {\omega_n}^2 \tag{35}$$

The PI speed controller coefficients can be determined by identification using the following equations:

$$\begin{cases} \frac{f+K_{pv}}{J} = 2\omega_n \xi \\ \frac{K_{iv}}{J} = {\omega_n}^2 \end{cases} \tag{36}$$

From Eq. 36, the parameters K_{pv} and K_{iv} of the speed controller are:

$$\begin{cases} K_{pv} = 2J\omega_n \xi - f \\ K_{iv} = J{\omega_n}^2 \end{cases} \tag{37}$$

In order to optimize the system's dynamic performance and stability, we choose a closed-loop ξ damping coefficient equal to 0.7. This value ensures a good compromise between controller response speed and system stability.

3.2 *Sliding Mode Control*

3.2.1 General Principles of a Sliding Mode Controller

The Sliding Mode Control (SMC) is a non-linear control with a variable structure, introduced in the 1950 [34]. This control is characterized by its fast dynamic response, excellent performance, and robustness to parametric variations and external disturbances [34]. Indeed, this approach is based on altering the functions of state variables, thus creating a specific surface in the state space, referred to as the sliding surface.

The SMC consists of two successive phases. The first is the convergence phase, during which the system transitions from an initial state to the sliding surface $s(x) = 0$. The second phase is the sliding phase, during which the system evolves along this surface to achieve the desired state, as illustrated in Fig. 6 [35].

The synthesis of SMC is carried out in three main steps:

- Choose the sliding surface.
- Establish the convergence conditions.
- Determine the control law.

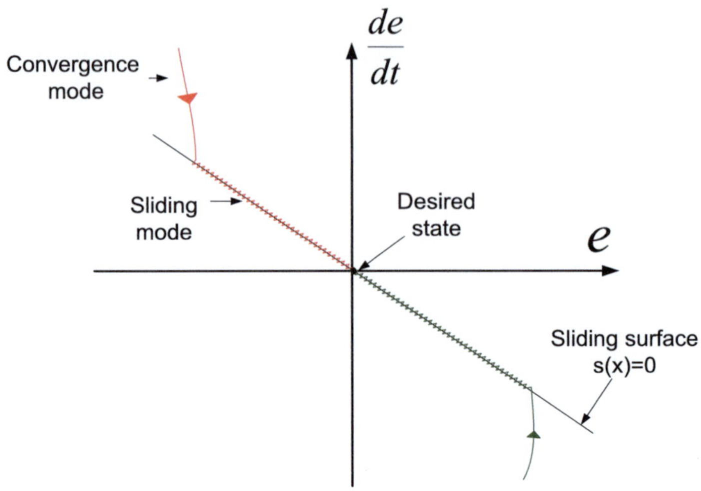

Fig. 6 SMC operating principle

3.2.2 Choice of Sliding Surface

In general, the state representation of a nonlinear system is determined by the system dynamics and desired performance, which is described by:

$$\dot{x}(t) = A(x, t).x(t) + B(x, t).u(t) \tag{38}$$

where A, B are non-linear functions, u is the control and x is the system state variable. The general formula of the sliding surface that ensures the convergence of a variable to its desired value is expressed by the following expression [36]:

$$s(x) = \left(\frac{\partial}{\partial t} + \lambda\right)^{n-1}.e(x) \tag{39}$$

with

- λ is a positive constant
- n is the relative degree of the system
- $e(x)$ is the tracking error determined by:

$$e(x) = x_{ref} - x \tag{40}$$

3.2.3 Establishing Convergence Conditions

To ensure the attractiveness of the chosen surface and to keep the state variable on this surface independently of perturbations, it is essential to impose additional conditions called sliding mode existence conditions. Indeed, these existence conditions are defined using Lyapunov's stability theory. We consider the following Lyapunov function to guarantee the stability of the system:

$$V(x) = \frac{1}{2}s(x)^2 \tag{41}$$

An attractiveness condition is guaranteed if the derivative of the chosen Lyapunov function is negative along the system's trajectories.

$$\dot{V}(x) = \dot{s}(x)s(x) < 0 \tag{42}$$

3.2.4 Determination of the Control Law

The structure of the SMC consists of two parts: the first is a continuous control ("equivalent control"), and the second is a discontinuous control. The equivalent control helps regulate the state vector's trajectory on the sliding mode surface. The expression of this control is determined by calculating the derivative of the sliding surface and assuming that:

$$s(x) = \dot{s}(x) = 0 \tag{43}$$

The SMC approach requires a control law U to drive the state variable of the system to the selected sliding surface while maintaining the convergence conditions. The SMC law is given by :

$$u = u_{eq} + u_d \tag{44}$$

where u_{eq} is the equivalent control and u_d is the discontinuous control. The attractiveness condition is guaranteed by the discontinuous control, which is defined by the following equation:

$$u_d = K sign(s) \tag{45}$$

where K is a negative constant and $sign(s)$ is the sign switching function, defined as follows:

$$sign(s) = \begin{cases} 1 & s > 0 \\ 0 & s = 0 \\ -1 & s < 0 \end{cases} \tag{46}$$

Fig. 7 illustrates the SM control law diagram.

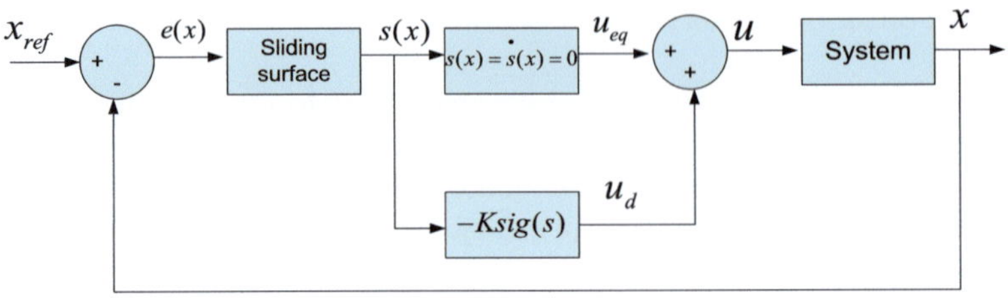

Fig. 7 Diagram of SM control law

3.2.5 Synthesis of SM Control

To maintain consistency in the comparison, the design of the SMC is justified as a control law equivalent to a PI controller. The choice of the sliding surface depends on the speed error, denoted as e_Ω, which is defined as follows;

$$s(t) = e_\Omega(t) + \lambda \int_{-\infty}^{t} e_\Omega(\tau)d\tau \tag{47}$$

where λ represents a positive constant. This selection ensures that the tracking error converges exponentially to zero. In other words, when $s = 0$, then $e_\Omega = 0$.

The mechanical equation of the SRM model can be rewritten as follows:

$$\dot{\Omega} = \frac{1}{J}T_e - \frac{f}{J}\Omega - \frac{1}{J}T_L = au - b\Omega - c \tag{48}$$

with: $a = 1/J, b = f/J, c = T_L/J$ et $u = T_e$.
The derivative of the Eq. 47 can be calculated as follows:

$$\begin{aligned} \dot{s}(t) &= \dot{e}_\Omega + \lambda e_\Omega \\ &= \dot{\Omega}_{ref} - \dot{\Omega} + \lambda e_\Omega \\ &= -au + b\Omega + c + \lambda e_\Omega \end{aligned} \tag{49}$$

To determine the equivalent control (u_{eq}), in our case the Eq. 49 reduces to $\dot{s} = 0$. We replace Eq. 48 in Eq. 49 and assume that $u = u_{eq}$.

$$u_{eq} = \frac{1}{a}\left[b\Omega + c + \lambda e_\Omega\right] \tag{50}$$

Finally, the SM control results from the combination of the two laws, i.e. the equivalent law and the discontinuous law.

$$T_{e-ref} = u = u_{eq} + u_d = \frac{1}{a}\left[b\Omega + c + \lambda e_\Omega\right] + K sign(s) \quad (51)$$

From the attractivity condition expressed in Eq. 42, we verify the stability condition by substituting the control law u into Eq. 42, we obtain the following equation:

$$s\dot{s} = s\left(-a\left[\frac{1}{a}\left[b\Omega + c + \lambda e_\Omega\right] + K sign(s)\right] + b\Omega + c + \lambda e_\Omega\right) = -asK sign(s) < 0 \quad (52)$$

3.3 *Fuzzy Logic Speed Control*

Artificial intelligence techniques are currently being used to address problems related to control and parameter estimation of electrical systems through conventional techniques. Fuzzy logic is one of the artificial intelligence techniques and is widely used for controlling nonlinear electrical systems. The principle of this technique resembles human reasoning, as the variables dealt with are linguistic variables rather than logical ones, closely related to natural language. These linguistic variables are manipulated using rules based on an understanding of the system's behavior. Fuzzy logic develops a set of fundamental concepts that serve to explain and support various fundamental principles.

3.3.1 General Principles of a Fuzzy Controller

Fuzzy logic control consists of four main parts: fuzzification, the rule base, inference, and defuzzification [37]. The flowchart detailing the operation of the fuzzy controller, summarizing the interaction between these different parts, is illustrated in Fig. 8.

- **Fuzzification**: This first part translates a numerical input, called the input variable, into a linguistic value using a membership function created by the fuzzy system designer. A membership function defines the degree of membership of a numerical value to a linguistic variable.
- **Rule Base**: This part contains the knowledge base for the targeted application area as well as the expected control outcome. It is composed of a database necessary to establish and manipulate the fuzzy control rules.
- **Inference Engine**: This is the core of fuzzy control as it emulates human decision-making processes and deduces control actions from the fuzzy implication rules. Inference is performed using various techniques, including the min-max inference method, the max-product inference method, and the sum-product inference method. The result is a control law formulated in terms of membership functions.

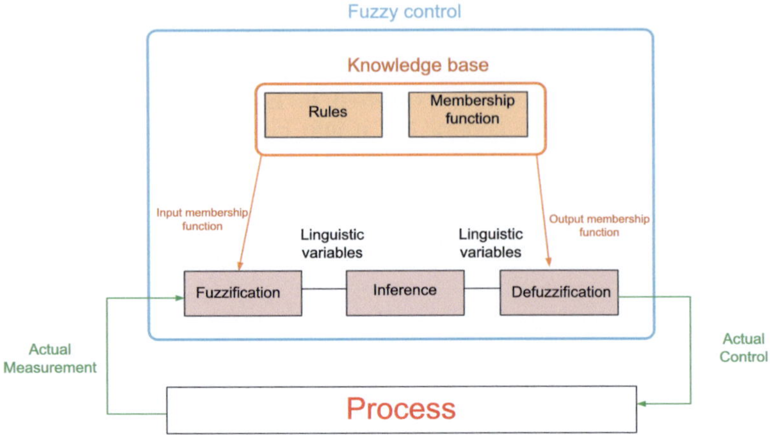

Fig. 8 Fuzzy control block diagram.

- **Defuzzification**: This final step involves merging and transforming the parameters from the previous step into numerical data to obtain a non-fuzzy control signal, as the system being controlled only receives deterministic values. Several methods exist for defuzzification. Among the most widely used are the method of the mean of the maxima and the centroid method.

Fuzzy control converts the deterministic input values into fuzzy values, processes them using fuzzy rules, and reconverts the control signal from fuzzy values back into deterministic values.

3.3.2 Fuzzy Regulation Synthesis for SRM Speed Control

To implement the fuzzy logic speed control algorithm for the SRM, the mechanical speed is measured using a mechanical sensor and then compared to the desired reference value. The error signal e_Ω and its derivative $\frac{de_\Omega}{dt}$ are used as inputs for fuzzy processing, as shown in Fig. 9. The output of the fuzzy controller generates the reference for the electromagnetic torque of the SRM.

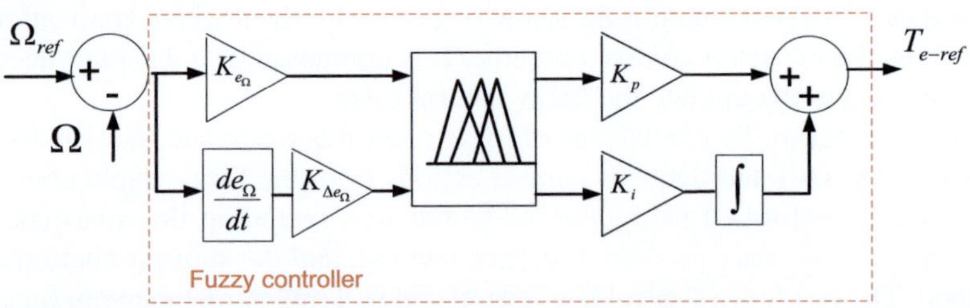

Fig. 9 Structure of the fuzzy-PI speed controller

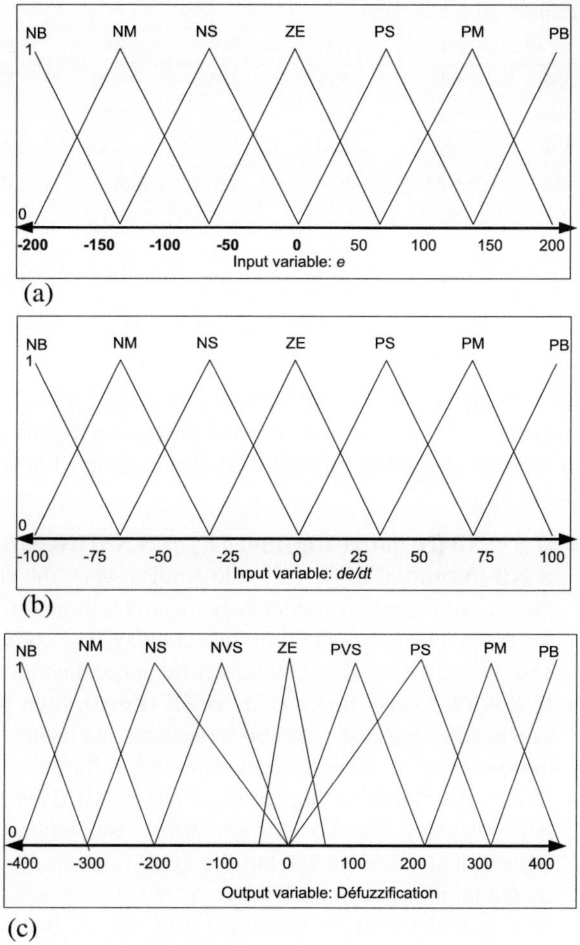

Fig. 10 Membership functions of the input and output variables of the fuzzy controller: **a** speed error, **b** derivative of the error, and **c** reference torque

Fuzzification strategies:
Figure 10 illustrates the membership functions of the input and output variables of the studied system. Triangular membership functions were chosen to represent the input and output variables. The speed error consists of seven fuzzy sets, namely Negative Big (NB), Negative Medium (NM), Negative Small (NS), Zero (ZE), Positive Small(PS), Positive Medium (PM), and Positive Big (PB). Similarly, the output variable of the inference system consists of nine fuzzy sets, namely Negative Big (NB), Negative Medium (NM), Negative Small (NS), Negative Very Small (NVS), Zero (ZE), Positive Very Small (PVS), Positive Small (PS), Positive Medium (PM), and Positive Big(PB).

Rule Base:
The number of rules and fuzzy sets is not exhaustive. It depends on the desired precision. The various possible combinations of seven fuzzy sets for the error and

Table 3 Rule base table.

E / dEdt	NB	NM	NS	ZE	PB	PM	PS
NB	NB	NB	NB	NM	NS	NVS	ZE
NM	NB	NB	NM	NS	NVS	ZE	PVS
NS	NB	NM	NS	NVS	ZE	PVS	PS
ZE	NM	NS	NVS	ZE	PVS	PS	PM
PB	NS	NVS	ZE	PVS	PS	PM	PB
PM	NVS	ZE	PVS	PS	PM	PB	PB
PS	ZE	PVS	PS	PM	PB	PB	PB

seven fuzzy sets for the derivative of the error form 49 rules in the inference system's base. This base reflects the knowledge acquired by the operator managing the process to be controlled. The linguistic control rules used in this work are shown in Table 3. For example:

- **If E is NB (Negative Big) and dE/dt is NM (Negative Medium), then the output is NB (Negative Big):** This rule applies when the error is large and negative, and the rate of change is moderately negative, indicating that the system is far below the desired state and moving further away. The control action in this case is strong and corrective (NB) to counteract the large deviation.
- **If E is ZE (Zero) and dE/dt is ZE (Zero), then the output is ZE (Zero)**: This rule handles the case when both the error and its derivative are negligible, meaning the system is near the desired state with minimal variation.
- **If E is PB (Positive Big) and dE/dt is NB (Negative Big), then the output is NS (Negative Small)**: This situation arises when the error is large and positive (above the desired state), but the system is rapidly correcting itself, as indicated by the large negative derivative.
- **If E is PM (Positive Medium) and dE/dt is NS (Negative Small), then the output is PVS (Positive Very Small)**: Here, the error is moderately positive, and the system is slowly correcting itself. The control action (PVS) is minimal to ensure smooth convergence to the desired state without overshooting.

This systematic approach ensures that each rule corresponds to a specific combination of E and dE/dt, providing an intuitive and predictable response. The symmetry in the rule base ensures that:

1. **Strong corrective actions** (e.g., NB or PB outputs) are applied when the error is large, and the derivative indicates further divergence.
2. **Mild corrective actions** (e.g., NS, PS outputs) are applied when the error is moderate, and the derivative indicates slight changes.
3. **Neutral actions** (e.g., ZE outputs) are applied when the error and its derivative suggest that the system is near stability.

Rule Inference: The $min - max$ inference method was selected to ensure the numerical processing of inference rules. It implements the AND operator using the

min function, the $THEN$ part of each inference rule using the min function, and links between the rules are handled using the OR operator via the max function.

Defuzzification: Defuzzification involves converting the fuzzy set into physically exploitable numerical variables. The center of gravity method was adopted to output the parameters K_p and K_I from the fuzzy controller.

4 Simulation Results

To illustrate the operation of the SRM, in this section, we perform a simulation using Matlab/Simulink for the DITC control of the SRM. It is worth noting that the SRM is powered by an asymmetric half-bridge converter, controlled by hysteresis. For the simulation, we have chosen four operating points ranging from 500 rpm to 1000 rpm, as illustrated in Fig. 11a. Additionally, we select a variable load torque ranging from 0 to 0.5 Nm, as shown in Fig. 11b.

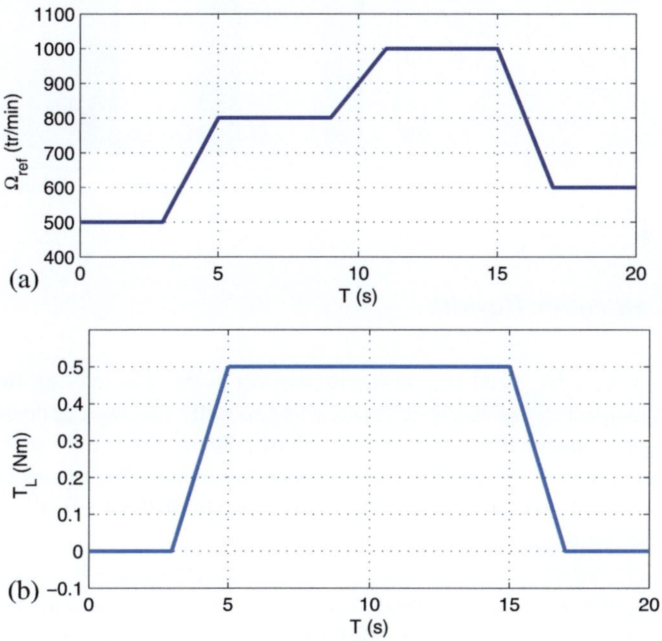

Fig. 11 Variation scenario: **a** load torque, **b** reference rotation speed

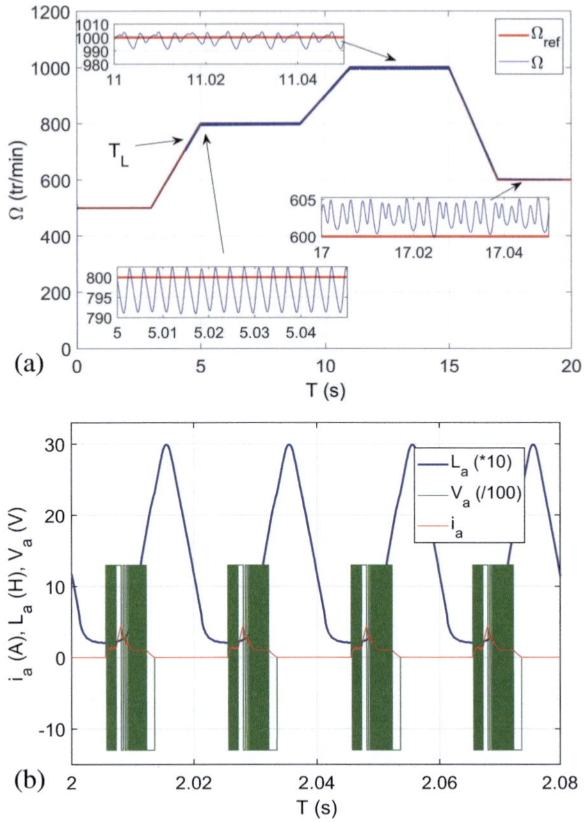

Fig. 12 **a** Speed response by the PI controller and **b** supply voltage, current, and inductance for one phase of the SRM with the switching angle $\theta_{off} = 20°$

4.1 PI Controller Results

The Fig. 12a shows the speed response with respect to the selected trajectory with PI controller. We observe that the speed response generally follows the desired profile, although ripples appear in each operating phase. These ripples are mainly caused by the nonlinearities of the model. Furthermore, we ensure that the speed response reaches the desired profile without significant overshoot thanks to the parameters chosen for the regulator.

Additionally, when the load torque varies, the speed ripples become more pronounced compared to those observed in steady-state conditions. This is due to the variation in load conditions and the additional constraints exerted on the machine, which can temporarily affect the stability of the speed response. Figure 12b illustrates the application of a 130 V supply voltage to the SRM. The SRM is powered according to a specific sequence that determines the operation of the machine's phases. Figure 13a shows the temporal evolution of the four stator currents of the SRM based on the selected speed profile.

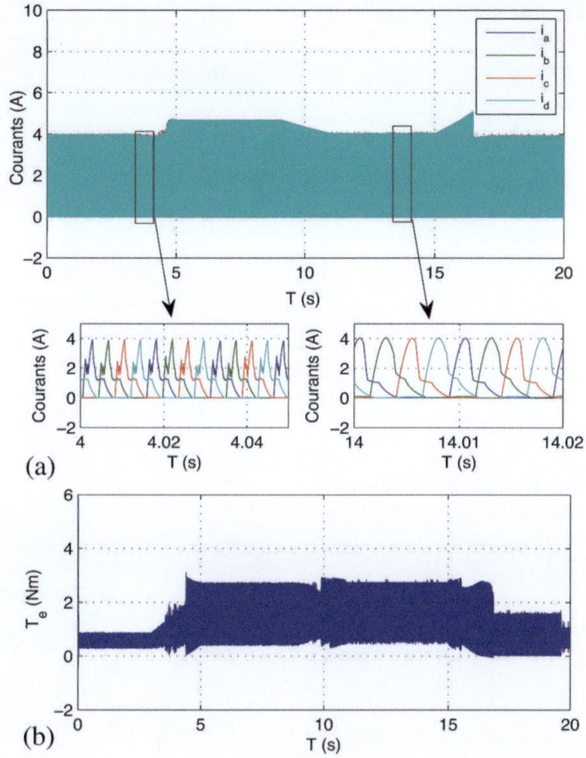

Fig. 13 **a** Current response with PI regulator and **b** total torque response with PI regulator

The currents are displayed for each phase of the machine. The zoomed-in views on this figure detail the sequential evolution of the currents, highlighting their phase shift and similar waveform. We observe that the stator currents are well-synchronized and phase-shifted relative to one another, which is essential for ensuring the proper operation of the SRM. This synchronization enables the creation of a rotating magnetic flux within the machine, thereby contributing to the generation of the torque necessary for mechanical motion. Figure 13b illustrates the evolution of the total torque generated by the SRM, resulting from the sum of the torques produced by each phase of the machine. It can be observed that the torque exhibits more significant ripples at high speeds. These fluctuations are mainly caused by the non-linearity of the SRM model and the hysteresis control. Indeed, these ripples are one of the major drawbacks of SRMs.

4.2 SMC Results

Maintaining the same speed profile and mechanical load, we conducted simulations using the previously developed speed controller. Figure 14 shows the speed curve

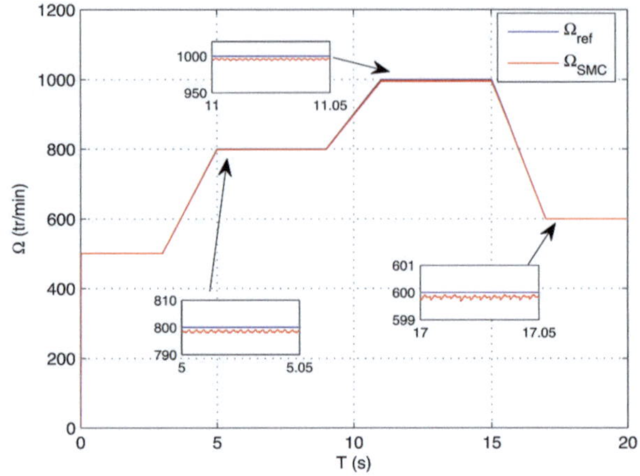

Fig. 14 SMC speed control response

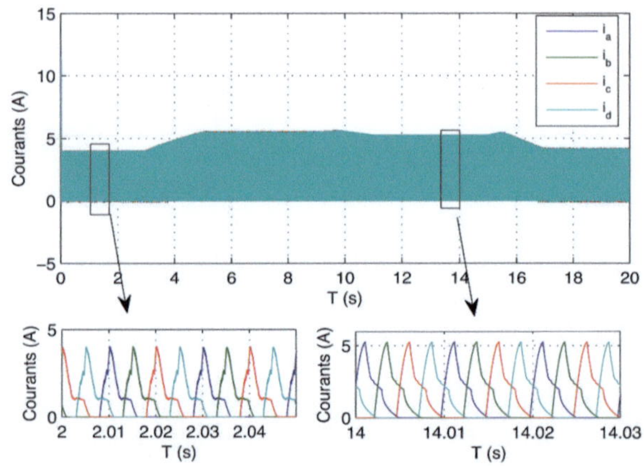

Fig. 15 Current response with SMC

of the SRM when the SMC controller was applied. A reduction in the steady-state error compared to that obtained with a conventional controller is observed, resulting in improved performance and robustness of the studied system, thereby enabling optimal tracking of reference speed fluctuations. The stator current responses shown in Fig. 15 are clearly improved by the SMC controller when the load torque varies. Regarding the torque response, Fig. 16 shows a significant reduction in torque ripples and tracking error for different speeds. These ripples have been considerably reduced compared to those obtained with the conventional PI controller.

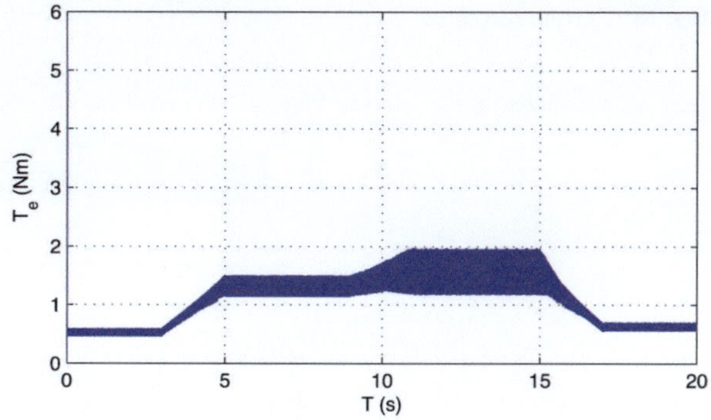

Fig. 16 Torque response with SMC

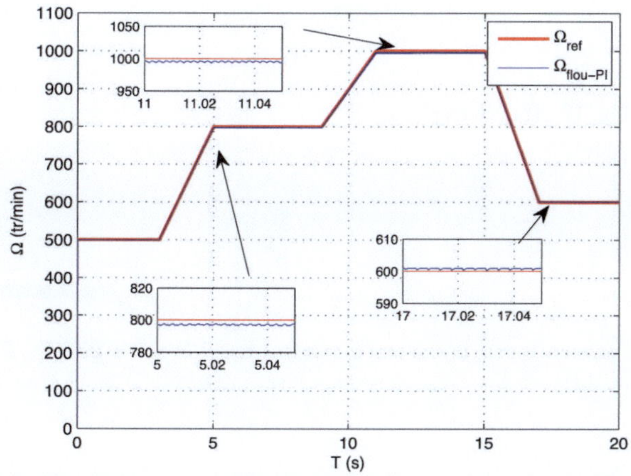

Fig. 17 Fuzzy-PI speed control response

4.3 *Fuzzy-PI Controller*

The fuzzy-PI-based controller we developed was applied to regulate the speed of the SRM. The speed and load torque reference profiles used are the same as those used for the PI and SMC controllers. Figure 17 shows the speed response, demonstrating excellent steady-state performance for the four operating points [500, 600, 800, and 1000] (rpm), with negligible ripple.

Moreover, regarding the transient behavior, the error is virtually nonexistent across the entire operating range, which is a significant improvement compared to the error observed when using the PI controller. The evolution of the stator currents for the four phases is shown in Fig. 18. The associated zooms in this figure reveal that the current ripples are more pronounced during the transient state due to the strong non-linearity of the system. Figure 19 illustrates the total torque response generated by the SRM using the FLC. We demonstrate that the torque fluctuations in steady-state at different speeds are lower compared to those obtained with the PI controller. This

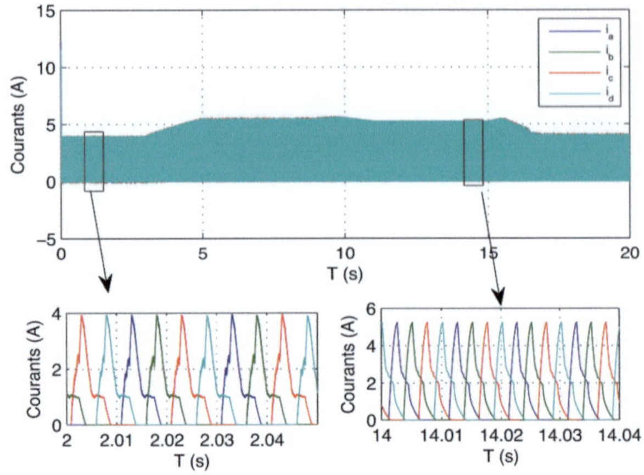

Fig. 18 Current response with FLC

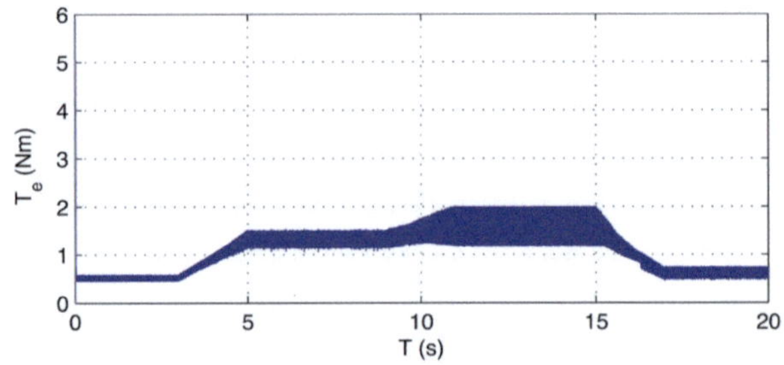

Fig. 19 Torque response with FLC

showcases the effectiveness of the fuzzy logic speed controller in reducing torque fluctuations and improving the SRM performance at various speeds.

4.4 Comparative Study

A comparative study was conducted to evaluate the performance of each controller presented earlier. Figure 20 shows the speed response of the three controllers for a speed profile varying from 0 to 800 rpm, with a constant mechanical load of 0.5 Nm applied at t=1.5 s. The zoom associated with this figure confirms the robustness of the nonlinear controllers, whether using sliding mode control or fuzzy-PI control, in the face of sudden changes in mechanical load torque compared to the performance of the conventional PI controller.

Figure 21 shows the total torque response for the three controllers. We observe that the torque fluctuations are significantly reduced in steady-state when using nonlinear controllers. To evaluate these ripples, we determined the ripple rate in steady-state using the following formula:

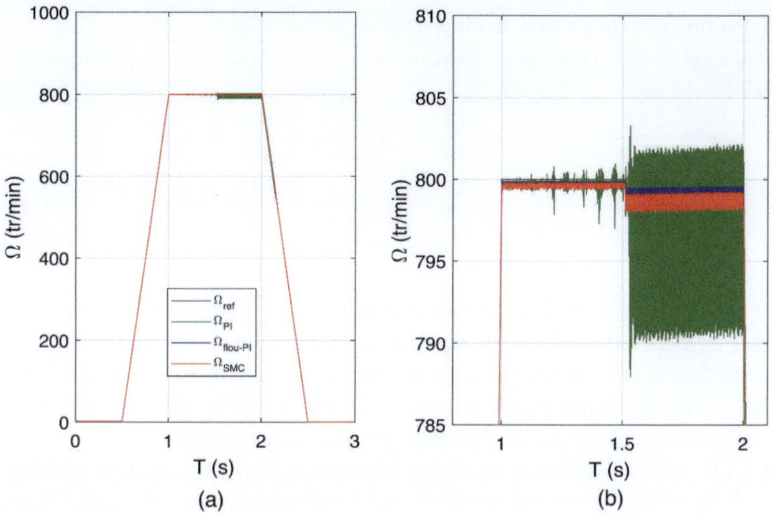

Fig. 20 **a** Speed response for all three controls and **b** zooming in when torque load varies

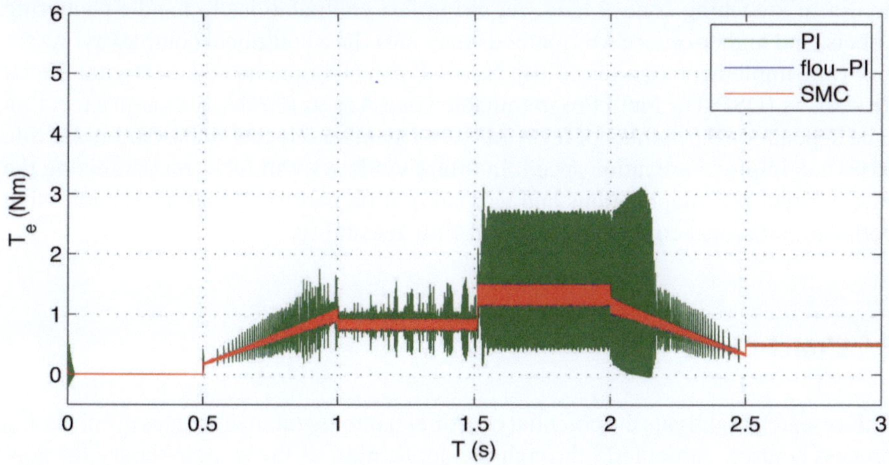

Fig. 21 Total torque with three controls

$$\Delta T_e(\%) = \frac{T_{e-\max} - T_{e-\min}}{T_{e-\text{moy}}} \tag{53}$$

where $T_{e-\min}$, $T_{e-\max}$ and $T_{e-\max}$ are the minimum, maximum and average torque values respectively

We note an increase in the torque ripple rate when the load torque is applied at t = 1.5 s compared to the no-load response. Specifically, the PI control caused an increase in the torque ripple rate from 34% in no-load to 85% under load. However, this ripple

Table 4 Performance comparison for different types of controller

Control method (with load)	Speed error (%)	RMS error	Response time (s)	Torque ripple (%)
PI	9.6	4.8	0.036	85
SMC	1.9	3.5	0.024	25
Fuzzy-PI	1.3	2.2	0.017	24

rate decreases further when nonlinear controllers are used. For example, the SMC controller reduces the load torque ripple rate to 25%, while the Fuzzy logic reduces it further to 24%. This demonstrates that nonlinear controllers, especially intelligent ones, help reduce torque ripple compared to the PI controller. Table 4 presents a comparative study of the results obtained by the three controls.

The FLC offers a balance between robustness and adaptability, making it an effective alternative to conventional PI and SMC approaches. However, it requires higher computational resources compared to PI control, which is computationally simpler and easier to implement in real-time. On the other hand, SMC involves nonlinear switching control laws, requiring fast computations to handle chattering effects, and higher-order SMC methods may introduce additional complexity. As for real-time implementation feasibility, the FLC can be implemented on Digital Signal Processors (DSPs) or Field Programmable Gate Arrays (FPGAs), though execution time depends on the number of fuzzy rules and membership functions. Optimized rule bases can improve execution speed. In future work, we will focus on optimizing the FLC for real-time applications and validating it on hardware platforms, addressing both computational complexity and real-time feasibility.

5 Conclusion

This research highlights the potential of artificial intelligence techniques in enhancing process control, particularly through the application of fuzzy logic theory for control of SRM drives. The chapter presents a self contained introduction to linear and nonlinear feedback control design for SRM. A comprehensive performance analysis of SRM drives is presented across a wide range of operating conditions, including load perturbation dynamics. Furthermore, the performance indices of the SRM drive system utilizing a FLC are compared with a conventional controller, highlighting the advantages of the proposed approach. The reference speed is consistently maintained across various load conditions without overshoot. Future research directions include the experimental validation of the proposed control system on real hardware to assess its practical feasibility and performance under real-world operating conditions. Additionally, exploring alternative optimization techniques could further enhance system efficiency, robustness, and adaptability.

References

1. Touati, Z., Mahmoud, I., Khedher, A.: Nonlinear modelling of switched reluctance machine. In: 2019 International Conference on Signal, Control and Communication (SCC), pp. 256–261. IEEE (2019). https://doi.org/10.1109/scc47175.2019.9116143
2. dos Santos Barros, T.A., dos Santos Neto, P.J., Nascimento Filho, P.S., Moreira, A.B., Ruppert Filho, E.: An approach for switched reluctance generator in a wind generation system with a wide range of operation speed. IEEE Trans. Power Electronics **32**(11), 8277–8292 (2017). https://doi.org/10.1109/tpel.2017.2697822
3. Xue, X., Cheng, K.W.E., Ho, S.L.: Correlation of modeling techniques and power factor for switched-reluctance machines drives. IEE Proc.-Electric Power Appl. **152**(3), 710–722 (2005). https://doi.org/10.1049/ip-epa:20049005
4. Touati, Z., Mahmoud, I., Khedher, A.: Modelling and electromagnetic characteristics for two mode functions of switched reluctance machine. In: 2019 International Conference on Signal, Control and Communication (SCC), pp. 114–120. IEEE (2019). https://doi.org/10.1109/scc47175.2019.9116181
5. Touati, Z., Mahmoud, I., Araújo, R.E., Khedher, A.: Fuzzy super-twisting sliding mode controller for switched reluctance wind power generator in low-voltage DC microgrid applications. Energies **17**(6), 1416 (2024). https://doi.org/10.3390/en17061416
6. Touati, Z., Araújo, R.E., Mahmoud, I., Khedher, A.: Analysis of skewing effects on radial force for different topologies of switched reluctance machines: 6/4 srm, 8/6 srm, and 12/8 srm. U Porto J. Eng. **9**(1), 55–71 (2023). https://doi.org/10.24840/2183-6493-009-001-001134
7. Li, S., Zhang, S., Habetler, T.G., Harley, R.G.: Modeling, design optimization, and applications of switched reluctance machines—a review. IEEE Trans. Ind. Appl. **55**(3), 2660–2681 (2019). https://doi.org/10.1109/tia.2019.2897965
8. Yan, W., Chen, H., Liao, S., Liu, Y., Cheng, H.: Design of a low-ripple double-modular-stator switched reluctance machine for electric vehicle applications. IEEE Trans. Transp. Electrification **7**(3), 1349–1358 (2021). https://doi.org/10.1109/tte.2021.3057133
9. Sun, X., Wu, J., Wang, S., Diao, K., Yang, Z.: Analysis of torque ripple and fault-tolerant capability for a 16/10 segmented switched reluctance motor in hev applications. COMPEL Int. J. Comput. Math. Electrical Electronic Eng. (2019). https://doi.org/10.1108/compel-11-2018-0477
10. Zarchi, H.A., Soltani, J., Markadeh, G.A.: Adaptive input—output feedback-linearization-based torque control of synchronous reluctance motor without mechanical sensor. IEEE Trans. Ind. Electronics **57**(1), 375–384 (2009). https://doi.org/10.1109/tie.2009.2026379
11. Cheng, H., Chen, H., Xu, S., Yang, S.: Adaptive variable angle control in switched reluctance motor drives for electric vehicle applications. J. Power Electronics **17**(6), 1512–1522 (2017). https://doi.org/10.6113/JPE.2017.17.6.1512
12. Mukhopadhyay, S., Mansani, S., Sekaran, S.: Torque ripple minimization and speed control of switched reluctance motor employing model predictive controller. In: Electrical engineeringm pp. 1–20 (2024). https://doi.org/10.1007/s00202-024-02425-5
13. Labiod, C., Srairi, K., Mahdad, B., Dib, A., Benchouia, M.T., Benbouzid, M.E.H.: Optimum performances for non-linear finite elements model of 8/6 switched reluctance motor based on intelligent routing algorithms. Adv. Electrical Electronic Eng. **15**(1), 1–11 (2017). https://doi.org/10.15598/aeee.v15i1.1906
14. Labiod, C., Srairi, K., Mahdad, B., Benchouia, M.T., Benbouzid, M.: Speed control of 8/6 switched reluctance motor with torque ripple reduction taking into account magnetic saturation effects. Energy Proc. **74**, 112–121 (2015). https://doi.org/10.1016/j.egypro.2015.07.530
15. Touati, Z., Pereira, M., Araújo, R.E., Khedher, A.: Comparative study of discrete pi and pr controller implemented in SRG for wind energy application: theory and experimentation. Electronics **11**(8), 1285 (2022). https://doi.org/10.3390/electronics11081285
16. Touati, Z., Pereira, M., Araújo, R.E., Khedher, A.: Improvement of steady state performance of voltage control in switched reluctance generator: Experimental validation. Machines **10**(2), 103 (2022). https://doi.org/10.3390/machines10020103

17. Kotb, H., Yakout, A.H., Attia, M.A., Turky, R.A., AboRas, K.M.: Speed control and torque ripple minimization of SRM using local unimodal sampling and spotted hyena algorithms based cascaded pid controller. Ain Shams Eng. J. **13**(4), 101719 (2022). https://doi.org/10.1016/j.asej.2022.101719
18. Vas, P.: Artificial-Intelligence-Based Electrical Machines and Drives: Application of Fuzzy, Neural, Fuzzy-Neural, and Genetic-Algorithm-Based Techniques, vol. 45. Oxford University Press (1999)
19. Zhao, Y., Collins, E.G.: Fuzzy pi control design for an industrial weigh belt feeder. IEEE Trans. Fuzzy Syst. **11**(3), 311–319 (2003). https://doi.org/10.1109/tfuzz.2003.812686
20. Uddin, M.N., Rahman, M.A.: High-speed control of ipmsm drives using improved fuzzy logic algorithms. IEEE Trans. Ind. Electronics **54**(1), 190–199 (2007). https://doi.org/10.1109/tie.2006.888781
21. Fnaiech, M.A., Betin, F., Capolino, G.A., Fnaiech, F.: Fuzzy logic and sliding-mode controls applied to six-phase induction machine with open phases. IEEE Trans. Ind. Electronics **57**(1), 354–364 (2009). https://doi.org/10.1109/tie.2009.2034285
22. Cheok, A.D., Ertugrul, N.: High robustness and reliability of fuzzy logic based position estimation for sensorless switched reluctance motor drives. IEEE Trans. Power Electronics **15**(2), 319–334 (2000). https://doi.org/10.1109/63.838105
23. Orlowska-Kowalska, T., Dybkowski, M., Szabat, K.: Adaptive sliding-mode neuro-fuzzy control of the two-mass induction motor drive without mechanical sensors. IEEE Transactions on Industrial Electronics **57**(2), 553–564 (2009). https://doi.org/10.1109/tie.2009.2036023
24. Mir, S., Elbuluk, M.E., Husain, I.: Torque-ripple minimization in switched reluctance motors using adaptive fuzzy control. IEEE Trans. Ind. Appl. **35**(2), 461–468 (1999). https://doi.org/10.1109/28.753642
25. Perry, A.G., Feng, G., Liu, Y.F., Sen, P.C.: A design method for pi-like fuzzy logic controllers for DC–DC converter. IEEE Trans. Ind. Electronics **54**(5), 2688–2696 (2007). https://doi.org/10.1109/tie.2007.899858
26. Koblara, T.: Implementation of speed controller for switched reluctance motor drive using fuzzy logic. In: 2008 11th International Conference on Optimization of Electrical and Electronic Equipment, pp. 101–105. IEEE (2008). https://doi.org/10.1109/optim.2008.4602464
27. Xiu, J., Xia, C.: An application of adaptive fuzzy logic controller for switched reluctance motor drive. In: Fourth International Conference on Fuzzy Systems and Knowledge Discovery (FSKD 2007), vol. 4, pp. 154–158. IEEE (2007). https://doi.org/10.1109/fskd.2007.141
28. Chowdhuri, S., Biswas, S.K., Mukherjee, A.: Performance studies of fuzzy logic based pi-like controller designed for speed control of switched reluctance motor. In: 2006 1st IEEE Conference on Industrial Electronics and Applications, pp. 1–5. IEEE (2006). https://doi.org/10.1109/iciea.2006.257304
29. Husain, I., Hossain, S.A.: Modeling, simulation, and control of switched reluctance motor drives. IEEE Trans. Ind. Electronics **52**(6), 1625–1634 (2005). https://doi.org/10.1109/TIE.2005.858710
30. Hedrick, J.K., Girard, A.: Control of nonlinear dynamic systems: theory and applications. Controllability Observability Nonlinear Syst **48**, 133–160 (2005)
31. Conte, G., Moog, C.H., Perdon, A.M.: Algebraic Methods for Nonlinear Control Systems. Springer Science & Business Media (2007)
32. Krishnan, R.: Switched Reluctance Motor Drives: Modeling, Simulation, Analysis, Design, and Applications. CRC Press (2017)
33. Pereira, M., Araújo, R.E.: Analysis and design of a speed controller for switched reluctance motor drive. U. Porto J. Eng. **5**(1), 46–58 (2019). https://doi.org/10.24840/2183-6493-005
34. Emel'yanov, S.: Theory of variable-structure control systems: inception and initial development. Comput. Math. Modeling **18**(4), 321–331 (2007). https://doi.org/10.1007/s10598-007-0028-6
35. Holkar, K., Waghmare, L.: Sliding mode control with predictive pid sliding surface for improved performance. Int. J. Comput. Appl. **78**(4) (2013). https://doi.org/10.5120/13474-0992

36. Slotine, J.J.E., Li, W., et al.: Applied Nonlinear Control, vol. 199. Prentice Hall, Englewood Cliffs, NJ (1991)
37. Driankov, D., Hellendoorn, H., Reinfrank, M.: An Introduction to Fuzzy Control. Springer Science & Business Media (2013)

Hidden Markov Prediction and Fuzzy Control Obstacle Avoidance for Submarine Robot

Johan M. Ruíz Holguín, Evgeni Magid, and Edgar A. Martínez-García

Abstract This paper introduces a novel obstacle avoidance strategy for an underwater robot, synergistically integrating Hidden Markov Chains (HMC) with a Mamdani Fuzzy inference controller. The HMC forecasts a vector of future states, and the maximum a posteriori state probabilities derived from this prediction directly inform the control actions of the fuzzy system. The proposed fuzzy logic controller employs five input variables: the HMC's predicted robot state (represented as crisp sets), instantaneous linear velocity, underwater depth, and yaw and pitch angular velocities. The fuzzy controller generates three output variables, which directly command the robot's actuators: the propulsion motor, the yawing motor, and the ballasting system, the latter operating via an integrated hydraulic piston mechanism. Furthermore, this work develops dynamics-based models for the robot's propulsion, steering, and ballasting subsystems, complemented by sensor fusion models designed to provide real-time control feedback. The paper also details the robot's platform and its underlying system architecture, engineered to support multi-threaded, real-time control operations. Experimental and simulation results are presented to validate the efficacy of the proposed obstacle avoidance strategy, demonstrating its robustness in challenging underwater locomotion scenarios.

Keywords Underwater robotics · Predictive planning · State-transition control · Collisionless hydrodinamics · Intelligent-locomotion · Dynamics-based model · Sensor-fusion · Real-time operations · Robot system architecture

---

J. M. Ruíz Holguín
Laboratorio de Robotica, Institute of Engineering and Technology, Universidad Autónoma de Ciudad Juarez, Chihuahua, Mexico

E. Magid
Institute of Information Technology and Intelligent Systems, Kazan Federal University, Kazan, Russia
e-mail: magid@it.kfu.ru

HSE University, Moscow, Russian Federation

E. A. Martínez-García (✉)
Laboratorio de Robotica, Institute of Engineering and Technology, Universidad Autonoma de Ciudad Juarez, Chihuahua, Mexico
e-mail: edmartin@uacj.mx

W. Pedrycz et al. (eds.), *Decision-Making in Computational Intelligence-Based Systems*, Studies in Systems, Decision and Control 628,
https://doi.org/10.1007/978-3-032-04056-5_9

1 Introduction

In recent years, the field of mobile robotics has seen significant advancements, particularly in the domain of underwater exploration. Autonomous Underwater Vehicles (AUVs) represent a cutting-edge class of mobile robots, uniquely designed to navigate the complex underwater environment with a wide range of intelligent functionalities. A review of underwater robotic functionalities such as localization, navigation and multiple sensing techniques can be found in [1]. AUVs are distinguished from Remotely Operated Vehicles (ROVs) by their ability to operate independently without direct human intervention. Unlike ROVs, which require a tethered connection to a surface vessel, AUVs are powered by their own onboard energy sources, allowing them greater freedom and versatility in their operations. The impetus for advancing AUV technology primarily stems from the need to collect vital oceanographic data, including measurements of temperature, salinity, and ocean currents. These vehicles play a crucial role in inspecting underwater infrastructures such as pipelines, submarine cables, and oil platforms. Beyond these applications, AUVs are instrumental in generating high-resolution bathymetric maps and aiding in the discovery and exploration of sunken vessels. They are invaluable assets in marine biological research, providing insights into marine life and underwater ecosystems. Additionally, they serve important purposes in surveillance and reconnaissance missions.

The development of UAVs in experimental environments with real-time performance and planning algorithms presents significant challenges. Achieving stability requires addressing critical factors such as real-time pathfinding algorithms, dynamic obstacle avoidance, path optimization, and sufficient processing speed, particularly when contending with hydrodynamic currents. A comprehensive review of these issues was presented by Li et al. [2]. The obstacle avoidance problem has been studied for robotic systems [3] in the presence of time-varying uncertainties with optimal obstacle-free coverage through Voronoi-based algorithm, and Spline-based avoidance navigation [4]. Kot et al. [5] reviewed two decades of research on collision avoidance and path planning for AUVs, comparing simulations with real-world implementations. They found that real-world applications predominantly rely on simpler, reactive behaviors.

This research presents an obstacle avoidance strategy for a robotic submarine that utilizes hydraulic ballasting for vertical motion and two asynchronous propellers for propulsion and yaw control. The primary contribution of this work is the integration of a Hidden Markov Chain (HMC) with a multi-input multi-output fuzzy logic controller (MIMO FLC) and model-based dynamics of the sensors and actuators. A key contribution is the perceptual schema of the sensing primitives, represented by a non-stationary sensing matrix that serves as the HMC's state transition matrix. This reduces the six-dimensional sensor input to a single value representing predicted motion, thereby decreasing the number of fuzzy inputs required by the FLC. Another contribution is the robot's acting primitive design, where the motor schema, a MIMO FLC, is directly interconnected with the HMC, effectively bypassing a distinct planning primitive. Part of this work's contribution is the proposal of a model-based control strategy for a customized hydraulic piston mechanism used for independent yet interconnected vertical motion and depth regulation, driven by one of the FLC's outputs. Furthermore, this research contributes a customized dynamic model of the robot's propulsion, yaw, pitch, and ballasting mechanisms. For the ballasting system, a Lyapunov-based stability analysis is provided.

The proposed fuzzy controller incorporates five inputs: the robot's state prediction quantized by crisp sets, instantaneous linear velocity, underwater depth, yaw velocity (angular azimuth), and pitch velocity (elevation angle), using a mix of Gaussian and sigmoid membership functions. Feedback information is obtained from ultrasonic sensors positioned at the six cardinal points (front, back, left, right, down, and up), a three-axis gyroscope, a three-axis accelerometer, and a liquid pressure sensor. The fuzzy controller generates three outputs, with a combination of Gaussian and sigmoid functions, one for each of the robot's actuators: the propulsion motor, steering motor, and a ballasting motor integrated into a hydraulic piston system. These outputs are interpreted by optical encoders that measure the motors' rotary information. Physics-based models for the propulsion, steering, and ballasting systems are developed, along with sensor fusion models that enable control feedback. Additionally, the paper introduces the robot's platform and system architecture. Experimental and numerical simulation results demonstrate the effectiveness of the proposed strategy in achieving robust obstacle avoidance for underwater navigation. As an initial validation, experimental results are presented using a prototype tested in a pool, assuming an ideal static fluid without external hydrodynamic forces.

Section 2 reviews relevant state-of-the-art works. Section 3 describes the architecture and design of the robotic platform architecture. Section 4 derives deterministic models for the robot's actuators and sensors. Section 5 describes the hidden Markov chain model that predicts the robot's state transitions for obstacle avoidance. Section 6 provides a comprehensive description of the fuzzy navigation control system. Section 7 deduces the dynamics of the robotic submarine's motion actuators and a stability analysis. Section 8 discusses the experimental validations. Finally, Sect. 9.

2 Related Work

The work reported by Cheng et al. [6] reviewed the growing importance of autonomous underwater vehicles (AUVs) in marine exploration, emphasizing path planning and obstacle avoidance. Their work provides a comprehensive overview of existing path planning methods, classifying them as global or local based on obstacle awareness, and offers valuable insights for new researchers and future research directions. In their work, Xu et al. [7] presented a comprehensive analysis of path planning technologies employed in autonomous underwater vehicle (AUV) missions. The authors reviewed a spectrum of algorithms, encompassing both traditional and intelligent methodologies, providing a comparative assessment of their respective advantages and limitations.

Chen et al. [8] conducted real-world experiments on underwater obstacle avoidance using a Remotely Operated Vehicle (ROV). Fuzzy logic controlled the ROV's depth and heading based on sensor data, with sonar aiding in heading adjustments. This approach ensured control stability and effective obstacle avoidance. Lin et al. [9] proposed a recurrent ANN with convolution for online obstacle avoidance in UUVs, focusing on feature extraction with reduced parameters. This method utilizes sonar and deep learning techniques. Bhopale [10] proposed a model-free reinforcement learning approach to adapt to environmental changes without manual tuning, using an ANN for state approximation for AUV obstacle avoidance in unknown environments. The stydy of Yang et al. [11] proposed a novel path planning solution for underwater robots using an N-step Priority Double DQN algorithm. This approach effectively addresses dynamic obstacle avoidance in realistic 3D underwater environments, incorporating real ocean current data. By introducing an experience screening mechanism, the algorithm enhances stability and adaptability without relying on prior experience.

Research by [12] proposed an obstacle avoidance approach for submarine robots centered on a 'motion balance point', aimed at long operation times and high energy efficiency. It combined local obstacle avoidance planning, control, and hydrodynamics, tracking dynamic obstacle avoidance through models of velocity, acceleration, and Singer acceleration. Zhang [13] discusses improving real-time obstacle avoidance in underwater robots using an enhanced artificial potential field method to address collision and path smoothness issues. The method employs a dynamic adjustment factor implemented in ROS, demonstrating effective real-time obstacle avoidance. The work of [14] introduced a 3-D obstacle avoidance algorithm for Autonomous Underwater Vehicles (AUVs) using the sphere cross-section method, employing Dubins-like curves for smooth trajectory planning. Yan et al. [15] proposed a real-time reactive obstacle avoidance algorithm for AUVs using forward-looking sonar. The algorithm simplifies obstacle outlines to convex polygons and introduces a memory-based solution for handling U-shaped obstacles. Wu [16] developed a dynamic obstacle avoidance method for AUVs within a velocity vector coordinate system, particularly for deep-sea exploration. This method addresses the complex relationship between velocity and displacement.

Yao [17] proposed a control method for AUVs to track 3D linear paths while avoiding obstacles. This method combines kinematic model predictive control with a dynamic controller utilizing sliding mode control to improve waypoint path tracking, reduce angle errors, and prevent saturation. Sun et al. [18] developed a control algorithm combining model predictive control for trajectory replanning and obstacle avoidance with sliding mode control for robust trajectory tracking in uncertain underwater environments, ensuring effective obstacle avoidance and stable tracking while reducing thruster saturation. Yu [19] proposes a robust nonlinear model predictive control scheme for high-speed waypoint tracking and real-time obstacle avoidance in large-scale, underactuated AUVs operating in unknown environments. It includes a dynamic sensing and collision avoidance scheme that transforms nonconvex obstacles into convex constraints. Li et al. [20] proposed a strategy that combines an enhanced A* algorithm with Model Predictive Control to improve path planning and obstacle avoidance for Unmanned Underwater Vehicles (UUVs). This approach uses a kinematic and dynamic UUV model for 3D path planning and supports real-time re-planning for dynamic obstacles.

Hidden Markov Models (HMMs) offer distinct advantages over alternative approaches in several key aspects. Their interpretable architecture with clearly defined transition and emission matrices provides transparency lacking in many modern machine learning methods. They demonstrate exceptional computational efficiency, leveraging optimized algorithms like Viterbi [21]. Unlike data-hungry neural networks [22–24], HMMs maintain robust performance with limited or noisy training data, thanks to their principled probabilistic framework. A particular strength lies in their explicit modeling of temporal dynamics through the Markov property, effectively capturing short-range sequential dependencies. The framework offers notable flexibility, seamlessly accommodating both discrete and continuous observations through appropriate emission distributions. Furthermore, their unsupervised learning capability via the Baum-Welch EM algorithm enables effective training when labeled data is scarce. These characteristics make HMMs particularly suitable for applications involving short-to-medium length structured sequences, such as gesture recognition, bioinformatics (gene prediction), or basic speech processing. They often outperform more complex alternatives like LSTMs or Transformers [25] in scenarios with limited training data [26], while avoiding the implementation complexity of POMDPs or reinforcement learning approaches [27, 28]. The Markov assumption inherently restricts modeling of long-range dependencies, and the framework requires careful discretization for continuous state spaces. However, when application requirements align with these constraints, HMMs remain a parsimonious, efficient, and mathematically elegant solution for sequential modeling tasks. Their combination of transparency, computational tractability, and reliable performance continues to make them relevant in the era of deep learning.

As comparative concern with traditional methods, the proposed hybrid approach presents several advantages over alternative decision-making techniques, including purely rule-based systems, classical PID control, and standalone machine learning models, primarily by offering enhanced robustness in uncertain underwater environments characterized by sonar noise and turbulence, where the Hidden Markov Model (HMM) probabilistically estimates optimal avoidance maneuvers by modeling hidden states of obstacle proximity trends, thereby exhibiting resilience to false readings, while the Fuzzy Logic Controller (FLC) effectively handles imprecise sensor data, unlike the brittle nature of crisp threshold-based methods or PID control under noise; furthermore, it enables adaptive decision-making without requiring explicit modeling of the complex and nonlinear robot hydrodynamics, allowing the FLC to generalize better through linguistic rules compared to traditional controllers needing precise tuning for each scenario; moreover, it facilitates predictive avoidance rather than purely reactive responses seen in systems like potential fields or simple threshold triggers that risk oscillations or local minima, as the HMM predicts future obstacle positions for smoother, preemptive maneuvers; additionally, the hybrid scheme naturally handles multi-sensor fusion, allowing the FLC to seamlessly integrate diverse inputs such as HMM predictions, raw sonar readings, and potential 3D data without

Table 1 Summarized comparison with other methods

Method	Pros	Cons
PID control	Simple, fast	Poor with nonlinearities, no prediction
Potential fields	Smooth paths	Local minima, reactive only
Neural networks	Learns complex behaviors	Needs big data, black-box
Rule-based (if-else)	Transparent	Brittle in dynamic environments
HMM + FLC (Proposed)	Predictive, robust, adaptive	Slightly higher computational load

explicit weighting, offering greater interpretability than neural networks without the need for large training datasets; finally, the HMM and FLC combination enhances motion efficiency by avoiding unnecessary thruster adjustments, such as reducing yaw corrections when an obstacle is predictably moving away, contrasting with the potential overreactions and energy waste of pure PID or potential-field methods to transient obstacles.

As evidenced in Table 1, the HMM+FLC hybrid approach demonstrates significant advantages in underwater robotics by proactively predicting obstacle trajectories rather than merely reacting to immediate sensor data. Its inherent adaptability to noise and environmental uncertainty, coupled with its efficiency in optimizing thruster commands, distinguishes it from purely reactive, rigidly defined, or data-intensive alternatives, rendering it particularly well-suited for the complexities of unpredictable underwater environments.

3 Robotic Platform

This section provides a comprehensive overview of the development and design of the underwater robotic platform's architecture deployed experimentally in this research. It outlines the construction of the mechanical physical system and electronic instrumentation, alongside the modular design of the robot's components. The software modules are elaborated at a multithread level to enable real-time operations. Additionally, the development of Autonomous Underwater Vehicles (AUVs), Remotely Operated Vehicles (ROVs), or Unmanned Underwater Vehicles (UUVs) can yield cost-effective robotic vehicles focused on robotics research, which may be achieved with a low-cost design approach, as referenced in [29].

Figure 1a depicts three main views of the proposed robot: rear, top, and lateral, along with their corresponding planes and Cartesian axes. It also shows the location of proprioceptive sensors such as an inertial measurement unit (IMU), a pressure sensor, and an exteroceptive sensor Red-Green-Glue Depth (RGBD) (not covered in this manuscript). Additionally, a graphical representation of other exteroceptive

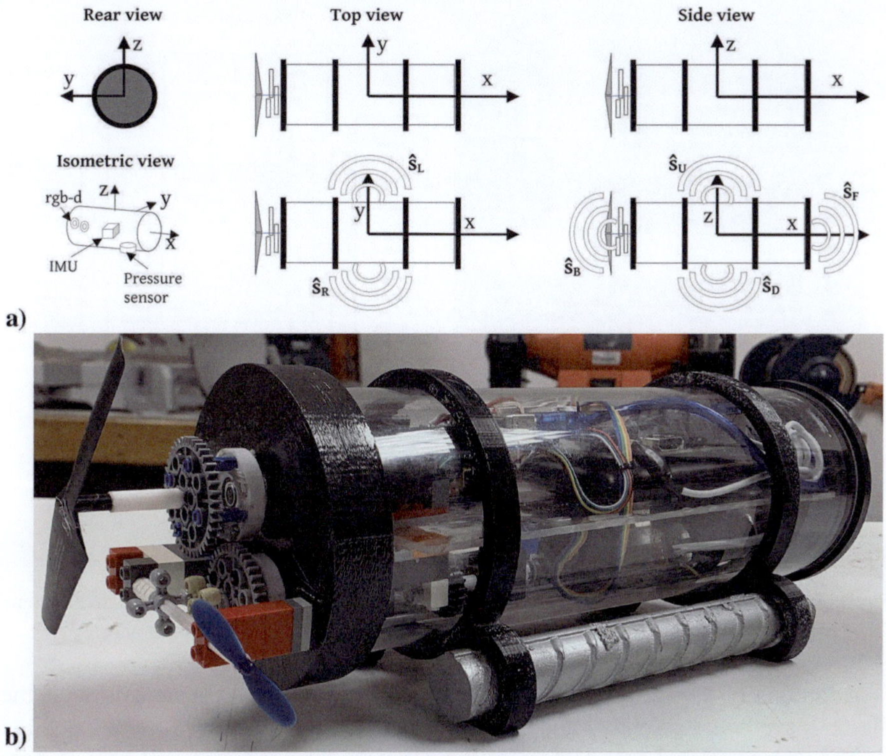

Fig. 1 Robot's reference system. **a** Multiview and sensors location. **b** Experimental prototype

sensors, such as the acoustic cones of ultrasonic sonars is illustrated, situated along the six cardinal axes: front and back ($\hat{s}_{F,B}$), up and down ($\hat{s}_{U,D}$), and left and right ($\hat{s}_{L,R}$).

In addition, Fig. 1b shows a photo of the robotic submarine that was developed to carry out the experimental issues of this research. The characteristics and physical parameters of the robot are detailed in Table 3, presented in Appendix 10.1.

Figure 2 illustrates the modular-level instrumentation diagram of the robotic architecture design (Fig. 2b). It includes two electronic boards: the Orange PI, a 64-bit, quad-core 1.5 GHz RISC mini-computer serving as the master system, and an Arduino ATmega328P control and sensing module card acting as the slave computer. The master computer primarily coordinates and synchronizes sensory data and control signals to the actuators.

The three actuators (propulsion, steering, and ballast) are DC motors controlled via pulse width modulation (PWM), with feedback provided by rotation data through pulse encoder integration. The master computer runs a 64-bit Linux operating system equipped with matrix calculation libraries and manages WiFi ports for TCP/IP communication.

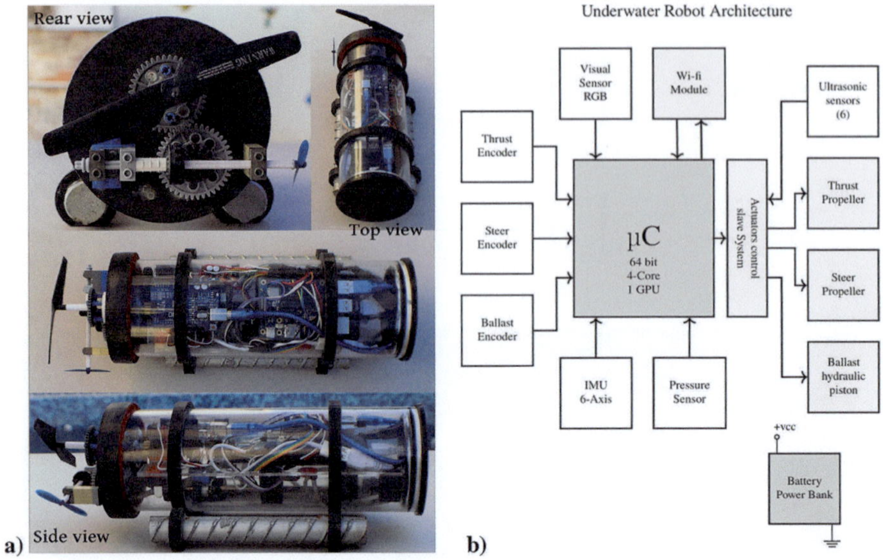

Fig. 2 Self-contained robotic submarine. **a** Multi-view prototype. **b** Modular system architecture

Additionally, Fig. 2a presents illustrations of the transparent chassis with the master-slave board assembly and other instrumentation devices.

The computational architecture of the robot is designed as a real-time system leveraging a multi-threaded framework to support concurrent execution of various processes. These processes are categorized into three primary classes:

a. **Sensing threads**: Responsible for acquiring and preprocessing data from the robot's sensors in real-time.
b. **Control and actuation threads**: Handle decision-making and command transmission to the robot's actuators, ensuring responsive and precise execution of tasks.
c. **Computation and planning threads**: It is purposed to perform intensive arithmetic and logical calculations, such as trajectory planning and optimization algorithms.

To enable seamless inter-thread communication and coordination, the system employs shared memory techniques coupled with mutual exclusion mechanisms to prevent data conflicts and ensure thread-safe operations. This structure allows for efficient real-time data flow and processing, crucial for the robot's functionality in dynamic environments.

A detailed diagram of the robot's computational organization, illustrating the interaction between threads and their respective responsibilities, is presented in Fig. 3.

The entire system, including all threads and routines, was implemented in C/C++ using the GNU Compiler Collection (GCC). To ensure a robust, efficient, and portable foundation for resource management and real-time operations, the implementation

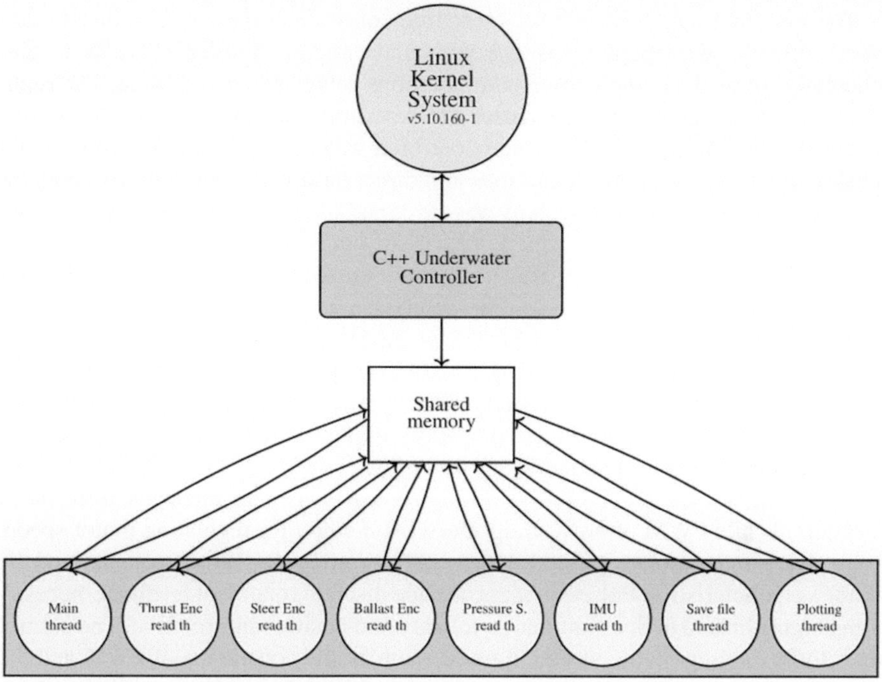

Fig. 3 Robot's real-time computational organization

adhered to the POSIX standard, leveraging features within the C++ standard library. This approach facilitated the development of a clean, well-structured, and maintainable codebase.

For computational tasks involving numerical linear algebra and matrix operations, the Armadillo C++ library was employed, providing a high-performance and user-friendly interface. Additionally, real-time visualization of experimental data during runtime was seamlessly integrated using the Sciplot library, enabling dynamic and interactive monitoring of the system's performance.

The algorithms executed within the threads operate at varying frequencies, tailored to the specific requirements of each process. These threads interact with reserved memory locations to read and write data, ensuring efficient data management and coordination. Synchronization mechanisms are employed to seamlessly integrate these operations, enabling precise low-level control of the actuators while simultaneously handling large volumes of data from the sensors. This architecture ensures real-time responsiveness and robust system performance, even in data-intensive and dynamic environments.

Given that this work's robot is a laboratory prototype, we identified computational resources as a potential area for enhancement. Currently, these limitations are mitigated by a real-time multi-threading design.

The system employs a relatively resource-constrained master computer (quad-core 1.5 GHz) and a separate sensing/control slave microcontroller. This architecture inherently limits the complexity of computations achievable in real-time. The separation of processing between the master and slave units indicates a strategy to offload specific tasks. The simpler slave microcontroller was selected for direct control and sensing due to its real-time capabilities and direct hardware interfacing. To address this, the computational architecture was designed as a real-time system, enabling task distribution across the available processor cores for concurrent execution of sensing, control, and planning. This design maximizes the utilization of the limited processing power to achieve real-time responsiveness. Sensing threads and control/actuation threads are likely managed, at least in part, by the slave computer for low-level, time-critical operations. The master computer handles higher-level coordination, synchronization, complex computations, and communication. This division of labor simplifies the algorithmic complexity required on each processor and optimizes performance based on hardware strengths.

The actuators are DC motors controlled via PWM, with feedback from pulse encoders. While PWM offers a straightforward method for regulating motor speed, its precision and responsiveness are inherently limited by the PWM frequency and the motor's characteristics. Pulse encoders provide discrete rotational feedback, necessitating algorithms to derive continuous velocity and position information. The control algorithms running on the master must account for the constraints of PWM and the resolution of the encoder feedback to achieve stable and accurate control.

Communication between the master (and potentially the slave) and external systems relies on WiFi and TCP/IP. Compared to direct hardware connections, wireless communication introduces latency and potential reliability issues, requiring algorithms designed to handle potential delays and packet loss. For inter-thread communication on the master, the algorithmic design incorporates shared memory and mutual exclusion mechanisms. This approach is efficient for real-time data sharing within a single processor but demands careful management to prevent race conditions, particularly concerning synchronized sensory data and control signals.

While the master computer possesses a capable processor, computationally intensive tasks such as planning and matrix operations can still strain its resources in real-time. The selection of a C++ library optimized for numerical linear algebra and matrix operations reflects an effort to leverage efficient implementations for these demanding tasks, maximizing the usable computational power of the hardware.

4 Sensor and Actuator Models

This section describes the sensor and actuator configuration relevant to the obstacle avoidance task. While the experimental robot prototype incorporates a comprehensive array of sensors, only those directly pertinent to the scope of this manuscript are presented herein. Specifically, the system employs optical incremental encoders for precise measurement of motor shaft angular displacement, an inertial measurement unit (IMU) integrating a three-axis gyroscope and a three-axis accelerometer for inertial navigation, and a pressure sensor for depth determination. The optical encoders operate as transducers, converting the continuous rotational motion of the motor shafts into a discrete pulse train at TTL voltage levels.

4.1 Observation Models

An observer model, commonly referred to as an observer, is a mathematical representation of a system designed to estimate its internal states based on observed inputs and outputs. In real-world robotics applications, such as the one discussed in this manuscript, it is often impractical or impossible to directly measure all internal states of the system.

Sensor measurements are typically noisy and influenced by various factors, such as environmental interference and inherent sensor inaccuracies, which can compromise the reliability of the readings. Observers address these challenges by leveraging a system model to estimate states that are either unmeasurable or significantly affected by noise. This approach enhances the overall accuracy and robustness of state estimation, enabling more effective system control and decision-making.

Therefore, we initially consider the angular velocity model derived by an optical encoder from the rotational motion of each motor, such that:

$$\hat{\phi}(\eta) = \frac{2\pi}{R}\eta_t, \tag{1}$$

Here, η_t represents the instantaneous number of pulses measured by the encoder, while R is a constant defining the encoder's resolution in pulses per revolution. Both parameters are dimensionless. Here, $\hat{\phi}$ represents the observation, rather than a measurement. Based on this, the following Proposition 1 is established.

Proposition 1 (Encoder's rotary velocity) *The numerical derivative is defined using the backward divided difference method, truncated through the Taylor series expansion. This approach allows for a high-precision calculation of the derivative as a function of time:*

$$\hat{\dot{\phi}}(\hat{\phi}, t) = \frac{3\phi_t - 4\phi_{t-1} + \phi_{t-2}}{2\Delta t} \tag{2}$$

By substituting the encoder model at different points in time, we derive an expression that depends on the number of pulses measured by the encoder and the corresponding time:

$$\frac{d\hat{\phi}}{dt} = \dot{\phi}(\hat{\eta}, t) = \left(\frac{\pi}{R\Delta t}\right)(-3\hat{\eta}_t - 4\hat{\eta}_{t-1} + \hat{\eta}_{t-2}). \tag{3}$$

Here, Δt represents the time interval used for the control and sampling cycle.

Therefore, Proposition 1 applies generally for any of the robot's rotary actuators: propulsion, steer and ballast.

The robot's proprioceptive motion measurements were conducted using an IMU that comprises a three-axis gyroscope and a three-axis accelerometer. These components are utilized for inferring observations across various physical units. Fundamentally, measurements of Euler angles and Cartesian accelerations were taken, with the IMU centrally positioned at the midpoint of the submarine.

The roll (α), pitch (β), and yaw (γ) angles, as well as the angular acceleration around the xyz axes, are inferred from the raw gyroscope data. To estimate any Euler rotation angle, the trapezoidal integration method was employed. This approach is a precise online numerical calculation method and becomes more accurate as the number of segments increases.

$$\phi_{\alpha,\beta,\gamma} = \int_{t_a}^{t_b} \hat{\dot{\phi}}_{\alpha,\beta,\gamma} dt = \frac{2n}{t_b - t_a} \left(\hat{\dot{\phi}}_{0_{\alpha,\beta,\gamma}} + 2 \sum_{i=1}^{n-1} \hat{\dot{\phi}}_{i_{\alpha,\beta,\gamma}} + \hat{\dot{\phi}}_{n_{\alpha,\beta,\gamma}} \right), \tag{4}$$

where n is the number of samples accumulated between t_b and t_a. $t_{a,b}$ are the time integration limits.

Besides, the angular accelerations observations are estimated by

$$\ddot{\phi}_{\alpha,\beta,\gamma} = \frac{d\hat{\dot{\phi}}_{\alpha,\beta,\gamma}}{dt} = \frac{3\hat{\dot{\phi}}_{t_{\alpha,\beta,\gamma}} - 4\hat{\dot{\phi}}_{t-1_{\alpha,\beta,\gamma}} + \hat{\dot{\phi}}_{t-2_{\alpha,\beta,\gamma}}}{2\hat{t}}, \tag{5}$$

where $\hat{t}$ is the measured time of data sampling.

Moreover, the linear accelerations $\hat{a}_{xyz}$ of the robot's body under inertial effects is measured by the accelerometer devices, and the linear velocity $\hat{\upsilon}$ is inferred by

$$\hat{\upsilon}_{xyz} = \int_{t_1}^{t_2} \hat{a}_{xyz} dt = \frac{2n}{t_2 - t_1} \left(\hat{\upsilon}_{o_{xyz}} + 2 \sum_{i=1}^{n-1} \hat{\upsilon}_{i_{xyz}} + \hat{\upsilon}_{n_{xyz}} \right), \tag{6}$$

Additionally, the pressure transducer, when subjected to a hydraulic load, requires a calibration model, such that

$$\rho_{abs} = \frac{\rho_{atm}}{\delta_{atm}} \hat{\delta}_\rho, \tag{7}$$

where ρ_{abs} is the absolute pressure. ρ_{atm} is the atmosphere pressure. δ_{atm} is a measurement of atmosphere pressure just above the water surface, and $\hat{\delta}_\rho$ is the the measurement of the transducer at an arbitrary underwater depth.

Therefore, the instantaneous underwater pressure is inferred by the following observer model,

$$\hat{\rho}_t = \frac{\rho_{atm}}{\delta_{atm}} \hat{\delta}_\rho - \rho_{atm} \tag{8}$$

The ultrasonic sonar, which measures the distance between the robot's main points and obstacles, is set within a unified coordinate system anchored to the robot's center. All obstacles are represented in Cartesian coordinates as s_x, s_y, s_z. According to Fig. 1a, the front ($\hat{s}_1 \equiv \hat{s}_F$) and back ($\hat{s}_2 \equiv \hat{s}_B$) sonars are longitudinally aligned along the x-axis. The left ($\hat{s}_3 \equiv \hat{s}_L$) and right ($\hat{s}_4 \equiv \hat{s}_R$) sonars are transversely aligned along the y-axis. The bottom ($\hat{s}_5 \equiv \hat{s}_D$) and top ($\hat{s}_6 \equiv \hat{s}_U$) sonars are vertically aligned along the z-axis.

All xy components measured by each sonar transducer on the horizontal xy-plane, are inferred by

$$s_{x_k} = l_k \cos(\psi_k) + \hat{s}_k \cos(\psi_k) \Big|_{k=1}^{6} \tag{9a}$$

and

$$s_{y_k} = l_k \sin(\psi_{xy}) + \hat{s}_k \sin(\psi_{xy}) \Big|_{k=1}^{6} \tag{9b}$$

Likewise, considering the vertical xz-plane the z component is obtained by

$$s_{z_k} = l_k \sin(\psi_{xz}) + \hat{s}_k \sin(\psi_{xz}) \Big|_{k=1}^{6} \tag{9c}$$

where, ψ is the orientation angle of each sonar sensor, such that $\psi_1 = 0$, $\psi_2 = \pi$, and $\psi_3 = \pi/2$, $\psi_4 = 3\pi/2$. In addition, over the xz-plane $\psi_5 = 3\pi/2$, $\psi_6 = \pi/2$.

4.2 Rotary Actuators

This section details the derivation of the kinematic model for the DC motors responsible for robot actuation. To address the influence of hydrostatic pressure on motor performance, particularly potential reductions in rotational power with increasing depth, a third-degree polynomial adaptive adjustment model is proposed. This model aims to dynamically compensate for these losses and maintain the desired motor speed.

The primary strategy is to develop an empirical model of each motor's kinematic behavior. This empirical model comprises experimental data as a tuple, including the motor's angular velocity $\dot{\varphi}\, s^{-1}$ correlated with a specified digital word δ (dimensionless), which is obtained through a controlled experiment. The digital word refers to a natural number linked to varying pulse width modulation (PWM) frequencies. By writing a digital word to the computer's PWM port, a desired motor velocity is set. Upon acquiring the empirical model, as illustrated in Fig. 4, a theoretical model is then derived.

This theoretical model is a mathematical function that closely replicates the real empirical model with higher precision. Additionally, the empirical model can be further analyzed mathematically.

The plot 4 illustrates the experimental performance of the propulsion and sterring motors angular velocity in relation to their digital control word (PWM).

Hence, approaches such as the regression model applied in robotics [30] that optimally approximates the empirical model is represented by the following system

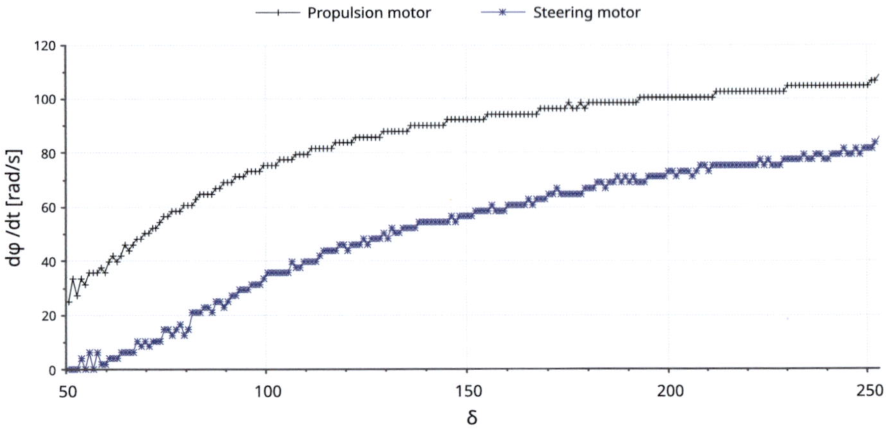

Fig. 4 Motors behavior model. **a** Propulsion/steer motor speed ($\dot{\varphi}_{D,S}$) versus PWM digital word (δ)

in Eq. 10, derived utilizing the least squares method.

$$\begin{pmatrix} \sum_i \delta_i \\ \sum_i \delta_i \dot{\varphi}_i \\ \sum_i \delta_i \dot{\varphi}_i^2 \\ \sum_i \delta_i \dot{\varphi}_i^3 \end{pmatrix} = \begin{pmatrix} n & \sum_i \dot{\varphi}_i & \sum_i \dot{\varphi}_i^2 & \sum_i \dot{\varphi}_i^3 \\ \sum_i \dot{\varphi}_i & \sum_i \dot{\varphi}_i^2 & \sum_i \dot{\varphi}_i^3 & \sum_i \dot{\varphi}_i^4 \\ \sum_i \dot{\varphi}_i^2 & \sum_i \dot{\varphi}_i^3 & \sum_i \dot{\varphi}_i^4 & \sum_i \dot{\varphi}_i^5 \\ \sum_i \dot{\varphi}_i^3 & \sum_i \dot{\varphi}_i^4 & \sum_i \dot{\varphi}_i^5 & \sum_i \dot{\varphi}_i^6 \end{pmatrix} \begin{pmatrix} a_0 \\ a_1 \\ a_2 \\ a_3 \\ a_4 \end{pmatrix}, \tag{10}$$

where n is the total number of experimental data. Thus, estimating the coefficient of interest, algebraically arrange for the algebraic solution,

$$\begin{pmatrix} a_0 \\ a_1 \\ a_2 \\ a_3 \\ a_4 \end{pmatrix} = \begin{pmatrix} n & \sum_i \dot{\varphi}_i & \sum_i \dot{\varphi}_i^2 & \sum_i \dot{\varphi}_i^3 \\ \sum_i \dot{\varphi}_i & \sum_i \dot{\varphi}_i^2 & \sum_i \dot{\varphi}_i^3 & \sum_i \dot{\varphi}_i^4 \\ \sum_i \dot{\varphi}_i^2 & \sum_i \dot{\varphi}_i^3 & \sum_i \dot{\varphi}_i^4 & \sum_i \dot{\varphi}_i^5 \\ \sum_i \dot{\varphi}_i^3 & \sum_i \dot{\varphi}_i^4 & \sum_i \dot{\varphi}_i^5 & \sum_i \dot{\varphi}_i^6 \end{pmatrix}^{-1} \cdot \begin{pmatrix} \sum_i \delta_i \\ \sum_i \delta_i \dot{\varphi}_i \\ \sum_i \delta_i \dot{\varphi}_i^2 \\ \sum_i \delta_i \dot{\varphi}_i^3 \end{pmatrix} \tag{11}$$

The polynomial representation for the digital control word δ for the motor's PWM output is formulated as a polynomial function of the angular velocity and is expressed as:

$$\delta(\dot{\varphi}) = a_0 + a_1\dot{\varphi} + a_2\dot{\varphi}^2 + a_3\dot{\varphi}^3, \tag{12}$$

This implies that the polynomial inverse solution is formulated in relation to a specified angular velocity in terms of the digital motor's control word or δ is

$$\begin{pmatrix} b_0 \\ b_1 \\ b_2 \\ b_3 \\ b_4 \end{pmatrix} = \begin{pmatrix} n & \sum_i \delta_i & \sum_i \delta_i^2 & \sum_i \delta_i^3 \\ \sum_i \delta_i & \sum_i \delta_i^2 & \sum_i \delta_i^3 & \sum_i \delta_i^4 \\ \sum_i \delta_i^2 & \sum_i \delta_i^3 & \sum_i \delta_i^4 & \sum_i \delta_i^5 \\ \sum_i \delta_i^3 & \sum_i \delta_i^4 & \sum_i \delta_i^5 & \sum_i \delta_i^6 \end{pmatrix}^{-1} \cdot \begin{pmatrix} \sum_i \dot{\varphi}_i \\ \sum_i \delta_i \dot{\varphi}_i \\ \sum_i \delta_i^2 \dot{\varphi} \\ \sum_i \delta_i^3 \dot{\varphi} \end{pmatrix}, \tag{13}$$

such that,

$$\dot{\varphi}(\delta) = b_0 + b_1\delta + b_2\delta^2 + b_3\delta^3. \tag{14}$$

5 Robot's State Prediction

This section describes the proposed Hidden Markov Chain (HMC) model for predicting seven process states of a submarine robot to avoid obstacles. Probabilities in each state q_i are calculated based on separate observations, which are the distances $\hat{s}_j$ measured by six proximity sensors positioned at the robot's cardinal points. Functioning as an intelligent and temporally aware sensor fusion layer, the Hidden Markov Chain (HMC) dynamically processes the sequential data streams from six distinct range sensors to generate a singular, high-level fuzzy prediction that encapsulates the most probable underlying environmental state. This abstraction not only significantly reduces the dimensionality of the input space for the controller, thereby simplifying its decision-making process, but also leverages the inherent temporal dependencies within the sensor readings to provide a more robust, contextually rich, and temporally consistent assessment of the surrounding environment, mitigating the impact of noisy or transient sensor fluctuations. The purpose of this approach is to predict the robot's next action tasks for obstacle evasion in real-time. The output of the HMC are inferred states that serve as critical input to one of the fuzzy sets in the fuzzy controller, which estimates numerical values for the robot's actuators (see Sect. 6). The proposed HMC features a non-stationary stochastic state transition matrix, where state transition probabilities change instantaneously according to probability observations as functions of obstacles proximity. During state prediction, the probabilities of hidden states are calculated, and the state with the highest probability at the current time is used as the input to the fuzzy controller.

A hidden Markov chain is a statistical model used to represent systems that transition between states in a Markov process, where the states themselves are not directly observable (hidden). Instead, you can observe a sequence of outputs or emissions, which are probabilistically related to the underlying hidden states. Hidden states $q_0, q_1, \ldots, q_6$ are the states the system transitions through, which are not directly visible. The observations $\hat{s}_F, \ldots, \hat{s}_U$ are the observable sensor data.

The transition probabilities $p(q_i | p_j)$ or for simplicity notation p_{ij}, define the likelihood of transitioning from one hidden state to another. The emission probabilities P_A, P_{CS} represent the probability of a particular hidden state.

The study by Brito et al. [31] presented a Markov chain-based approach to model the sequential steps of autonomous underwater vehicle (AUV) missions, from

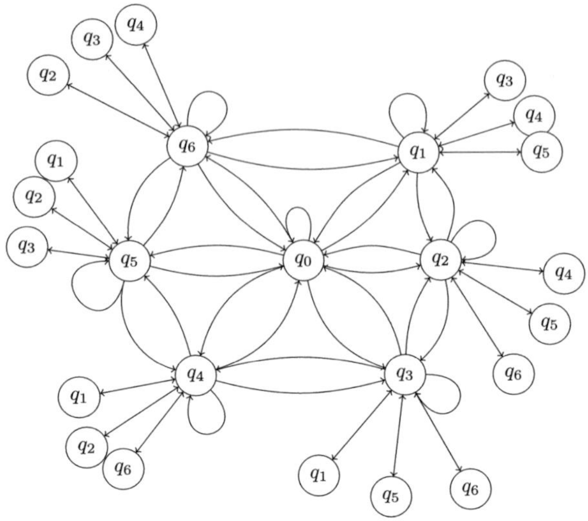

Fig. 5 Simplified planar representation of the robot's state process Markov chain

Table 2 Markov's elements description

Current state	State meaning	Activated sensor	Observation likelihood	States probability	Transition probability
q_0	Stop	All sensors	$p(c_S) = 0$	$p(q_0) = \frac{\sum_i p(q_i)}{6}$	$p_{i0} = \frac{\sum_i p(c_i)}{6}$
q_1	Forward	Back-side sensor	$p(c_F) = \lambda_F e^{-\lambda_F \hat{s}_F}$	$p(q_1) = p(c_B)$	$p_{11} = 1 - p(c_F)$
q_2	Backward	Front-side sensor	$p(c_B) = \lambda_B e^{-\lambda_B \hat{s}_B}$	$p(q_2) = p(c_F)$	$p_{22} = 1 - p(c_B)$
q_3	Left-turn	Right-side sensor	$p(c_L) = \lambda_L e^{-\lambda_L \hat{s}_L}$	$p(q_3) = p(c_R)$	$p_{33} = 1 - p(c_L)$
q_4	Right-turn	Left-side sensor	$p(c_R) = \lambda_R e^{-\lambda_R \hat{s}_R}$	$p(q_4) = p(c_L)$	$p_{44} = 1 - p(c_R)$
q_5	Sink	Up-side sensor	$p(c_D) = \lambda_D e^{-\lambda_D \hat{s}_D}$	$p(q_5) = p(c_U)$	$p_{55} = 1 - p(c_D)$
q_6	Rise	Down-side sensor	$p(c_U) = \lambda_U e^{-\lambda_U \hat{s}_U}$	$p(q_6) = p(c_D)$	$p_{66} = 1 - p(c_U)$

The observation probability indicates likelihood of collision. p_{i0} means transition probability from any state q_i to q_0 (*stop*)

prelaunch to recovery, with the goal of identifying high-risk states and transitions. The model outlines event sequences and assigns transition probabilities across 11 discrete states, incorporating distance-dependent survival statistics. Additionally, the study by Ali et al. [32] introduced a deep learning approach utilizing a network for real-time state estimation of underwater maneuvering objects within a Markov chain framework. It evaluates the network's performance in bearings-only tracking under various ocean conditions and noise levels.

The elements depicted in Fig. 5 represent all the states of the chain and their respective transitions, as detailed in Table 2. Define $\mathbf{q} = (q_0, q_1, \ldots, q_6)^\top$ to denote the transitions between the robot's states. The functions $p(q_i)$ represent the probability of each state occurring, while the probabilities $p(C_i)$ indicate the likelihood of distinct output observations. Additionally, p_{ij} represents the probability of transitioning between states q_i and q_j.

Definition 1 *(Transition from any q_i to q_0)* Let $p_A \equiv p_{i0}$ that represent the probability model describing transitions from any of the states $q_1, q_2, \ldots, q_6$ toward the state q_0. The model p_A serves as an alternative to the inverse exponential probabilities $p_{10} = p_{20} = p_{30} = p_{40} = p_{50} = p_{60}$ and is defined as:

$$p_A(\hat{\mathbf{s}}, \lambda_k) \doteq \sum_{k=1}^{6} \lambda_k e^{-\lambda_k \hat{s}_k},$$

where $\hat{\mathbf{s}} = (s_1, s_2, s_3, s_4, s_5, s_6)^\top$ are the distances measured to an obstacle by the kth ranging sensor, and λ_k is an adjustment parameter that determines the amplitude and range of the probability based on the proximity of obstacles to the robot. This definition captures the scenario in which all sensors detect obstacles in their respective directions (cardinal points), influencing the robot to transition toward the state q_0.

In addition, there is another important Definition 2.

Definition 2 *(Probability of current state transition p_{ii})* Let $p_{CS} \equiv p_{ii}$ represent the inverse exponential probability model describing transitions from current state q_i toward the same state q_i. The model p_{CS} serves as an alternative to the inverse exponential probabilities $p_{11}, p_{22}, p_{33}, p_{44}, p_{55}, p_{66}$ and is defined as:

$$p_{cs}(\hat{s}_k, \lambda_i) \doteq 1 - \lambda_i e^{-\lambda_i \hat{s}_k},$$

where $\hat{s}_k$ is the distance measured to an obstacle by the kth ranging sensor, and λ_i is an adjustment parameter that determines the amplitude and range of the probability of state q_i based on the proximity of obstacles to the robot. This definition describes a scenario in which the robot remains in its current state, q_i, when no obstacles influence that state. In this case, the complement of the transition probability is inversely proportional to the presence of obstacles, resulting in a high likelihood for the robot to stay in the same state.

The Markov state transition stochastic matrix $\mathbf{P}$, as referenced in Definitions 1 and 2, assumes a probability $p(\hat{s}_k) = 1$ when the measured obstacle range is approximately zero, $\hat{s}_k \approx 0$. Therefore, we have:

$$1 = \lambda_k e^{\lambda_k(0)}, \qquad \therefore \lambda_k = 1 \tag{15}$$

For analysis purposes, let us assume $\lambda_F = \lambda_B = \lambda_L = \lambda_R = \lambda_S = 1$. Figure 6 illustrates the proposed function trend, where the maximum distance for obstacle detection is 5 m.

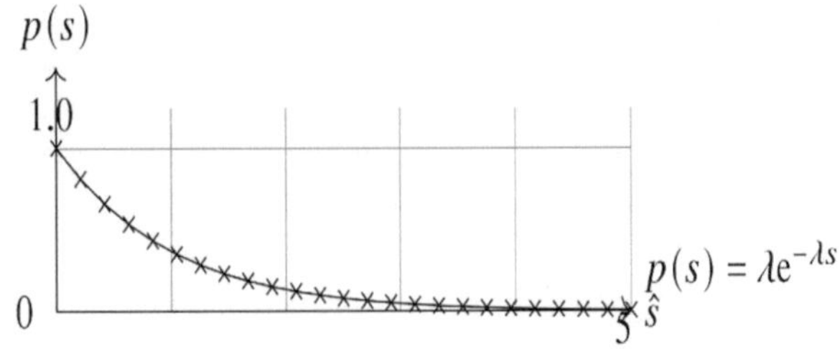

Fig. 6 Probability function $p(s) = \lambda e^{-\lambda \hat{s}}$ and $\lambda = 1$

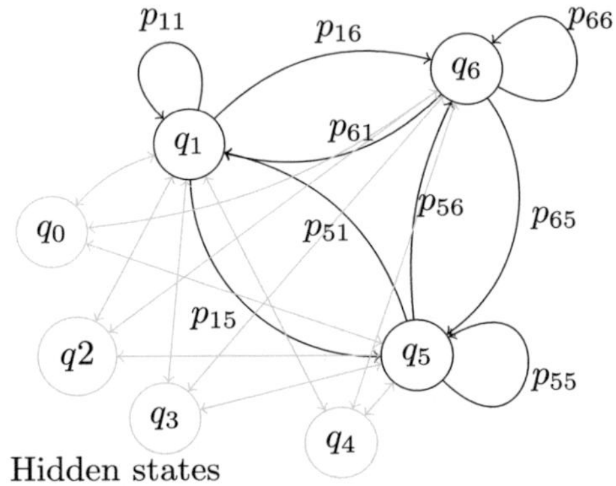

Fig. 7 Markov chain illustrating state transitions involved in upward collision while moving both forward and upward

The Markov state transition stochastic matrix is a non-stationary matrix model that is updated at each control loop based on range sensor measurements. Consequently, its dynamic elements change according to the following inverse exponential functions:

$$
\mathbf{P} = \begin{pmatrix}
q_0 & q_1 & q_2 & q_3 & q_4 & q_5 & q_6 & \\
\hline
p_A & p_A & p_A & p_A & p_A & p_A & p_A & q_0 \\
e^{-\hat{s}_B} & p_{cs}(\hat{s}_F) & e^{-\hat{s}_B} & e^{-\hat{s}_B} & e^{-\hat{s}_B} & e^{-\hat{s}_B} & e^{-\hat{s}_B} & q_1 \\
e^{-\hat{s}_F} & e^{-\hat{s}_F} & p_{cs}(\hat{s}_B) & e^{-\hat{s}_F} & e^{-\hat{s}_F} & e^{-\hat{s}_F} & e^{-\hat{s}_F} & q_2 \\
e^{-\hat{s}_R} & e^{-\hat{s}_R} & e^{-\hat{s}_R} & p_{cs}(\hat{s}_L) & e^{-\hat{s}_R} & e^{-\hat{s}_R} & e^{-\hat{s}_R} & q_3 \\
e^{-\hat{s}_L} & e^{-\hat{s}_L} & e^{-\hat{s}_L} & e^{-\hat{s}_L} & p_{cs}(\hat{s}_R) & e^{-\hat{s}_L} & e^{-\hat{s}_L} & q_4 \\
e^{-\hat{s}_U} & e^{-\hat{s}_U} & e^{-\hat{s}_U} & e^{-\hat{s}_U} & e^{-\hat{s}_U} & p_{cs}(\hat{s}_D) & e^{-\hat{s}_D} & q_5 \\
e^{-\hat{s}_D} & e^{-\hat{s}_D} & e^{-\hat{s}_D} & e^{-\hat{s}_D} & e^{-\hat{s}_D} & e^{-\hat{s}_D} & p_{cs}(\hat{s}_U) & q_6
\end{pmatrix} \tag{16}
$$

where $\hat{s}_k$ are range measurements, such that $\hat{s}_B$ is a measurement of the back-side sensor, $\hat{s}_F$ is from the front-side sensor, $\hat{s}_R$ and $\hat{s}_L$ are from the right-hand and the left-hand sensors, and $\hat{s}_U$, $\hat{s}_D$ are measurements from up-side and down-side sensors.

Example 1

In an aquatic scenario, the robot senses an obstacle directly above it at a distance of 0.3 m. The robot is in two initial states when detecting the obstacle above: moving forward ($q_1 = 1$) and simultaneously surfacing ($q_6 = 1$). The Markov state transition prediction process is illustrated by Fig. 7, showing only the main involved states for simplicity.

The prediction at discrete time $k + 1$ for the state vector $\mathbf{q}_0 = (0, 0.99, 0, 0, 0, 0, 0.99)^\top$ is established

$$\mathbf{q}_1 = \mathbf{P} \cdot \mathbf{q}_0, \tag{17}$$

such that, the normalized state transition vector becomes

$$\mathbf{q}_1 = \begin{pmatrix} 0.123 & 0.123 & 0.123 & 0.123 & 0.123 & 0.123 & 0.123 \\ 0 & 1 & 0 & 0 & 0 & 0 & 0 \\ 0 & 0 & 1 & 0 & 0 & 0 & 0 \\ 0 & 0 & 0 & 1 & 0 & 0 & 0 \\ 0 & 0 & 0 & 0 & 1 & 0 & 0 \\ 0.74 & 0.74 & 0.74 & 0.74 & 0.74 & 1 & 0.74 \\ 0 & 0 & 0 & 0 & 0 & 0 & 0.259 \end{pmatrix} \cdot \begin{pmatrix} 0 \\ 1 \\ 0 \\ 0 \\ 0 \\ 0 \\ 1 \end{pmatrix} = \begin{pmatrix} 0.082 \\ 0.33 \\ 0 \\ 0 \\ 0 \\ 0.495 \\ 0.086 \end{pmatrix} \begin{matrix} q_0 \\ q_1 \\ q_2 \\ q_3, \\ q_4 \\ q_5 \\ q_6 \end{matrix} \tag{18}$$

where the prediction states that as the robot detects an obstacle above of it, then now the Markov prediction process establishes that there is now a 33% to continue forward (q_1) and a 49.5% to move downward in order to safety avoid the above obstacle. Nevertheless, for a second state prediction, at $k + 1$ the state transition vector is

$$\mathbf{q}_2 = \begin{pmatrix} 0.123 & 0.123 & 0.123 & 0.123 & 0.123 & 0.123 & 0.123 \\ 0 & 1 & 0 & 0 & 0 & 0 & 0 \\ 0 & 0 & 1 & 0 & 0 & 0 & 0 \\ 0 & 0 & 0 & 1 & 0 & 0 & 0 \\ 0 & 0 & 0 & 0 & 1 & 0 & 0 \\ 0.74 & 0.74 & 0.74 & 0.74 & 0.74 & 1 & 0.74 \\ 0 & 0 & 0 & 0 & 0 & 0 & 0.259 \end{pmatrix} \cdot \begin{pmatrix} 0.082 \\ 0.33 \\ 0 \\ 0 \\ 0 \\ 0.495 \\ 0.086 \end{pmatrix} = \begin{pmatrix} 0.091 \\ 0.247 \\ 0 \\ 0 \\ 0 \\ 0.643 \\ 0.016 \end{pmatrix} \begin{matrix} q_0 \\ q_1 \\ q_2 \\ q_3 \\ q_4 \\ q_5 \\ q_6 \end{matrix} \tag{19}$$

The Markov state transition process for the submarine robot indicates that there is a significant likelihood of 64.3% for the robot to 'sink' in order to avoid the obstacle above. Additionally, there is a 24.7% chance to continue moving forward and safely avoid the obstacle. Hence, the most critical state to consider as an input for the fuzzy set of tasks is the 'sink' prediction.

Example 2

In a typical scenario, the underwater robot initially advances with current state 'forward'($q_1 = 1$) and detects an obstacle obstructing its frontal area a few meters ahead, primarily on the left (0.25m) and near the water surface, a fraction of meter above (0.3m). As a result, the robot maneuvers by diving and subtly turning to the right, allowing the obstacle to remain on the surface.

The Markov chain state transitions for diving towards the right side are shown in Fig. 8.

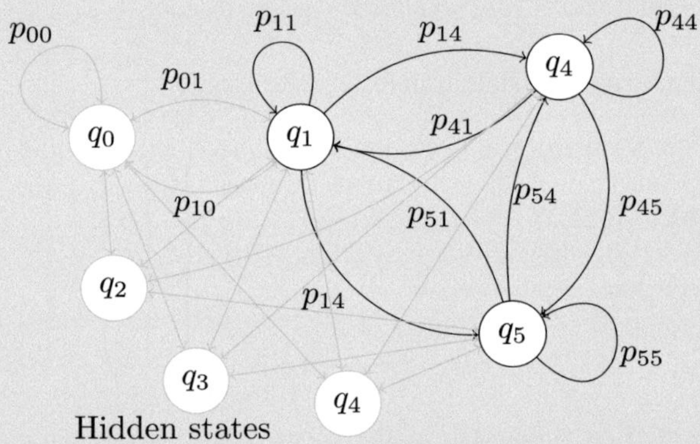

Fig. 8 Markov chain illustrating state transitions primarily affected to prevent simultaneous downward and rightward movements

Initially, the stochastic transition state vector is given by $\mathbf{q}_0 = (0, 0.99, 0, 0, 0, 0, 0)^\top$. Consequently, the subsequent prediction is obtained at the first computation as $\mathbf{q}_{k+1} = \mathbf{P} \cdot \mathbf{q}_k$.

$$\mathbf{q}_1 = \begin{pmatrix} 0.253 & 0.253 & 0.253 & 0.253 & 0.253 & 0.253 & 0.253 \\ 0 & 0.221 & 0 & 0 & 0 & 0 & 0 \\ 0.778 & 0.778 & 1 & 0.778 & 0.778 & 0.778 & 0.778 \\ 0 & 0 & 0 & 1 & 0 & 0 & 0 \\ 0 & 0 & 0 & 0 & 1 & 0 & 0 \\ 0.74 & 0.74 & 0.74 & 0.74 & 0.74 & 1 & 0.74 \\ 0 & 0 & 0 & 0 & 0 & 0 & 0.259 \end{pmatrix} \cdot \begin{pmatrix} 0 \\ 0.99 \\ 0 \\ 0 \\ 0 \\ 0 \\ 0 \end{pmatrix} = \begin{pmatrix} 0.12 \\ 0.11 \\ 0.39 \\ 0 \\ 0 \\ 0.37 \\ 0 \end{pmatrix} \begin{matrix} q_0 \\ q_1 \\ q_2 \\ q_3 \\ q_4 \\ q_5 \\ q_6 \end{matrix} \tag{20}$$

Thus, the state transition probabilities were estimated as follows: $q_0 = 12\%$ (*stop*), $q_1 = 11\%$ (*forward*), $q_2 = 39\%$ (*backward*), and $q_5 = 37\%$ (*sink*) during the first computation loop, which corresponds to a time period of 3 s. The Markov chain indicates that the robot should perform a backward propulsive motion ($q_2 = 0.39$) to avoid the front-side obstacle at 0.25m, and initiate a descent ($q_5 = 0.37$) to evade the obstacle located 0.3m above its path. Subsequently, at discrete time $k = 2$, the state predictions evolve as follows:

$$\mathbf{q}_2 = \begin{pmatrix} 0.253 & 0.253 & 0.253 & 0.253 & 0.253 & 0.253 & 0.253 \\ 0 & 0.221 & 0 & 0 & 0 & 0 & 0 \\ 0.778 & 0.778 & 1 & 0.778 & 0.778 & 0.778 & 0.778 \\ 0 & 0 & 0 & 1 & 0 & 0 & 0 \\ 0 & 0 & 0 & 0 & 1 & 0 & 0 \\ 0.74 & 0.74 & 0.74 & 0.74 & 0.74 & 1 & 0.74 \\ 0 & 0 & 0 & 0 & 0 & 0 & 0.259 \end{pmatrix} \cdot \begin{pmatrix} 0.12 \\ 0.11 \\ 0.39 \\ 0 \\ 0 \\ 0.37 \\ 0 \end{pmatrix} = \begin{pmatrix} 0.12 \\ 0.01 \\ 0.43 \\ 0 \\ 0 \\ 0.42 \\ 0 \end{pmatrix} \begin{matrix} q_0 \\ q_1 \\ q_2 \\ q_3 \\ q_4 \\ q_5 \\ q_6 \end{matrix} \tag{21}$$

The Markov state transition process for the submarine robot indicates a 43% probability of moving backward to avoid the nearest obstacle in front. Similarly, there is a 42% likelihood of diving to evade an overhead obstacle. Consequently, the 'backward' prediction is the most crucial state to consider as input for the next stage, which involves the fuzzy set of tasks.

6 Fuzzy Motion Control

This research proposes a fuzzy controller for adjusting the three motors speeds (drive, steer and ballast) taking as inputs five distinct fuzzy inputs, that include the Markov chain robot's avoidance prediction (crisp sets), the robot's linear and angular velocities, as well underwater depth, azimuth, and elevation angles. This approach allows the fuzzy controller to dynamically adjusting the temporal parameters to govern the movements of the underwater robot. In other reported works, David et al. [33] introduced a fuzzy logic-based controller for autonomous navigation and heuristic obstacle avoidance in unmanned underwater vehicles (UUVs) equipped with six thrusters for precise motion control and heading correction. Similarly, Peng et al. [34] developed a fuzzy logic guidance scheme for an autonomous underwater vehicle (AUV), utilizing multi-directional thrusters and adaptively adjusting the navigation coefficient (Fig. 9).

The proposed controller is a Mamdani Fuzzy inference controller system characterized as a multiple-input (five) multiple-output (three) governing system. The robot can execute seven distinct motion tasks to avoid obstacles, which are categorized by the first input fuzzy set (Fig. 10a): stop, move forward, move backward, turn left, turn right, dive (sink), and surface (rise). For this first fuzzy variable, we present Definition 3, which elaborates on the parametric structure of the input sets.

Definition 3 *(Input fuzzy tasks)* The output of the Markov network is mapped to a fuzzy input denoted as $q_t = \max_{q_i} \mathbf{q}$. This q_t represents the state transition prediction, maximizing the Markov network's probabilistic calculation. These values are constrained to the discrete set $\mathcal{C} = [0, 1, \ldots, 6]$, and their membership in this crisp set is denoted by $\mu(q_t)$.

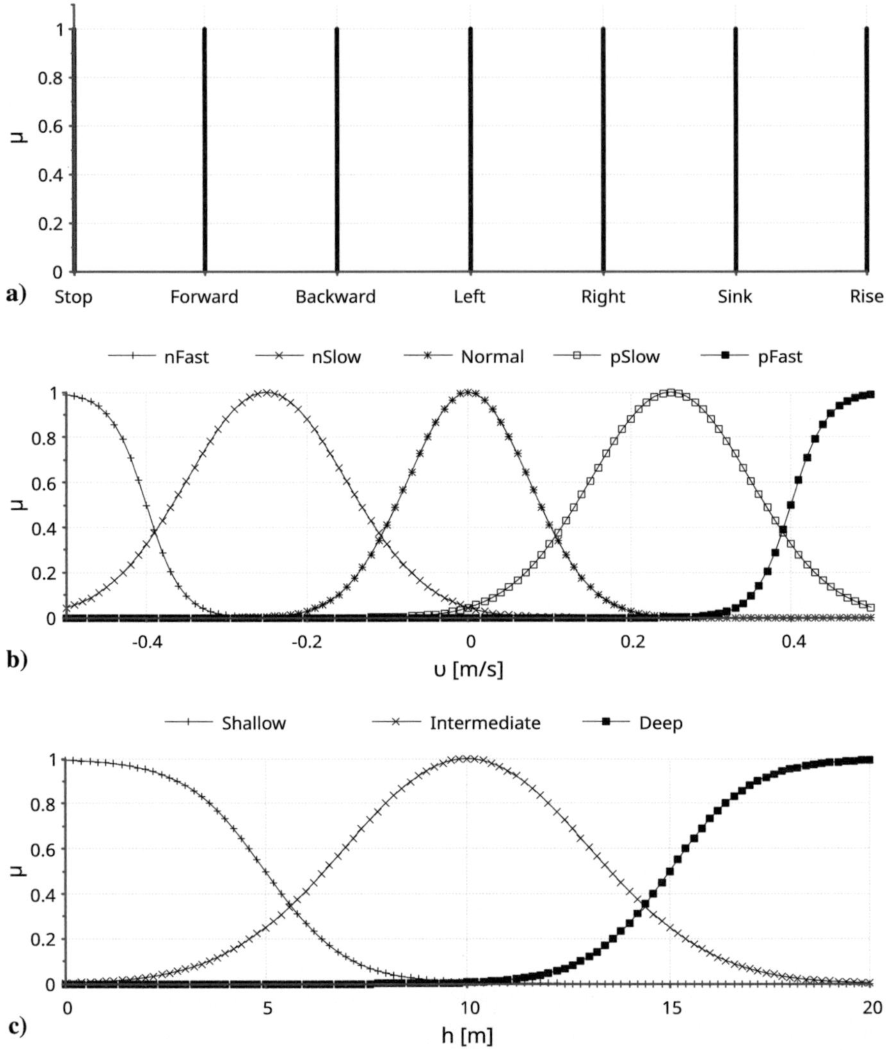

Fig. 9 Robot's input fuzzy sets. **a** Action task (HMC's output state). **b** Robot's thrusting velocity υ. **c** Robot's underwater depth h

$$\mu(q_t) = \begin{cases} 0, & q_t \notin \mathcal{C} \\ 1, & q_t \in \mathcal{C}. \end{cases} \tag{22a}$$

The submarine robot's sensor observations utilize sigmoid membership functions to model the S-shaped sets of both thrust speed υ [cm/s] and angular speed ω [rad/s]. This consistent modeling approach captures the extreme characteristics of both input variables.

Let $\mu(\upsilon)$ denote the sets categorized as 'normal', 'slow', and 'fast'. These sets are prefixed with either 'n' for negative or 'p' for positive direction, corresponding to the push velocity υ, defined by:

$$\mu_{\upsilon,h,\omega,\zeta} = \frac{1}{1+\mathrm{e}^{\pm x \mp a}}, \tag{22b}$$

where x is any of the physical variables $(\upsilon, h, \omega, \zeta)$ and a is an adjustment parameter.

Similarly, the sigmoid functions with respect to the input physical variable with adjustment parameter b is modeled as:

$$\mu_{\upsilon,h,\omega,\zeta} = \frac{1}{1+\mathrm{e}^{\pm x \mp b}} \tag{22c}$$

Moreover, the remaining intermediate sets can be described as any of the kth Gaussian membership functions corresponding to the robot's kinematic variables, characterized by a parametric mean value $\bar{x}_k$ and a standard deviation σ_k.

$$\mu_k = \mathrm{e}^{-\left(\frac{(\bar{\upsilon}-\upsilon)^2}{2\sigma_{\bar{x}_k}^2}\right)}, \tag{22d}$$

Consequently, the reasoning rules articulated within the inference engine have been designed by applying the fuzzy inputs, specifically tailored to produce desired outputs that correspond to the underwater motion variables of the underwater robot (refer to Fig. 12). These inference rules form the most basic core decision-making engine for obstacle avoidance. However, still numerous additional rules can be included to sophisticate the avoidance planning.

Specifically, for the crisp set of avoidance tasks, the rules are abbreviated as follows: S (stop), F (forward), B (backward), L (left turn), R (right turn), D (sink or dive), and U (rise or surface). Moreover, the crisp input q^{max}, a scalar component of the Markov state prediction vector, is evaluated based on the maximal value for certain inference rules, whereas other rules utilize the two or three largest values.

Figure 11 illustrates the most general obstacle avoidance scenarios, with the exception of the robot's stopping action. Sub-figures 2 through 5 delineate the fundamental fuzzy rules governing these scenarios, which are listed subsequently. It is pertinent to note that the total number of fuzzy rules can be significantly expanded to accommodate more complex obstacle arrangements arising from the combination of multiple scenarios.

1. **if** $q_1^{max} = \mathrm{S} \wedge \upsilon = \text{any} \wedge h = \text{any} \wedge \dot{\theta} = \text{any} \wedge \dot{\zeta} = \text{any}$ **then** $\dot{\varphi}_D = \text{Stop} \wedge \dot{\varphi}_S = \text{Stop} \wedge \dot{\varphi}_B = \text{Stop}$.
2. **if** $q_1^{max} = \text{Dive} \wedge \upsilon = \text{Normal} \wedge h = \text{Shallow} \wedge \dot{\theta} = \text{Straight} \wedge \dot{\zeta} = \text{Gentle}$ **then** $\dot{\varphi}_D = \text{Stop} \wedge \dot{\varphi}_S = \text{Stop} \wedge \dot{\varphi}_B = \text{Fast}^-$.
3. **if** $q_1^{max} = \text{Surface} \wedge \upsilon = \text{Normal} \wedge h = \text{Intermediate} \wedge \dot{\theta} = \text{Straight} \wedge \dot{\zeta} = \text{Gentle}$ **then** $\dot{\varphi}_D = \text{Stop} \wedge \dot{\varphi}_S = \text{Stop} \wedge \dot{\varphi}_B = \text{Fast}^+$.

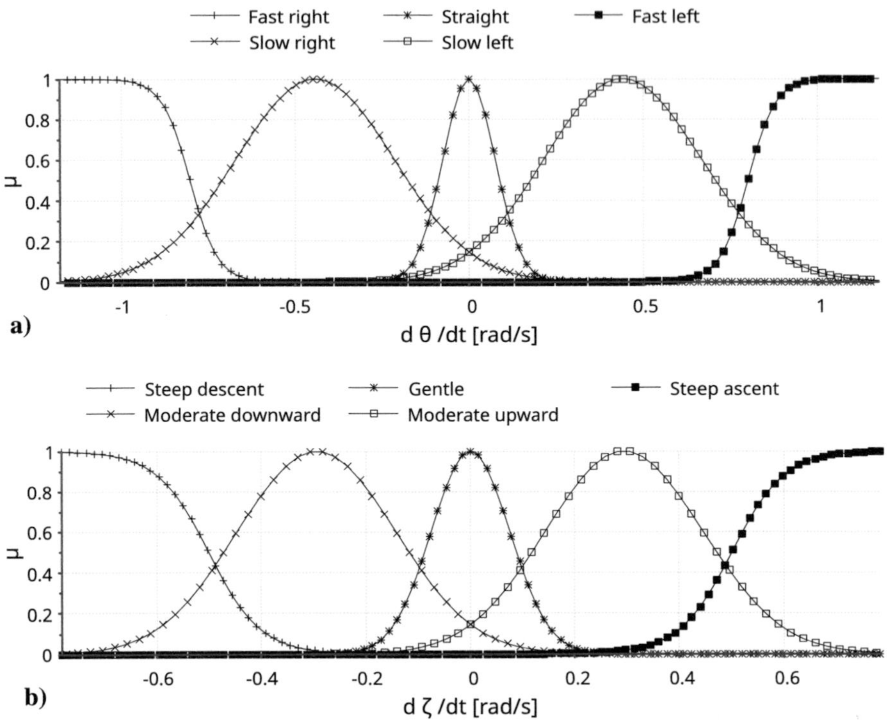

Fig. 10 Additional robot's input fuzzy sets. **a** Robot's yaw velocity $\dot{\theta}$. **b** Robot's pitch velocity $\dot{\zeta}$

4. **if** q_1^{max} = Forward $\wedge\ \upsilon$ = Normal $\wedge\ h$ = Shallow $\wedge\ \dot{\theta}$ = Straight $\wedge\ \dot{\zeta}$ = Gentle **then** $\dot{\varphi}_D$ = Fast$^+$ $\wedge\ \dot{\varphi}_S$ = Stop $\wedge\ \dot{\varphi}_B$ = Stop.
5. **if** q_1^{max} = Backward $\wedge\ \upsilon$ = Normal $\wedge\ h$ = Shallow $\wedge\ \dot{\theta}$ = Straight $\wedge\ \dot{\zeta}$ = Gentle **then** $\dot{\varphi}_D$ = Fast$^-$ $\wedge\ \dot{\varphi}_S$ = Stop $\wedge\ \dot{\varphi}_B$ = Stop.
6. **if** q_1^{max} = LeftTurn $\wedge\ \upsilon$ = Normal $\wedge\ h$ = any $\wedge\ \dot{\theta}$ = Straight $\wedge\ \dot{\zeta}$ = Gentle **then** $\dot{\varphi}_D$ = Slow$^+$ $\wedge\ \dot{\varphi}_S$ = Fast$^+$ $\wedge\ \dot{\varphi}_B$ = Stop.
7. **if** q_1^{max} = RightTurn $\wedge\ \upsilon$ = Normal $\wedge\ h$ = any $\wedge\ \dot{\theta}$ = Straight $\wedge\ \dot{\zeta}$ = Gentle **then** $\dot{\varphi}_D$ = Slow$^+$ $\wedge\ \dot{\varphi}_S$ = Fast$^+$ $\wedge\ \dot{\varphi}_B$ = Stop.

Therefore, the output fuzzy sets outline the rotary speeds of the robotic submarine's propels and ballasting actuators, as illustrated in Fig. 12.

The inference engine rules define the application of fuzzy operators used to evaluate the terms involved in fuzzy decision-making. For any of the input fuzzy sets, the membership value is the one that maximizes

For the avoidance task the crisp set chosen is

$$\mu_q^{max} = \max_{\mu_k \in q_k} \left(q_S, q_F, q_B, q_L, q_R, q_D, q_U \right), \tag{23a}$$

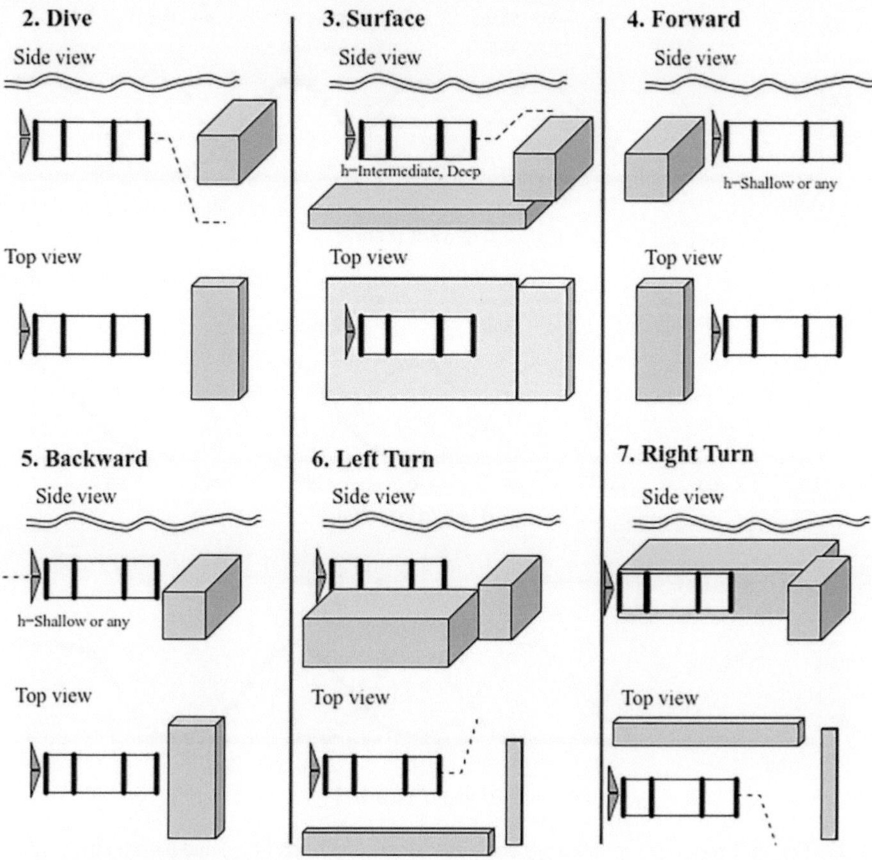

Fig. 11 The six most fundamental fuzzy rules, outlining responses to the most general obstacle avoidance scenarios

as for the linear speed

$$\mu_{\upsilon}^{max} = \max_{\mu_k \in \upsilon} \left(\mu_{nfast}, \mu_{nslow}, \mu_{normal}, \mu_{pslow}, \mu_{pfast} \right), \tag{23b}$$

likewise, the depth value membership is the one that maximizes a given set,

$$\mu_{h}^{max} = \max_{\mu_k \in h} \left(\mu_{shallow}, \mu_{intermediate}, \mu_{deep} \right) \tag{23c}$$

similarly for the robot's yaw speed

$$\mu_{\dot{\theta}}^{max} = \max_{\mu_k \in \dot{\theta}} \left(\mu_{fastR}, \mu_{slowR}, \mu straight, \mu_{slowL}, \mu_{fastR} \right) \tag{23d}$$

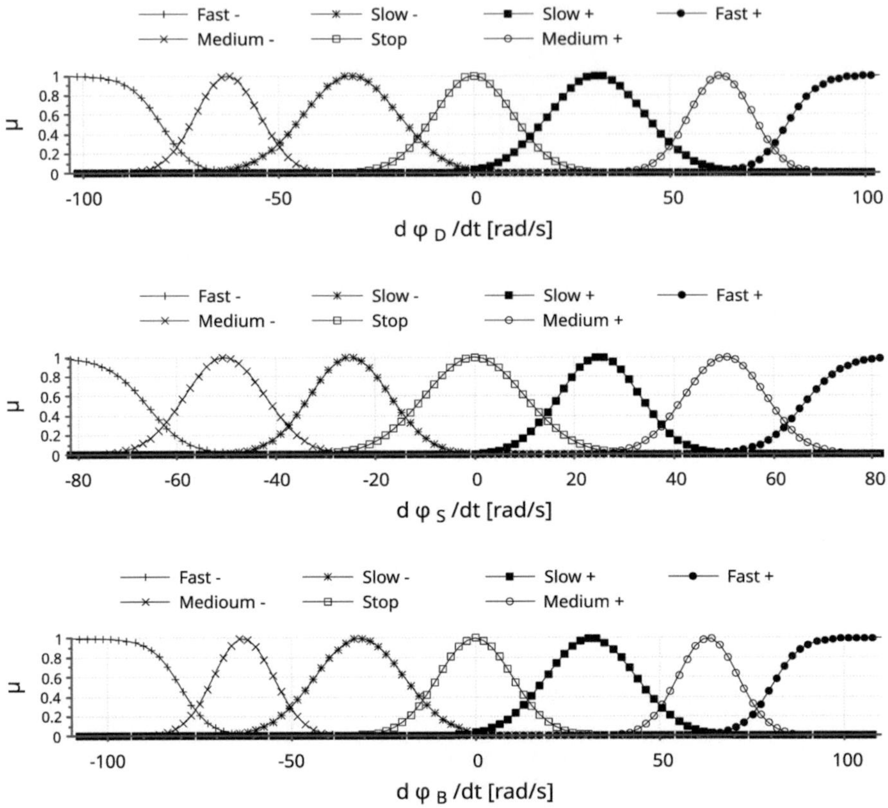

Fig. 12 Output fuzzy sets for the rotary actuators: thrust ($\dot{\varphi}_D$), steer ($\dot{\varphi}_S$) and ballast ($\dot{\varphi}_B$)

and for the robot's pitch speed

$$\mu_{\dot{\zeta}}^{max} = \max_{\mu_k \in \dot{\zeta}} \left(\mu_{steepDesc}, \mu_{modDown}, \mu_{gentle}, \mu_{modUp}, \mu_{steepAsc} \right) \tag{23e}$$

Based on the logical premise of a given proposition rule, the application of the subsequent fuzzy operator is global for the input sets,

$$\mu^{max} = \max_{\mu_k \in q \cup \upsilon \cup \ldots \zeta} \left(\mu_q^{max}, \mu_\upsilon^{max}, \mu_h^{max}, \mu_\theta^{max}, \mu_\zeta^{max} \right).$$

Proceeding to the defuzzification process, the primary objective is to derive an inverse solution. The fuzzy output sets, however, are categorized into two types of distributions: Gaussian and sigmoid, as outlined in Definition 4. The functional form of Gaussian distributions is also formally specified in Definition 4.

Definition 4 *(Output fuzzy sigmoid variables)* The membership functions for any of the sigmoid output sets are generally defined as:

$$\mu(\dot{\varphi}) = \frac{1}{1 + \mathrm{e}^{\pm\dot{\varphi}_k \mp c_k}}. \quad (24)$$

In this expression, $\dot{\varphi}_k$ [rad/s] denotes the angular velocity of one of the propulsion motors, with the sign indicating the direction of rotation. The parameter c_k represents a displacement factor, and k is the numerical index corresponding to a specific set.

Furthermore, the membership functions for the Gaussian output sets are generalized for any of the outputs as follows:

$$\mu(\dot{\varphi}) = \mathrm{e}^{-\left(\frac{(\bar{\dot{\varphi}}_k - \dot{\varphi})^2}{2\sigma^2_{\dot{\varphi}_k}}\right)}, \quad (25)$$

Here, $\bar{\dot{\varphi}}_k$ represents the mean value, while $\sigma_{\dot{\varphi}_k}$ denotes the standard deviation of the set k, where k serves as the numerical index for a specific set.

According to Definition 4, each μ_k yields a normalized membership value within the interval $\mu_k \in [0, \ldots, 1]$. The inverse sigmoid membership function, denoted as $T_k \in \mathbb{R}\ \forall\ \mu_k$, is given by the following general inverse expression:

$$\dot{\varphi}(\mu_k) = \mp \ln\left(\frac{1}{\mu_k} \pm c_k\right). \quad (26)$$

Similarly, the inverse Gauss function membership $v\dot{a}rphi_k \in \mathbb{R}\ \forall\ \mu_k$, is defined for by the following inverse solution:

$$\dot{\varphi}(\mu_k) = \mu_k - \sqrt[2]{-2\sigma_k^2 \ln(\mu_k)}. \quad (27)$$

Consequently, the Centroid method is utilized for defuzzification, specifically for the jth category of output fuzzy sets activated by the pertinent fuzzy inference rule for estimating $\dot{\varphi}_k$, using the following expression:

$$\dot{\varphi}_{D,S,B} = \frac{\sum_j \mu_j^* \dot{\varphi}_j(\mu_j^*)}{\sum_j \mu_j^*}, \quad (28)$$

or more specifically expressed $f^{\pm}$ (positive/negative fast), $m^{\pm}$ (positive/negative medium), $s^{\pm}$ (positive/negative slow) and $st^{\pm}$ (stop). such that,

$$\dot{\varphi}_{D,S,B} = \frac{\mu_f^* \dot{\varphi}(\mu_f^*) + \mu_m^* \dot{\varphi}(\mu_m^*) + \mu_s^* \dot{\varphi}(\mu_s^*) + \mu_{st}^* \dot{\varphi}(\mu_{st}^*) + \mu_{ts}^*(\mu_{ts}^*)}{\mu_f^* + \mu_m^* + \mu_s^* + \mu_s^* + \mu_{ts}^*}, \quad (29)$$

Assuming that for the defuzzification solution, either the inverse sigmoid (26) or the Gauss (27) applies depending on the type of probabilistic set.

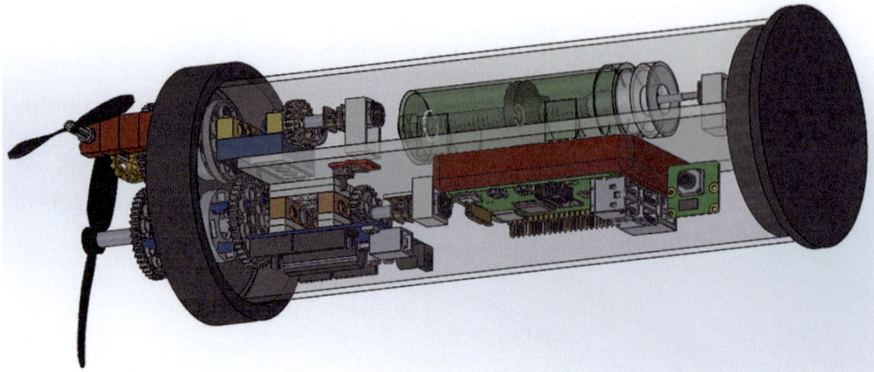

Fig. 13 Underwater robot's inner elements

7 Propulsion, Steer and Ballast Controllers

This section details the dynamic model of the robot, with an emphasis on the propulsion and ballast mechanisms, which are governed by the numerical outputs of the fuzzy controller. In contrast to conventional underwater biorobots that feature bionic designs and underactuated mechanisms [35], the robot introduced here employs a unique design centered on geared speed reducers and an electric piston, as illustrated by Fig. 13. Although it does not replicate the appearance of a biorobot, this innovative design preserves the underactuated functionality, combining mechanical efficiency with precision control.

7.1 Propulsion Dynamics

Consider φ_D as the controlled fuzzy output variable, transmitted to a gear with a radius of r_D which is the DC motor pinion or global conductor gear. Gears ϕ_1 and ϕ_2, both having the same radius $r_1 = r_2$, are connected in parallel to the same shaft, as depicted in Fig. 14.

Consequently, the transmission of rotary motion is accomplished by

$$\phi_1 = \phi_2 = -\frac{r_D}{r_1}\varphi_D \tag{30}$$

and subsequently, external gear ϕ_D fixes the outside propelling element and is connected in parallel by a set of magnets throughout the chassis with inner gears ϕ_3 and ϕ_4, such that

$$\phi_D = \phi_4 = \phi_3 = \frac{r_D r_2}{r_1 r_3}\varphi_D. \tag{31}$$

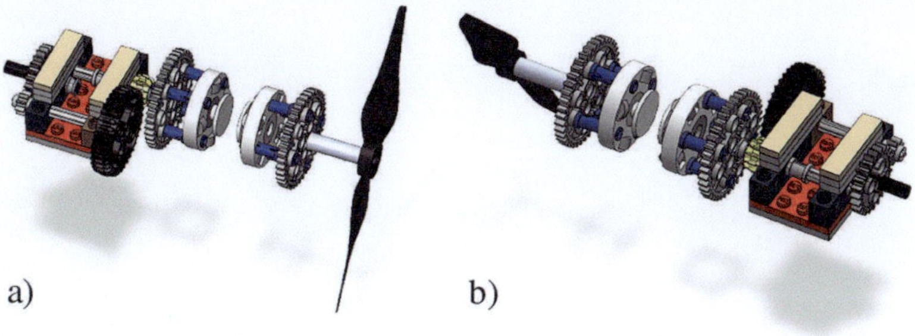

Fig. 14 Robot's propulsion mechanism employing permanent magnets to transmit motion through the chassis wall

Furthermore, the angular moment the propelling element reaches is disclosed by the following dynamic function given in terms of the numerical controlled fuzzy output for thrusting φ_D and $\ddot{\varphi}_D$:

$$\tau_D = \left(\frac{m_D r_D^2 r_D r_2}{r_1 r_3}\right)\ddot{\varphi}_D + m_D g r_D \cos\left(\frac{r_D r_2}{r_1 r_3}\varphi_D\right), \tag{32}$$

where m_D is the mass of the propelling gear D, g is the gravity acceleration and τ_D is the propelling torque τ_D [Nm].

To derive the underwater linear velocity v and thrust force F_D of the robotic submarine deploying a double-helix propeller, this work proposed to model a relationship between the propeller's angular velocity $\dot{\varphi}_D$ and the resulting motion. Thus, the submarine when moving in the water (viscous fluid), the robot experiencing drag. When the double-helix propeller generates thrust along the longitudinal axis, the system reaches a steady-state velocity where thrust balances drag.

The thrust force from the propeller F_D is generated by the propeller that is related to its fuzzy controller angular velocity $\dot{\varphi}_D$ (rad/s), which is modeled as

$$F_D = K_T \rho \dot{\varphi}_D^2 \ell_D^4, \tag{33}$$

where, K_T is a dimensionless thrust coefficient that depend on propeller geometry. ρ is the water density (approx. 1000 kg/m^3), and ℓ_D is the thrusting propeller diameter. Alternatively, for a helical propeller, thrust can be expressed as

$$F = C_T \dot{\varphi}_D^2, \tag{34}$$

where C_T is a lumped thrust coefficient that depends on its propeller pitch P (distance advanced per revolution), its blade geometry and fluid properties.

In this work, the robot's dragging force is assumed at steady state, thrust F_D equals the drag force F_g and are counteracted which means constant velocity of equilibrium condition, and it is modeled as:

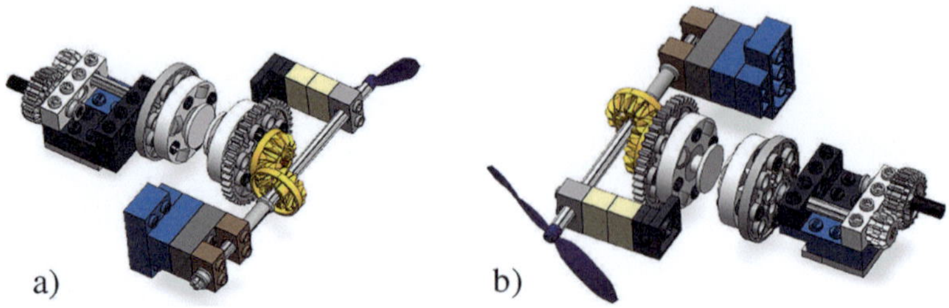

Fig. 15 Robot's steering mechanism employing permanent magnets to transmit motion through the chassis wall

$$F_g = \frac{1}{2}\rho C_D A \upsilon^2 \tag{35}$$

where C_D is a drag coefficient (depending on shape). A is the robot's cross-sectional area, and υ_t submarine's instantaneous linear velocity (m/s).

Therefore, the steady-state velocity at equilibrium point $F_g = F_D$ is established as

$$C_T \dot{\varphi}_D^2 = \frac{1}{2}\rho C_D A \upsilon_t^2 \tag{36}$$

Thus, solving for υ_t,

$$\upsilon_t = \dot{\varphi}_D \sqrt[2]{\frac{2C_T}{\rho C_D A}}. \tag{37}$$

This shows that velocity scales linearly with angular velocity $\dot{\varphi}_D$ under ideal conditions, but is limited by drag. Concerning drag effects, when the robotic submarine operates in deeper water, remote from boundaries such as the surface, drag forces may be reduced (Fig. 15).

7.2 *Steering Dynamics*

Similarly, as depicted in Fig. 16 the steering mechanism includes another smaller propelling element pointing out laterally, mechanically connected to external gear ϕ_S and conducted by the global conductor gear that is governed by the controlled fuzzy output φ_S,

$$\phi_S = \frac{r_4 r_S}{r_5 r_1}\varphi_S, \tag{38}$$

and essentially the dynamic function for torque steering τ_S [Nm] is obtained and stated in terms of the controlled fuzzy output for the robot's steering motion.

$$\tau_S = \left(\frac{m_5 r_5^2 r_4 r_0}{r_5 r_1}\right) \ddot{\varphi}_S + m_5 g r_5 \cos\left(\frac{r_4 r_0}{r_5 r_1} \varphi_S\right). \tag{39}$$

In order to derive the robot's body yaw rate or angular velocity $\dot{\theta}$ and the yaw's angular momentum M_Y (around z-axis) due to the lateral dual-helix steering propeller with fuzzy controlled rotary speed $\dot{\varphi}_S$, the torque generated by the steering propeller and the resulting rotational dynamics is analyzed.

The robot has a cylindrical body with mass m and moment of inertia I_z about the yaw z-axis (vertical axis). In addition, the steering propeller is positioned at a distance r from the center of mass (near the back-side perimeter). The steering propeller is purposed to generate a lateral thrust force F_s that induces an angular moment M_Y for yaw rotation. Besides, it is assumed the system reaches a steady-state yaw rate where the torque balances rotational drag.
In order to perform a steering thrust force, the lateral thrust force F_s from the steering propeller is modeled acting perpendicular to the robot's longitudinal axis, such that,

$$F_s = C_{T_s} \dot{\varphi}_s^2, \tag{40}$$

where, C_{T_s} is the thrust coefficient of the steering propeller that depend on blade geometry, fluid density ρ, and propeller diameter D_s.

The robot's yaw torque generation expressed by τ_Y concerns the steering propeller, assuming it is located at the center of mass, located at a distance r, then

$$\tau_Y = F_s \cdot r = C_{T_s} \dot{\varphi}_s^2 \cdot r, \tag{41}$$

this torque causes the robot to rotate about its yaw axis. Furthermore, the yaw motion rotational dynamics involves the yaw acceleration that follows from Euler's rotation equation,

$$\tau_Y = I_z \ddot{\theta}, \tag{42}$$

where I_z is the moment of inertia of the robot about the yaw axis (for a cylinder shape, $I_z \approx \frac{1}{2} m R^2$ if mass is evenly distributed, where R is the cylinder radius), and $\ddot{\theta}$ is the robot's angular acceleration.

At steady state, the torque τ_Y is balanced by the rotational drag torque τ_g,

$$\tau_g = -C_{g_r} \dot{\theta}, \tag{43}$$

where C_{g_r} is the rotational drag coefficient. Hence, setting $\tau_Y = \tau_g$,

$$C_{T_s} \omega_s^2 r = C_{d_r} \dot{\psi}, \tag{44}$$

and solving for the steady-state yaw rate $\dot{\theta}$:

$$\dot{\theta} = \frac{C_{T_s} r}{C_{g_r}} \dot{\varphi}_s^2. \tag{45}$$

Moreover, the robot's body angular momentum M_Y is stated by

$$M_Y = I_z\dot{\theta} = I_z\left(\frac{C_{T_s}r}{C_{g_r}}\dot{\varphi}_s^2\right). \tag{46}$$

Therefore, the robot's yaw rate as a function of the steering propeller's angular velocity is

$$\dot{\theta} = \left(\frac{C_{T_s}r}{C_{g_r}}\right)\dot{\varphi}_s^2 \tag{47}$$

The yaw rate scales with the square of the steering propeller's angular velocity $\dot{\varphi}_s^2$. In addition, by increasing the lever arm length or the thrust coefficient C_{T_s} improves maneuverability. Higher rotational drag C_{g_r} reduces yaw rate.

Therefore, the robot's angular momentum (yaw) is essentially described by the following expression:

$$M_Y = I_z\dot{\theta} = I_z\left(\frac{C_{T_s}r}{C_{g_r}}\right)\dot{\varphi}_s^2, \tag{48}$$

as well as the torque produced by steering propeller around the robot's z-axis,

$$\tau_Y = C_{T_s}r\dot{\varphi}_s^2. \tag{49}$$

Moreover, in practice these authors have found the effects of the cross-coupling. The main propeller's thrust may induce fluid swirl, affecting steering efficiency, and a non-symmetric wake could require compensation in control. In addition, the rotational drag coefficient C_{g_r} depends on the robot's shape and surface roughness. Concerning the effects of transient dynamics, before reaching steady-state $\dot{\theta}$, the robot undergoes angular acceleration, given by

$$\ddot{\theta} = \frac{\tau_Y}{I_z} = \frac{C_{T_s}r}{I_z}\dot{\varphi}_s^2. \tag{50}$$

Thus, in order to achieve a controlled yaw rate $\dot{\theta}_{t+1}$, we must solve for φ_s,

$$\varphi_s = \sqrt{\frac{C_{g_r}\dot{\theta}_{t+1}}{C_{T_s}r}} \tag{51}$$

7.3 Pitch Dynamics

Moreover, to derive the robot's pitch rate $\dot{\zeta}$ and pitch angular momentum M_P, the combined effects of both propellers are considered as strategy in this research. The main (longitudinal) propeller (thrust F_D) at the rear center and the steering (lateral) propeller (thrust F_s) offset from the centerline by distance r. The propellers induce pitching moments due to their thrust vectors and positions. The robot geometry is assumed of length l, with center of mass (CoM) located at the geometric center.

The thrust forces tangential and around the y-axis is given by previously stated $F_D = C_T \dot{\varphi}_D^2$ (thrust propeller). Likewise, the steering propeller $F_s = C_{T_s} \dot{\varphi}_s^2$ that acts laterally at position. Further, the linear damping torque $C_{g_p} \dot{\zeta}$) opposes pitch motion. The moment of inertia is I_y about the pitch axis. Given that the robot's shape is a cylinder, $I_y \approx \frac{1}{4} m R^2 + \frac{1}{12} m l^2$.

Therefore, the pitch torque τ_P arises from the longitudinal thrust F_t. If the main propeller is offset vertically by h_x (e.g., due to buoyancy), it creates a moment arm h_x,

$$\tau_P = F_D \cdot h_x = C_T \dot{\varphi}_D^2 \cdot h_x. \tag{52}$$

Moreover, the lateral thrust F_s produced by the steering propeller's lateral force F_s acts creating a moment arm x about the CoM:

$$\tau_s = F_s \cdot h_y = C_{T_s} \omega_s^2 \cdot h_y \tag{53}$$

Hence, in three-dimension Cartesian space, the combined total pitching torque is established by

$$\tau_Y = \tau_P + \tau_s = C_T h_x \dot{\varphi}_D^2 + C_{T_s} h_y \dot{\varphi}_S^2. \tag{54}$$

Thus, it follows that the robot's pitch acceleration is described by the expression

$$\tau_Y = I_y \ddot{\zeta} \tag{55}$$

Therefore, at steady-state, the torque balances rotational drag $C_{g_p} \dot{\zeta}$:

$$C_T h_x \dot{\varphi}_D^2 + C_{T_s} h_y \dot{\varphi}_s^2 = C_{g_p} \dot{\zeta} \tag{56}$$

Thus, algebraically establishing an analytical solution for the pitch rate $\dot{\zeta}$:

$$\dot{\zeta} = \frac{C_T h_x \dot{\varphi}_D^2 + C_{T_s} h_y \dot{\varphi}_s^2}{C_{g_p}} \tag{57}$$

Likewise, the pitch angular momentum about the pitch axis is deduced as

$$M_P = I_y \dot{\zeta} = I_y \left(\frac{C_T h_x \dot{\varphi}_D^2 + C_{T_s} h_y \dot{\varphi}_s^2}{C_{g_p}} \right) \tag{58}$$

From previous statements, we met a main propeller dominance, if $h_x \neq 0$, $\dot{\varphi}_D$, then it induces pitch even when $\dot{\varphi}_s = 0$ (e.g., during straight-line motion). The steering propeller impact is provided by the term $C_{T_s} h_x \dot{\varphi}_s^2$, which allow active pitch control by varying $\dot{\varphi}_s$. Besides, a positive h_x (main propeller below CoM) causes upward

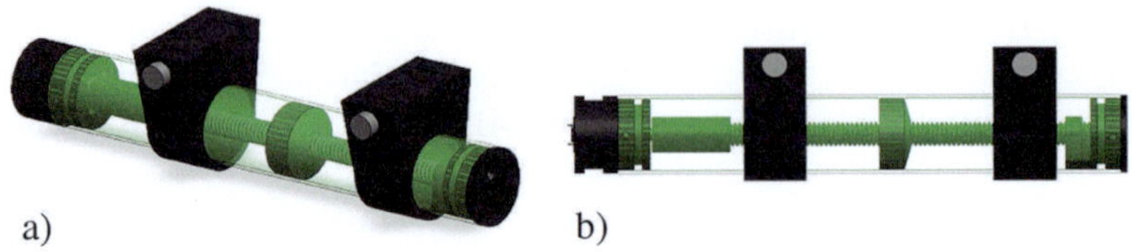

Fig. 16 Robot's ballasting mechanism employing a screw-driven hydraulic piston to regulate water intake and expulsion for controlling diving and surfacing motions

pitch $\dot{\zeta} > 0$ for forward thrust.
In addition, when the robot is buoyant, by adding a restoring torque $-\kappa\zeta$ (e.g., $\kappa = mgz_{\text{CB}}$, then z_{CB} is the center of buoyancy height).

Concerning, the transient response, before steady-state, the instantaneous pitch angle ζ_t evolves as

$$I_y\ddot{\zeta} + C_{g_p}\dot{\zeta} = C_T h_x \dot{\varphi}_D^2 + C_{T_s} h_y \dot{\varphi}_s^2. \tag{59}$$

Thus, in order to achieve a desired controlled pitch rate $\dot{\zeta}_{t+1}$, then solve for φ_D and φ_s, such that

$$C_T h_x \varphi_D^2 + C_{T_s} h_x \varphi_s^2 = C_{g_p}\dot{\zeta}_{t+1} \tag{60}$$

From the proposed approach, the pitch rate is proportional to the weighted sum of φ_D^2 and φ_s^2, and pitch angular momentum scales with I_y and the net torque.

7.4 *Ballasting Dynamics*

In this study, the ballasting system garnered special focus due to its unique characteristics, setting it apart from the thrusting and steering mechanisms. It operates using a hydraulic piston equipped with pressure measurement feedback, which precisely controls the vertical depth movement. The study by [36] presents an EMG-based cybernetic control framework for a robotic fish processed by a neural network incorporated to a fuzzy-based oscillation pattern generator to produce swimming motions. While their proposed ballasting system may exhibit a few implementation differences from the current context, the dynamic model deduction exhibits significant similarities.

Regarding the screw mechanism, let L [m] be a constant parameter known as the advance coefficient, provided by the screw manufacturer. Additionally, let φ_B represent the controlled fuzzy output for the ballasting DC motor, such that the hydraulic piston kinematic displacement, s_B, is estimated by the equation:

$$s_B = L \cdot \varphi_B \tag{61}$$

and the piston screw's torque is established as

$$\tau_B = m_B r_B^2 \ddot{\varphi}_B + m_B g r_B \cos(\varphi_B), \tag{62}$$

with screw mass m_B and radius r_B, and gravity acceleration constant g. Therefore, the general piston's push/pulling force f_B during the ballasting task is expressed by

$$f_B = m_B L \ddot{\varphi}_B - m_1 g \tag{63}$$

The schematic diagram on the lower side of Fig. 13 illustrates the fundamental components of a hydraulic piston, which are essential for control modeling.

The primary operational functions of the ballast device involve filling the chamber of its container with water to achieve submersion or expelling water from the container to attain buoyancy. Both actions entail the application of a linear force to manipulate a cylindrical hydraulic piston or plunger, thereby controlling the flow of water through suction or expulsion. Consequently, the volume of the liquid mass fluctuates over time, dependent on a control reference or desired level denoted as H, coupled with the quantification of a filling rate $u(t)$ and the measurement of the actual liquid level $h(t)$.

Therefore, we can characterize the filling rate $u(t)$ as the rate of change of volume V with respect to time, expressed as

$$\frac{dV(t)}{dt} = u(t), \tag{64}$$

assuming a cylindrical fluid chamber of radius r and circular area $A = \pi r^2$, the volume is expressed as

$$V(t) = Ah(t), \tag{65}$$

In here, $h(t)$ represents the real position of the plunger due to volume of incoming hydraulic mass. Thus, in consequence the filling rate may also be expressed by

$$u(t) = k(H - h(t)). \tag{66}$$

Let k represent an adjustment coefficient and H the reference or desired fill level, with $h(t)$ denoting the instantaneous longitudinal fill level. Substituting the preceding expressions into the initial equation (64), we arrive at the following first-order linear differential equation:

$$A\frac{dh(t)}{dt} = k(H - h(t)). \tag{67}$$

To solve the mentioned equation, we employ the method of integrating factor, so that

$$\dot{h}(t) + \frac{k}{A}h(t) = \frac{k}{A}H. \tag{68}$$

In this case, the integrating factor is determined as follows:

$$\mu(t) = e^{\int \frac{k}{A} dt} = e^{\frac{kt}{A}}. \tag{69}$$

Consequently, application of the integrating factor effectively reduces the order of the derivatives in the following steps.

$$e^{\frac{kt}{A}} \dot{h}(t) + e^{\frac{kt}{A}} \frac{k}{A} h(t) = \frac{k}{A} H e^{\frac{kt}{A}}, \tag{70}$$

Algebraic simplification of the left-hand side of the aforementioned expression yields the following result.

$$\left(h(t) e^{\frac{kt}{A}}\right)' = \frac{k}{A} H e^{\frac{kt}{A}}. \tag{71}$$

Subsequently, upon integrating both sides of the equation with respect to time,

$$\int_t \left(h(t) e^{\frac{kt}{A}}\right)' dt = \int_t \frac{k}{A} H e^{\frac{kt}{A}} dt, \tag{72}$$

where the expression on the left-hand side transforms into

$$h(t) e^{\frac{kt}{A}} = \frac{k}{A} H \int_t e^{\frac{kt}{A}} dt, \tag{73}$$

and the right side of the equation, once solved, transforms into

$$h(t) e^{\frac{kt}{A}} = H e^{\frac{kt}{A}} + c. \tag{74}$$

Now, to obtain the solution for $h(t)$, we isolate by rearranging the term $e^{\frac{kt}{A}}$:

$$h(t) = H + c e^{-\frac{kt}{A}} \tag{75}$$

Based on the initial condition $h(t_0) = 0$, which signifies that the piston is fully retracted within the chamber at the initial time $t_0 = 0$s the integration constant c is subsequently evaluated as

$$0 = H + c e^0. \tag{76}$$

Therefore, the value of c is $c = -H$, and by substituting it into the previously obtained solution,

$$h(t) = H(1 - e^{-\frac{kt}{A}}). \tag{77}$$

Moreover, given that the required piston force f_e is expressed as:

$$f_e = m(t) \frac{dv}{dt} + f_k + \rho_a A, \tag{78}$$

where f_k denotes the frictional force acting on the piston within the hydraulic cylinder, and $\rho_a A$ represents the hydrostatic pressure exerted at that depth upon the piston's inlet area. The instantaneous mass accounts for both the piston mass, m_e, and the mass of the inflowing water m_a:

$$m(t) = m_e + m_a, \tag{79}$$

where the density of the water is given by $\delta_a = \frac{m_a}{V_a}$ y and the volume of water is expressed as $V_a = \pi r^2 h(t)$, thereby completing the mass model:

$$m(t) = m_e + \delta_a \pi r^2 H(1 - e^{-\frac{kt}{A}}). \tag{80}$$

Consequently, the force required for actuation of the piston device is governed by the control law given by

$$f_e = \left(m_e + \delta_a \pi r^2 H(1 - e^{-\frac{kt}{A}})\right) \frac{dv}{dt} + f_k + \rho_a A. \tag{81}$$

7.5 Stability Analysis

To ascertain the equilibrium point and stability characteristics of the robotic submarine's motion control system, the ballasting task was subjected to analysis. Empirical observations by the authors revealed that the vertical diving and surfacing motion exhibited moderate to significant instability, particularly during rapid ascent, frequently leading to uncontrolled buoyancy oscillations (porpoising). Furthermore, during buoyancy-pitch coupled controlled maneuvers, rapid alterations in depth induced substantial pitch moments, especially in instances of misalignment between the center of buoyancy and the center of mass. Similarly, near-surface diving operations were susceptible to wave disturbances and ventilation effects arising from fluctuating hydrodynamic conditions. Moreover, a prevalent phenomenon encountered was ballast delay, wherein sluggish ballast systems contributed to overshooting desired depth targets due to the persistent influence of gravitational and hydrostatic pressure forces.

The vertical motion of a robotic submarine can be modeled by a second-order linear differential equation that accounts for buoyancy, damping, and control forces.

$$m\frac{d^2z}{dt^2} + b\frac{dz}{dt} + kz = F_t, \tag{82}$$

where $z(t)$ is the vertical displacement (depth) of the submarine at time t. m is the effective mass of the robotic submarine (including added mass due to water displacement). Let b be the damping coefficient (due to water resistance). Let k be the buoyancy stiffness (related to restoring force from buoyancy). Let F_t be any external control force (e.g., from thrusters or ballast adjustment).

The inertia term $m\frac{d^2z}{dt^2}$ represents the submarine's resistance to acceleration. Likewise, the damping term $b\frac{dz}{dt}$ essentially models the drag/friction from water. The restoring term kz represents the buoyancy trying to return the submarine to equilibrium depth. Eventually, the forcing term F_t is a general external control force input.

Consequently, in the event of instantaneous ballasting-induced free motion occurring at a specific underwater depth, the robot is required to maintain a stable equilibrium at that precise depth.

When $F_t = 0$, the equation becomes:

$$m\frac{d^2z}{dt^2} + b\frac{dz}{dt} + kz = 0 \tag{83}$$

This describes the robotic submarine's natural oscillatory motion (damped or undamped, depending on b and k.

Moreover, deriving the state-space representation from the given second-order linear differential equation for the robotic submarine's vertical motion.

Rewriting the differential equation as a system of first-order ordinary differential equation, the state variables are defined as position and velocity respectively,

$$x_1 = z, \quad x_2 = \dot{x}_1 = \dot{z}, \quad \dot{x}_2 = \ddot{z}. \tag{84}$$

Therefore, the second-order equation can be split into two first-order equations in thye form of state variables:

$$m\dot{x}_2 + bx_2 + kx_1 = F_t \tag{85}$$

separating the highest derivative implies that

$$\dot{x}_2 = -\frac{b}{m}x_2 - \frac{k}{m}x_1 + \frac{1}{m}F_t. \tag{86}$$

Thus, state-space matrix representation in its standard state-space form is

$$\dot{\mathbf{x}} = \mathbf{A}\mathbf{x} + \mathbf{B}u, \quad y = \mathbf{C}\mathbf{x} + \mathbf{D}u, \tag{87}$$

where $\mathbf{x} = [x_1, x_2]^T$ is state vector, $u = F_t$ is an input signal, and y is the output (here, we state that $y = z = x_1$ to measure depth). Thus,

$$\dot{\mathbf{x}} = \begin{bmatrix} 0 & 1 \\ -\frac{k}{m} & -\frac{b}{m} \end{bmatrix}\mathbf{x} + \begin{bmatrix} 0 \\ \frac{1}{m} \end{bmatrix}u, \tag{88}$$

$$y = \begin{bmatrix} 1 & 0 \end{bmatrix}\mathbf{x} + 0 \cdot u. \tag{89}$$

The state matrix **A** encodes the robotic submarine's natural dynamics (oscillations and damping). The input matrix **B** shows how the control force u affects acceleration. Likewise, the output matrix **C** selects depth x_1 as the measurable output.

As previously stated, considering the unforced system $u = 0$, such that $\dot{\mathbf{x}} = A\mathbf{x}$.

The system is asymptotically stable if all eigenvalues of the state matrix **A** have negative real parts. Thus, the characteristic equation is

$$\det(s\mathbf{I} - \mathbf{A}) = s^2 + \frac{b}{m}s + \frac{k}{m} = 0. \tag{90}$$

The eigenvalues are

$$s = \frac{-\frac{b}{m} \pm \sqrt{\left(\frac{b}{m}\right)^2 - 4\frac{k}{m}}}{2}. \tag{91}$$

If $b > 0$ and $k > 0$, then the real parts of s are negative or stable for any physically realistic parameters.
In addition, if $b = 0$, the system is marginally stable and oscillates indefinitely. For linear systems, a quadratic Lyapunov function $V(\mathbf{x})$ is sufficient:

$$V(\mathbf{x}) = \mathbf{x}^T\mathbf{P}\mathbf{x}, \tag{92}$$

where **P** is a symmetric positive definite matrix ($P \succ 0$). To prove stability, we require:

$$\dot{V}(\mathbf{x}) = \mathbf{x}^T(\mathbf{A}^T\mathbf{P} + \mathbf{P}\mathbf{A})\mathbf{x} < 0 \quad \forall \mathbf{x} \neq 0. \tag{93}$$

Consequently, in solving the Lyapunov equation, it is required that the matrix **Q** be positive definite ($\mathbf{Q} \succ 0$, e.g., $\mathbf{Q} = \mathbf{I}$) and subsequently solve for the matrix **P** in the following equation:

$$\mathbf{A}^T\mathbf{P} + \mathbf{P}\mathbf{A} = -\mathbf{Q}. \tag{94}$$

For the state matrix **A**, let $\mathbf{Q} = \mathbf{I}$ be stated, such that:

$$\begin{bmatrix} 0 & -\frac{k}{m} \\ 1 & -\frac{b}{m} \end{bmatrix} \mathbf{P} + \mathbf{P} \begin{bmatrix} 0 & 1 \\ -\frac{k}{m} & -\frac{b}{m} \end{bmatrix} = -\begin{bmatrix} 1 & 0 \\ 0 & 1 \end{bmatrix}. \tag{95}$$

Let us assume the matrix **P**. Thus, by substituting and solving the linear system for **P**. To solve for the symmetric positive definite matrix P in the Lyapunov equation of the robotic submarine's state-space system,

$$\mathbf{P} = \begin{pmatrix} p_{11} & p_{12} \\ p_{12} & p_{22} \end{pmatrix} = \begin{pmatrix} \frac{b}{2k} & \frac{m}{2k} \\ \frac{m}{2k} & \frac{m}{2b}\left(1 + \frac{m}{k}\right) \end{pmatrix}. \tag{96}$$

For **P** to be positive definite: $p_{11} = \frac{b}{2k} > 0$ (true if $b, k > 0$). For $\det(\mathbf{P}) = p_{11}p_{22} - p_{12}^2 > 0$:

$$\det(\mathbf{P}) = \left(\frac{b}{2k}\right)\left(\frac{m}{2b} + \frac{m^2}{2bk}\right) - \left(\frac{m}{2k}\right)^2 = \frac{m}{4k} > 0. \tag{97}$$

It is always true for $m, k > 0$. The terms p_{11}, p_{12}, p_{22} depend on $\frac{b}{k}$ and $\frac{m}{k}$, reflecting how damping and buoyancy ratios influence stability. Thus, for Larger b (damping) or smaller k (stiffness) increases diagonal terms of **P**, implying a stiffer Lyapunov function.

The solved **P** gives an explicit energy-like function $V(\mathbf{x}) = \mathbf{x}^T\mathbf{P}\mathbf{x}$ that decays over time.

1. if $b > 0$ and $k > 0$, this yields a unique $P \succ 0$, proving asymptotic stability.
2. if $b > 0$ and $k > 0$, the system is asymptotically stable (damped oscillations).
   a. $b > 0$: Water drag dissipates energy, stabilizing the submarine.
   b. $k > 0$: Buoyancy provides restoring force.
3. if $b = 0$ and $k > 0$, the system is marginally stable (undamped oscillations).

Lyapunov analysis confirms that the submarine's natural dynamics are stable when $b, k > 0$.

8 Experimental Validation

This section discusses the experimental findings derived from the implementation of the six most general obstacle avoidance scenarios within real-world aquatic environments, concurrently addressing the challenges encountered during these trials. The robotic submarine underwent an evaluation of its obstacle avoidance capabilities across fundamental operational paradigms. A principal objective herein is to provide a concise synthesis of the control performance, achieved through a comparative analysis of the commanded variables generated by the fuzzy logic controller (specifically, drive, steer, and ballast) against the empirically observed states of the electromechanical actuators.

The fuzzy controller's multiple output variables are evaluated, specifically the angular velocities governing the propulsion, steering, and hydraulic piston motors, across evasion maneuvers. The validation protocol encompasses the following evasive experiments: forward and backward propulsion, port and starboard turns, and ballast system actuation for submersion and emersion.

The analysis focused on comparing the real (observed) motors angular velocities with the predicted values of the control variables, quantifying their absolute and relative errors, which produced differences attributed to the influence of gravitational forces, inertial effects, ambient underwater pressure, and external hydrodynamic flow conditions.

8.1 Evasion maneuvers: Forward and backward motion

Forward-Backward experimental evasive tasks
The underwater robot navigated at a shallow depth, maintained between 0.3 m and 0.5 m below the water surface. Two distinct objects were strategically positioned: one trailing the robot, inducing forward motion, and another situated ahead, both at a range of ∓ 0.5 m, prompting backward locomotion. Experimental data is showed in Figure 17.

Figure 17, in its initial plot, illustrates the propelling motor's angular velocity, derived from the model (3) as described in Proposition 1. Subsequent to the initial 100 seconds, the propelling motor's rotation direction reverses, propelling the robot forward following the detection of an obstacle astern. Fuzzy rule number 4 is activated for approximately 20 seconds as the robot increases its distance from the rear obstacle. Subsequently, the propulsion is deactivated, and the robot, continuing its motion due to inertial forces, abruptly detects another obstacle directly ahead. Correspondingly, the first plot, depicting the observed angular velocity $\omega_P = \dot{\varphi}_D$ of the same motor, now indicates rotation in the positive direction, coinciding with the activation of a specific fuzzy rule. It is noteworthy that in this experiment, the fuzzy control system sequentially activates fuzzy rules 4 and 5. These rules, in turn, deactivate the steering motors, denoted as $\omega_D = \dot{\varphi}_S$, and the ballast hydraulic piston motor, represented as $\omega_L = \dot{\varphi}_B$, a response that aligns with anticipated behavior. In the aforementioned experiment, the second plot within Figure 17 presents the acquired sensory data from the submersible pressure transducer. Several transient pressure peaks are discernible, attributable to the compounded influence of inherent sensor noise and minute pressure fluctuations induced by subtle aquatic turbulence. It is noteworthy that a discernible reduction in pressure readings correlates with the temporal intervals during which the robot executes both forward and backward evasive maneuvers. Furthermore, the pressure sensor registers a comparatively elevated hydrodynamic drag pressure when the robot is engaged in backward locomotion. The third row of graphical representations in Figure 17 depicts the sensed angular velocities corresponding to the roll, pitch, and yaw rotations of the robotic platform's body. Within the temporal interval spanning 50 to 100 seconds, a hydrodynamic turbulence was detected, predominantly perturbing the robot's orientation in yaw and roll, with a negligible effect observed on its pitch dynamics. This minimal susceptibility to pitch perturbations is attributed to the robot's longitudinal axis presenting a greater resistance to such aquatic undulations. Concurrently, the angular velocities associated with roll and yaw exhibit consistency with the executed forward and subsequent backward evasive maneuvers, particularly correlating with the initial acceleration phase of the forward evasion.

The fourth graphical representation in Fig. 17 illustrates the sensed accelerations of the robotic platform's motion. A consistent detection of the turbulent flow and the resultant accelerative magnitudes imparted to the robot is observed within the tempo-

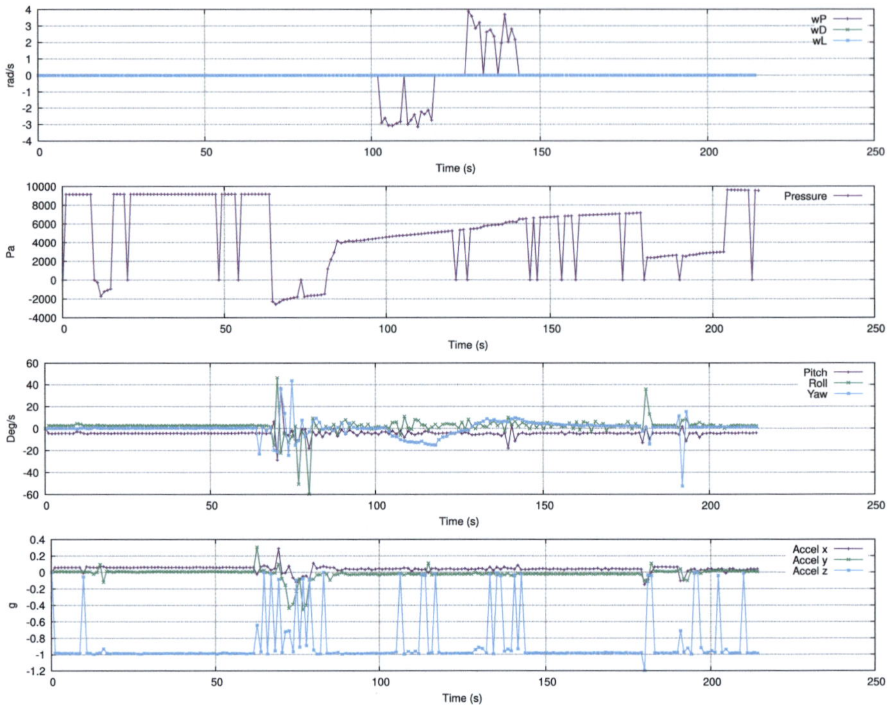

Fig. 17 Experimental robot's sensing system during forward-backward propulsion

ral interval of 50–100 s. Notably, the vertical undulating effects of the aquatic medium are registered by the vertical accelerometer, frequently accompanied by spurious positive readings manifesting as transient peaks approaching zero. The baseline value for the z-axis accelerometer corresponds to 1 standard gravity (9.81 m/s^2), consistently detecting the influence of the Earth's gravitational field. Within this graphical depiction, the most pertinent element is the acceleration along the x-axis, as this axis is aligned with the longitudinal axis of the robot's structure, corresponding to forward and backward motion. Only minimal accelerations were detected along this axis, a consequence of the fuzzy control system's action in maintaining a regulated angular velocity of the propulsion motor in both directions of rotation.

8.2 Evasion Maneuvers: Left and Right Turning Motion

Furthermore, a set of evasion experiments employing left and right yawing maneuvers were performed under analogous conditions. In Fig. 18, the initial graphical representation of the data pertains to the observed angular velocities of the robot's three actuators. For this specific experimental paradigm, only the actuator associated with the steering propeller, denoted as $\omega_D = \dot{\varphi}_S$ was activated.

Forward-Backward experimental evasive tasks
Initially, a lateral obstruction was positioned proximal to the robot, at a distance of approximately 1 meter on its starboard (right) side. Subsequently, upon the robot's yawing motion sufficiently clearing this obstruction from its sensor field, a second obstruction was detected on its port (left) side, resulting in the activation of the same steering actuator but with a reversal of its rotational direction.

The angular velocity of the steering actuator, while governed sequentially by fuzzy rules six and seven, demonstrating operational consistency, also exhibits transient responses to abrupt cessations in propeller rotation. These interruptions are attributed to mechanical load effects arising from hydrostatic pressure and hydrodynamic drag forces. The period of actuator governance spans the temporal interval of 30–47 s, followed by a directional change occurring between 50 and 68 s. The third graphical representation in Fig. 18 illustrates the angular velocities observed by three accelerometers, corresponding to roll, pitch, and yaw orientations, respectively. As anticipated, the most significant perturbation in angular velocity is registered along the yaw axis.

8.3 *Evasion Maneuvers: Submergence and Emergence Motion*

Following the two preceding sets of controlled evasion experiments, the concluding experimental paradigm investigates the vertical dynamics of the robotic submarine's ballast system, specifically its controlled ascent and descent within the water pool. In one of the experiments, an obstruction was introduced at the limit of the water surface, positioned directly above the robotic platform's body. Simultaneously, the physical beneath obstacle constituted by the pool's floor was also considered as a limiting condition, and acting as an obstruction within the operational space. Figure 19 provides a visual illustration of the aquatic environment and the robot's spatial locations across a range of submersion depths.

Ballasting system experimental data
The submarine descended from the surface to the bottom of a pool with a depth of 0.85m and later resurfaced. The ballasting system operated by suctioning water between $t = 103$s and $t = 146$s and expelling it between $t = 147$s and $t = 191$s of Fig. 20. Figure 20 displays the experimental results.

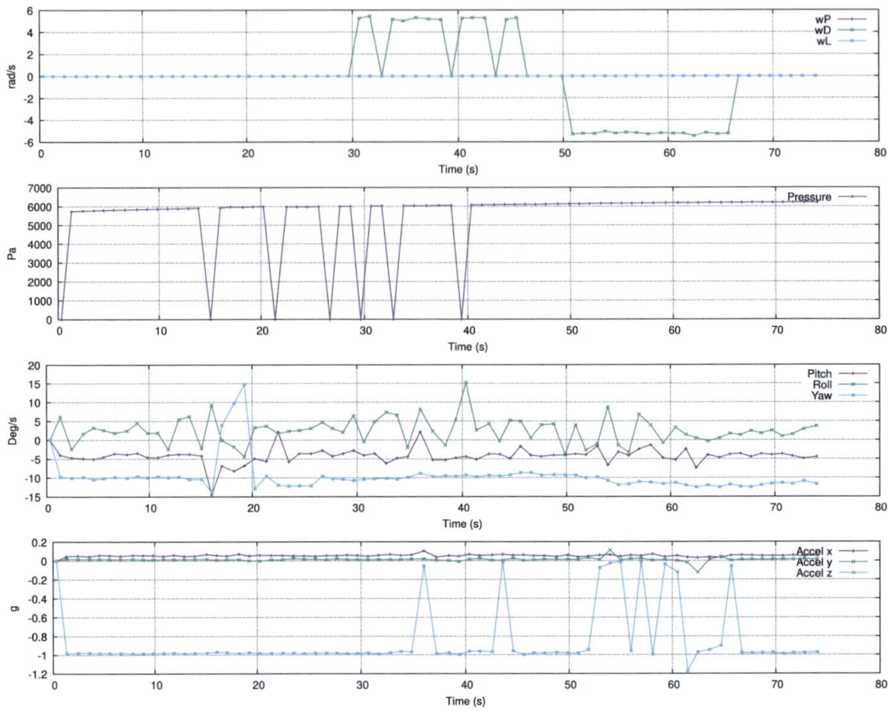

Fig. 18 Experimental robot's sensing system during left-right steering

The initial graphical representation in Fig. 20 illustrates the activation profile of the hydraulic actuator motor. Initially, the motor exhibits rotation in the positive direction, governed by the activation of fuzzy rule 2. Subsequently, upon near-proximal detection of the pool's floor, fuzzy rule 3 is triggered, resulting in an immediate reversal of the ballast motor's rotational direction, denoted as $\dot{\varphi}_B$.

Due to the mass of internal air trapped inside the robot's chassis, a certain amount of water mass was required within the hydraulic piston for the robot to commence its descent, albeit gradually. This was observed between $t = 143$ seconds and $t = 191$ seconds as it reached the bottom of the test pool. Subsequently, it began to ascend, surfacing at approximately $t = 240$ seconds. Although the piston had expelled some water mass, the robot continued to sink until the remaining water was insufficient to keep it submerged. The ascent was noticeably slower than the descent due to forces counteracting gravity and the pressure of the surrounding liquid.

In its present configuration, the system experienced instances of multiple sensor malfunctions; however, the HMC-Fuzzy controller demonstrated robust responsiveness to these anomalies. Beyond inherent sensor noise, several such sensor failures are evident in Figs. 17, 18, and 20, particularly in the graphical representations depict-

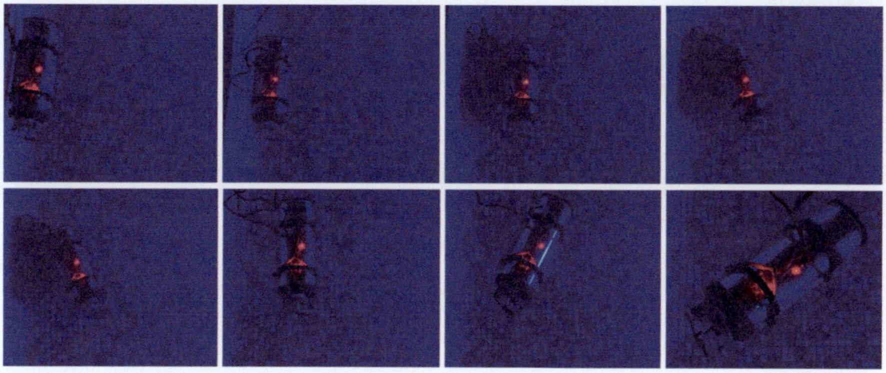

Fig. 19 Experimental underwater controlled scenario. Robotic structure deployed at different depths

ing motor encoders (first plot), depth inferred from pressure sensors (second plot), and acceleration sensors (fourth plot). These failures manifested as abrupt, transient peaks. Figure 21 illustrates the efficacy of the fuzzy controller in mitigating the impact of these noisy sensor signals, as applied in [37].

8.4 Fuzzy Controller Output Variables

This section discusses the findings pertaining to a core aspect of this research: the behavioral response of the robotic submarine's actuators when encountering obstacles. This study summarizes the results of controlled experiments designed to evaluate the fundamentals evasion maneuvers: forward and backward translation, port and starboard yawing, and submergence and emergence. The authors have consistently integrated the pertinent fuzzy inference rules governing these basic modes of locomotion to provide a clear illustration of the controller's operational characteristics for the reader. However it is worth mentioning that a larger of combined fuzzy rules can be added that involve a mix of scenario's configurations. A particularly salient attribute of this approach is the inherent consistency exhibited by the fuzzy logic controller, which demonstrated robust performance and resilience in the face of environmental discontinuities and nonlinearities. Figure 21 presents a synthesis of the primary analytical metric: the observed angular velocity of the actuators juxtaposed with the commanded velocity values computed by the fuzzy logic controller. The first row of graphical representations illustrates the temporal intervals during which the fuzzy controller predicts the ideal evasive velocity for the propulsion motor in the forward (left column) and backward (right column) directions. The commanded velocity is generated at significantly higher frequencies than

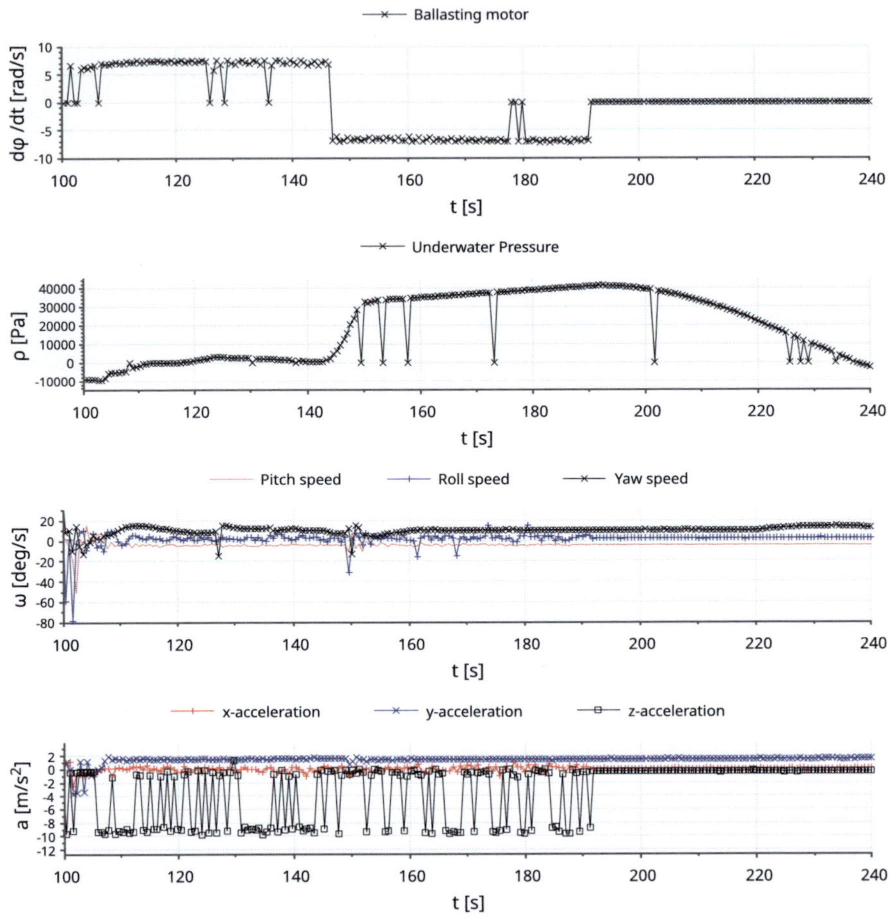

Fig. 20 Experiment of the robot's vertical evasion behavior

the physically manifested velocity within the robot's mechanical system. Hydrodynamic drag forces, resulting from hydraulic pressure, constitute the primary factor impeding the physical velocity of the actuator. Owing to pressure fluctuations on the propeller surfaces induced by micro-wave disturbances, the encoder observations register commutated peaks indicative of velocity variations. Nevertheless, the graphical data demonstrate the controller's persistent action in enforcing the maintenance of a predefined desired velocity. The second row of graphical representations in Fig. 21 illustrates the observed and commanded velocities of the steering actuator during yawing maneuvers executed for lateral evasion. Notably, the sign reversals in the scales of each column correspond to the opposing rotational directions of the motor. During leftward yawing, instances of spurious negative readings were detected, attributed to mechanical vibrations affecting the encoder mounted on the steering motor's shaft. However, the commanded velocity maintains consistent val-

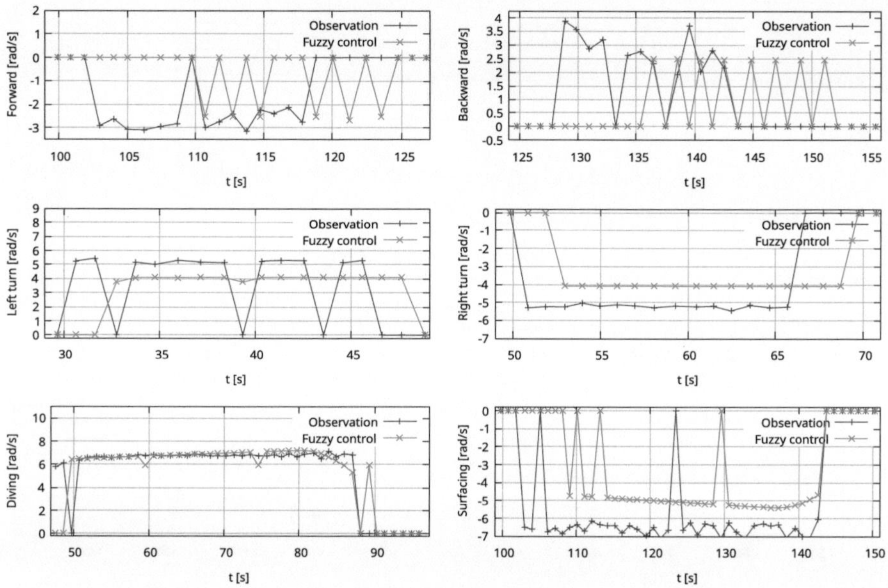

Fig. 21 Motors' controlled velocity versus observed velocities

ues, exhibiting a temporal lead with respect to the observed velocity, which requires a finite duration to attain the commanded values due to the mechanical loads imposed upon it. A similar effect is evident in the case of rightward yawing, although with a near absence of perturbations. The depicted temporal windows correspond to the duration required for the robot to complete the yawing motion and cease sensing the obstacle that triggered its evasive response.

The third row of graphical representations in Fig. 21 illustrates the observed and commanded velocities of the hydraulic piston motor during the ballast system's filling and emptying cycles, corresponding to the robot's evasive maneuvers of submergence and emergence. In this instance, the minimal numerical discrepancies between the observed and commanded velocities were detected during robot submergence (left column). This can be attributed to the gravitational force facilitating the robot's descent, while external hydrodynamic forces appear to exert a comparatively reduced influence on the vertical impulse. Conversely, during evasive vertical ascent locomotion (right column), the robot's propulsive force must overcome not only the hydrodynamic drag forces but also the opposing force of gravity. Consequently, the ascent maneuver exhibits a slower dynamic response, as evidenced by the graphical data wherein the observed physical velocity of the robot lags behind the commanded velocity. Achieving congruence with the commanded velocity necessitates a greater velocity and torque output from the actuator.

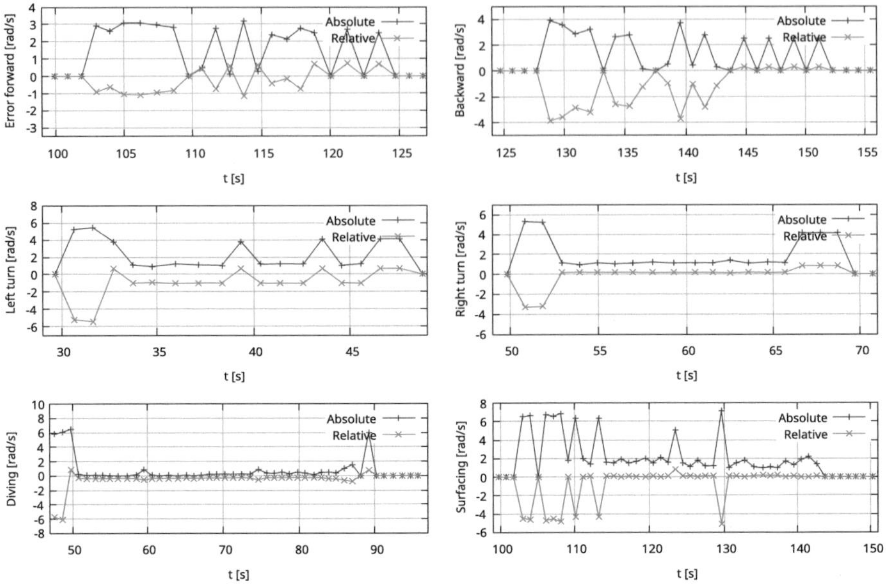

Fig. 22 Absolute and relative errors overtime between real motors' speed and controlled speeds

Finally, Fig. 22 presents the absolute and relative errors derived from the six experimental trials, comparing the actual observed velocities of the three actuators with their corresponding commanded control velocities. The absolute error serves as a direct metric of imprecision, quantifying the difference between the true value (the observation) and the control value. A smaller absolute error signifies a greater degree of proximity between the commanded and observed values, thus indicating enhanced precision in absolute terms. Furthermore, the relative error provides a proportional measure of the temporal discrepancy between the observed and commanded values. This metric offers more comprehensive insight into the stability and convergence performance of the control system with respect to the observed system behavior over time.

Analysis of the relative error indicates that the evasion locomotion experiments involving backward translation, rightward yawing, submergence, and emergence exhibit the most pronounced convergence over time. This signifies that the control system effectively drives the observed physical velocities to progressively approximate the commanded fuzzy control values.

9 Conclusion

The robotic platform underwent extensive development to ensure both operability and reliability in underwater tasks, tested in a medium-sized pool with a depth of 2 m. One of the most challenging aspects, requiring significant precision and development time, was achieving seamless multi-thread synchronization within the master-slave computational system. This was particularly critical for ensuring proper coordination and synchronization during access to I/O ports, with special emphasis on the I2C serial communication ports. The propulsion and steering mechanisms were designed using plastic gear systems, comprising two components: one housed inside the cylindrical body of the robot and the other positioned externally in the aquatic environment. Both components were constructed with plastic materials for lightweight and durability. To avoid compromising the integrity of the robot's chassis with perforations, the two gear systems were mechanically coupled using neodymium permanent magnets, effectively replicating the functionality of a rotational shaft. This innovative design proved highly effective, ensuring smooth operation with minimal friction between the magnets and the surface of the chassis.

The Markovian network excels at making accurate predictions. It utilizes a non-stationary stochastic matrix, with probability values dynamically estimated based on sensor-measured distances. The Markov chain forecasts the robot submarine's next state, such as stop, move forward, move backward, turn left, turn right, dive, or surface-based on the task at hand. Experimentally validated, the chain employs transitions between ideal states, selecting the most suitable predictions as the input value for the fuzzy set in the robot's fuzzy controller.

The fuzzy controller exhibited outstanding reliability and precision in determining the motors speed values. Upon implementation, the routine-designed using Gaussian and sigmoid input and output functions emerged as an exceptionally efficient numerical method. Its ability to deliver accurate results with remarkable speed, requiring essentially only a single iteration, highlights its effectiveness and computational efficiency.

In contrast to the relatively straightforward design of the propulsion and steering mechanisms, the ballasting system–developed as a laboratory prototype–posed considerable engineering challenges. Its complexity stemmed from the operation of a screw-driven piston designed to manipulate a liquid mass for precise buoyancy control. Moreover, the ballasting system necessitated the development of a dedicated controller, leveraging advanced physical models of the hydraulic piston to achieve accurate depth regulation. This control strategy relied on real-time water pressure measurements as a feedback mechanism, ensuring precise and responsive operation.

Integrating HMC with Fuzzy logic control (FLC) enables robotic submarines to anticipate dynamic underwater complex scenarios. HMCs predict environmental changes, like approaching obstacles, or system failures, while FLC converts these predictions into real-time motor adjustments. This hybrid approach foresight and adaptability, where sudden turbulence risks collisions may happen such as offshore oil rig inspections. Navigate in fragile ruins amid sediment clouds at deep-sea archaeology, or where hidden threats demand probabilistic path planning such as military mine neutralization. By combining HMCs' predictive intelligence with FLC's robust execution, these underwater robots robustly operate in long-term tasks like in monitoring missions.

The proposed robotic architecture, while functional, operates with limited computational resources. To achieve real-time responsiveness, it employs a multi-threaded framework for concurrent task execution. However, computationally intensive tasks like matrix operations can still strain the system. The use of optimized C++ libraries for these operations helps maximize the limited processing power. In essence, the architecture balances functionality with hardware constraints through efficient software design.

Future endeavors within this research project entail the transposition of the developed HMC-Fuzzy evasive control strategy to a bio-inspired robotic platform featuring a caudal fin propulsion mechanism. We posit that a bio-inspired design offers enhanced efficiency in terms of mechanical energy utilization and, more significantly, the potential for superior locomotive performance. This transition will involve a reduction in the complement of ultrasonic sonar sensors, coupled with the integration of an RGB-D visual sensor to facilitate environmental mapping and the utilization of 3D point clouds for navigation in cluttered environments. The objective is to capture natural landmarks for autonomous route planning and, leveraging these features, to implement exteroceptive self-localization capabilities.

Acknowledgements The corresponding author acknowledges the support of Laboratorio de Robótica. The second author acknowledges the support of the Kazan Federal University Strategic Academic Leadership Program ("PRIORITY-2030").

10 Appendix

10.1 Platform's Features

Table 3 offers a detailed overview of several features, organized into categories: physical attributes, electrical specifications, dynamic performance, sensory sensitivity, and computing capabilities.

Table 3 Robotic platform's physical features

Materials	Polycarbonate, PLA and Carbon steel
Weight	2.5 kg
Length	32.5 cm
Diameter	15.7 cm
Water resistance	IP8
Operating voltage	5 VDC
Underwater operation time	45 min
Computing power	Quad-core 1.5 Hhz 64-bit
Operating system	Linux kernel 5.10.160-1 × 86 64bits
Developing language	GNU C/C++
Acceleration sensitivity	16g
Gyro speed sensitivity	2000 deg/s
Euler angles sensitivity	$\pi/8$ rad
Underwater pressure range	0 to 40 kPa
Ultrasonic range	3m
Ultrasonic acoustic cone	$5^o - 10^o$m
Ulstrasonic sensor waterproof level	IP68
Drive motor max speed	105 rad/s
Steer motor max speed	80 rad/s
Ballast motor max speed	75 rad/s
Drive torque	5 Nm
Steer torque	4.5 Nm
Ballast torque	7.5 Nm
Thrust velocity	0.15 m/s
Angular velocity	$\pi/8$ rad/s
Dive velocity	0.04 m/s
Surfacing velocity	0.045 m/s

References

1. Wu, Y., Ta, X., Xiao, R., Wei, Y., An, D., Li, D.: Survey of underwater robot positioning navigation. Appl Ocean Res **90**, 101845 (2019). https://doi.org/10.1016/j.apor.2019.06.002
2. Li, D., Wang, P., Du, L.: Path planning technologies for autonomous underwater vehicles-a review. IEEE Access **7**, 9745–9768 (2019). https://doi.org/10.1109/access.2018.2888617
3. Bai, Y., Wang, Y., Svinin, M., Magid, E., Sun, R.: Adaptive multi-agent coverage control with obstacle avoidance. IEEE Cont. Syst. Lett. **6**, 944–949 (2022). https://doi.org/10.1109/lcsys.2021.3087609
4. Lavrenov, R., Matsuno, F., Magid, E.: Modified spline-based navigation: guaranteed safety for obstacle avoidance, pp. 123–133. Springer International Publishing (2017). https://doi.org/10.1007/978-3-319-66471-2_14
5. Kot, R.: Review of collision avoidance and path planning algorithms used in autonomous underwater vehicles. Electronics **11**(15), 2301 (2022). https://doi.org/10.3390/electronics11152301

6. Cheng, C., Sha, Q., He, B., Li, G.: Path planning and obstacle avoidance for auv: a review. Ocean Eng. **235**, 109355 (2021). https://doi.org/10.1016/j.oceaneng.2021.109355
7. Xu, C., Tan, F.: Research on autonomous underwater vehicles path planning method and its obstacle avoidance solution. In: 2024 5th International Conference on Artificial Intelligence and Electromechanical Automation (AIEA), pp. 977–982. IEEE (2024). https://doi.org/10.1109/aiea62095.2024.10692552
8. Chen, S., Lin, T., Jheng, K., Wu, C.: Application of fuzzy theory and optimum computing to the obstacle avoidance control of unmanned underwater vehicles. Appl. Sci. **10**(17), 6105 (2020). https://doi.org/10.3390/app10176105
9. Lin, C., Wang, H., Yuan, J., Yu, D., Li, C.: An improved recurrent neural network for unmanned underwater vehicle online obstacle avoidance. Ocean Eng. **189**, 106327 (2019). https://doi.org/10.1016/j.oceaneng.2019.106327
10. Bhopale, P., Kazi, F., Singh, N.: Reinforcement learning based obstacle avoidance for autonomous underwater vehicle. J. Mar. Sci. Appl. **18**(2), 228–238 (2019). https://doi.org/10.1007/s11804-019-00089-3
11. Yang, J., Ni, J., Xi, M., Wen, J., Li, Y.: Intelligent path planning of underwater robot based on reinforcement learning. IEEE Trans. Automation Sci. Eng. **20**(3), 1983–1996 (2023). https://doi.org/10.1109/tase.2022.3190901
12. Lu, X., Liu, D.: Design of obstacle avoidance algorithm for submarine intelligent robot. J. Coastal Res. **97**(sp1), 162 (2019). https://doi.org/10.2112/si97-023.1
13. Zhang, J., Lu, Z., Yu, C., Shang, Y., Chen, Y.: The underwater obstacle avoidance method based on ros. J. Phys. Conf. Ser. **2562**(1), 012052 (2023). https://doi.org/10.1088/1742-6596/2562/1/012052
14. Cai, W., Wu, Y., Zhang, M.: Three-dimensional obstacle avoidance for autonomous underwater robot. IEEE Sens. Lett. **4**(11), 1–4 (2020). https://doi.org/10.1109/lsens.2020.3034309
15. Yan, Z., Li, J., Zhang, G., Wu, Y.: A real-time reaction obstacle avoidance algorithm for autonomous underwater vehicles in unknown environments. Sensors **18**(2), 438 (2018). https://doi.org/10.3390/s18020438
16. Wu, S.: Dynamic obstacle avoidance method for 2d robot with underwater actuator in velocity vector coordinate system. Int. J. Multiphys. **18**(2s), 11–22 (2024). https://themultiphysicsjournal.com/index.php/ijm/article/view/933
17. Yao, X., Wang, X., Wang, F., Zhang, L.: Path following based on waypoints and real-time obstacle avoidance control of an autonomous underwater vehicle. Sensors **20**(3), 795 (2020). https://doi.org/10.3390/s20030795
18. Sun, B., Zhang, W., Song, A., Zhu, X., Zhu, D.: Trajectory tracking and obstacle avoidance control of unmanned underwater vehicles based on mpc. In: 2018 IEEE 8th International Conference on Underwater System Technology: Theory and Applications (USYS), pp. 1–6. IEEE (2018). https://doi.org/10.1109/usys.2018.8779160
19. Yu, L., Qiao, L., Shen, C.: High-speed obstacle avoidance of a large-scale underactuated autonomous underwater vehicle under a finite field of view. IEEE Trans. Automat. Sci. Eng. 1–10 (2024). https://doi.org/10.1109/tase.2024.3373607
20. Li, X., Yu, S., Gao, X.z., Yan, Y., Zhao, Y.: Path planning and obstacle avoidance control of uuv based on an enhanced a* algorithm and mpc in dynamic environment. Ocean Eng. **302**, 117584 (2024). https://doi.org/10.1016/j.oceaneng.2024.117584
21. Yuan, S., Chen, J.: Submarine navigation and communication model based on hmm and tda. In: 2024 IEEE 4th International Conference on Data Science and Computer Application (ICDSCA), pp. 298–310. IEEE (2024). https://doi.org/10.1109/icdsca63855.2024.10860138
22. Khan, A., Fouda, M.M., Do, D.T., Almaleh, A., Alqahtani, A.M., Rahman, A.U.: Underwater target detection using deep learning: methodologies, challenges, applications, and future evolution. IEEE Access **12**, 12618–12635 (2024). https://doi.org/10.1109/access.2024.3353688
23. Sasi, J., Nidhi Pandagre, K., Royappa, A., Walke, S., G, P., L, N.: Deep learning techniques for autonomous navigation of underwater robots. In: 2023 10th IEEE Uttar Pradesh section International Conference on Electrical, Electronics and Computer Engineering (UPCON), pp. 1630–1635. IEEE (2023). https://doi.org/10.1109/upcon59197.2023.10434865

24. Wang, Y., Wang, H., Liu, Y., Wu, J., Lun, Y.: 6-dof uav path planning and tracking control for obstacle avoidance: a deep learning-based integrated approach. Aerospace Sci. Technol. **151**, 109320 (2024). https://doi.org/10.1016/j.ast.2024.109320
25. Lin, C., Cheng, Y., Wang, X., Yuan, J., Wang, G.: Transformer-based dual-channel self-attention for uuv autonomous collision avoidance. IEEE Trans. Intell. Veh. **8**(3), 2319–2331 (2023). https://doi.org/10.1109/tiv.2023.3245615
26. Zheng, H., Sun, Y., Zhang, G., Zhang, L., Zhang, W.: Research on real time obstacle avoidance method for auv based on combination of vit-dpfn and mpc. IEEE Trans. Inst. Meas. **73**, 1–15 (2024). https://doi.org/10.1109/tim.2023.3348890
27. Zhang, B., Xing, H., Zhang, Z., Feng, W.: Autonomous obstacle avoidance decision method for spherical underwater robot based on brain-inspired spiking neural network. Expert Syst. Appl. **274**, 127021 (2025). https://doi.org/10.1016/j.eswa.2025.127021
28. Fang, Z., Jiang, D., Huang, J., Cheng, C., Sha, Q., He, B., Li, G.: Autonomous underwater vehicle formation control and obstacle avoidance using multi-agent generative adversarial imitation learning. Ocean Eng. **262**, 112182 (2022). https://doi.org/10.1016/j.oceaneng.2022.112182
29. Yar, G.N.A.H., Ahmad, A., Khurshid, K.: Low cost assembly design of unmanned underwater vehicle (uuv). In: 2021 International Bhurban Conference on Applied Sciences and Technologies (IBCAST), pp. 829–834. IEEE (2021). https://doi.org/10.1109/ibcast51254.2021.9393178
30. Bakirci, M., Toptas, B.: Kinematics and autoregressive model analysis of a differential drive mobile robot. In: 2022 international congress on human-computer interaction, optimization and robotic applications (HORA), pp. 1–6. IEEE (2022). https://doi.org/10.1109/hora55278.2022.9800071
31. Brito, M.P., Griffiths, G.: A markov chain state transition approach to establishing critical phases for auv reliability. IEEE J. Oceanic Eng. **36**(1), 139–149 (2011). https://doi.org/10.1109/joe.2010.2083070
32. Ali, W., Li, Y., Raja, M.A.Z., Khan, W.U., He, Y.: State estimation of an underwater Markov chain maneuvering target using intelligent computing. Entropy **23**(9), 1124 (2021). https://doi.org/10.3390/e23091124
33. David, K.K.A., Vicerra, R.R., Bandala, A.A., Gan Lim, L.A., Dadios, E.P.: Unmanned underwater vehicle navigation and collision avoidance using fuzzy logic. In: Proceedings of the 2013 IEEE/SICE International Symposium on System Integration, pp. 126–131. IEEE (2013). https://doi.org/10.1109/sii.2013.6776715
34. Peng, X.F., Zhang, J.J., Li, P.: Fuzzy logic guidance of a control-configured autonomous underwater vehicle. In: 2018 37th Chinese control conference (CCC), pp. 3989–3993. IEEE (2018). https://doi.org/10.23919/chicc.2018.8484224
35. Martínez-García, E.A., Lavrenov, R., Magid, E.: Robot fish caudal propulsive mechanisms: a mini-review. AI, Comput. Sci. Robot. Technol. **2022**, 1–17 (2022). https://doi.org/10.5772/acrt.09
36. Montoya Martínez, M.A., Torres-Córdoba, R., Magid, E., Martínez-García, E.A.: Electromyography-based biomechanical cybernetic control of a robotic fish avatar. Machines **12**(2), 124 (2024). https://doi.org/10.3390/machines12020124
37. Bakirci, M.: Data-driven system identification of a modified differential drive mobile robot through on-plane motion tests. Electrica **23**(3), 619–633 (2023). https://doi.org/10.5152/electrica.2023.22164

Ensemble-Based DDoS Attack Detection Model for Software-Defined Networks

Ermias Tsegu and Ayalew Belay Habtie

Abstract In this modern era, networking has advanced rapidly. The need for businesses to incorporate numerous dynamic applications, services, physical objects, and machines is increasing. Consequently, networking infrastructure has be- come more complex in terms of devices and resource utilization. The traditional networking paradigm has proven ineffective and inefficient in handling these demands. As a result, Software Defined Networking (SDN) has garnered significant attention from researchers and practitioners. Unlike the traditional paradigm, SDN offers efficient resource utilization, simplified network management, improved performance, network virtualization, and programmability. However, it faces security challenges such as network tampering, unauthorized access, flow rule conflicts, poor controller deployment, and Distributed Denial of Service (DDoS) attacks. Among these, DDoS is one of the most threatening attacks on SDN. In response, several studies have been conducted to detect DDoS in SDN networks using statistical approaches, traditional machine learning (ML) techniques, and state-of-the-art methods like deep learning (DL). Traditional ML methods, though effective, are less efficient than advanced DL approaches, which, in turn, are computationally complex. To address this issue, we proposed an ensemble-based DDoS detection model for SDN using a flow-based dataset. The experiment was conducted with the InSDN dataset, which includes normal traffic, metasploitable-2 attacks targeting a server, and attacks targeting an Open Virtual Switch (OVS) machine. Since the dataset contains various attacks such as DoS, DDoS, web attacks, R2L, malware, probe, and U2R, it was necessary to isolate the DDoS data for our purposes. The dataset was split into 70% for training and 30% for testing. Results indicated that the adaptive boosting ensemble technique achieved the highest accuracy at 100%. However, in terms of latency, the gradient boosting algorithm performed best, with a latency of 60.6 ms, while the KNN_DT-based stacking algorithm had the highest latency at 119,431.5 ms.

Keywords Software defined network (SDN) · Distributed denial of service attack (DDoS) · Machine learning (ML) · Ensemble-based algorithms

---

E. Tsegu · A. B. Habtie (✉)
Department of Computer Science, Addis Ababa University, Addis Ababa, Ethiopia
e-mail: ayalew.belay@aau.edu.et

W. Pedrycz et al. (eds.), *Decision-Making in Computational Intelligence-Based Systems*,
Studies in Systems, Decision and Control 628,
https://doi.org/10.1007/978-3-032-04056-5_10

1 Introduction

In this technological era, the rapid growth of networking has been a continuous trend. AL-Badrany and Al-Somaidai [1] stated that networking is growing exponentially in terms of scalability and complexity. As a result, adopting an effective approach is essential to address the challenges posed by these rapid advancements in computer networks and related technologies. To this end, the authors argue that Software Defined Networking (SDN) offers elegant solutions to network complexity. Consequently, SDN has recently garnered significant attention from the network research community.

Zhai et al. [2] describe SDN as a new networking architecture that is dynamic, flexible, manageable, and adaptable. According to Kyaw et al. [3], SDN is composed of three layers: the infrastructure layer, control layer, and application layer. The infrastructure layer, also known as the data plane, includes devices like switches and routers. These devices forward or drop incoming packets based on the flow table configured by the control layer. The control layer's primary function is to install flow rules that determine whether traffic should be forwarded or dropped. The application layer, also referred to as the management plane, contains applications that leverage the functionalities provided by the northbound interface.

The core philosophical concept behind SDN is the separation of the control plane from the data plane. Additionally, the implementation of a logically centralized scheme for centralized control and management of network activities is a key contribution to SDN architecture [4]. This design enables a global view of network activities, allowing SDN to program networks via an application programming interface and integrate necessary functions [5]. Moreover, SDN applications can manage and control networks, ensuring efficient load balancing, access control, and routing. These features represent the most significant advantages of this innovative networking paradigm [5].

Despite SDN's numerous benefits and innovations tailored to current technological advancements, one significant drawback of the architecture is its centralization, which makes it particularly vulnerable to Distributed Denial of Service (DDoS) attacks. A DDoS attack aims to render resources unavailable by overwhelming the target with fake packets, often initiated by multiple connected devices. According to Bavani et al. [6], DDoS attacks are categorized into volumetric, protocol, and application layer attacks. Volumetric attacks aim to overwhelm the victim's network bandwidth by injecting large amounts of traffic. Protocol attacks exploit vulnerabilities in the network and transport layers' protocol stacks, while application layer attacks target weaknesses in the application layer's protocol stack. Anagha et al. [7] noted that SDN architecture is particularly prone to DDoS attacks, where malicious users can deplete resources in the data plane and disrupt services in the control plane. Such malicious activities pose a significant DDoS threat to SDN, making it a real challenge for this networking paradigm.

In recent years, numerous research efforts have focused on detecting and mitigating the devastating impact of DDoS attacks. From the literature survey conducted for this study, four broad categories of approaches for detecting DDoS attacks in SDN environments were identified: statistical approaches [6, 8, 9], traditional ML techniques [3, 10], DL-based state-of-the-art methods [11], and ensemble-based techniques [12].

Statistical approaches for detecting DDoS attacks in SDN-based networks rely on a threshold value compared with a calculated entropy value, but their effectiveness is limited by the subjective nature of the threshold, which depends on researchers' experience. Consequently, these methods often struggle to accurately identify DDoS attacks. In comparison to traditional machine learning techniques such as Support Vector Machine (SVM) and K-Nearest Neighbour (KNN), statistical approaches are generally regarded as less efficient and effective in practice.

Traditional machine learning (ML) techniques generally exhibit lower accuracy compared to state-of-the-art approaches like deep learning (DL) [13]. Many studies in this area rely on outdated datasets, such as KDD1999, which do not reflect current network protocols or the complexities of modern networking scenarios. Additionally, these datasets are primarily derived from conventional networks, lacking representation of the OpenFlow protocol used in SDN, making them unsuitable for developing models applicable to contemporary SDN architectures. Furthermore, a significant number of researchers including [9, 13] focus solely on detecting UDP flood attacks, neglecting other types of flooding attacks like TCP and ICMP floods, which limits the effectiveness of their proposed models to only addressing UDP flood attacks in DDoS scenarios.

Recent studies employing state-of-the-art deep learning (DL) techniques demonstrate significantly improved accuracy compared to statistical and traditional machine learning approaches. However, they come with substantial training and testing times as well as high computational costs [12]. As highlighted by Mbasuva and Zoid, while these advanced DL techniques enhance learning accuracy, they are often resource-intensive, requiring over 100 GB of RAM and GPU acceleration for effective implementation. Additionally, many of these studies still rely on outdated and non-SDN representative datasets, which hampers their efficiency and overall performance, indicating a critical need for improvement in research methodologies and data selection.

Ensemble-based approaches for detecting DDoS attacks on Software Defined Networking (SDN) have seen limited exploration so far, with some studies focusing on combinations of deep learning (DL) techniques. However, these methods often introduce significant time, space, and computational overhead compared to single DL techniques [12, 14]. Although there have been efforts to apply boosting methods, the choice of datasets remains a crucial concern that affects their effectiveness. Therefore, there is a clear need for further research into ensemble-based techniques to fill these gaps. Hence, this research aims to address these issues by developing an ensemble-based DDoS detection model for SDN that utilizes a flow-based dataset generated from a real SDN networking environment, incorporating up-to-date traffic patterns and protocols.

2 The Proposed Ensemble-Based DDoS Attack Detection Model

The proposed architecture for the ensemble-based DDoS attack detection model in SDN, as shown in Fig. 1, is grounded in insights from the literature survey conducted in the field. This architecture includes several essential components: data preparation, data pre-processing, train-test split, and model fitting and evaluation. Each component plays a critical role in ensuring the accuracy and efficiency of the detection model, making it a robust framework for addressing the challenges posed by DDoS attacks in SDN environments.

Data preparation component creates context-aware raw data that can be used for training and evaluating the proposed models. The InSDN dataset, being a comprehensive SDN dataset originally designed as a generic IDS dataset, requires modifications to align with the specific goals of this research. It contains normal traffic and various attack categories relevant to SDN architecture. DDoS attack data from the metasploitable-2 and OVS groups in the InSDN dataset are segregated and combined to form a unified DDoS attack dataset. This combined DDoS dataset is then merged with the normal traffic group, resulting in a raw dataset that is ready for training and testing the proposed ensemble-based DDoS detection model for SDN networks. The process is described using Algorithm 1.

Algorithm. Data preparation algorithm

Input: normal_dataset, metasploitable_attack, ovs_attack;
Output: prepared_dataset;

START
Normal_dataset ← read normal dataset file metasploitable_attack ← read metasploitable attack file ovs_attack ← read OVS attack file
select DDoS attack from metasploitable_attack select DDoS attack from ovs_attack DDoS_ attack ← combine the two DDoS attacks
prepared_dataset ← combine normal_dataset and ddos_attack write prepared_dataset to file
END

The dataset pre-processing component is responsible for cleaning and removing or filling missing values, as well as eliminating unwanted and duplicate information before feeding the data into the learning model. One important task is to drop unnecessary variables from the original InSDN dataset, which contains 83 features. Although all these features can be used for classification, their importance varies. Therefore, it is essential to identify and select the most valuable feature sets for the specific problem. Additionally, we encode the labels for binary classification, classifying the input traffic as either normal or malicious. For this purpose, all DDoS attack classes are considered as part of the "attack" category. The string values representing normal and attack traffic are then encoded into binary values: 0 for normal traffic and 1 for attack traffic. This step is elaborated using Algorithm 2.

Algorithm 2. Data pre-processing algorithm

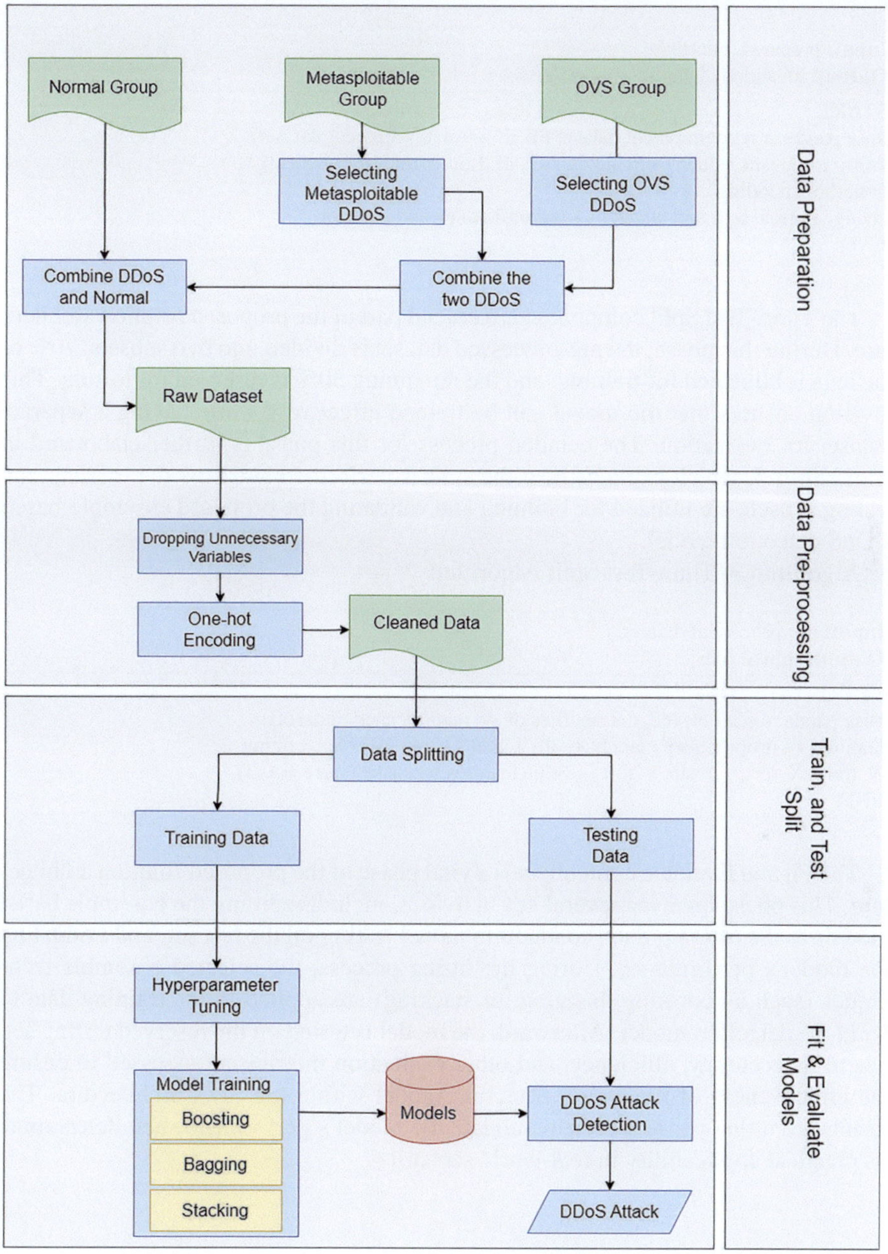

Fig. 1 The proposed ensemble-based DDoS attack detection model

Input: prepared_dataset
Output: cleanded_dataset;

```
START
#use panda to read prepared_dataset file df ← read(prepared_dataset)
#drop irrelevant values from the dataset. df.drop([irrelevant_values])
#one-hot encoding
change normal to 0 and attack to 1 write file to cleand_dataset
END
```

The Train-Test Split component is a crucial part of the proposed solution architecture. During this phase, the pre-processed dataset is divided into two subsets: 70% of the data is allocated for training, and the remaining 30% is reserved for testing. This division ensures that the model can be trained effectively while having a separate dataset for evaluation. The detailed process for this phase is further elaborated in **Algorithm 3**, which explains how the split is performed and how the training and testing subsets are utilized for building and validating the proposed ensemble-based DDoS detection model.

Algorithm 3. Train-Test Split Algorithm

Input: pre-processed dataset
Output: splited files;

```
START
#use panda to read cleand_dataset files df ← read(cleaned_dataset)
features ← drop('Label') labels ← df['Label'] #split 70% for training
X_train, X_test, y_train, y_test ← split(features,labels,test_size = 0.3)
END
```

The Fit and Evaluate component is a vital phase of the proposed solution architecture. This phase involves several key activities, including fitting the ensemble based models to the training data, conducting model testing on the test set, and evaluating the model's performance. During the fitting process, the selected ensemble techniques (such as boosting, bagging, or stacking) are applied to the training data to build the detection model. Afterward, the model is tested on the reserved testing set, where its accuracy, efficiency, and other evaluation metrics are assessed to ensure the effectiveness of the DDoS detection model within the SDN architecture. The results from this phase will help validate the model's performance and deter- mine its practical applicability in real-world scenarios.

3 Experimentation and Evaluation

The dataset used for the experiment is InSDN, which was generated by [15] and reflects contemporary Internet attacks that can be launched in current SDN networks. Designed for Intrusion Detection Systems (IDS), it encompasses various attack types, including DoS, DDoS, Web Attacks, R2L, Malware, Probe, and U2R (Exploitation). The dataset is organized into three groups based on traffic types and target machines: the first group consists solely of normal traffic, the second group contains attack traffic targeting the Metasploitable-2 server, and the third group focuses on attacks against the OVS machine. In total, it comprises 343,939 instances, with 68,424 instances of normal traffic, 136,743 instances of attack traffic for the Metasploitable-2 group, and 138,772 instances for the OVS group. The dataset features are with 83 attributes categorized into various types, including network identifier attributes (such as IP address and port number), packet-based attributes (total number of packets in both forward and backward directions), bytes-based attributes (total bytes in both directions), interarrival time attributes, flow timers attributes (active and inactive timing), flag attributes (like SYN and RST flags), flow descriptors attributes (number of packets and bytes), and subflow descriptors attributes.

For the experimentation process, the dataset underwent thorough preparation and pre-processing, resulting in a final dataset comprising 121,942 instances of attack samples and 68,424 instances of normal traffic. According to [15, 16], the dataset features can vary significantly across networks, and attackers can generate both normal and attack traffic from the same IP address. Consequently, training the model using socket features could lead to overfitting, as the model may become biased towards the socket information. After removing unnecessary variables, we ultimately selected 47 features from the raw dataset using the Scikit-Learn Python library. This comprehensive dataset serves as the foundation for training and testing the ensemble-based DDoS detection models specifically designed for SDN environments.

The dataset is then split into 70% for training and 30% for testing. The model training incorporates various ensemble techniques, including adaptive boosting, gradient boosting, decision tree (DT)-based bagging, and K-Nearest Neighbors (KNN) combined with decision tree (DT)-based stacking models. This diverse approach aims to enhance the robustness and accuracy of the DDoS detection models tailored for SDN environments.

The hyperparameters used during the experiment are detailed in Table 1. This table outlines the specific values and configurations for each model's hyperparameters, which are essential for optimizing performance and ensuring effective training of the ensemble-based DDoS detection models.

3.1 Evaluation Result

The proposed work develops four ensemble-based models to compare and contrast their performance: adaptive boosting, gradient boosting, decision tree (DT)-based bagging, and KNN-DT-based stacking techniques. These models are meticulously designed and evaluated using a testing dataset. The performance of the proposed ensemble-based DDoS detection models for SDN networks is assessed using various metrics, including accuracy, precision, recall, F1-score, and latency (prediction time), which are detailed in Table 2.

Table 1 Hyperparameters used during the training

Ensemble-Algorithm	Parameters	Values	Best Values
Adaptive boosting	n_estimators	[10,50,100]	10
	learning_rate	[1,10,20]	1
	Base estimator	DT	–
Gradient boosting	n_estimators	[10,50,100]	10
	learning_rate	[1,10,20]	1
	max_depth	[10,50,100]	5
	Base estimator	DT	–
Bagging	n_estimators	[10,50,100]	10
	random_state	[1,10,20]	1
	Base estimator	DT	–
Stacking	knn n_neighbors	[5,7]	5
	dt random_state	[0,3]	3
	Base estimator	KNN and DT	–
	final_estimator	LR	–

Table 2 performance of the proposed ensemble-based models

Evaluation metrix	Ensemble-Techniques			
	Adaptive boosting	Gradient boosting	DT_Baggi ng	KNN_DT stacking
TN	20,426	20,426	20,426	20,426
FP	0	1	1	1
FN	0	0	1	1
TP	36,684	36,684	36,684	36,684
Accuracy	1.0	0.999982	0.999964	0.999964
Precision	1.0	0.999972	0.999972	0.999972
Recall	1.0	1.0	0.999972	0.999972
F1-score	1.0	0.999985	0.999971	0.999971
Latency (PREDICTIONTIME)	501.6 ms	60.6 ms	145.5 ms	119, 431.5 ms

The results, as shown in Table 1, indicate that the adaptive boosting model achieves exceptional performance, attaining 100% accuracy, precision, recall, and F1-score, with zero false positive and false negative values. Similarly, the gradient boosting ensemble technique follows closely, achieving 99.99% in accuracy, precision, and F1-score, while attaining a perfect recall of 100%. The DT-based bagging and KNN_DT-based stacking ensemble techniques also excel, achieving 99.99% in accuracy, precision, recall, and F1-score. In terms of latency, the gradient boosting ensemble-based technique demonstrates the lowest prediction latency at 60.6 ms, whereas the KNN_DT-based stacking model exhibits the highest prediction time, with a significant value of 119,431.5 ms.

The confusion matrix serves as a valuable tool for assessing the performance of the models. Figure 2a–d illustrate the confusion matrix results for each ensemble-based technique on the testing set. Notably, the adaptive boosting ensemble technique successfully classifies all instances without any errors. In contrast, the gradient boosting model misclassifies one instance out of 20,426 normal instances as an attack. For the DT-based bagging and DT-KNN-based stacking ensemble techniques, there are misclassifications where one instance is incorrectly classified as an attack when it is actually normal, and another instance is misclassified as normal when it is indeed an attack.

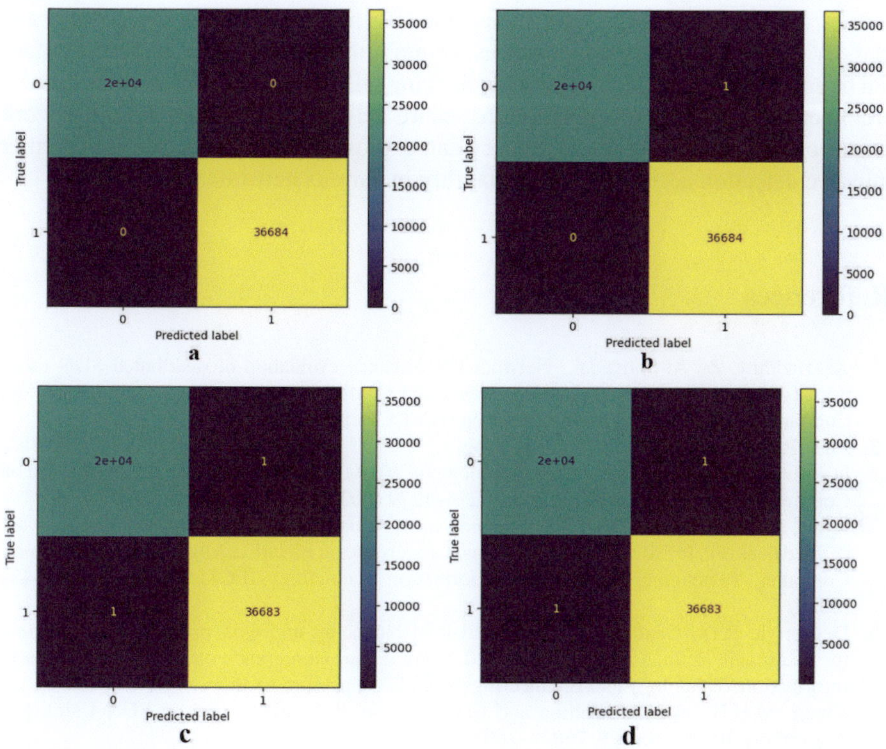

Fig. 2 **a** Adaptive boosting, **b** gradient boosting, **c** bagging classifier, **d** stacking classifier

4 Conclusion

In conclusion, this research successfully develops and evaluates four ensemble-based DDoS detection models tailored for Software-Defined Networking (SDN) environments, utilizing a comprehensive dataset from the InSDN collection. The models—adaptive boosting, gradient boosting, DT-based bagging, and KNN-DT-based stacking—demonstrated exceptional performance metrics, particularly the adaptive boosting model, which achieved 100% accuracy, precision, recall, and F1-score, alongside zero false positives and negatives. The gradient boosting technique also performed admirably, with a marginally lower accuracy but the fastest prediction latency. In contrast, the KNN-DT-based stacking model exhibited the highest prediction time, indicating potential areas for optimization. The confusion matrix analysis further validated the robustness of these models, highlighting their effectiveness in classifying normal and attack instances. Overall, the findings underscore the potential of ensemble-based approaches in enhancing DDoS detection capabilities in SDN architectures, paving the way for more secure network environments. Future work may focus on optimizing model latency and exploring additional datasets to further improve detection accuracy and adaptability in various network scenarios.

References

1. AL-Badrany, Z., Al-Somaidai.: Network performance evaluation of distributed SDN using floodlight controllers. In: 2019 2nd International Conference on Electrical, Communication, Computer, Power and Control Engineering (ICECCPCE), pp. 34–39 (2019)
2. Zhai, P., Song, Y., Zhu, X., Cao, L., Zhang, J., Yang, C.: Distributed denial of service defense in software defined network using OpenFlow. In: 2020 IEEE/CIC International Conference on Communications in China (ICCC), pp. 1274–1279 (2020)
3. Kyaw, A.T., Zin Oo, M., Khin, C.S.: Machine-learning based DDOS attack classifier in software defined network. In: 2020 17th International Conference on Electrical Engineering/Electronics, Computer, Telecommunications and Information Technology (ECTI-CON), pp. 431–434 (2020)
4. Devapriya, R.D., Gandhi, S.I.: Enhanced load balancing and QoS provisioning algorithm for a software defined network. In: 2020 International Conference on Emerging Trends in Information Technology and Engineering (ic-ETITE), pp. 1–5 (2020)
5. Khairi, M.H.H., et al.: Detection and Classification of Conflict Flows in SDN Using ML Algorithms. IEEE Access **9**, 76024–76037 (2021)
6. Bavani, K., Ramkumar, M.P., Selvan, E.G.S.R.: Statistical approach based detection of distributed denial of service attack in a software defined network. In: 2020 6th International Conference on Advanced Computing and Communication Systems (ICACCS), pp. 380–385 (2020)
7. Anagha, M.J., Lepakshi, R., Goutham, V., Thavish, V., Keerthan Kumar, T.G.: Packet injection and Dos attack controller software (PDACS) module to handle attacks in software defined network. In: 2020 Fourth International Conference on Computing Methodologies and Communication (ICCMC), pp. 966–970 (2020)
8. Ahalawat, A., Dash, S.S., Panda, A., Babu, K.S.: Entropy based DDoS detection and mitigation in OpenFlow enabled SDN. In: 2019 International Conference on Vision Towards Emerging Trends in Communication and Networking (ViTECoN), Vellore, India, pp. 1–5 (2019)

9. Duy, P.T., Hien, D.T.T., Pham, V.-H.: A role-based statistical mechanism for DDoS attack detection in SDN. In: 2018 5th NAFOSTED Conference on Information and Computer Science (NICS), Ho Chi Minh City, Vietnam, pp. 177–182 (2018)
10. Sahoo, K.S., Iqbal, A., Maiti, P., Sahoo, B.: A machine learning approach for predicting DDoS traffic in software defined networks. In: 2018 International Conference on Information Technology (ICIT), Bhubaneswar, India, pp. 199–203 (2018)
11. Nadeem, M.W., Goh, H.G., Aun, Y., Ponnusamy, V.: Detecting and Mitigating Botnet Attacks in Software-Defined Networks Using Deep Learning Techniques. IEEE Access **11**, 49153–49171 (2023)
12. Haider, S et al.: A Deep CNN ensemble framework for efficient DDoS attack detection in software defined networks. IEEE Access **8**, 53972–53983 (2020)
13. Gadri, S.: Developing an efficeint predictive model based on ML and DL approaches to detect diabetes. Inform. , Int. J. Comput. Inform. **45**(3), 433–440, 2021
14. Mbasuva, U., Zodi, G.-A.L.: Designing ensemble deep learning intrusion detection system for DDoS attacks in software defined networks. In: 2022 16th International Conference on Ubiquitous Information Management and Communication (IMCOM), 2022, pp. 1–8 (2022)
15. Elsayed, M.S., Le-Khac, N.-A., Jurcut, A.D.: InSDN: A novel SDN intrusion dataset. IEEE Access **8**, 165263–165284 (2020)
16. Elsayed, M.S., Le-Khac, N.-A., Dev, S., Jurcut, A.D.: DDoSNet: a deep-learning model for detecting network attacks. In: 2020 IEEE 21st International Symposium on "A World of Wireless, Mobile and Multimedia Networks" (WoWMoM), Cork, Ireland, pp. 391–396 (2020)

Leveraging Computational Intelligence for Decision-Making in Healthcare and Environmental Science

Hina Bansal, Kritika Shukla, and Ashish Kumar Karn

Abstract Decision-making driven by the CI (Computational Intelligence) has revolutionized healthcare, environmental sciences, finance, business management, robotics, and gaming. Computational Intelligence (CI) is a subfield of artificial intelligence that focuses on developing intelligent systems by mimicking natural biological processes. These processes include evolution and swarm behavior, utilizing techniques like neural networks, fuzzy logic, and evolutionary computation to solve complex problems. It might be difficult to tackle with traditional computing methods, which makes it highly relevant for applications requiring adaptive learning and decision-making in uncertain environments. Using AI for decision-making significantly increases accuracy efficiency, fastens decision-making, reduces biases, and helps in better predictive analysis and risk management. Artificial Intelligence is transforming disease diagnosis by analyzing large-scale medical data with the help of image analysis and pattern recognition to identify different diseases, which will further lead to making personalized treatment plans. AI models can also analyze patient data, identifying individuals at high risk of getting readmitted even after discharge and permitting healthcare providers to step in with preventative measures. Similarly, AI is crucial to ecosystem monitoring and management in environmental science. It analyses the data from sources like camera traps, satellite images and sensor networks, enabling faster and more accurate identification of environmental changes, threats, and trends. Conservation organizations utilize AI to analyze camera trap images to track individual elephants, monitor population dynamics, and identify poaching hotspots. Ethical considerations in environmental management are important for the responsible use of AI. This chapter delves into the implications of computational intelligence in the healthcare and environmental sciences. With the help of advanced techniques such as neural networks and fuzzy logic, CI not only addresses complex challenges but also leads to adaptive learning in unknown

H. Bansal (✉) · K. Shukla · A. K. Karn
Centre for Computational Biology and Bioinformatics, Amity Institute of Biotechnology, Amity University, Noida, Uttar Pradesh, India
e-mail: hinabansal@gmail.com

K. Shukla
e-mail: kritika.shukla1@s.amity.edu

W. Pedrycz et al. (eds.), *Decision-Making in Computational Intelligence-Based Systems*, Studies in Systems, Decision and Control 628,
https://doi.org/10.1007/978-3-032-04056-5_11

environments. The applications discussed in this chapter demonstrate how AI-driven systems can enhance disease diagnosis and ecosystem management. The insights of this chapter provide a foundation for understanding how CI can be used to navigate the complexities of modern society, paving the way for innovative solutions that enhance individual and collective well-being. The chapter also explains the real-world examples, like how neural networks help in diagnosing diseases through medical imaging, or how fuzzy logic and evolutionary algorithms support smarter farming and environmental decisions. These stories reflect both the powerful impact and the moral questions that come with including CI in our daily lives.

Keywords Computational intelligence · Environmental sciences · Healthcare · Artificial intelligence

1 Introduction to Computational Intelligence

Computational intelligence (CI) is an interesting combination of advanced technology and real-world applications. At its core, CI targets replicating human-like thinking processes, such as learning, reasoning, and problem-solving, by mixing elements of artificial intelligence (AI), machine learning (ML), neural networks, and evolutionary algorithms. These skills improve decision-making across various sectors, notably healthcare and environmental sciences, where CI delivers generative advances.

The journey of computational intelligence may be traced back to early AI research initiatives. The main focus was on constructing machines capable of doing activities formerly associated with human intelligence. Significant events, such as the development of expert systems in the 1970s and the advent of neural networks in the 1980s, paved the way for today's faster CI systems. These complex systems use enormous datasets and strong algorithms to detect patterns, make judgments, and dynamically adapt to changing conditions.

In the medical industry, computational intelligence (CI) is revolutionizing disease diagnosis and patient care approaches. The most notable application of CI is the capacity to break down complex medical data, such as medical images, patient histories, and genetic data, for disease diagnosis with remarkable accuracy and efficiency. For example, CNNs accurately identify anomalies, diagnose tumors, and segment organs in medical pictures by capturing complex patterns and structures [1].

The CI has also helped doctors reshape care of their patients with the help of predictive analysis and personalized treatment plans, which happens by integrating data from various sources like clinical trials and more electronic data found on the system about the patient.

This strategy increases the likelihood of therapeutic success while reducing its side effects. One of the best examples of artificial intelligence in medicine is the Watson for Oncology (WFO) support system. It uses natural language processing (NLP) and machine learning (ML) to make a unique patient treatment plan based on their characteristics by integrating big datasets from clinical trials and electronic data of individual patients [2].

Computational intelligence has a significant impact beyond the medical field, such as environmental science, where it plays an important role in monitoring and managing the ecosystem [3]. CI systems work with large amounts of environmental data, which helps the researchers identify ecological shifts, forecast climate fluctuations, and address biodiversity issues. Nowadays, Machine learning algorithms detect deforestation patterns, which helps climate fluctuations conservationists take timely actions against threatened habitats [4]. For instance, climate change modelling also uses CI to analyze past climatic data and give information that can help in the prevention of global warming effects.

Sustainable resource management is another area that CI has revolutionized. It plays an important role in conserving our natural resources by analyzing previous patterns, predicting the available resources, and recommending sustainable practices. CI-driven models are helping to optimize resource use effectively. Example—Precision farming: It helps farmers estimate crop yields, adjust irrigation, and evaluate soil health [5]. With the help of data analysis, these developments improve farmer's decision-making regarding their crop productivity, boost sustainable agriculture, and increase global food security. It can also help determine the right amount of nutrients and water needed for a particular crop.

In summary, CI is transforming the ways in how we make decisions in the healthcare and environmental science. We're not only protecting our ecosystem but also promoting sustainability. By keeping in mind the ethical challenges, we can make sure to make our future healthier and more sustainable.

2 Importance of Decision-Making in Computational Intelligence

Decision making involves choosing between alternatives based on values, goals, and available information. Several well-established concepts and frameworks guide decision-making across various fields. At the heart of computer intelligence lies decision making—the ability of system to process and distill a vast volume of data into actionable insights. This capability plays a pivotal role across diverse domains, including healthcare, finance, and beyond. In this section, we explored the transformative impact that CI-driven decision-making has had over the years.

Decision-making is a mental process that involves identifying a problem, gathering information, and evaluating options. The process is governed by a variety of factors, including:

- *Mental processes*: Biases, reason, emotions, and memories all play a role in the decision-making process.
- *Culture, values, and beliefs*: The decision-making process is based on the decision-maker's culture, values, beliefs, personality, and knowledge.

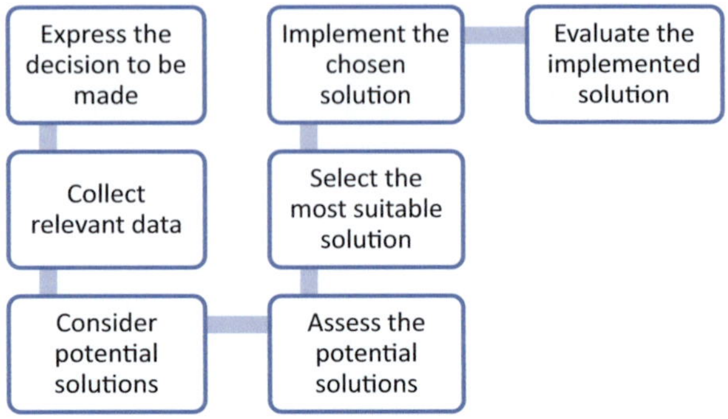

Fig. 1 Steps of decision-making in CI

- *Brain regions*: The prefrontal cortex (working memory) and hippocampus (long-term memory) are known to communicate during the decision-making process.

Decision-making in CI works similarly. It first collects all the data, processes the information, and then evaluates the outcomes based on the processed data with the help of techniques like machine learning, fuzzy logic, and neural networks [6]. Though it seems simple, it is a complex process with many steps shown in the Fig. 1. In this section, we will learn about how it happens and how it has impacted various fields. The ability to make decisions effectively is considered the defining characteristic of intelligence, whether in humans or machines.

The 7 steps of the decision-making process:

(a) Express the decision to be made: The decision-making team should explicitly state what decision it is attempting to make. Everyone on the team should understand the circumstances underlying the choice and what they hope to achieve by making it. The problem or scenario should be unambiguous. The more accurately the team can describe the decision [7] and why it should be taken, the easier it will be to move forward with the rest of the decision-making process, and the better the result.
(b) Collect relevant data: Once a choice has been made, the team should collect all relevant information. It may search historical data on the issue or locate information on previous activities in the organization that have either succeeded or failed. It should collect important data from sources outside the organization, such as studies or market research. The team could also benefit from speaking with external peers in similar situations or engaging a consultant.
(c) Consider potential solutions: Provided with the relevant facts, the team may begin considering viable solutions. For example, the team may need to decide how to increase storage capacity over the next year. Possible solutions include purchasing new storage systems, moving into the cloud, repurposing existing systems, leasing equipment, and other choices. This is an excellent moment for creative thinking and brainstorming, while the final list of answers should only include the most plausible options.

(d) Assess the potential solutions: The team should carefully assess each solution, recognizing its strengths and drawbacks. It should also seek internally and externally for instances of how similar solutions have performed. Sometimes, the team may reject some choices due to apparent issues or overriding challenges. The team's evaluation should also assess how a specific solution affects key stakeholders and implementation needs.
(e) Select the most suitable solution: After assessing the potential options, the team should select which best meets its requirements. Sometimes, the choice is obvious, and one solution has arisen. In other circumstances, the decision will be more complex, and the team must consider the benefits and drawbacks of numerous promising options. However, the team may determine that combining several choices into a single overarching solution is the best option.
(f) Implement the chosen solution: Once you have decided, it is time to take action. The team should create a clear implementation strategy that outlines duties specifically. Every action has implications; the planning should account for potential challenges and establish a procedure for dealing with unanticipated setbacks. Open communication is crucial, as are clearly stated responsibilities and expectations.
(g) Evaluate the implemented solution. Once the solution is deployed, the team should assess its effectiveness and decide whether to choose that option. The team should ensure that the solution addresses the original problem. If it does not, the team may have to repeat some of the decision-making steps. The team should consider whether another option would have worked better.

2.1 Techniques Used in Decision-Making in Computational Intelligence

2.1.1 Neural Networks

A neural network is a model that is inspired by the human brain, in which the information is processed with the help of neurons that are interconnected nodes [8]. They work in layers by first identifying the patterns and then learning from the data, which enables tasks like complex decision making. Neural Networks consist of three layers of neurons [9], input layer—data receiving, hidden layer- performing complex problems, output layer—gives the final results. They learn in a process called training, where they perceive vast amounts of data and then adjust the weights (the connection between the neurons), which helps in improving the accuracy and efficiency of the results [9]. Some daily life applications of neural networks are image recognition, speech recognition, NLP (Natural Language Processing), and medical diagnosis.

2.1.2 Fuzzy Logics

Fuzzy logic is a type of mathematical reasoning that helps computers process information that is not always true or false but is somewhere between the two, just like how humans interpret problems with uncertain information. Fuzzy logic uses membership functions to assign a degree of truth to the statements that range from completely false (0) to completely true (1), unlike other binary logics. It can handle terms like sweet, sour, and a little spicy, which are ambiguous in traditional computer systems. It uses the if–then rules to make decisions that incorporate linguistic variables, which helps in more nuanced interpretations of input data. It works by a process called fuzzification in which the initial input data is converted into fuzzy sets by assigning membership degrees to different values based on relevant linguistic variables. The fuzzy rules are then applied to the fuzzified input, which in turn generates a fuzzy output. Fuzzy output is then converted back into a crisp value that can be used by the system. Hence, Fuzzy Logic is a form of logic used by computer Intelligence that allows reasoning in a way that mimics human decision-making, unlike traditional binary logic (0 or 1), fuzzy logic allows for a degree of truth.

Example: In medicine, many decisions involve vague or imprecise information-symptoms are not always clearly present or absent, and severity can vary. Fuzzy logic helps model this uncertainty to support better diagnosis, treatment, and monitoring. Like in diabetes risk assessment, a Fuzzy system evaluates risk using multiple fuzzy variables such as blood sugar level, BMI, age, family history, etc. if each input is partially true then the system uses fuzzy rules like: "IF blood sugar is elevated AND BMI is overweight AND age is middle-aged THEN diabetes risk is moderate to high."

2.1.3 Evolutionary Algorithms

Evolutionary algorithms are a subset of optimization techniques that copy the process of natural selection to find optimal solutions to complex decision-making problems. By iteratively improving a population of potential solutions [10] through operations like selection, crossover, and mutation, essentially simulating the evolutionary process of "survival of the fittest" to arrive at the best possible result. Evolutionary algorithms take inspiration from biological evolution. Individuals with favorable traits are more likely to survive and reproduce [11], which then passes on those beneficial traits to the next generation. For instance, in cancer treatment, EAs can help tailor chemotherapy regimens based on a patient's genetic makeup, improving treatment efficacy while minimizing side effects. Similarly, EAs are used in clinical decision support systems to analyse patient data and recommend the most effective interventions, enhancing healthcare outcomes. Evolutionary algorithms work with a lot of potential solutions, allowing them to explore a wider range of possibilities and avoid getting stuck. There are some common types of evolutionary algorithms shown in Fig. 2.

Common types of evolutionary algorithms:

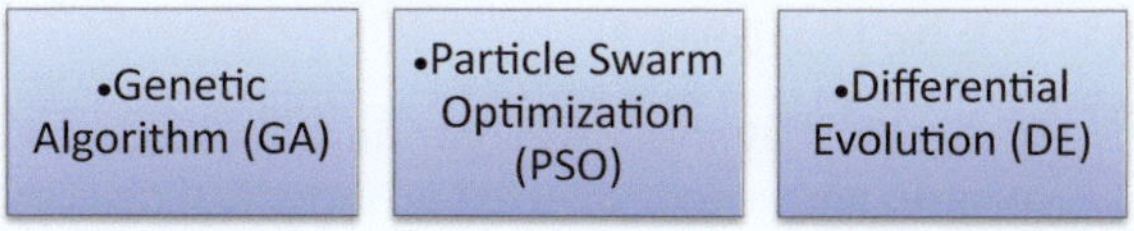

Fig. 2 Types of evolutionary algorithms

- Genetic Algorithm (GA):

  The most well-known evolutionary algorithm, widely used for optimization problems. Here, solutions are represented as strings of genes and manipulated using crossover and mutation operators.
- Particle Swarm Optimization (PSO):

  This is inspired by the bird's flocking behaviour; here, each particle represents a solution. Then this particle updates its position based on its own best experience of the swarm.
- Differential Evolution (DE):

  It utilizes a difference vector between population members to help the search process, often effective for complex problems.

2.1.4 Reinforcement Learning (RL)

In computational intelligence, reinforcement learning is a machine learning technique that helps autonomous agents to make decisions by learning through the trial-and-error method in an environment in which they essentially mimic how humans learn by receiving feedback (rewards or penalties) for their actions. This allows them to improve their behaviour over time, which helps in achieving the best possible outcome in complex decision-making scenarios.

An RL system consists of an agent that interacts with its environment by taking actions and receiving feedback (rewards) based on those actions [12], allowing it to adjust its behavior to maximize rewards in the future.

The applications of RL include personalized treatment plans in dosage recommendations, adaptive prosthetics and robotic surgery control, optimizing resource allocation in hospitals, and building healthcare simulators for safe training without real-world risk.

3 Applications in Healthcare

Computational Intelligence includes a range of techniques that allow computers to perform tasks that typically require human intelligence like learning and problem-solving [13]. In the healthcare industry, CI has been an evolutionary force which enhanced a lot of aspects like patient care, diagnostics and treatment plans.

A. Disease Diagnosis

In disease diagnosis, AI algorithms look into medical data like (X-rays, MRIs), records, and genetic data to find the actual disease with great accuracy. This helps healthcare professionals in making more reliable and faster diagnostic decisions, especially in complex and early-stage detections. Data Analysis is one of the key characteristics of CI as AI models collect and analyze large amounts of medical data and identify the patterns to indicate specific diseases. Deep Learning models like convolutional neural networks, are good at medical imaging like CT scans and MRIs to detect abnormalities (tumors and fractures). After the disease detection, AI can provide diagnosis and treatment recommendations for the patients which can help the professionals in easy treatment. For example, cancer detection by imaging analysis, cardiovascular disease detection of heart failures, and coronary artery diseases by ECGs. In emergency departments, AI algorithms can prioritize patients based on severity, helping clinicians make quicker decisions. The use of AI comes with some challenges like large and well-annotated data requirements, bias in data, and privacy concerns because of sensitive patient data.

B. Robotics in Surgery

In Robotics, AI algorithms can be utilized to analyze real-time surgical data, which may help surgeons make more informed decisions while performing the procedures and enhancing precision which results in identifying complex structures and optimizing surgical techniques acting as a smart assistant to the surgeons. AI algorithms help in identifying important anatomical structures within the field by live imaging, which allows them precise navigation around the target tissues. It can also identify the steps and with experience provide the surgeons with the next action to be performed. By using the patient data, AI can predict potential complications and suggest optimal surgical strategies according to the patient. Adaptive control is also another aspect in which AI algorithms can adjust robotic arm movements in the surgical field. CI in robotic surgery can provide greater accuracy in tissue dissection in comparison to traditional laparoscopic surgery, it can help in smaller incisions which results in less recovery time.

C. Healthcare Administration

In Healthcare Administration, CI can help in making more informed decisions regarding allocations, treatment plans, and operational efficiency with the help of AI techniques like machine learning. High-risk patients can be identified by analyzing their history and current health and by intervening with treatment plans and preventive methods. Forecasting patient volumes to optimize staffing levels, bed capacity, and resource allocation based on historical data and seasonal trends. Patient data can be analyzed to get the most effective treatment suggestions which are cost effective as well. Detection of insurance and billing fraud is another way in which AI algorithms can help. It enhances the quality of care, cost reduction, and data-driven decision-making.

D. Telemedicine

Telemedicine is the remote delivery of healthcare services over the telecommunications infrastructure [14]. Telehealth helps healthcare providers to evaluate, diagnose, and inform patients without an in-person visit about their condition. With the help of a large amount of patient data from wearable devices or remote monitoring systems, AI can analyze and identify different patterns and predict health risks. Virtual consultations can also be done by providing diagnosis and treatments with the help of AI. Chatbots and voice-activated systems can answer basic health-related questions and provide medications. During the remote monitoring of vital signs like blood pressure and glucose levels for diabetes patients, AI can alert healthcare providers if the patient is facing any complications. AI can also be used to analyze medical images like X-rays or MRIs remotely, aiding in diagnosis by radiologists [15]. The advantages of telemedicine is that people who live in rural areas also get better access to healthcare facilities and it saves both doctors and patients time. In healthcare, the use of AI raises significant ethical concerns, including bias in medical algorithms, which can lead to unequal treatment outcomes across different demographics. Additionally, patient data privacy is a critical issue, as AI systems often require access to sensitive health information, necessitating stringent safeguards to protect individuals' rights and confidentiality. Addressing these concerns is essential to ensure that AI applications in healthcare are both effective and equitable.

3.1 AI-Driven Diagnosis and Prognosis

AI has proven to be a tremendous help in predicting patient outcomes. From identifying diseases at an earlier stage to identifying treatment options, that may have been missed due to its ability to analyse and interpret complex patterns in medical data, including imaging scans and electronic health records. Despite significant advances in medicine, diagnosing diseases effectively remains a global challenge. Developing tools for early diagnosis is particularly difficult because of the complexity of disease mechanisms and the variety of symptoms they present. However, artificial intelligence (AI) has the potential to transform healthcare, especially in the area of diagnosis [16]. Machine learning (ML) is a part of artificial intelligence (AI) that uses data to get better at tasks [17]. This makes it a great tool for helping with disease diagnosis. How well ML works depends on the quality and amount of data it uses. Deep learning on the other hand adds layers by using CNN and data mining as well for pattern recognition. This can help in identifying the disease or even predicting the diseases like Alzheimer's.

The Role of AI in Diagnosis

- Medical imaging analysis

  It is the use of AI/ML in analyzing medical images like XRAYS, and MRIs to get meaningful information for diagnosis, treatment planning and disease monitoring by identifying patterns in the data. It divides medical images into pieces and then analyses the results. These pieces are then classified into different patterns and then indicate a particular result. For example, in cancer detection-cancerous lesions can be identified via CT scans or MRIs, and Bone Fractures can be assessed by X-rays. Registration of images of the same person at different times of different organs can also be done.

- Pathology

  Use of machine learning techniques to analyze digital pathology images which allows automated interpretation of tissue samples to diagnose diseases, predict prognoses, and discover new biomarkers. AI analyses the cell patterns better than the traditional methods and enhances accuracy as well. It helps in identifying various cancers by differentiating between benign and malignant cancer with their grading severity. It can be done by digitalization of tissue slides using slide scanners which create high-resolution images that can be used in analysis. For example, identification and classification of cancerous tissue, grading tumors, and detection of tumor margins Estimating the likelihood of disease progression or patient survival based on tissue analysis.

3.2 Treatment Optimization in Healthcare

Artificial intelligence (AI) is crucial for dose optimization and adverse pharmaceutical event prediction to enhance patient safety and treatment outcomes. By employing AI algorithms to predict potential adverse drug events and optimize medication dosages for each patient, healthcare providers may reduce risks and improve patient care. In a study that sought to create a decision support system for optimizing the dosage of warfarin maintenance and an AI-based prediction model for the prothrombin time international normalized ratio (PT/INR) [18], The algorithm beat expert clinicians with substantial differences in predicting future PT/INRs, and the generated customized warfarin dose was dependable, according to the authors' analysis of data from 19,719 inpatients across three institutions. Conversely, a brand-new dose optimization system called CURATE.AI is an AI-powered platform that uses patient data to dynamically optimize chemotherapy dosages. To validate this method as an open-label, prospective trial, three distinct chemotherapy regimens were administered to patients with advanced solid tumors. In pilot investigations, clinicians accepted virtually all (97%) of CURATE.AI's dosage recommendations, with some patients receiving doses that were roughly 20% lower than usual levels to reduce potential side effects. While these first findings are intriguing, they highlight the need for further confirmation via rigorous randomized clinical studies [19].

3.3 *Predictive Analysis*

Predictive analytics is used in identifying and guiding health initiatives because of the increase in health management risks. predictive analytics is a discipline that significantly utilizes modelling, data mining, AI, and ML. To anticipate the future, it analyses past and current data. Using machine learning algorithms and other technologies, data can be analyzed, and prediction models are then developed to improve patient outcomes and reduce costs. Finding those who are at risk of developing chronic illnesses like heart or endocrine issues can be aided by predictive analytics. Hospital readmissions can also be predicted using predictive analytics. By looking at patient demographics, medical history, and social health determinants, predictive models can pinpoint treatments to prevent readmissions and identify patients who are more likely to experience hospital readmissions. By enhancing the accuracy and efficacy of predictive models, artificial intelligence can promise enhanced healthcare [20]. AI systems can analyze large amounts of data and identify patterns and connections that human analysts would overlook, improving the accuracy of predictive models and enabling patients to receive the best care possible. With AI, some public health management tasks, such as care coordination and patient outreach, can also be automated. It ensures patients receive timely and appropriate care, which improves patient outcomes and reduces healthcare expenditures.

It is important to note that the quality of data and the technology infrastructure necessary to construct and operate prediction models show how well predictive analytics works in public health management. Furthermore, for patients who are at risk, humans are required to monitor to ensure the suitability and efficacy of treatments. In conclusion, predictive analytics is becoming increasingly important in population health [21]. Using machine learning, healthcare companies can develop prediction models that identify individuals at risk of chronic illness or hospital readmission. In addition, artificial intelligence is required to overcome vaccine production issues and supply chain constraints. It can be difficult to test algorithms for real-time vaccination supply chains. To address this, research and development must be prioritized to develop robust algorithms capable of properly predicting and optimizing vaccination supply chains. Edge analytics can also detect anomalies and anticipate Disease X episodes, as well as the related dangers to the healthcare system. Sehaa, a big data analytics tool in Saudi Arabia, uses Twitter data to diagnose diseases. It was discovered that dermatological diseases, heart diseases, hypertension, cancer, and diabetes are the top five diseases in the kingdom [22].

3.3.1 Key Applications of Predictive Analytics in Healthcare

- Early disease detection and diagnosis: predicting the likelihood of developing chronic conditions like diabetes, heart disease or cancer based on past history, genetics, and lifestyle data.

- Hospital readmission prediction: identifying patients at high risk of being readmitted within 30 days of discharge.
- Personalised treatment plans: using patient data to forecast how individuals will respond to certain treatments, like chemotherapy.
- Clinical decision support: alerting clinicians to potential diagnoses or complications based on patterns in lab results and vital signs.
- Resource and staffing optimization: predicting ER visit volume based on seasonality and local events.
- Chronic disease management: monitoring wearable data to predict deterioration in patients with heart failure or COPD.
- Sepsis and infection risk prediction: using real-time data to identify early signs of sepsis in hospitalised patients
- Mental health risk assessment: using patient behaviour, speech pattern or social determinants to predict risk of depression or suicide.

The overall benefits of predictive analytics in healthcare includes improves patient outcomes, reduce avoidable hospitalisation and complications, optimises treatment and resource allocation, supports evidence based personalised care and cut down healthcare costs [23].

3.4 Case Study in Healthcare

Artificial intelligence (AI) is transforming the way we approach [24] public health, which makes it easier to address complicated issues and results in the enhancement of healthcare services around the world. AI assists us in finding diseases earlier, anticipating epidemics, and tailoring treatment to individual requirements by rapidly processing enormous amounts of data. This trend is affecting both emergency responses to health crises and routine medical care.

One of the most striking aspects of AI is its capacity to detect diseases early and anticipate when epidemics will occur. For example, BlueDot, an AI platform, identified the COVID-19 outbreak in Wuhan before it was officially disclosed by health institutions. Similarly, HealthMap monitored the spread of the Ebola virus, allowing health authorities to respond more quickly. Tools such as Nextstrain track how viruses change over time. AI can also help with the management of chronic diseases. IBM Watson Health, for example, analyzes medical studies and patient data to assist clinicians in diagnosing and curing cancer [25]. Other systems, such as PathAI, use sophisticated image recognition to identify illness markers in pathology samples. Wearable health gadgets and telemedicine have enabled continuous patient monitoring, enabling timely interventions for illnesses such as diabetes and high blood pressure. During the COVID-19 epidemic, telemedicine allowed people to receive care without having to attend crowded medical facilities.

Furthermore, AI has altered vaccine delivery procedures, ensuring equitable access to immunization. During the epidemic, organizations such as the Rockefeller Foundation and Gavi used artificial intelligence to improve vaccination distribution. These technologies forecasted demand and identified potential obstacles in distributing vaccines to underserved areas, reducing waste and ensuring that those in greatest need received vaccines on schedule. AI is also helping to eliminate healthcare gaps. For example, the Aetna Health Equity Index uses artificial intelligence to identify gaps in healthcare access, assisting politicians in the development of tailored remedies. Maven Clinic provides AI-powered maternal care solutions that give pregnant women virtual consultations in locations where essential services are absent. These initiatives demonstrate how AI may help make healthcare more accessible.

However, there are challenges to implementing AI in healthcare [26]. Health data is sensitive and subject to manipulation; therefore, privacy and security are critical issues. Furthermore, biases in AI systems can worsen existing inequality by disproportionately affecting underrepresented groups. The high costs of adopting AI technology, along with poor infrastructure in some areas, can limit their application. Furthermore, public distrust of AI's role in healthcare decisions limits its implementation. Despite these obstacles, artificial intelligence has a lot to offer in the field of public health. To properly address these issues, cooperation between governments, IT companies, healthcare facilities, and communities is crucial. The establishment of ethical regulations and a global regulatory framework will ensure that AI is used in an ethical manner that benefits everyone equitably. Looking ahead, AI in Clinical Informatics has huge potential. Imagine a future where AI helps doctors create personalized treatment plans just for you, based on your genetic makeup. With wearables that track your health 24/7, AI will be able to help spot potential problems before they even happen, keeping you out of the hospital. In the next decade, AI might just become the backbone of healthcare, making care more personalized and efficient.

Case Study: Leveraging Predictive Analytics to Reduce Sepsis Mortality at Johns Hopkins Hospital

Sepsis, a severe and often fatal response to infection, remains a leading cause of hospital mortality. Its early detection is challenging due to symptom overlap with other conditions, and delayed treatment significantly increases the risk of death. To address this, Johns Hopkins Hospital developed TREWS (Targeted Real-Time Early Warning System), a machine learning-based tool that analyzes real-time patient data to identify early signs of sepsis.

Integrated directly into the hospital's electronic health record (EHR) system, TREWS continuously monitors lab results, vital signs, medications, and clinician notes. It applies predictive modeling to generate sepsis risk scores and alerts clinicians when a patient's risk crosses a critical threshold.

The implementation of TREWS led to early identification in 76% of sepsis cases and contributed to a reduction in sepsis-related mortality by 18–20%. Due to its success, TREWS is being adopted by other hospitals and stands as a leading example of how predictive analytics can enhance clinical decision support systems (CDSS) and improve patient outcomes.

Table 1 Key benefits of CI in ecosystem monitoring

Application	Description	Example
Species Identification	Automated recognition of flora and fauna using AI algorithms	CNNs for species ID
Threat Detection	Identifying poaching activities and environmental threats via sensor data	Satellite and drone usage
Habitat Analysis	Monitoring habitat changes due to urbanization or climate change	Remote sensing and GIS AI
Ecological Forecasting	Predicting disruptions like invasive species or habitat degradation	Temporal AI predictions

4 Applications in Environmental Sciences

4.1 Ecosystem Monitoring and Management

Computational Intelligence offers a different approach to the observation and management of ecological environments. Using data from satellites, sensors, and camera traps, we can detect environmental threats and track ecosystem changes through biomonitoring. These can lead to improved conservation methods such as in the application of CNN, successfully deployed in monitoring the populations as well as recognizing species' characteristics from images with their movements that may involve mass migration. Such findings have proven crucial for biodiversity conservation since they can detect endangered species and habitats under threat. It increases early warning mechanisms because it will forecast a potential ecological disruption, say the outbreak of invasive species or degradation of a habitat [27]. Integrating real-time data streams with historical records by AI tools brings about actionable insight for conservation action that enables action well before something undesirable has occurred, hence reducing the time to intervene and increasing intervention efficacy. AI technologies can also integrate multiple sources of data through one platform so that comprehensive knowledge about ecosystems can be established. Such platforms will include GIS mapping, temporal data analysis, and predictive modeling to achieve a holistic view of ecological health and vulnerabilities. As shown in Table 1 CI applications provide significant advantages in species identification, threat detection, habitat analysis, and ecological forecasting, enhancing conservation efforts through automated monitoring and predictive analytics.

4.2 Climate Change Prediction

AI and CI are among the most sophisticated tools that can theoretically be employed to predict and mitigate the impacts of climate change [28]. These serve as a revolutionary means of acquiring and analyzing enormous datasets including atmospheric measurements, oceanic behavior, and historical records of climate variations. After computing these datasets using Weighted Neural Networks, Deep Learning systems, and other advanced algorithms, sophisticated patterns are disclosed that could hardly have been detected. Such an impressive ability of data processing is deeply needed for bringing a comprehensive understanding of the dynamics of climate change and for preempting proactive countermeasures [29]. It is the ability to simulate the impacts of greenhouse gas emissions and predict the future pathway of global climate change that lays the more essential strengths of AI and CI approaches in building strong adaptation and mitigation strategies. Information of this nature is critical in aiding the formulation of potent adaptation and mitigation strategies [30].

Such systems allow for modeling of what would happen if certain interventions—such as reductions in carbon emissions, deforestation, or renewable energy adoption—were implemented, thus assisting policymakers in creating evidence-based constructive approaches toward environmental protection. Events of great interest simulated by AI models include critical scenarios that factor into how countries build resilience against climate change, such as sea-level rise, extreme weather conditions, and changes in agricultural productivity. Among other applications, AI and CI tools are used to study complex environmental systems with multiple logistics such as changing weather patterns, human interventions, and natural processes. The fuzzy systems are some of the tools that can handle these uncertainties very effectively about seasonal or climate predictions. Fuzzy logic systems work with incomplete or imprecise data by showing a range of possible outcomes rather than a single deterministic solution [31]. This makes them particularly valuable for addressing the uncertainty that naturally surrounds climate change projections because such variables as human behavior, policy changes, and incidents in the natural world pose a communications challenge when it comes to a single clear answer. Fuzzy logic systems also play an essential role in modeling a range of outcomes that allow decision-makers to consider various scenarios to motivate decision-makers toward more resilient policies and solutions. In addition, AI and CI-based tools facilitate more efficient and informed decision-making in areas where interdependent agents interact on complex variables, which are likewise linked with each other [32]. Climate change adaptation strategies have for a long time been involved in a trade-off between economic development, environmental protection, and social equity. With real-time data at their disposal, AI and CI tools will allow the policymakers to weigh those competing interests among each other through simulations showing how a multitude of policies may affect various sectors of society. With this information integrated into their decision-making, governments and most organizations will see to it that the strategies they implement are effective in both scientific and economically viable terms, as well as socially acceptable. One of the primary benefits brought in by AI and CI in climate change prediction and adaptation is convergence among stakeholders for communication. As one of the challenges in climate change response, elaborate technical scientific data often pose difficulties in their coming across to non-specialist-oriented decision-makers [33]. AI and CI-powered tools present advanced visualization schemes that frame the presentation in terms more accessible to audiences by simplifying and communicating rather complicated climatic patterns and their variables. These tools enable stakeholders-from local community representatives to international organizations to understand and even quantify potential climate change impact and the various adaptation and mitigation options open to them [34]. Such tools not only increase collaborations for planning and resource allocation, for instance, local government, NGOs, and other involved parties may collaborate maximizing the usefulness of interventions while avoiding any duplication of efforts by creating such visualizations of climate risks and mitigation strategies. Also, visualizing explains climate data which can have the power to engage local communities in the planning processes while ensuring the inspecting policies respond to the distinctive needs and vulnerabilities of the populations [35]. Hopefully, greater coordination will lead to more efficient use of resources, as well as better strategies for protecting vulnerable ecosystems and communities. The application of AI and CI in climate change also provides more equitable and inclusive solutions for issues akin. By harvesting and analyzing data on social and economic disparities traveling from such twenty-first century tools can indeed enable policymakers to identify the most vulnerable populations requiring their resource allocation. In turn, through wider inclusion of stakeholders during the decision-making processes, AI and CI foster more inclusionary approaches towards climate change adaptation besides providing means for engaging the public [36]. In this regard, it would be sufficient to say that the introduction of AI and CI in climate change prediction and response brings great opportunities for the improvement of resilience at both local and global levels. With advanced data collection, simulation, and decision-making capabilities, these tools provide insights that are relevant to help policymakers better anticipate and mitigate against climate change effects. The introduction of fuzzy logic systems and high-end visualizations largely allows the stakeholders to make informed choices, cutting across any silo divides in increasing collaboration and resource mobilization for priority programs. In conclusion, while the relationship between the improvement of community resilience with better adapting to climate policies, some contributions of AI and CI could still constitute a strong cable to fight the climate change challenges and ensure a sustainable future for generations.

4.3 Sustainable Resource Management

When implementing CI for managing sustainable resources, there is a need to focus on the optimal use of natural resources in such a way that environmental degradation is minimized. This all happens is thus aided by improved machine learning algorithms that predict the demands of the resource and render efficient use of resources over various fields such as agriculture, water, and energy shown in Fig. 3. For instance, a CI algorithm is directed at the optimization of the irrigation schedule [37]. The aim is to conserve water with this new approach, which does not waste water on the land but uses water scientifically and efficiently according to the prevailing tendencies at the moment. Evolutionary algorithms serve as feasible recommendations regarding layout patterns and pest control in sustainable agriculture [38].

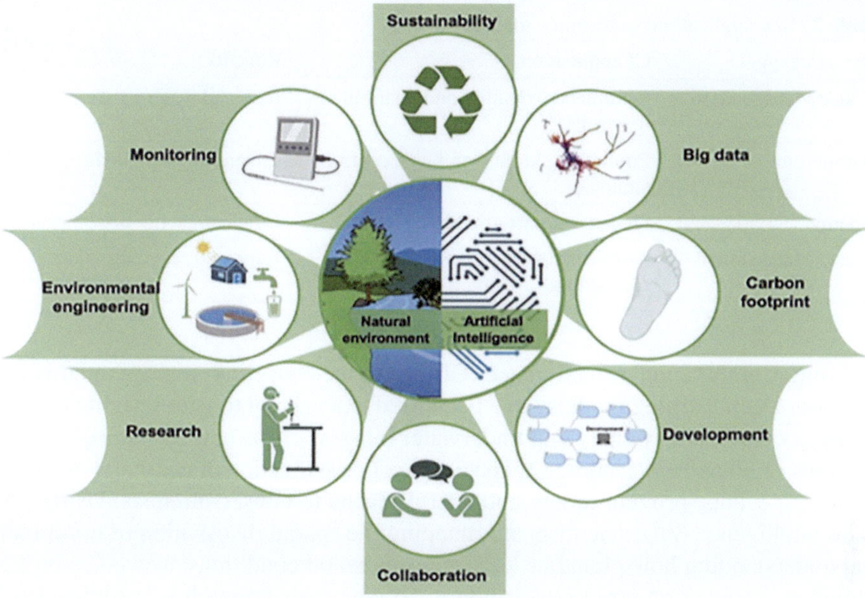

Fig. 3 AI and the natural environment integration for sustainability

Besides that, from the soil, weather data, and plant health, this algorithm maximizes productivity and environment-friendliness in all crops it studies as shown in Table 2. This way, AI-powered smart grid designs could also help enable the use of renewable energy resources in balancing the distribution of energy besides reducing the total energy loss transmitted. It could allow more effective real-time demands-based dynamic balancing of resource distribution, thus reducing wastage and enhancing the resilience of the grid. Similarly, artificial intelligence has a broad scope of applications, such as in waste management. Here, machine learning models will classify and sort the waste, which results in recycling and protecting landfill resources. Additionally, using predictive analytics to analyze and predict waste production patterns can assist municipalities in devising where they may enhance collection and disposal strategies.

4.4 Case Study in Environmental Sciences

The use of Computational Intelligence (CI) in environmental science offers significant advantages, including the ability to manage nonlinear, uncertain, and high-dimensional data; enable real-time monitoring and adaptive system management; support evidence-based policy-making and efficient resource optimization; and foster sustainability through data-driven decision-making.

Table 2 Key applications in resource management

Resource type	CI application	Benefits
Water Management	Optimized irrigation and distribution networks	Reduced water waste
Agriculture	Predictive planting and pest control strategies	Enhanced crop yields
Energy	Dynamic smart grid allocation	Efficient energy distribution
Waste Management	Automated waste classification and forecasting	Improved recycling rates

Using Computational Intelligence (CI) in marine ecosystem management is one of those. Deep learning models have been used extensively to gain insights into the status of coral reefs by analyzing underwater images. These new technologies, with upwards of 90 percent accuracy at species-level classification of coral and detecting bleaching events, provide timely and useful results to conservationists. Using artificial intelligence (AI), scientists are mapping the spatial distribution of coral reefs and understanding how global or local environmental conditions, as well as human activities, impinge on coral reef habitats. To take their research to a higher level, they have combined satellite images and real-time underwater surveys using a geek-chic method. These integrated approaches have enabled the formulation of predictive models for areas that are at risk of degradation or that are expected to degrade. This hardly meant anything toward targeted conservation actions but proved instrumental by simply predicting the establishment of marine protected areas and coral transplantation methods to restore damaged reefs. These not only helped in protecting fragile ecosystems but also provided scalable solutions for managing marine conservation issues in every corner of the world [39]. The engagement of community members has been integral to the success of the engagements. There is full community involvement. Communities were able to engage directly in conservation initiatives through community members trained to do real-time monitoring and reporting of data; examples include coral bleaching assessments and illegal fishing activities noted on AI-powered mobile applications. The bottom-up approach not only streamlined data collection but also nurtured a more meaningful relationship between local stakeholders and marine conservation efforts. As community members became more informed and invested in their environment, environmental education moved forward considerably in ways that fostered a more sustainable practice at the local level. In one of the attempts, this innovative aspect would include fusing artificial intelligence tools with drone technology. Drones, embedded with artificial intelligence capabilities, were sent for aerial reconnaissance, thereby allowing scientists difficult access to remote or hard-to-reach environments for observational and research purposes. Such technology combinations, therefore, developed further perspectives for research consequently providing wider information about marine ecosystems. They proved time after time effective in mapping big coral reef systems, directing intervention on processes of degradation, and gauging effectiveness through the continuum of conservation efforts [40]. Furthermore, the CI-based systems assisted in designing the allocation of resources with real-time visualization of data and insights on prediction. Also, such tools could be put to better priorities by policymakers and conservationists for action and resource allocation and be useful for tracking project progress. This high-level technology stated the need for interdisciplinary cooperation as it required the interactions of personnel with expertise in marine biology, the IT field, and environmental policy towards the achievement of shared goals. All these holistic measures indicate the great potential of CI in ameliorating marine ecosystem protection. This initiative perfectly demonstrates how a combination of high technology, involvement of the community, and sustainable practices may serve to transform CT into an instrument for accomplishing multi-scale environmental challenges. As this kind of project continues to blossom and grow, much more may be expected in raising the benchmark for conservation action to ensure continued livelihood.

5 Challenges and Ethical Considerations

The rapid advancement and widespread adoption of Computational Intelligence (CI) across critical sectors—such as healthcare, finance, defense, and environmental science has introduced a complex array of ethical challenges. These issues arise from the inherent complexity, autonomy, and lack of transparency in intelligent systems, combined with their far-reaching societal implications. The problem of biased AI models is of great concern due to the unworthy or unbalanced training data. The main problem in environmental sciences is that the scores in an environment are far from assessment, and they need prediction. Challenges in CI Applications are shown in Table 3. Generally, training and deployment typically use massive computational resources, and AI models are tied to direct-consuming ecological resources, back to major carbon emissions, which is in the opposite direction from the environmental goals [41]. There are also important ethical discussions that go together with CIs. Deployment of CI systems should respect data privacy, particularly satellite and drone imagery, while not infringing on the land and resources of indigenous communities. Informed consent by indigenous communities must be solicited before the implementation of CI solutions. Algorithms that are completely clear and all-inclusive services will help enable a CI solution that benefits all stakeholders evenly.

Below is a detailed overview of key ethical concerns emerging from real-world applications of CI:

- Bias in Medical Diagnosis Tools: CI systems trained on unbalanced datasets may produce biased diagnostic outcomes, exacerbating health disparities and undermining equitable care delivery.
- Opacity in Financial Decision-Making: Automated loan approval systems often lack explainability, making it difficult for applicants to understand or challenge decisions—eroding transparency and accountability.

Table 3 Challenges in CI applications

Challenge	Description	Mitigation strategy
Data Bias	Incomplete or skewed datasets affecting accuracy	Inclusive dataset design
Energy Consumption	High computational needs increase carbon emissions	Adoption of green AI practices
Ethical Concerns	Risks to data privacy and community rights	Policy frameworks and regulations
Accessibility	Limited access to technology in remote areas	Promotion of affordable solutions

- Privacy Concerns in Health Monitoring Technologies: Wearable devices and health apps frequently collect sensitive data without clear consent, posing serious risks to privacy and patient confidentiality.
- Autonomous Decision-Making in Defense Systems: The use of CI in autonomous weapons raises moral and legal dilemmas, especially when life-or-death decisions are made without human oversight.
- Environmental Impact of Model Training: The computational demands of training large-scale CI models contribute significantly to energy consumption and carbon emissions, raising sustainability concerns.
- Labor Displacement in Smart Agriculture: Automation driven by CI in agricultural systems can displace rural labor forces, leading to economic disruption without adequate retraining or support mechanisms.
- Algorithmic Bias in Law Enforcement: Predictive policing tools often reflect and reinforce systemic biases, resulting in the disproportionate targeting of marginalized communities.
- Misdiagnosis in Public Health Forecasting: Inaccurate CI-based predictions of disease outbreaks—due to insufficient validation or poor data quality—can cause misallocation of resources and public fear.

6 Future Trends

The future of the CI within environmental sciences can be considered as the development of emerging technologies and new methods of cooperation. The special place of quantum computing is also pointed out as a promising future development in improving the efficiency of CI systems, enabling the paranormal innovation of highly computationally intensive environmental models [42]. All these would create entirely new frontiers in modeling climate change, resource management, and biodiversity conservation. Edge computing allows real-time data analysis in remote areas. By processing data closer to the data source, edge computing reduces latencies and power consumption same principles that enable sustainability and effectiveness in CI solutions. Joining CI with citizen science brings participatory monitoring to a completely new level. The actual installation of AI in mobile apps that will allow reporting and documenting the changes in one's surroundings can create decentralized and community-based datasets. This kind of democratization of data collection energizes environmental monitoring and advances the process of public engagement. The area of explainable AI should facilitate the interpretability of complex CI models and should earn the trust of wider groups of stakeholders. Transparency in CI is expected to lead to its adoption in wider areas of policy and community activities [43]. The future of CI in healthcare and environmental sciences lies in intelligent, integrated, and ethically aligned systems that can adapt to complex, evolving challenges. These technologies will not only improve outcomes but also enhance sustainability, equity, and resilience in both human and natural systems.

7 Conclusion

This chapter underscores the transformative potential of Computational Intelligence (CI) in redefining decision-making processes across healthcare and environmental sciences. By leveraging core techniques such as neural networks, fuzzy logic, evolutionary algorithms, and reinforcement learning, CI provides a powerful and adaptive toolkit capable of addressing real-world challenges that often surpass the limitations of traditional computational models.

In healthcare, CI is enabling early disease detection, customizing treatment plans, and optimizing surgical procedures. In environmental science, it enhances precision agriculture, improves climate forecasting, and supports biodiversity conservation. These innovations point toward a future that is more responsive, efficient, and informed.

Real-world examples discussed throughout the chapter illustrate how CI bridges the gap between raw data and meaningful action. For instance, intelligent systems now support medical image diagnostics and monitor coral reef degradation using AI-powered drones. These examples reflect the broader trend of combining computational tools with domain expertise to unlock new pathways for sustainable development and human well-being.

However, the growing influence of CI brings forth critical ethical and practical challenges. Issues like data privacy, algorithmic bias, and equitable access must be addressed through the development of robust ethical frameworks and transparent, inclusive algorithms. As CI continues to evolve, its integration with emerging technologies, such as quantum computing, edge AI, and citizen science, will further expand its reach and democratize its benefits.

One of the central challenges CI faces, particularly in environmental applications, is scalability. Environmental data is often vast, fast-changing, and computationally intensive. Meeting acceptable performance thresholds under these constraints remains difficult. Yet, several breakthroughs, such as using specialist algorithms and hybrid systems, have shown promising results. Notably, recent studies have demonstrated that social media data, when integrated effectively, can enrich environmental datasets and enhance CI-based analyses.

These insights offer a practical roadmap for expanding CI's role in both domains. By aligning technological innovation with empathy, sustainability, and global collaboration, we can ensure that Computational Intelligence not only advances scientific frontiers but also contributes meaningfully to a more equitable and resilient world.

References

1. Li, M., Jiang, Y., Zhang, Y., Zhu, H.: Medical image analysis using deep learning algorithms. Frontiers Public Health **11** (2023). https://doi.org/10.3389/fpubh.2023.1273253
2. Liu, C., Liu, X., Wu, F., Xie, M., Feng, Y., Hu, C.: Using Artificial intelligence (Watson for oncology) for treatment recommendations amongst Chinese patients with lung cancer: feasibility study. J. Med. Internet Res. **20**(9), e11087–e11087 (2018). https://doi.org/10.2196/11087
3. Chisom, O.N., et al.: Reviewing the Role of AI in Environmental Monitoring and Conservation: A Data-Driven Revolution for our Planet. ResearchGate (2024). https://doi.org/10.30574/wjarr.2024.21.1.2720
4. Raihan, A.: Artificial intelligence and machine learning applications in forest management and biodiversity conservation. Natural Resources Conservation Res. **6**(2), 3825–3825 (2023). https://doi.org/10.24294/nrcr.v6i2.3825
5. Bongiovanni, R., et al.: Precision Agriculture and Sustainability. ResearchGate (2025). https://doi.org/10.1023/B:PRAG.0000040806.39604.aa
6. Russell, S., Norvig, P.: Artificial Intelligence A Modern Approach, 3rd edn (2010). Retrieved from https://people.engr.tamu.edu/guni/csce421/files/AI_Russell_Norvig.pdf
7. Hosen, M.S.: An influence of data visualization on the decision-making process in the business world. Int. J. Recent Innov. Trends Comput. Commun. **12**(2), 262–269 (2024). Retrieved from https://ijritcc.org/index.php/ijritcc/article/view/10566
8. Yap, M.H.: Enhanced Algorithms for Lesion Detection and Recognition in Ultrasound Breast Images. Figshare %22 (2008)
9. Goh, A.T.C.: Empirical Design in Geotechnics Using Neural Networks. ResearchGate (2023). https://doi.org/10.1680/geot.1995.45.4.709
10. Rajan, S.: Genetic Algorithms and its Use-Cases in Machine Learning (2021). Retrieved 31 Jan 2025, from Analytics Vidhya website: https://www.analyticsvidhya.com/blog/2021/06/genetic-algorithms-and-its-use-cases-in-machine-learning/
11. Lippi, G., Longo, U.G., Maff Lippi, G., Longo, U.G., Maffulli, N.: Genetics and sports. Br. Med. Bull. **93**(1), 27–47 (2009). https://doi.org/10.1093/bmb/ldp007
12. Day9's Introduction to Machine Learning—reason.town.: https://reason.town/day9-machine-learning/
13. Home—Jomstay Digital Technology (M) Sdn. Bhd.: (2022). Retrieved 31 Jan 2025, from Jomstay Digital Technology (M) Sdn. Bhd. https://jomstaydigital.com/
14. Martinez, J.A., Parikh, P.D., Wong, R.W., Harper, C.A., Dooner, J.W., Levitan, M., Ghafoori, S.D.: Telemedicine for diabetic retinopathy screening in an urban, insured population using fundus cameras in a primary care office setting. Ophthalmic Surgery Lasers Imaging Retina **50**(11) (2019). https://doi.org/10.3928/23258160-20191031-14
15. Schork, N.J.: Artificial intelligence and personalized medicine. In: Cancer Treatment and Research (2019), pp. 265–283. https://doi.org/10.1007/978-3-030-16391-4_11
16. Gebretensae, Y.A., Asmelash, D.: Trend analysis and forecasting the spread of COVID-19 pandemic in Ethiopia using box-Jenkins modeling procedure. Int. J. General Med. **14**, 1485–1498 (2021). https://doi.org/10.2147/ijgm.s306250

17. Saqib, S., Ahmad, F.: An Integrative Decision Support Model for Smart Agriculture Based on Internet of Things and Machine Learning. ResearchGate (2021). https://doi.org/10.54060/JIEEE/002.02.007
18. (2025) Retrieved 31 Jan 2025, from Jwatch.org: https://www.jwatch.org/na53712/2021/06/09/novel-method-monitoring-anticoagulation-with-warfarin
19. Alowais, S.A., Alghamdi, S.S., Alsuhebany, N., Alqahtani, T., Alshaya, A.I., Almohareb, S.N., Albekairy, A.M., et al.: Revolutionizing healthcare: the role of artificial intelligence in clinical practice. BMC Med. Educ. **23**(1) (2023). https://doi.org/10.1186/s12909-023-04698-z
20. Research, Evidence-Based Practice, and Clinical Improvement/Innovation Posters. AORN J. **97**(6), 654–672 (2013). https://doi.org/10.1016/j.aorn.2013.04.004
21. Reliable Group—AI software development services company in USA (2023, March 14). Retrieved 31 Jan 2025, from Reliablegroup.com: https://www.reliablegroup.com/blog/why-artificial-intelligence-is-the-future-of-hollywood/
22. Persistent Launches Population Health Management Solution Using Microsoft Azure OpenAI Service (2024, February 8). Retrieved 31 Jan 2025, from Persistent Systems: https://www.persistent.com/media/press-releases/persistent-launches-population-health-management-solution-using-microsoft-azure-openai-service/
23. Alowais, S.A., et al.: Revolutionizing Healthcare: The Role of Artificial Intelligence in Clinical Practice. ResearchGate (2023). https://doi.org/10.1186/s12909-023-04698-z
24. B2B International. (2024, February 15). Master B2B Marketing | B2B International. Retrieved 31 Jan 2025, from B2B International: https://www.b2binternational.com/publications/b2b-marketing/
25. Bharadwaj, R.S., et al.: Revolutionizing Dental Health Care: An In-Depth Exploration of Technological Advancements. ResearchGate (2024). https://doi.org/10.1055/s-0043-1776321
26. The Impact and Potential of Artificial Intelligence in Healthcare: A Critical Review of Current Applications and Future Directions | Request PDF. ResearchGate (2023). https://doi.org/10.22214/ijraset.2023.55537
27. Adger, W.N., Barnett, J., Chapin, F.S., Ellemor, H.: This Earth of mankind: climate change, resilience, and human security. Ambio **35**(6), 441–447 (2011). https://doi.org/10.1007/s13280-011-0213-8
28. Boulton, C.A., Lenton, T.M., Boers, N.: Pronounced loss of Amazon rainforest resilience since the early 2000s. Nat. Clim. Chang. **12**(3), 271–278 (2022). https://doi.org/10.1038/s41558-022-01287-8
29. Brooks, N., Adger, W.N., Kelly, P.M.: The determinants of vulnerability and adaptive capacity at the national level and the implications for adaptation. Glob. Environ. Chang. **15**(2), 151–163 (2005). https://doi.org/10.1016/j.gloenvcha.2004.12.006
30. Byers, E., et al.: Global exposure and vulnerability to multi-sector development and climate risks. Nat. Clim. Chang. **8**(7), 545–551 (2018). https://doi.org/10.1038/s41558-018-0190-4
31. Chen, J., Dirmeyer, P.A.: The interdependence of land cover changes, climate variability, and ecosystem health. Environ. Res. Lett. **17**(3), 034001 (2022). https://doi.org/10.1088/1748-9326/ac5130
32. Creutzig, F., et al.: Cities: the core of climate mitigation efforts. Nat. Clim. Chang. **9**(12), 773–777 (2019). https://doi.org/10.1038/s41558-019-0606-2
33. Dickson, B., et al.: Monitoring biodiversity and ecosystems: a global framework for data integration and analysis. Nature Ecol. Evol. **6**(4), 391–400 (2022). https://doi.org/10.1038/s41559-022-01760-2
34. Erb, K.-H., et al.: Land management: data availability and modeling. Glob. Environ. Chang. **39**, 1–6 (2016). https://doi.org/10.1016/j.gloenvcha.2016.04.009
35. Fawzy, S., Osman, A.I., Doran, J., Rooney, D.W.: Strategies for mitigation of climate change. Environ. Chem. Lett. **18**(6), 2069–2094 (2020). https://doi.org/10.1007/s10311-020-01059-w
36. IPCC: Climate Change 2022: Impacts, Adaptation, and Vulnerability. Contribution of Working Group II to the Sixth Assessment Report. Cambridge University Press (2022). https://doi.org/10.1017/9781009325844

37. Kumar, P., et al.: Climate change and ecosystem services. Ecol. Ind. **98**, 19–27 (2019). https://doi.org/10.1016/j.ecolind.2018.10.059
38. Levin, K., Cashore, B., Bernstein, S., Auld, G.: Overcoming the tragedy of super wicked problems: Constraining our future selves to mitigate climate change. Policy Sci. **45**(2), 123–152 (2012). https://doi.org/10.1007/s11077-012-9151-0
39. Lobell, D.B., Asseng, S.: Climate adaptation in agriculture: crops, cropping systems, and cultivars. Environ. Res. Lett. **12**(3), 033001 (2017). https://doi.org/10.1088/1748-9326/aa6793
40. Mansouri, R., Safari, H., Goodarzi, S.: AI models to forecast drought and resilience. J. Hydrol. **598**, 125747 (2021). https://doi.org/10.1016/j.jhydrol.2021.125747
41. McKinney, M.L.: Urbanization, biodiversity, and conservation. Bioscience **52**(10), 883–890 (2002). https://doi.org/10.1641/0006-3568(2002)052[0883:UBAC]2.0.CO;2
42. Meier, S., Kim, J.: Precision conservation using computational intelligence. Front. Environ. Sci. **9**, 718392 (2021). https://doi.org/10.3389/fenvs.2021.718392
43. Pörtner, H.-O., et al.: Biodiversity-climate feedbacks: a roadmap for integrating global biodiversity targets. Nature **593**(7859), 551–557 (2021). https://doi.org/10.1038/s41586-021-03415-6

Selection of Vital Signs Monitors and ECGs with Automatic Diagnosis of Cardiac Arrhythmias Using a Decision Maker with Topsis

Alberto Ochoa-Zezzatti, Roberto Contreras-Masse, Irma Hernandez-Baez, and Victor López-Sánchez

Abstract Cardiac arrhythmias pose a significant global health concern, necessitating early diagnosis for timely intervention. While automatic cardiac arrhythmia detection has been integrated into vital sign monitors, selecting the most suitable device for individual needs remains a complex task involving multiple conflicting criteria. This research addresses this challenge by developing a novel software tool that assists users in selecting the optimal vital sign monitor. The software employs the TOPSIS (Technique for Order of Preference by Similarity to Ideal Solution) multi-criteria decision-making method to evaluate and rank available devices based on user-defined preferences. Compared to simpler decision-making approaches that may rely on subjective evaluations or single-criterion assessments, the TOPSIS-based tool provides a quantitative and systematic framework for device selection. It achieves this by calculating a closeness coefficient for each device, representing its relative distance to the ideal and negative-ideal solutions across multiple weighted criteria. This allows for a more objective and comprehensive ranking of devices, considering trade-offs between different factors such as clinical performance, cost, and usability. Furthermore, the software has been expanded to incorporate crucial considerations such as compliance with Mexican NOM-035 regulations for psychological risk assessment and carbon footprint analysis, enabling users to make informed decisions that prioritize both health and environmental sustainability. This research contributes to improved patient care and a more sustainable healthcare system by empowering individuals and healthcare providers with a robust and comprehensive tool for selecting the most appropriate vital sign monitoring devices.

A. Ochoa-Zezzatti
UACJ, Ciudad Juarez, Mexico
e-mail: alberto.ochoa@uacj.mx

R. Contreras-Masse (✉)
TecNM Campus ITCJ, Ciudad Juárez, Mexico
e-mail: rcontreras@itcj.edu.mx

I. Hernandez-Baez · V. López-Sánchez
Universidad Politécnica del Estado de Morelos, Cuernavaca, Mexico

W. Pedrycz et al. (eds.), *Decision-Making in Computational Intelligence-Based Systems*, Studies in Systems, Decision and Control 628,
https://doi.org/10.1007/978-3-032-04056-5_12

Keywords Multi-criteria decision making (MCDM) · TOPSIS method · Cardiac arrhythmia detection · Vital signs monitoring · Medical device evaluation · Environmental sustainability · Psychosocial risk assessment · Clinical decision support systems

1 Introduction

According to the World Health Organization, approximately 17 million people die each year from cardiovascular diseases, with cardiac arrhythmias being a significant contributor requiring early diagnosis for timely intervention. While automatic cardiac arrhythmia detection has been integrated into vital sign monitors, selecting the most suitable device for individual needs remains a complex task involving multiple conflicting criteria. To address this challenge, this research develops a novel software tool that assists users in selecting the optimal vital sign monitor. The software employs the TOPSIS (Technique for Order of Preference by Similarity to Ideal Solution) multi-criteria decision-making method to evaluate and rank available devices based on user-defined preferences. TOPSIS is particularly well-suited for this application due to its ability to handle multiple criteria simultaneously, its clear and logical process for identifying the best alternative by considering both the ideal and negative-ideal solutions, and its computational efficiency. Unlike some other MCDM methods, TOPSIS provides a cardinal ranking of alternatives, indicating the relative performance of each device across all considered criteria, which is valuable for informed decision-making in complex medical device selection. Furthermore, the software has been expanded to incorporate crucial considerations such as compliance with Mexican NOM-035 regulations for psychological risk assessment and carbon footprint analysis, enabling users to make informed decisions that prioritize both health and environmental sustainability. This research contributes to improved patient care and a more sustainable healthcare system by empowering individuals and healthcare providers with a robust and comprehensive tool for selecting the most appropriate vital sign monitoring devices.

Given the significant global impact of arrhythmias, a vital signs monitor capable of automatically detecting them could contribute substantially to early diagnosis and, especially, to the reduction of complications or mortality resulting from them.

Since the early 2000s, advancements in electrocardiographic monitoring and data storage systems have enabled more efficient and accurate arrhythmia identification.

Currently, vital signs monitors are available that can store cardiac monitoring data and compare it with a database of arrhythmia models to identify any matches. The specifications of each monitor with this capability typically vary by model. Some may achieve more efficient data comparison, resulting in high-quality automatic diagnoses. Therefore, it is important to select the most effective vital signs monitor, considering various criteria depending on the purpose of the comparison.

2 Literature Review

The application of the Technique for Order of Preference by Similarity to Ideal Solution (TOPSIS) in decision-making processes has gained significant attention across various fields, including healthcare, materials selection, and supplier management. This literature review synthesizes relevant studies that highlight the methodology's effectiveness, advantages, and areas of application, particularly in the context of selecting vital signs monitors and ECGs for automatic diagnosis of cardiac arrhythmias.

TOPSIS is a multi-criteria decision-making (MCDM) technique that ranks alternatives based on their geometric distance from an ideal solution. It is particularly valued for its simplicity and computational efficiency, making it suitable for various applications. For instance, Yang et al. demonstrated the effectiveness of TOPSIS in materials selection, where it provided a clear ranking of alternatives based on multiple criteria, showcasing its practical applicability in engineering contexts [1]. Similarly, Sucahyo et al. emphasized the advantages of TOPSIS in decision support systems, noting its high accuracy and efficiency in selecting Wi-Fi modems, which reflects its versatility across different domains [2].

In the healthcare sector, TOPSIS has been utilized to assess and prioritize various medical devices, including in the selection of mechanical ventilators. Kaya's study illustrated how TOPSIS and fuzzy-TOPSIS methods were employed to evaluate mechanical ventilators based on multiple criteria, demonstrating the method's robustness in medical device selection [3]. Furthermore, Abdel-Basset et al. applied a neutrosophic TOPSIS approach for the selection of smart medical devices, highlighting the method's adaptability to complex decision-making scenarios involving uncertainty and imprecision [4]. This adaptability is particularly relevant in the context of selecting ECGs and vital signs monitors, where multiple conflicting criteria must be considered.

The integration of TOPSIS with other decision-making frameworks, such as the Analytic Hierarchy Process (AHP), has also been explored in the literature. For example, Chen proposed a model that combines AHP with TOPSIS for building material supplier selection, emphasizing the importance of subjective and objective weighting in decision-making [5]. This combination can enhance the decision-making process in selecting medical devices by providing a structured approach to evaluate and rank alternatives based on both qualitative and quantitative criteria.

However, the current study focuses on the application of TOPSIS and does not incorporate AHP. The integration of AHP to determine criteria weights was considered but ultimately left for future research due to the scope and objectives of the current work. Future studies could explore the benefits of using AHP to establish the relative importance of the criteria and compare the results with the weighting approach used in this study.

Moreover, the application of TOPSIS in healthcare extends to performance assessment in teaching hospitals, where it has been used to prioritize hospitals based on various performance indicators [6]. This illustrates the method's potential for broader applications in healthcare decision-making, including the selection of diagnostic tools like ECGs and vital signs monitors.

The literature demonstrates that TOPSIS is a valuable tool for multi-criteria decision-making, particularly in the selection of medical devices and diagnostic tools. Its ability to handle multiple conflicting criteria, coupled with its computational efficiency, makes it an ideal candidate for evaluating vital signs monitors and ECGs in the context of automatic cardiac arrhythmia diagnosis. The integration of TOPSIS with other decision-making frameworks further enhances its applicability, providing a comprehensive approach to addressing complex healthcare decisions.

While the literature review provides a useful overview of studies employing the TOPSIS method, it currently lacks a critical appraisal of the method's potential limitations. This omission could present an incomplete picture of the strengths and weaknesses of TOPSIS in the context of multi-criteria decision-making.

One of TOPSIS's limitations is regarding results sensitivity. TOPSIS results can be highly sensitive to the weights assigned to the criteria. Even small changes in the weights can significantly alter the ranking of alternatives. In the context of vital signs monitor selection, if the weight for 'Patient Safety' is slightly increased, it could drastically change the ranking compared to when 'Cost' is given a slightly higher weight. This sensitivity necessitates a robust and well-justified weighting scheme, as discussed earlier.

Another limitation is the normalization. The normalization method used in TOPSIS can also influence the results. Different normalization techniques can lead to different rankings. For instance, if vector normalization is used instead of linear normalization, the ranking of monitors might change, especially if they have significantly different performance ranges across criteria.

In conclusion, the application of TOPSIS for vital signs monitor selection, while offering a structured approach to multi-criteria decision-making, requires careful consideration of its inherent limitations. The method's sensitivity to criteria weighting underscores the critical importance of employing a robust and well-justified weighting scheme to ensure the reliability and stability of the resulting rankings. As demonstrated, even slight variations in weights, particularly for crucial factors like 'Patient Safety' versus 'Cost,' can significantly alter the outcomes. Furthermore, the potential influence of the chosen normalization method on the final rankings highlights the need for transparency and justification in the data preprocessing steps. The selection of an appropriate normalization technique, such as vector or linear normalization, must be carefully considered and its potential impact acknowledged, especially when dealing with devices that exhibit diverse performance characteristics. Therefore, users of the TOPSIS method in this context should be cognizant of these sensitivities and exercise due diligence in the parameter selection and data preparation phases of the analysis.

Table 1 Key criteria for devices selection and devices in scope

Criteria	Devices
Monitored parameters	D_1 Philips intelliVue MX450/ MX550
Operating modes	D_2 GE healthcare CARESCAPE B450/B65 0/B850
Automatic ECG Interpretation Capability	D_3 Mazy beneVision
Custom alarms	D_4 Dräger infinit y delta series
Display and interface	D_5 Mortar (Hillrom) ely series
Connectivity	D_6 Nihon Kohden life scope
Storage capacity and trends	D_7 Welch allyn connex spot monitor
Portability and battery	
Dimensions and weight	
Certifications and compliance	
Ease of maintenance	
Access to spare parts	

2.1 What Criteria Should Be Taken Into Account?

For decision making, different criteria must be taken into account according to the purpose of use of the equipment that is being qualified. In this study, 12 criteria will be taken into account, which are important according to the objective that is required to be met. Vital signs monitors can be qualified for different functions that contain image resolution, measurement functions of medical parameters such as oximetry, blood pressure, electrocardiogram, and even specific characteristics of each monitor such as size or portability, due to the fact that there are Holter devices that are portable monitors for monitoring cardiac activity. Table 1 shows the twelve criteria for our study. With the help of the program that was developed, decision-making will be carried out by means of topsis, selecting the value of the criteria to be qualified with the Likert scale, the criteria previously provided based on research of the teams with common characteristics, which will be ordered according to the weight of importance assigned based on the objective to be met. Each of the criteria has a weight that indicates how important the criterion is for the fulfillment of the objective and the lower the weight, the less importance it has, these criteria and ratings are later shown on the screen in the form of a list in order to better analyze the information and verify which of the values of the criteria you want to change or if not, allows you to select the option to graph the data obtained.

The literature supports the weighting of individual selection criteria for medical devices by emphasizing the critical role each plays in ensuring patient safety, diagnostic accuracy, and operational efficiency. One of the most fundamental criteria is the ability to monitor key physiological parameters accurately and comprehensively [7]. Devices must reliably measure vital signs such as heart rate, oxygen saturation, and respiratory rate, especially in critical care settings. The significance of each parameter may vary depending on medical specialty, underscoring the importance of aligning device capabilities with clinical needs.

Operating modes contribute to device versatility by allowing for different functionalities such as diagnosis, therapy, or monitoring. Effective mode selection by users is essential to patient safety, making the clarity and usability of mode interfaces a priority. Similarly, devices with automated ECG interpretation capabilities enhance diagnostic speed and accessibility, particularly when used by non-specialists [8]. While AI-driven interpretations offer notable advantages, they must be validated rigorously, and reliance on them should be tempered by clinical oversight.

Customizable alarm features are vital in combating alarm fatigue and ensuring that alerts are clinically meaningful. Devices that allow alarm thresholds to be tailored to individual patients improve the relevance and timeliness of alerts [9]. The design of the device's display and user interface is equally crucial; intuitive, readable, and user-friendly interfaces reduce cognitive load and prevent user errors, particularly in high-stress environments.

Connectivity is another essential criterion, enabling seamless integration with hospital information systems, facilitating remote monitoring, and supporting data analytics. Devices must also meet strict data security and privacy standards. Adequate storage capacity and the ability to analyze trends from stored data allow for retrospective evaluation and early identification of clinical issues, enhancing long-term patient care strategies.

Portability and reliable battery power increase device flexibility, especially in emergency or resource-limited settings. Device dimensions and weight also affect usability and ergonomics, with smaller, lighter devices being preferred for their ease of handling and integration into various environments [10]. Certifications and regulatory compliance are non-negotiable, reflecting a device's safety, quality, and market readiness. These credentials should carry the highest weight in the selection process.

Ease of maintenance is important to ensure device uptime and reduce costs, requiring accessible design and clear documentation. Finally, consistent access to spare parts is necessary to prevent service disruptions and support timely repairs. Taken together, these criteria form a comprehensive framework for evaluating and selecting medical devices, each contributing to the effectiveness, safety, and sustainability of healthcare delivery [11].

2.2 *Why Is a Vital Signs Monitor Used to Detect Cardiac Arrhythmias?*

The utilization of vital signs monitors, particularly electrocardiograms (ECGs), for the detection of cardiac arrhythmias is a critical component of modern cardiology. This literature review synthesizes recent studies that elucidate the importance of these devices in identifying arrhythmias, the technological advancements that enhance their efficacy, and the implications for patient care.

Cardiac arrhythmias, characterized by irregular heartbeats, can lead to severe complications such as stroke and sudden cardiac death. The ability of ECGs to capture the electrical activity of the heart makes them indispensable for diagnosing these conditions. As noted by Deng, cardiac arrhythmias significantly elevate the risk of adverse cardiovascular events, and their detection through ECG monitoring is crucial for timely intervention [12]. The ECG provides a continuous record of the heart's electrical activity, allowing for the identification of abnormal rhythms that may not be apparent during a standard clinical examination.

Recent advancements in wearable technology have further enhanced the capability of vital signs monitors in detecting arrhythmias. For instance, Choi et al. demonstrated that watch-type ECG devices can reliably detect paroxysmal cardiac arrhythmias, outperforming traditional photoplethysmography-based devices by accurately capturing P waves and differentiating QRS complex morphologies [13]. This capability is essential for effective arrhythmia detection, as the morphology of the QRS complex can provide critical information about the underlying cardiac condition.

Moreover, the integration of machine learning algorithms into ECG analysis has revolutionized arrhythmia detection. Lakshmanadoss highlighted the role of deep learning techniques, such as convolutional neural networks (CNNs) and recurrent neural networks (RNNs), in automating the diagnosis of various cardiac arrhythmias from ECG data [14]. These advanced algorithms can analyze large datasets to identify patterns indicative of arrhythmias, thereby improving diagnostic accuracy and reducing the burden on healthcare professionals.

The effectiveness of continuous monitoring systems, such as the Zio patch, has also been substantiated in the literature. Yenikomshian et al. conducted a systematic review that revealed higher detection rates of atrial fibrillation (AF) and other arrhythmias among patients monitored with the Zio system compared to traditional Holter monitors [15]. This finding underscores the importance of prolonged monitoring in capturing transient arrhythmias that may be missed during shorter monitoring periods.

Furthermore, the feasibility of automated systems for arrhythmia detection has been explored. For example, the study by Seidel et al. demonstrated that vital signs monitoring could predict cardiac instability in pediatric patients, suggesting that similar methodologies could be applied to adult populations for early arrhythmia detection [16]. The ability to detect short-term instabilities is crucial, as it may serve as an early warning system for more severe arrhythmias.

Vital signs monitors, particularly ECGs, play a pivotal role in the detection of cardiac arrhythmias. Their ability to provide continuous, real-time monitoring, coupled with advancements in wearable technology and machine learning, enhances their diagnostic capabilities. As the understanding of cardiac arrhythmias evolves, the integration of these technologies into clinical practice will likely lead to improved patient outcomes through timely and accurate diagnosis.

While the integration of automated cardiac arrhythmia detection into vital signs monitors offers significant advantages for early diagnosis, it's important to acknowledge the inherent challenges and potential limitations associated with these systems. One key concern is the possibility of false positives, where the algorithm identifies an arrhythmia that is not actually present. This can lead to unnecessary anxiety for patients and potentially trigger further, invasive investigations. Conversely, false negatives, where a genuine arrhythmia is missed by the automated system, pose a more serious risk by delaying necessary treatment.

Several factors can contribute to these inaccuracies. Artifacts in the ECG signal, caused by patient movement, muscle tremors, or electrical interference, can be misinterpreted by the algorithms. The complexity and variability of cardiac signals, particularly in patients with multiple comorbidities or less common arrhythmias, can also challenge the detection capabilities of current algorithms. Furthermore, the sensitivity and specificity of different automated detection algorithms can vary significantly between manufacturers and models.

Acknowledging these limitations is crucial for responsible implementation and interpretation of automated ECG analysis. Healthcare professionals must be aware of the potential for both false positives and negatives and should always correlate the automated interpretations with the patient's clinical presentation and their own expert judgment. Future research should focus on improving the robustness and accuracy of these algorithms, reducing the impact of artifacts, and developing more sophisticated methods for handling complex cardiac signals. Continuous monitoring and validation of these systems in real-world clinical settings are also essential to ensure their reliability and safety.

The Mexican Official Standard NOM-035-STPS-2018 plays a crucial role in enhancing the psychosocial safety of workers in various environments. This standard mandates the identification, analysis, and prevention of psychosocial risk factors (PRFs) that can adversely affect the mental and emotional well-being of employees. The implementation of NOM-035 is essential for fostering a positive work environment, which is linked to improved organizational resilience and productivity [17, 18]. The standard aligns with international frameworks such as ISO 45003:2021, which emphasizes the importance of recognizing and mitigating psychosocial risks in the workplace [18].

In the context of decision-making processes, the integration of NOM-035 into multicriteria decision-making frameworks, such as the Technique for Order Preference by Similarity to Ideal Solution (TOPSIS), can facilitate the evaluation of various workplace environments based on their adherence to psychosocial safety standards. TOPSIS allows for the ranking of alternatives based on their distance to an ideal solution, which can incorporate factors related to employee well-being, job satisfaction, and overall workplace safety [17, 18]. By employing such methodologies, organizations can make informed decisions that prioritize the mental health of their workforce while also complying with regulatory requirements.

Moreover, the inclusion of carbon footprint as a criterion in the selection of devices and technologies is increasingly recognized as vital in contemporary decision-making processes. The carbon footprint reflects the total greenhouse gas emissions associated with a product or service throughout its lifecycle, which is critical for assessing environmental sustainability [17, 18]. As organizations strive to reduce their environmental impact, evaluating devices based on their carbon footprint can lead to more sustainable practices and contribute to corporate social responsibility goals [17, 18].

Incorporating carbon footprint assessments into decision-making frameworks, such as TOPSIS, allows organizations to balance environmental considerations with other operational factors, such as cost and efficiency. This multifaceted approach not only aids in selecting environmentally friendly technologies but also aligns with broader sustainability objectives [17, 18]. The integration of these criteria into decision-making processes is essential for fostering a culture of sustainability within organizations, ultimately leading to enhanced corporate reputation and compliance with environmental regulations.

The reviewed literature shows how NOM-035 serves as a foundational standard for promoting psychosocial safety in the workplace, while the consideration of carbon footprint in device selection underscores the importance of environmental sustainability. The application of multicriteria decision-making methods, such as TOPSIS, can facilitate the effective integration of these critical factors into organizational practices, ensuring a holistic approach to workplace safety and environmental responsibility.

3 Methodology

In this study, 12 important criteria were chosen for the selection of the best alternative in order to generate an automatic diagnosis of cardiac arrhythmia, which are presented below:

- **Monitored parameters**: Specifies the parameters that the vital signs monitor can measure, depending on the model, parameters include ECG, SpO2, NIBP, CO2 measurement, among others.
- **Operating modes**: The operating modes are those modalities that the vital signs monitor has and its configuration change as it is for adult, pediatric or neonatal patients.
- **Automatic ECG interpretation capability**: The ECG monitor's ability to generate advanced diagnosis and recognition of cardiac arrhythmias stands out. Some of these monitors have highly accurate ECG analysis capability and have an arrhythmia detection system.
- **Personalized alarms**: This criterion indicates the customization that alarms have, because in some devices they are completely configurable and advanced to alert any anomaly detected either in patient parameters or equipment operation.
- **Screen and interface**: This criterion includes all the characteristics of the screen because some of the monitors have advanced settings to customize the interface, have a touch screen and allow you to configure the appearance of visual elements.

- **Connectivity**: This criterion is important due to the limitations that may be had in terms of connectivity. Monitors usually have an ethernet, wi-fi, HL7 or even bluetooth connection for easier data storage.
- **Storage capacity and trends**: This criterion depends on how much data can be stored in order to analyze it. Some monitors have 96 h of trend storage while others manage to reach 240 h of storage.
- **Portability and battery**: Portability and battery is an important feature because the optimal operation of the equipment must be guaranteed even in blackouts or places where there is no access to electricity connection. Portability guarantees that the equipment can be transported from one place to another without difficulties, this to cover the needs where it is required, taking into account the mobility of the equipment within the hospital or even in the patient's home.
- **Dimensions and weight**: Like portability and battery, dimensions and weight are important to meet the goal of making the equipment portable and easy to transport.
- **Certifications and regulatory compliance**: As it is a medical device in contact with the patient and that works with electricity, it must have all the permits and certifications, as well as its user manual and compliance with regulations, thus ensuring safe use by the patient.
- **Ease of maintenance**: This is another important criterion due to the complexity that maintenance can have. Most monitors need simple maintenance or with trained personnel.
- **Access to spare parts**: This last criterion is important and goes hand in hand with the previous one, mentioning the ease of maintenance because in certain countries it is not accessible to obtain spare parts for the equipment.

On the side of the alternatives that are going to be evaluated, seven vital signs monitors with somewhat similar characteristics were chosen to be able to determine by means of TOPSIS which of them is the one that best meets the objective. The seven alternatives are presented below, and the image of each device is included in Fig. 1.

D_1 **Philips IntelliVue MX450**. This device monitors a wide range of parameters such as ECG, SpO2, NIBP, CO2, has advanced algorithms for ECG anomaly recognition, has fully configurable alarms, a customizable touch screen, is portable and has a long battery life, has very complete connectivity and up to 96 h of storage, and is adaptable to various modes of use: adult, pediatric, and neonatal.

D_2 **GE Carescape B450**. It also handles a wide range of parameter measurement, contains Marquette 12SL for advanced ECG analysis, alarms are highly configurable, has a configurable touch screen, is portable with a battery change option, has advanced high connectivity, up to 72 h of storage, and also has adult, pediatric, and neonatal operation modes.

D_3 **Mindray BeneVision N12-N15**. This device monitors a similar variety of parameters, has arrhythmia detection and advanced analysis, alarms are configurable with advanced customization, has a high-resolution touch screen, is portable with a long-lasting battery, wide connectivity, can store up to 120 h, and has different modes such as adult, pediatric, and neonatal.

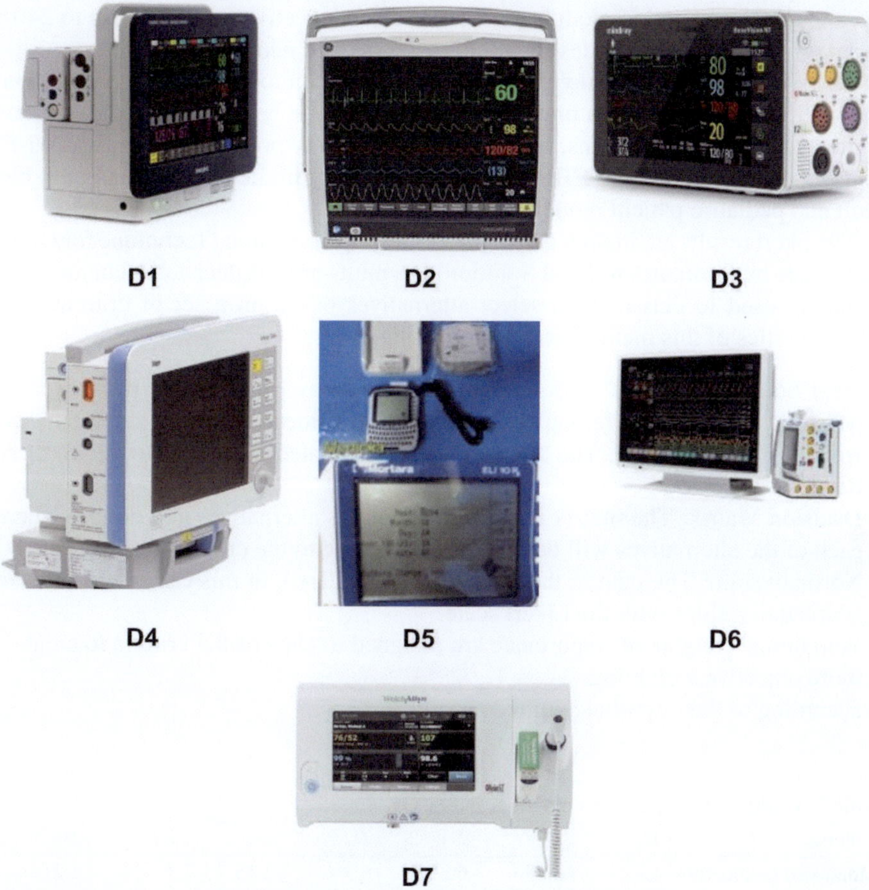

Fig. 1 Devices selected in scope for TOPSIS method

D_4 **Drager Infinity Delta Series**. It is capable of monitoring multiple parameters including PPIs and BIS, features advanced ECG interpretation, fully customizable alarms, an advanced multi-color touchscreen, is portable with battery options, has full connectivity, stores up to 96 h, and has adult, pediatric, and neonatal patient mode.

D_5 **Mortara ELI Series**. It only has ECG measurement, highly accurate ECG interpretation, no alarms because it is not applicable, specialized but less complete color display, it is compact and easy to transport, it has limited connectivity, only ECG centered, stores only ECG data just like the modes only managed focused on ECG.

D_6 **Nihon Kohden Life Scope G9**. It monitors a full range of parameters, has advanced capabilities for ECG analysis, alarms that are configurable and advanced, has a touch screen and is modular, has advanced connectivity, achieves up to 240 h of data storage, and has modes for adult, pediatric, and neonatal patients.

D_7 **Welch Allyn Connex Spot Monitor**. It monitors essential parameters, although less complete than previous models, has basic heart rhythm analysis, has limited customization options, basic touch screen, very portable design, has Wi-Fi connectivity, bluetooth and EHR integration, has very limited storage and only has adult and pediatric patient mode.

The likert results are included in Table 2. The Topsis method (Technique for Order Preference by Similarity to Ideal Solution) is a multi-criteria decision technique that is mainly used to classify and select alternatives based on a set of criteria. The characteristics of this method are:

- Ideal Solutions: In this method, two ideal points are defined, the positive ideal point that is the best alternative, and the negative ideal point that represents the worst option to choose, The alternatives are evaluated according to the distance between those points.
- Decision Matrix: The matrix is built that contains alternatives and criteria where each of the alternatives will be qualified according to the criteria.
- Normalization: The data in the matrix are normalized, in this case the data were adapted to ratings with the Likert scale.
- Weighting: Weights of importance are assigned to each of the criteria to create a more objective weighting
- according to the importance of the criteria.

Table 2 Evaluation of vital signs monitors and ECGs

Criteria	D_1	D_2	D_3	D_4	D_5	D_6	D_7
Monitored parameters	6.32	6.29	6.31	6.3	2.15	7	4.21
Operating modes	7	7	6.99	7	2.17	4.23	4.23
Automatic ECG interpretation	6.34	6.85	6.32	6.37	6.3	4.95	4.95
Custom alarms	7.17	6.89	6.99	7	2.15	7	4.21
Display and interface	6.23	6.3	6.9	6.28	2.81	6.3	4.93
Connectivity	6.99	6.99	6.95	6.27	4.21	7	6.31
Storage capacity	5.62	4.9	6.9	6.3	2.82	7	3.55
Portability and battery	6.32	5.6	6.31	5.6	4.91	6.3	7
Dimensions and weight	4.94	4.2	4.9	4.2	6.31	4.9	6.98
Certifications	7	7	7	7	5.6	7	5.62
Ease of maintenance	6.34	6.3	5.6	6.3	2.49	5.6	4.91
Access to spare parts	7	7	5.6	5.6	4.9	5.6	4.95

Fig. 2 Metaverse application to interact with the users to do the device evaluation and visualize the alternatives

- Distance calculation: The distances between each alternative and the ideal points are calculated in order to determine how close or far each alternative is located from the ideal points.
- Final Classification: Each alternative is classified according to its proximity to the positive ideal point and its distance to the negative point, allowing the best option to be selected.

A hospital environment was created in Unity. The project, featuring a TOPSIS decision-making tool, is presented through a Vroid-designed avatar (Fig. 2). This avatar utilizes facial mapping technology via a laptop camera to mirror user movements and incorporates voice integration.

Python code, using the Tkinter library, was implemented to design the interactive user interface (the software was written in Spanish, therefore the screenshots show elements in that language.). This interface serves as the main menu, displaying ten primary options. The first option, "Show Monitors," presents all data from the database of monitors being evaluated.

The second menu option, "Change Criteria," allows for the modification of evaluation criteria. Clicking the corresponding button opens a window with two selection lists:

- Monitor Selection: This list presents the seven monitor alternatives, enabling the user to choose the specific device they wish to modify.
- Criteria Selection: This list displays the twelve evaluation criteria. The user can then select the specific criterion they want to adjust.

This feature provides flexibility in the evaluation process, allowing users to customize the criteria according to their specific needs and priorities. Figure 3 shows the screenshots for this process.

Given the focus of this paper on the application of TOPSIS within the developed software tool for vital signs monitor selection and the integration of additional criteria like NOM-035 and ecological footprint, the methodology section emphasizes

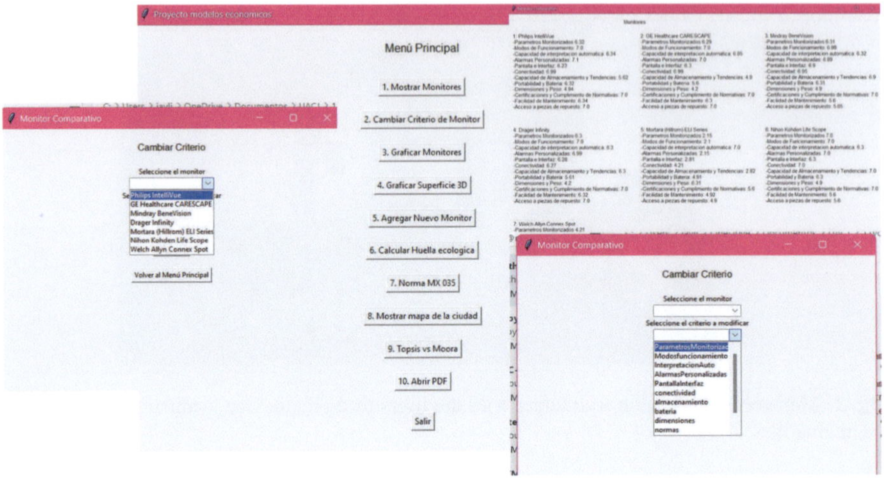

Fig. 3 Graphical user interface to enter devices grades

the adaptation of TOPSIS to this specific context rather than providing a generic calculation example.

The development of the user interface is a key component of the decision-making system, as it directly impacts the user's ability to interact with the software and utilize its functionalities. However, this study did not include a formal usability evaluation of the interface. Usability testing, involving representative users such as clinicians and biomedical engineers, is crucial for assessing the interface's intuitiveness, efficiency, ease of learning, and overall user satisfaction. Future work should prioritize conducting usability studies to gather user feedback on the interface design, navigation, data input methods, and the clarity of information presentation. Incorporating user feedback-driven improvements will enhance the tool's user-friendliness and ensure its practical applicability in real-world healthcare settings, thereby increasing confidence in its effectiveness.

The user interface allows for dynamic modification of the evaluation criteria. Selecting the "Change Criteria" option from the main menu opens a window. Here, the user first selects the desired monitor from a list of seven alternatives. Subsequently, another list presents the twelve evaluation criteria, allowing the user to choose the specific criterion to be modified. The user then enters a new value for the chosen criterion. Upon clicking the "Update" button, the system saves the new value, updates the database, and displays a success message confirming the change.

The "Graph Monitors" option provides visual representations of the evaluation data. Clicking this option opens a screen with two choices: "Bar Graph" and "Return to Main Menu." The "Bar Graph" option generates a bar chart. The X-axis represents the evaluation criteria, while the Y-axis displays the corresponding scores for each alternative. Each alternative is assigned a unique color, making it easy to distinguish between the different devices.

The "Graph 3D Surface" option, accessible from the main menu, opens a screen with options to generate a 3D surface graph or return to the main menu. When the "3D Graph" button is clicked, a 3D surface graph is generated. This graph visualizes the performance of the seven alternatives across all twelve evaluation criteria. Each criterion is represented by an axis, and the score for each alternative is color-coded, providing a comprehensive overview of the evaluation results.

While bar charts and 3D surface plots were employed to visualize the results and facilitate understanding of the multi-criteria decision-making process, the current study does not include direct feedback from stakeholders regarding the effectiveness and usability of these visualizations. Future research should incorporate user response analysis and formal usability studies to evaluate how well these visual representations aid in the interpretation of the device rankings and the underlying criteria trade-offs. Gathering insights from the end-users, including healthcare professionals and decision-makers, would provide valuable information for refining the visualization techniques and ensuring the practical utility of the developed software tool in real-world scenarios.

4 Additional Criteria Taken in Consideration

The system also incorporates an ecological footprint calculator. Accessible through the sixth menu option, this feature presents a user interface for inputting relevant data for the calculation. Upon clicking "Calculate," a new window displays a graph illustrating the calculated ecological footprint alongside the date and time of the calculation.

The ecological footprint is a valuable metric that quantifies the environmental impact of human activities. It represents the biologically productive land and water area required to support a population's consumption of resources and absorb the waste it generates.

Calculating the ecological footprint is crucial for assessing sustainability. It helps determine whether a region can withstand the impact of human activities without causing environmental damage. Furthermore, it raises awareness about the environmental consequences of our actions, encouraging individuals and societies to adopt more sustainable practices. By providing data on resource consumption and environmental impact, ecological footprint calculations can inform the development of effective environmental policies, promoting the conscious use of resources and the implementation of regulations to mitigate environmental damage.

The inclusion of an ecological footprint calculator in the device selection tool underscores the importance of a life cycle perspective. Selecting a cardiac monitoring device involves more than just considering its purchase price and immediate functionality. A comprehensive evaluation includes assessing the environmental impact throughout the device's entire life cycle. This encompasses the extraction of raw materials, the energy-intensive manufacturing processes, transportation, usage, and eventual disposal or recycling. Devices with a smaller ecological footprint often reflect more efficient manufacturing, reduced energy consumption during use, and designs that minimize waste. By integrating ecological footprint as a selection criterion, we encourage manufacturers to adopt more sustainable practices and enable healthcare providers to make informed decisions that minimize the environmental consequences of their procurement choices.

The system incorporates a module for assessing compliance with NOM-035-STPS-2018, the Mexican Standard for the Identification, Analysis, and Prevention of Psychosocial Risks in the Workplace. This module utilizes a survey-based approach. Employees answer a series of questions through a selection of options, reflecting their experiences and perceptions regarding their work environment. Upon completion of the survey, clicking the "Calculate Result" button triggers an analysis. The system then displays the results in a color-coded box: green for low risk, orange for moderate risk, and red for high risk.

A green result indicates a generally positive work environment. Employees likely report low levels of stress, a positive work atmosphere, and good relationships with colleagues. An orange result suggests moderate risk. While the work environment may generally be positive, there are indications of potential stress, dissatisfaction, or unfavorable work conditions. A red result signifies high psychosocial risk, indicating a significant level of stress, dissatisfaction, or other negative factors in the work environment. This may manifest as excessive workload, poor working relationships, insufficient rest time, or other factors that negatively impact employee well-being.

NOM-035-STPS-2018 is a crucial legal framework in Mexico, mandating the identification, analysis, and prevention of psychosocial risks in the workplace. Its primary objective is to foster a healthy and safe work environment for all employees. This standard encompasses a wide range of factors, including stress, workplace violence, and lack of social support. By requiring employers to assess and address these risks, NOM-035-STPS-2018 aims to improve employee well-being, productivity, and overall job satisfaction. Key measures include implementing preventive and corrective actions, conducting regular assessments of the work environment, and providing employee training on maintaining a healthy work environment.

While the primary function of an ECG monitor is to accurately diagnose cardiac arrhythmias, the selection process must also account for the organizational context in which the device will be used. NOM-035 highlights the importance of psychosocial factors in the workplace, including those that can affect efficiency and error rates. A well-designed ECG monitor, with intuitive alarms and a clear display, can reduce the cognitive load on healthcare providers, minimizing the risk of errors caused by stress or fatigue. Furthermore, adequate training and support for using the device, as emphasized by NOM-035, can improve staff competence and confidence, leading to better patient outcomes. By considering NOM-035 criteria in the selection process, we aim to choose devices that not only meet the clinical requirements but also promote a safer and more productive work environment, indirectly enhancing the quality of patient care and reducing potential risks.

5 Conclusions and Future Direction

The development of a GUI-based software for selecting vital signs monitors using the TOPSIS method offers several significant advantages. Firstly, it streamlines the decision-making process. By providing a user-friendly interface for data input, analysis, and visualization, the GUI eliminates the need for manual calculations and data entry, saving time and effort for healthcare professionals. Secondly, it enhances accessibility. The GUI makes the TOPSIS methodology more accessible to users without extensive technical expertise in multi-criteria decision analysis, democratizing the decision-making process and empowering clinicians, biomedical engineers, and other stakeholders to make informed choices. Moreover, the GUI enhances transparency by providing a clear and transparent view of the decision-making process. Users can easily see how the criteria weights and device performance scores contribute to the final ranking, increasing trust in the results. Additionally, the GUI facilitates efficient data management and storage by integrating with a database, allowing for easy tracking of device evaluations, comparison of different models, and analysis of trends over time.

Furthermore, the GUI enhances data visualization. The graphical representations, such as bar charts and 3D surface plots, provide a visual and intuitive understanding of the evaluation results. This aids in identifying key strengths and weaknesses of different devices and facilitates the comparison of performance across various criteria. These visualizations can also be effectively used to communicate the evaluation results to stakeholders, such as hospital administrators or purchasing committees, in a clear and concise manner.

Moreover, the GUI significantly improves the user experience. The intuitive design, with features like drag-and-drop, easy navigation, and clear instructions, enhances usability and minimizes the learning curve. The interactive nature of the GUI, with features such as 3D visualizations and the ability to dynamically adjust criteria weights, can increase user engagement and make the evaluation process more enjoyable.

Finally, the GUI can be further enhanced through integration with other tools. For example, integration with hospital information systems or inventory management software can further enhance the efficiency and effectiveness of the decision-making process. Additionally, integrating machine learning algorithms could further enhance the decision-making process by identifying patterns in historical data and predicting future performance trends. Incorporating modules for risk assessment, such as the NOM-035-STPS-2018 compliance module, can provide a more holistic evaluation of device selection, considering factors beyond technical specifications. Gathering user feedback on the software and incorporating customization options can further improve its usability and relevance to specific user needs.

The development of a GUI-based software for selecting vital signs monitors using the TOPSIS method offers numerous advantages. By enhancing efficiency, improving data visualization, and providing a user-friendly interface, this software empowers decision-makers to select the most suitable devices for their specific needs, ultimately improving patient care and healthcare outcomes.

Also, psychosocial risks in the workplace can have significant negative consequences for both employees and the organization. These consequences can include:

- Mental health issues: Exposure to stressors such as workplace bullying can contribute to serious mental health problems, including anxiety, depression, and in extreme cases, even suicide.
- Decreased productivity: High levels of stress and other risk factors can negatively impact employee productivity. This can lead to poor performance, lower work efficiency, and missed deadlines.
- Increased absenteeism: Employees experiencing mental health issues due to work-related stress are more likely to be absent from work, affecting the continuity of operations and productivity within the organization.
- High staff turnover: A negative and stressful work environment can increase employee turnover rates. This leads to increased costs associated with recruitment, hiring, and training new employees, impacting the organization's bottom line.
- Deteriorated labor relations: Psychosocial risks can contribute to interpersonal conflicts and a hostile work environment, damaging teamwork and employee morale.
- Economic impact: The cumulative effects of increased absenteeism, high staff turnover, and decreased productivity due to psychosocial risks can have a significant negative impact on the organization's overall economic performance.

Several mitigation measures can be implemented to address psychosocial risks in the workplace:

- Regular risk assessment: Conducting regular assessments to identify and evaluate psychosocial risk factors is crucial. This involves analyzing work content, working conditions, and interpersonal relationships within the organization.
- Training and awareness programs: Providing employees with training on stress management techniques and strategies for coping with psychosocial stressors can empower them to better manage their mental well-being.
- Promoting open communication: Fostering open and effective communication channels within the organization allows employees to express their concerns and feel heard and supported.
- Establishing clear protocols: Implementing clear protocols for addressing issues such as harassment and workplace violence is essential to create a safe and supportive work environment.
- Employee participation and engagement: Involving employees in decision-making processes and organizing social events can improve employee morale, engagement, and overall well-being, thereby mitigating the negative impact of psychosocial risks.

- Promoting professional development: Providing opportunities for employee growth and advancement can enhance job satisfaction, reduce feelings of stagnation, and contribute to a more positive work environment.

By implementing these measures, organizations can effectively address psychosocial risks, create a healthier and more productive work environment, and improve overall employee well-being.

As a future development for the multi-criteria decision-making system, we propose to incorporate another MCDM method to compare and validate the results obtained from TOPSIS. The Multi-Objective Optimization on the basis of Ratio Analysis (MOORA) method has been selected for this comparison. MOORA offers a distinct approach to multi-criteria decision-making compared to TOPSIS, primarily in how it normalizes data and aggregates criteria values. While TOPSIS relies on the concepts of ideal and negative-ideal solutions and calculates Euclidean distances, MOORA utilizes ratio analysis to compare the performance of alternatives relative to each criterion. This difference in methodology can provide a valuable cross-check, potentially revealing nuances in the ranking of alternatives that might not be apparent with a single method. Furthermore, MOORA has been reported to be effective in various selection problems, demonstrating its capability to handle complex decision scenarios, similar to the selection of medical devices. Comparing the results from TOPSIS and MOORA will provide a more comprehensive evaluation of the selected devices and enhance the robustness of the decision-making tool.

While other robust MCDM methods such as VIKOR and ELECTRE are available, MOORA was specifically chosen for this comparative analysis with TOPSIS due to several key reasons. Unlike VIKOR, which focuses on minimizing the group regret and the individual regret of alternatives, MOORA offers a more direct and computationally efficient approach by utilizing ratio analysis and aggregating performance scores based on weighted criteria. This straightforward methodology allows for a clearer comparison of the underlying ranking mechanisms when contrasted with the distance-based approach of TOPSIS. Furthermore, while ELECTRE excels at outranking relations and dealing with incomparability, its complexity in parameter selection and the potential for generating a partial preorder rather than a complete ranking make MOORA a more suitable choice for a direct, point-by-point comparison of the final results. The goal is to identify if a fundamentally different aggregation logic (ratio-based vs. distance-based) yields consistent or divergent rankings, and MOORA's clear and direct calculation facilitates this objective.

The difference between the TOPSIS (Technique for Order of Preference by Similarity to Ideal Solution) and MOORA (Multi-Objective Optimization on the basis of Ratio Analysis) methods is based on the approach that each of these techniques has and the procedure for multi-criteria decision-making.

In the field of biomedical engineering, it can be very useful to choose the best medical equipment for a purpose, since there is equipment from different brands and with a wide variety of functions. It is also used to evaluate projects based on criteria such as profitability, technical feasibility and sustainability, helping organizations to make better decisions and prioritize investments.

References

1. Yang, W.C., Chon, S.H., Choe, C.M., Yang, J.Y.: Materials selection method using topsis with some popular normalization methods. Eng. Res. Exp. (2021). https://doi.org/10.1088/2631-8695/abd5a7
2. Sucahyo, N., Fitriansyah, A., Suhanda, Y.: Decision support system for wi-fi modem selection using multiple attribute decision making with topsis. J. Riset. Informatika. (2022). https://doi.org/10.34288/jri.v4i4.388
3. Kaya, G.K.: Rational decision-making in purchasing medical devices (2021). https://doi.org/10.46254/eu04.20210384
4. Abdel-Basset, M., Manogaran, G., Gamal, A., Smarandache, F.: A group decision making framework based on neutrosophic topsis approach for smart medical device selection. J Med Syst (2019). https://doi.org/10.1007/s10916-019-1156-1
5. Chen, C.H.: A novel multi-criteria decision-making model for building material supplier selection based on entropy-ahp weighted topsis. Entropy (2020). https://doi.org/10.3390/e22020259
6. Jafari, M., Seyedjavadi, M., Zaboli, R.: Assessment of performance in teaching hospitals: Using multicriteria decision-making techniques. J. Educ. Health Promot. (2020). https://doi.org/10.4103/jehp.jehp_89_20
7. Shbool, M.A., Arabeyyat, O.S., Al-Bazi, A., AlAlaween, W.H.: An integrated multi-criteria decision-making framework for a medical device selection in the healthcare industry. Cogent. Eng. **8**(1), 1968741 (2022). https://doi.org/10.1080/23311916.2021.1968741
8. Roldán, Ú.B., et al.: Multi-criteria decision analysis as a decision-support tool for drug evaluation: a pilot study in a pharmacy and therapeutics committee setting. Int. J. Technol. Ass. Health Care. **34**(5), 519–526 (2018). https://doi.org/10.1017/s0266462318001378
9. Badia, X., et al.: Patient involvement in reflective multicriteria decision analysis to assist decision making in oncology. Int J Technol Ass Health Care **35**(1), 56–63 (2019). https://doi.org/10.1017/S0266462318003641
10. Tavakol, M., et al.: The application of multicriteria decision analysis methods in health care: a literature review. Med. Dec. Making **42**(2), 262–274 (2022). https://doi.org/10.1177/0272989x211019040
11. Shbool, M.A., Arabeyyat, O.S., Al-Bazi, A., AlAlaween, W.H.: An integrated multi-criteria decision-making framework for a medical device selection in the healthcare industry. Cogent. Eng. **8**(1), 1968741 (2022). https://doi.org/10.1080/23311916.2021.1968741
12. Deng, J.: An energy-efficient ecg processor based on hdwt and a hybrid classifier for arrhythmia detection. Appl. Sci. (2023). https://doi.org/10.3390/app14010342
13. Choi, J.I., Kim, H., Min, K., Choi, Y.Y., Shim, J., Son, H.S., Kim, Y.H.: A watch-type electrocardiography is a reliable tool for detecting paroxysmal cardiac arrhythmias. J. Clin. Med. (2022). https://doi.org/10.3390/jcm11123333
14. Lakshmanadoss, U.: Practical applications of electrocardiogram (2020). https://doi.org/10.5772/intechopen.77737
15. Yenikomshian, M., Jarvis, J., Patton, C., Yee, C., Mortimer, R., Birnbaum, H.G., Topash, M.: Cardiac arrhythmia detection outcomes among patients monitored with the zio patch system: a systematic literature review. Current Med. Res. Opinion (2019). https://doi.org/10.1080/03007995.2019.1610370
16. Seidel, G., Murthy, S., Peters, C., Rostalski, P., Görges, M .: Feasibility of automated vital sign instability detection in children admitted to the pediatric intensive care unit (2019). https://doi.org/10.22489/cinc.2019.111
17. Cano-Gutiérrez, J.C., Pérez-Morán, J.C., Bernal-Baldenebro, B., Arenas-Meneses, D., Vazquez-Lira, R., Olguín Tiznado, J.E.: Factor structure and measurement invariance of the psychosocial risk factors inventory of nom-035-stps-2018. Front Psychol. (2023). https://doi.org/10.3389/fpsyg.2022.1022707
18. Cuello-Cuello, Y.: Study of mental workload in public administration managers. Dyna (2024). https://doi.org/10.15446/dyna.v91n232.112592

Adaptability of Neuro-fuzzy Inference System for Clinical Decision-Making in Heart Failure Death

Debajyoti Bora, Juri Borah, and Manash Pratim Barman

Abstract Early detection of health-related events through a cost-effective predictive model is crucial in guiding treatment decisions and preventive measures. Machine learning techniques have been widely utilized in healthcare research. This study explores the effectiveness of a neuro-fuzzy inference system in diagnosing heart failure based on three key variables: age (in years), ejection fraction (percentage), and serum creatinine levels (mg/dL). The model is developed using a publicly available dataset comprising medical records of 299 heart failure patients from the Kaggle database. The dataset is divided into training and testing sets. A first-order Sugeno inference system with six layers is employed for model development. The Kolmogorov-Smirnov test is utilized to assess the normality of the estimated output values, while the Mann-Whitney test assesses whether the mean of estimated values significantly differs between patients who survived and those who did not. The model's performance is measured using the area under the receiver operating characteristic curve (AUC), correlation coefficient, mean squared error (MSE), and mean squared logarithmic error (MSLE). Additionally, the extracted data and corresponding estimated outputs are analyzed individually. The development and performance assessment of the model are conducted using MATLAB and Python. The results indicate an AUC of 0.880, a correlation coefficient of 0.649, an MSE of 0.147, and an MSLE of 0.069. This study demonstrates the potential of a neuro-fuzzy inference system for predicting heart failure risk.

Keywords Decision-making · Heart failure · Sugeno inference system · Machine learning

D. Bora (✉)
Department of Statistics, Pandit Deendayal Upadhyaya Adarsha Mahavidyalaya, Eraligool, Assam, India
e-mail: debu.bora19@gmail.com

J. Borah
Department of Anthropology, Gurucharan University, Silchar, Assam, India

M. P. Barman
Department of Statistics, Dibrugarh University, Dibrugarh, Assam, India
e-mail: manash_stats@dibru.ac.in

W. Pedrycz et al. (eds.), *Decision-Making in Computational Intelligence-Based Systems*, Studies in Systems, Decision and Control 628,
https://doi.org/10.1007/978-3-032-04056-5_13

1 Introduction

The decision-making process plays a crucial role in healthcare systems and medical practice. Healthcare is a complex and dynamic sector involving multiple stakeholders with diverse preferences and values that influence decisions [1]. While the inclusion of different stakeholders enhances the sustainability of solutions, it also makes consensus-building more challenging, as each stakeholder may prioritize different values and interests [2]. Medical decision-making is intricate due to uncertainties in diagnosis and treatment, patient preferences, and cost considerations [3].

In healthcare, decision-making involves selecting the most appropriate course of action from multiple available options to address a specific medical issue [4]. It is vital at all levels, from individual patient care to hospital management and policy development [5–7]. A major challenge in the healthcare industry is making decisions that benefit both patients and the system as a whole [8]. These decisions impact various aspects, including diagnosis, treatment plans, resource distribution, risk analysis, and healthcare policy development [9–11]. Physicians rely on their expertise, scientific knowledge, evidence-based guidelines, and patient preferences to make well-informed decisions that enhance patient health outcomes [12]. Unlike routine decisions, many medical decisions carry significant consequences and involve substantial uncertainties, including those related to diagnosis, the accuracy of tests, disease progression, and treatment efficacy [3].

The field of decision-making science is continually evolving, introducing new strategies to enhance accuracy and effectiveness. Healthcare decision-makers consider numerous variables, from selecting optimal treatments to ensuring efficient resource allocation [12]. To overcome the limitations of traditional methods, innovative approaches have been introduced to facilitate more precise decision-making. For decades, researchers have employed diverse techniques to address real-world healthcare challenges [2]. The adoption of advanced methodologies has significantly transformed the healthcare sector. Multi-criteria decision-making (MCDM) methods have become increasingly popular, as they incorporate both qualitative and quantitative data, enabling more reliable and consensus-driven decisions even in cases of conflicting evidence. Glaize et al. [13] proposed MCDM tools for improving various aspects of healthcare services.

Over the past few decades, artificial intelligence (AI) and machine learning (ML) have played a crucial role in clinical decision-making by addressing uncertainties and enhancing decision accuracy. These advanced technologies facilitate early diagnosis and assist in treatment planning, leading to better patient outcomes. Moreover, they are cost-effective and simplify complex data analysis, making healthcare insights more accessible to the general public. The primary objective of this study is to develop a cost-effective, interpretable, and accurate clinical decision-support system for the early detection of critical heart failure outcomes. This is achieved by integrating the interpretability of fuzzy logic with the adaptive learning capabilities of neural networks. The robustness and validity of the proposed system have been rigorously evaluated through comprehensive statistical analysis, incorporating multiple performance metrics and nonparametric tests.

1.1 Related Work

Various machine learning approaches have been widely used in medical research for predictive analysis, including Logistic Regression (LR), Decision Trees (DT), Artificial Neural Networks (ANN), k-Nearest Neighbors (KNN), Linear Discriminant Analysis (LDA), Support Vector Machines (SVM), Random Forests (RF), and Fuzzy Inference Systems (FIS). However, most of these studies primarily assess model performance using classification accuracy alone, without estimating risk scores or providing error metrics such as MSE or correlation coefficients with actual outcomes. Moreover, many do not incorporate individual case-level analysis, limiting the interpretability and clinical applicability of the results for patient-specific decision-making.

Agrawal [14] developed a cancer prediction model using a Support Vector Machine (SVM), Logistic Regression, Naïve Bayes, and K-Nearest Neighbors (KNN). The dataset was split into 75% for training and 25% for testing, and the results indicated that the KNN algorithm had the highest accuracy (0.98). Omondiagbe et al. [15] compared SVM, Artificial Neural Networks (ANN), and Naïve Bayes for breast cancer diagnosis using the Wisconsin Diagnostic Breast Cancer dataset. Their findings led to the development of a hybrid model incorporating Linear Discriminant Analysis (LDA) with SVM, achieving an accuracy of 98.82%. Lazuwardi [16] analyzed the factors influencing infant mortality in East Java Province using logistic regression. His study found that education and parity were significant predictors, whereas antenatal care, birth assistance, low birth weight, and employment were not. Ye et al. [17] utilized Gradient Boosted Decision Trees, Adaptive Boosting (AdaBoost), Light Gradient Boosting, Logistic Regression, and Random Forest to predict gestational diabetes mellitus. Their findings showed that Gradient Boosted Decision Trees performed the best (AUC 0.74, 95% CI 0.71–0.76). Rani [18] developed an early diabetes prediction model using KNN, Logistic Regression, Random Forest, SVM, and Decision Tree, concluding that the Decision Tree (DT) had the highest accuracy (99%). Soni and Varma [19] applied multiple ML algorithms to predict diabetes, with Random Forest (RF) demonstrating the highest accuracy. Kishore et al. [20] assessed diabetes prediction models using the Pima Indians Diabetes dataset, reporting an accuracy of approximately 75% for Random Forest. Khanam and Foo [21] tested various classification methods and found that Neural Networks performed best, achieving 86% accuracy. Further studies by Abaker and Saeed [22] on diabetic patients at Alsukari Hospital identified logistic regression (LR) as the most accurate model (81%). Llaha and Rista [23] evaluated classification techniques and determined that the Decision Tree (DT) performed best. Butt et al. [24] compared Random Forest, Multilayer Perceptron, and Logistic Regression for diabetes classification, finding Multilayer Perceptron to be the most effective (86.08% accuracy). Srivastava et al. [25] conducted a retrospective analysis of infant mortality using Cox regression and Kaplan-Meier survival curves, concluding that female infants had a higher mortality risk. Nyoni and Nyoni [26] applied artificial neural networks to predict Morocco's infant mortality rate and found their model to be stable. Hajipour et al. [27] utilized various ML techniques to determine infant mortality predictors and found that modern modeling approaches were more effective than traditional methods. Wegbom et al. [28] used logistic regression (LR) to analyze short birth intervals in Nigeria, identifying significant factors such as maternal age, region, education, and the number of children. Kazemi and Mohammadi [29] developed a fuzzy diagnostic model for diabetes based on factors such as age, weight, and physical activity. Singh et al. [30] combined fuzzy logic-driven image segmentation with an advanced deep-learning model to enhance melanoma detection. Apasiya et al. [31] proposed a mobile-based fuzzy expert system aimed at improving breast cancer prognosis and diagnosis. In cardiovascular research, Ali et al. [32] achieved 98.08% accuracy using a fuzzy system that analyzed seven clinical variables including HbA1c, HDL, LDL, and blood pressure. El-Ibrahimi et al. [33] further demonstrated the strength of fuzzy expert systems in early coronary artery disease prediction, reporting 99.61% accuracy, 100% sensitivity, and 99.20% specificity based on critical risk factor identification.

1.2 Background

AI and ML have revolutionized healthcare decision-making by providing cost-effective and highly accurate predictive models. Their application continues to improve diagnosis, treatment, and overall healthcare efficiency, making them indispensable tools in modern medicine. Recently, hybrid systems have gained increasing attention for their application in diagnosing health outcomes. Among these, the Neuro-Fuzzy Inference System (NFIS) stands out as a powerful approach that integrates the principles of artificial neural networks (ANN) and fuzzy inference systems (FIS) [34, 35]. By combining the learning capabilities of neural networks with the interpretability of fuzzy logic, NFIS enhances decision-making in complex medical modeling tasks. This hybrid approach leverages the strengths of both techniques [34], improving accuracy and adaptability in healthcare diagnostics. A brief overview of the Neuro-Fuzzy Inference System is provided below.

1.2.1 Neuro-fuzzy Inference System

The Neuro-Fuzzy Inference System (NFIS) effectively combines the key attributes of both neural networks and fuzzy logic systems [35]. It possesses the inference capabilities of fuzzy systems while incorporating the adaptive learning features of neural networks. Nodes within the NFIS serve distinct roles and are not necessarily fully connected. Some nodes act as output variables, while others contribute to defining fuzzy rules.

Typically, neuro-fuzzy systems are constructed using fuzzy systems based on prior knowledge or expert insights. The foundation of NFIS is derived from the Takagi-Sugeno-Kang (TSK) model, commonly referred to as the Sugeno-type fuzzy model. In this model, an inference rule R_j is formulated as follows:

$$R_j : \text{IF } \mu_{M_i}(\text{x}) \text{ AND } \mu_{N_i}(\text{y}) \text{ THEN f} = p_j\text{x} + q_j\text{y} + r_j$$

where, the jth rule, denoted as R_j, consists of fuzzy membership functions M_i and N_i, which can take various forms, such as triangular, Gaussian, or trapezoidal, and are represented by μ in the antecedent part of the rule. The parameters p_j, q_j, and r_j in the consequent section are linear coefficients that influence the rule's outcome. The learning process involves optimizing both the antecedent parameters in the membership functions and the consequent parameters to enhance accuracy.

The neuro-fuzzy inference system closely resembles the Mamdani inference method, with the primary distinction being that Sugeno-type inference replaces fuzzy sets in the consequent part with a mathematical function of input variables.

1.2.2 Architecture of Neuro-fuzzy Inference System

A typical neuro-fuzzy inference system consists of six layers, incorporating two types of neurons: fixed and adaptable. While neurons in the antecedent and consequent layers are adjustable, the remaining neurons remain static. Below is an overview of the six layers:

Input Layer: This is the first layer of the neuro-fuzzy inference system. It consists of input nodes representing system parameters, which merely pass values to the next layer without performing any computations. Each node corresponds to a single input variable, and the link weight in this layer is set to unity [36].

Membership Function (Fuzzification) Layer: Each node in this layer represents a linguistic label (e.g., "good" or "very good") associated with an input variable. A suitable membership function determines the degree to which an input value belongs to a fuzzy set, producing a membership value. These membership functions are fine-tuned during the learning process, with their parameters known as premise parameters. The output of this layer consists of the membership values of the premise part.

Rule Layer: Each node in this layer corresponds to the antecedent part of a fuzzy rule. A T-norm operator, typically the logical AND operation, determines the firing strength of each rule. The rule base is formed by combining all possible variable categories. If there are n inputs and each input is divided into c categories, the total number of rules is c^n [37]. The conjunction of rule antecedents is calculated using a product operator, yielding the output. The mathematical formulation is defined by Eq. (1).

$$y_i = \prod_{j=1}^{k} x_{ij} \tag{1}$$

where x_{ij} are the inputs and y_i is the output of rule neuron i in Layer 3.

Normalization Layer: This layer calculates the normalized firing strength of each rule by dividing the firing strength of an individual rule by the sum of all rule firing strengths. This ratio determines the contribution of each rule to the final output. The mathematical formulation is defined by Eq. (2).

$$\overline{w_j} = \frac{w_j}{w_1 + w_2 + \cdots + w_n}, \quad j = 1, 2, \ldots, n \tag{2}$$

where $\overline{w_j}$ is the normalized weight of the jth rule, w_j is the weight of the jth rule [31].

Defuzzification Layer: This layer applies a mathematical function to determine the final output. Parameters in this layer, referred to as consequent parameters, adjust based on learning. Each neuron in this layer connects to the respective normalization neuron and receives the original input values x_1 and x_2. The mathematical formulation is defined by Eq. (3).

$$\overline{w_i} f_i = \overline{w_i}(p_i x_1 + q_i x_2 + r_i) \tag{3}$$

where $\overline{w_i}$ is the output of the normalization layer, and p_i, q_i, r_i are the parameters.

The weighted average method is typically used for defuzzification in Sugeno-type inference, particularly for symmetrical output membership functions. This method closely resembles the centroid method and its mathematical formulation is defined by Eq. (4).

$$X^* = \frac{\sum_{j=1}^{k} \mu_C(\overline{x}) \cdot \overline{x}}{\sum_{j=1}^{k} \mu_C(\overline{x})} \tag{4}$$

where X^* is the defuzzified output and C represents an output fuzzy set.

Output Layer: The final layer consists of a single node that aggregates all incoming signals to generate the system's final output. The mathematical formulation of the overall output [37] is defined by Eq. (5).

$$Overall\ output = \sum_i \overline{w_i} f_i = \frac{\sum_i w_i f}{\sum_i w_i} \tag{5}$$

1.3 Heart Failure Death

Heart failure is a prevalent condition that affects individuals with underlying health issues such as hypertension, insomnia, and cardiovascular diseases. It occurs when the heart is unable to pump sufficient blood to meet the body's needs, often due to conditions like high blood pressure, diabetes, or other cardiac disorders [38]. This condition is categorized based on ejection fraction, which measures the proportion of blood ejected from the heart with each contraction. A normal ejection fraction falls between 50 and 75%, whereas a value below 40% indicates heart failure with reduced ejection fraction (HFrEF), also known as systolic heart failure [39]. Conversely, heart failure with preserved ejection fraction (HFpEF), or diastolic heart failure, arises when the left ventricle contracts properly during systole but fails to relax adequately during diastole [40–45].

The onset of heart disease is frequently linked to unhealthy lifestyle choices, including inadequate physical activity, poor diet, and excessive consumption of alcohol and tobacco [46, 47]. Implementing healthier habits such as incorporating more fruits and vegetables into the diet, reducing salt intake, engaging in regular physical activity, and avoiding smoking and excessive alcohol consumption can significantly lower the risk of developing heart-related conditions [48]. Research by Wang et al. [49] and Zhenya and Zhang [50] suggests that heart failure is a growing health concern worldwide. Early detection is essential for effective treatment, and clinical decision support systems (CDSS) play a crucial role in diagnosing heart conditions by analyzing patient data from hospitals and clinics. These systems enhance diagnostic accuracy, streamline treatment processes, and reduce the need for unnecessary medical tests, saving both time and healthcare costs [51, 52]. Additionally, studies have demonstrated that CDSS improves clinical decision-making, enhances patient outcomes, and supports preventive healthcare strategies [50, 53].

Several research studies have compared different methodologies for heart failure risk assessment. Ahmad et al. [54] utilized Cox regression models to predict mortality in heart failure patients and employed Kaplan-Meier plots to analyze survival trends. Weng et al. [55] assessed the accuracy of the American College of Cardiology's traditional cardiovascular risk prediction algorithm over 10 years. Their study compared this algorithm with machine learning techniques such as random forest, logistic regression, gradient boosting machines, and neural networks, revealing that these advanced models significantly enhance predictive accuracy. Zahid et al. [56] developed a semi-parametric proportional hazards Cox model to identify key risk factors associated with heart failure mortality.

In another study, Jeyaganesan et al. [57] evaluated heart disease prediction using machine learning algorithms, including Decision Trees, Logistic Regression, Random Forest, and Naïve Bayes. They found that Random Forest achieved the highest accuracy, exceeding 90%. Similarly, Oguntade and Ajayi [58] employed logistic regression modeling to identify independent risk factors contributing to hypertensive heart failure. Wang [59] conducted a comparative analysis of 18 machine learning models, including K-Nearest Neighbors, Decision Trees, Support

Vector Machines (SVM), Multi-Level Perceptron, Ridge Classifier, Random Forest, AdaBoost, and Gradient Boosting. By applying z-score and min-max normalization techniques to clinical variables, the study determined that z-score normalization provided superior accuracy in predicting heart failure.

Research on gender-related risk factors in heart failure patients has yielded varying results [60, 61]. Some studies suggest that women are at a higher risk than men of developing heart failure [62, 63]. Women, particularly older individuals, are more likely to experience congestive heart failure with preserved ejection fraction [64]. Kajimoto et al. [65] investigated the relationship between anemia, left ventricular ejection fraction, and survival in patients with acute decompensated heart failure, identifying significant gender-based differences. Although men over 40 in Western populations have a higher prevalence of heart failure, this trend reverses after the age of 80, with women being more affected [66]. However, there is limited data on the prevalence and incidence of heart failure in South Asian populations. McKeigue et al. [67] found that people from the Indian subcontinent have unusually high rates of heart disease, which highlights the need for more region-specific research.

Drawing from these extensive reviews, which examine diverse demographic and health variables, this chapter presents a neuro-fuzzy inference system designed to predict heart failure mortality using real-world data. The model aims to assess the likelihood of death due to heart failure, a leading cause of mortality and disability worldwide [68] by incorporating key risk factors associated with the condition.

2 Data of the Study

This study examined the medical records of 299 heart failure patients (105 females and 194 males), aged 40–95 years, who received treatment at the Faisalabad Institute of Cardiology and Allied Hospital in Faisalabad, Punjab, Pakistan, between April and December 2015 [38, 54]. The dataset was sourced from the Kaggle repository (available at https://www.kaggle.com/datasets). All patients in the dataset had left ventricular systolic dysfunction and a history of prior heart failure episodes. Among the participants, 96 patients were recorded as deceased, while 203 were alive.

The dataset comprises 12 features: age, anaemia, creatinine phosphokinase, diabetes, ejection fraction, high blood pressure, platelets, serum creatinine, serum sodium, sex, smoking, and the time of heart failure death. Filter method of feature selection was applied to identify the most relevant and informative variables before constructing the neuro-fuzzy inference system for heart failure diagnosis. The results from the feature selection process identified three key variables: age, ejection fraction, and serum creatinine, as having the highest information (greater than 0.07) concerning heart failure outcomes. These three variables were therefore selected to build the prediction model. For analysis, only patient records containing complete and accurate data for the selected input variables and the output variable (mortality due to heart failure) were included.

Table 1 Descriptive statistics of the dataset

Measures	Dead			Alive		
	Age (in years)	Ejection fraction (in %)	Serum creatinine (in mg/dL)	Age (in years)	Ejection fraction (in %)	Serum creatinine (in mg/dL)
Mean	65.22	33.47	1.84	58.76	40.27	1.18
Standard deviation	13.21	12.53	1.47	10.64	10.86	0.65
Minimum	42	14	0.60	40	17	0.50
Maximum	95	70	9.40	90	80	6.10

Table 1 provides descriptive statistics for the selected data, detailing the patients' age (in years), ejection fraction percentages, and serum creatinine levels (measured in mg/dL).

Several studies have identified age, ejection fraction, and serum creatinine as significant predictors of mortality in patients with heart failure. Wannamethee et al. [69] reported that elevated serum creatinine levels have been strongly correlated with an increased risk of stroke and have been identified as a significant contributor to mortality. Advanced age is also recognized as a critical predictor of complications and in-hospital death in heart failure patients [70, 71]. Age consistently emerges as an essential clinical risk factor for cardiovascular diseases, including coronary artery disease and stroke [72, 73]. As the global population ages, the prevalence of heart failure is expected to rise significantly [68, 74]. Additionally, age has been identified as a key factor in heart failure cases involving preserved left ventricular ejection fraction [54]. Studies by Kajimoto et al. [65] and Zahid et al. [56] provide valuable insights into the relationship between ejection fraction, serum creatinine levels, and survival outcomes in heart failure patients. According to Chicco and Jurman [38], serum creatinine and ejection fraction are critical clinical markers for predicting survival in heart failure cases. Age, being a non-modifiable risk factor, remains a significant influence on cardiovascular diseases across various demographic groups [58, 75].

3 Research Methodology

3.1 Categorization of Input and Output Variable for the Model

To categorize the input variables patient age, ejection fraction, and serum creatinine levels various literature sources were reviewed. For adult males with normal kidney function, serum creatinine levels typically range between 0.6 and 1.2 mg/dL, while adult females usually fall within 0.5 to 1.1 mg/dL [76, 77]. The reference range for serum creatinine in the general male population is 0.7–1.3 mg/dL [78], with standard serum creatinine levels for males and females being 0.8–1.3 and 0.6–1.3 mg/dL, respectively [79]. Elderly individuals face an increased risk of developing cardiovascular disease [80, 81]. An ejection fraction above 55% is typically regarded as normal by both the American Society of Echocardiography and the European Society of Echocardiography [82]. Reports from Sweden and Denmark highlight the growing prevalence of heart failure, particularly among younger populations [83, 84]. Lam et al. [85] introduced a new cut off for left ventricular ejection fraction, setting 55% for males and 60% for females as indicative of "normal" in the general population. Tromp et al. [86] noted that individuals aged 65 and older bear the largest burden of heart failure and classified age groups for assessing the impact of risk factors: young (< 55 years), middle-aged (55–64 years), old (65–74 years), and elderly (≥ 75 years).

Based on these findings, the variable age has been divided into three categories for both male and female datasets. Ejection fraction (EF) for both male and female datasets has been classified into two categories. Lastly, serum creatinine (SC) levels are divided into three linguistic labels for both male and female datasets.

3.2 *Model Development*

A two-step methodology is implemented to develop the model using real-world datasets, which are divided into training and testing subsets. For model training, 70% of the total data comprising 136 male and 74 female patients is utilized, while the remaining 30% including 58 male and 31 female patients is reserved for testing to validate the model's performance. The model is built using the Neuro-Fuzzy Inference System feature available in MATLAB (R2013a). The modeling process consists of the following stages:

3.3 *Model Construction*

The neuro-fuzzy inference system is based on a first-order Sugeno inference model. A six-layered architecture is employed to analyze the heart failure dataset. The training process integrates both forward and backward learning algorithms, utilizing least squares estimation and gradient descent techniques to optimize the model's network.

3.3.1 Structure of the Neuro-fuzzy Inference System

The structure of the neuro-fuzzy inference system is illustrated in Fig. 1. The model takes three input variables:

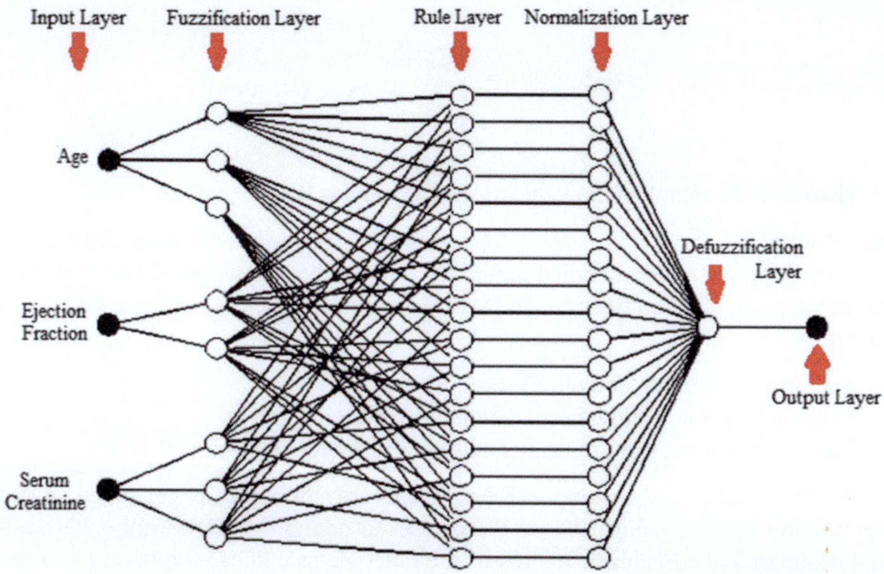

Fig. 1 Structure of neuro-fuzzy inference system used in the study

Input 1: Patient's age (in years), categorized into three membership functions: young, middle-aged, and old.
Input 2: Ejection fraction (percentage), categorized into two membership functions: below normal and normal.
Input 3: Serum creatinine (mg/dL), categorized into three membership functions: low, normal, and high.

The output variable, which represents the risk of heart failure-related mortality, is expressed through a single linear membership function. The model consists of six interconnected layers, each performing a specific function in the inference process. The layers are described below.

Input Layer: This layer consists of three input nodes (see Fig. 1) representing the patient's age, ejection fraction, and serum creatinine levels. Each node corresponds to a single input variable and directly transmits the input values to the next layer. The link weights in this layer are set to 1 [36].

Fuzzification Layer: In this layer, a triangular membership function is applied to associate the input nodes with their respective linguistic labels. The degree of membership for each input variable is determined using triangular fuzzy numbers. For example, the ejection fraction input is mapped to the labels below normal and normal based on its membership degree.

A Triangular fuzzy number is a fuzzy number represented by three points. If young age is a fuzzy number with points of 15 years, 25 years, and 35 years, then the mathematical formulation [87] of its membership function is defined by Eq. (6).

$$\mu_{(Young\ Age)}(m) = \begin{cases} 0, & m < 15 \\ \frac{m-15}{25-15}, & 15 \le x \le 25 \\ \frac{35-m}{35-25}, & 25 \le x \le 35 \\ 0, & m > 35 \end{cases} \tag{6}$$

Therefore the membership value at age 20 years is 0.5.

Rule Layer: This layer constructs a Sugeno-type fuzzy rule base. Logical AND operations are applied to the fuzzified inputs to compute the rule weights. The three key predictors, age, ejection fraction, and serum creatinine are classified into predefined categories:

Age: Young, middle, and old (3 categories).
Ejection Fraction: Below normal and normal (2 categories).
Serum Creatinine: Low, normal, and high (3 categories).

Based on these classifications, 18 fuzzy rules are automatically generated during the training process using 70% of the dataset to define the relationships between inputs and outputs. Each rule expresses the outcome as a linear combination of the input variables. Table 2 presents the set of rules generated by MATLAB for this model.

For example: Rule 1 states that if the first input patient's age belongs to the young category, the second input ejection fraction percentage falls under the below normal category, and the third input serum creatinine level is classified as low, then the output risk of heart failure related mortality is very low.

Normalization Layer: This layer processes input values from all neurons in the rule layer and computes the normalized weights of the rules. The weight of each rule is determined as a ratio of its individual weight to the sum of all rule weights. The mathematical formulation is defined by Eq. (7).

$$\overline{v_j} = \frac{v_j}{v_1 + v_2 + \cdots + v_n}, \quad j = 1, 2, \ldots, n \tag{7}$$

where $\overline{v_j}$ is the normalized weight of the jth rule, v_j is the weight of the jth rule.

Defuzzification Layer: In this layer, each neuron is connected to both the corresponding normalization neuron and the original input variables, such as the patient's age, ejection fraction, and serum creatinine levels. The rule output in this layer is determined by a linear node function. The mathematical formulation of the output of the jth rule is defined by Eq. (8).

$$\overline{v_j}\left(p_j y_1 + q_j y_2 + s_j y_3 + r_j\right) = \overline{v_j} f_j \tag{8}$$

where $\overline{v_j}$ represents the output from the previous layer, p_j, q_j, s_j, and r_j are the parameters, and y_i, $(i = 1, 2, 3)$ are the values of three input variables. The unknown parameters are estimated using the least squares method. For defuzzification, the

Table 2 The inference rules generated by MATLAB

Rule No.	Input			Output
	Age	Ejection fraction	Serum creatinine	Risk of heart failure death
1	Young	Below normal	Low	Very low
2	Young	Below normal	Normal	Low
3	Young	Below normal	High	Moderate
4	Young	Normal	Low	Very low
5	Young	Normal	Normal	Low
6	Young	Normal	High	Moderate
7	Middle	Below normal	Low	Moderate
8	Middle	Below normal	Normal	Moderate
9	Middle	Below normal	High	High
10	Middle	Normal	Low	Moderate
11	Middle	Normal	Normal	Moderate
12	Middle	Normal	High	High
13	Old	Below normal	Low	Moderate
14	Old	Below normal	Normal	High
15	Old	Below normal	High	Very high
16	Old	Normal	Low	Medium
17	Old	Normal	Normal	High
18	Old	Normal	High	Very high

weighted average (Wtaver) approach is used to compute the weighted consequent value of each rule.

Output Layer: The final output of the neuro-fuzzy inference system is obtained by summing all incoming signals at a single node. The mathematical formulation is defined by Eq. (9).

$$Overall\ output = \sum_j \overline{v_j} f_j = \frac{\sum_j v_j f_j}{\sum_j v_j}; \quad j = 1, 2, \ldots, n \tag{9}$$

3.3.2 Model Training and Testing

The neuro-fuzzy inference system is trained using a hybrid optimization technique combined with a backpropagation algorithm. Training is conducted over 100 iterations with an error tolerance threshold of 0.00001. Grid partitioning is used to initialize the fuzzy inference system, and as training progresses, the parameters of the membership functions are adjusted accordingly. This technique helps determine the number of membership functions that define the premise of the rules.

Once training is complete, the model is tested on the remaining 30% of the dataset, consisting of 89 patients. The testing dataset includes the same variables as the training data, allowing the model to predict the risk of heart failure-related mortality based on the given input variables.

3.4 Normality Test of Output Values

The output values generated by the neuro-fuzzy inference system are tested for normality using the Kolmogorov-Smirnov test on the testing datasets. Since the results indicate that the model's output values do not follow a normal distribution, the Mann-Whitney test is employed to determine whether the mean estimated values significantly differ between heart failure deaths and survivors.

3.5 Performance Evaluation of the Model

The accuracy and effectiveness of the developed model are evaluated in three distinct steps using the same input parameters. As an initial step, a Receiver Operating Characteristic (ROC) curve is generated, and the area under this curve (AUC) is calculated using the formula provided by Zhu et al. [88]. The mathematical formulation is defined by Eq. (10).

$$AUC = \int_0^1 ROC(t)dt \tag{10}$$

where t = (1 − specificity) and ROC (t) is sensitivity.

Subsequently, statistical metrics such as the Karl Pearson correlation coefficient, Mean Square Error (MSE), and Mean Squared Logarithmic Error (MSLE) are computed. The mathematical formulation of the correlation coefficient, MES and MSLE are defined by Eqs. (11), (12) and (13) respectively.

$$Correlation\,coefficient = \frac{\sum_{i=1}^{n}(O_i - \overline{O})(C_i - \overline{C})}{\sqrt{\sum_{i=1}^{n}(O_i - \overline{O})^2 \sum_{i=1}^{n}(C_i - \overline{C})^2}} \tag{11}$$

$$Mean\,Square\,Error = \sum_{i=1}^{n} \frac{(O_i - C_i)^2}{n} \tag{12}$$

Table 3 Mann-Whitney test statistic and mean ranks

Mann-Whitney		Mean rank	
Statistic	p-value	Alive	30.94
231.0	0.0001	Dead	64.76

$$Mean\,Squared\,Logarithmic\,Error = \frac{1}{n}\sum_{i=1}^{n}(\log(1+O_i) - \log(1+C_i))^2 \quad (13)$$

where O_i is the original output of the dataset, C_i is the computed output from the dataset, $\overline{O}$ and $\overline{C}$ are the mean values of O_i and C_i respectively, n is the number of data and log is the natural logarithm (base e).

Finally, the output values generated by the neuro-fuzzy inference system for the testing dataset are individually compared with the corresponding original observation i.e. heath failure (death or alive). To estimate the 95% confidence interval for each performance metric, bootstrap simulations are performed 1000 times. All performance analyses are conducted using Python 3.

4 Results

4.1 Normality of Output Values

The Kolmogorov-Smirnov test conducted on the output values yielded a test statistic of 0.514 with a significant p-value of 0.0001. This result indicates that the estimated output values for heart failure-related mortality do not follow a normal distribution. The results of the Mann-Whitney test are presented in Table 3.

4.2 Receiver Operating Characteristic (ROC) Curve

Figure 2 presents the ROC curves for the estimated risk values of the testing dataset from the developed model. The area under the curve (AUC) is calculated as 0.880, with a 95% confidence interval (CI) ranging from 0.807 to 0.939.

4.3 Statistical Measures

Table 4 presents the estimated values of the Karl Pearson correlation coefficient, Mean Squared Error (MSE), and Mean Squared Logarithmic Error (MSLE) between

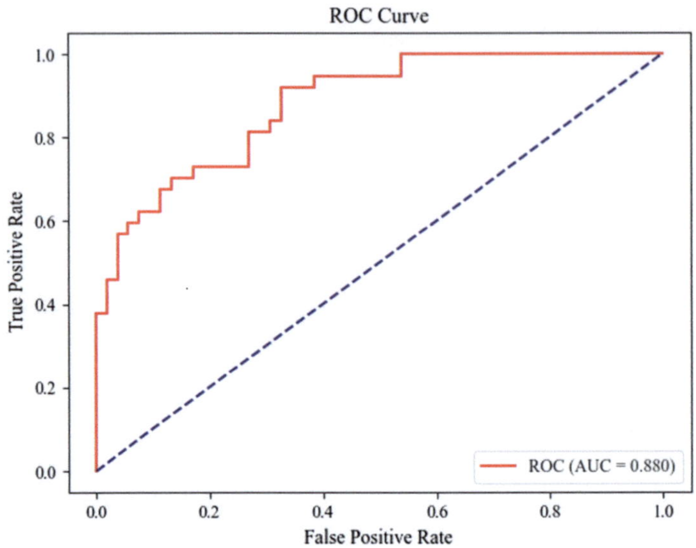

Fig. 2 ROC curves of the testing dataset

Table 4 Estimated values of statistical measures

Measure	Estimated value	Lower bound (95%)	Upper bound (95%)
Correlation coefficient	0.649	9.369e−17	0.742
Mean square error	0.147	0.112	0.186
Mean squared logarithmic error	0.069	0.054	0.085

the actual and predicted outcomes of the testing dataset, as assessed by the fuzzy logic model.

4.4 *Individual Observation*

Table 5 presents the individual observations of the actual output alongside the individual risk predictors from 40% of the testing dataset along with the risk of heart failure death estimated by the model.

Table 5 Individual observation with estimated risk value based on individual-level risk predictors

Sl. No.	Age of the patient	Ejection fraction	Serum creatinine	Heart failure death	Estimated risk value
1	82	30	1.2	Yes	0.567409
2	94	38	1.83	Yes	0.850525
3	85	45	3	Yes	0.750179
4	69	35	3.5	Yes	0.763399
5	60	30	2.3	No	0.370675
6	70	20	1.83	Yes	0.609747
7	72	50	1	No	0.129231
8	51	25	0.9	Yes	0.310371
9	60	40	0.6	No	0.112643
10	49	20	1.1	Yes	0.457683
11	72	20	1.3	Yes	0.560794
12	45	25	1	Yes	0.323114
13	72	25	1	Yes	0.413565
14	58	35	0.9	No	0.195133
15	65	50	1.3	No	0.151025
16	42	40	1.2	No	0.066763
17	67	50	1.18	No	0.126976
18	51	50	0.7	No	0.068658
19	55	35	1.1	No	0.18786
20	65	60	0.8	No	0.193815
21	60	38	2.2	Yes	0.77178
22	50	40	2.3	Yes	0.89720
23	45	80	1.18	No	0.14968
24	72	45	2.5	Yes	0.82399
25	80	38	1.3	Yes	0.63976
26	55	35	0.8	No	0.08052
27	73	40	0.9	No	0.22602
28	50	30	0.5	No	0.07041
29	70	35	1.2	No	0.34422
30	42	15	1.3	Yes	0.53320
31	70	60	0.8	No	0.28898
32	55	45	0.9	No	0.08429
33	46	35	0.9	No	0.06347
34	72	25	1.2	No	0.37099
35	60	30	1.7	Yes	0.61361
36	55	35	1.2	No	0.187112

5 Discussion

The results of this study underscore the importance of using neuro-fuzzy inference systems in clinical decision-making, particularly for predicting mortality risks in heart failure patients. By integrating variables such as age, ejection fraction, and serum creatinine levels, the neuro-fuzzy inference system demonstrated its capability to manage uncertainty and imprecision, which are common challenges in clinical decision-making. This ability positions the neuro-fuzzy inference system as a valuable tool in healthcare, where accurate outcome predictions can have a direct impact on patient care and resource allocation.

A key advantage of the neuro-fuzzy inference system is its hybrid nature, combining artificial neural networks and fuzzy inference systems. It can mimic human-like reasoning through neural network learning, while the inclusion of fuzzy inference rules allows the system to handle the vagueness inherent in clinical data.

This characteristic is particularly important in healthcare, where decisions often rely on qualitative judgments that traditional computational models might struggle to process.

Beyond its technical capabilities, the Mann-Whitney test results support the model's accuracy in predicting heart failure death. The estimated mean ranks of the model's outputs show that it provides high estimates (64.76) for patients who died due to heart failure and low estimates (30.94) for those who survived. The model also demonstrated significant predictiveness, with a good area under the curve (AUC) of 0.880, a positive correlation of 0.649, and minimal errors (MSE = 0.147, MSLE = 0.069). Additionally, individual case observations show that the model assigns high-risk values to patients who died and low-risk values to survivors.

The broader implications of this study suggest that neuro-fuzzy inference systems have great potential in healthcare. Several studies have employed neuro-fuzzy systems for cardiovascular disease prediction across different datasets, regions, and experimental settings. Bhuvaneswari [89] used a heart disease dataset with Principal Component Analysis (PCA) for dimensionality reduction and ANFIS for classification. They achieved an accuracy of 93.2%, though specific input variables and dataset size were not disclosed. Ziasabounchi et al. [90] applied ANFIS to a 303-patient dataset to classify the degree of heart disease. Their approach, based on conventional clinical inputs, yielded 92.3% accuracy. The data source and preprocessing methods, however, were not deeply described. Zabbah et al. [91] used a 200-subject dataset to compare pure artificial neural networks with neuro-fuzzy systems for coronary artery disease diagnosis. They reported that the MSE dropped from 0.25 (ANN) to 0.007 (neuro-fuzzy), indicating a substantial performance gain when combining learning with fuzzy logic. Opeyemi and Justice [92] designed a patient-facing neuro-fuzzy system using a hybrid dataset from the UCI repository, incorporating eight clinical inputs (e.g., blood pressure, chest pain, cholesterol). The dataset was split 80/20 for training/testing, and they reported > 90% accuracy. Their model was validated for usability in real-world settings, suggesting potential for deployment. Feng et al. [93]

enhanced ANFIS with a genetic algorithm to tune parameters. Using the UCI Cleveland dataset, split into 253 training and 50 testing instances, and 13 clinical features, they achieved 91.25% testing accuracy. Their setup closely resembles many benchmark studies, making their results widely comparable. Rajabzadeh et al. [94] used a large cohort (n = 7385) from the Mashhad population with 20 variables initially considered and 18 selected for modeling. Data were split 70/30 for training/testing. Their ANFIS model achieved 83.4% accuracy, outperforming logistic regression and discriminant analysis. Importantly, they followed a structured feature selection process using non-parametric tests.

Neuro-fuzzy inference systems (NFIS) have also been successfully applied to diagnose a range of diseases beyond cardiovascular conditions. Polat et al. [95] combined ANFIS with linear and quadratic discriminant analysis to classify diabetes, achieving an accuracy of 89.47%. Sancar et al. [96] evaluated ANFIS in predicting Body Mass Index (BMI), reporting low RMSE values, indicating biological reliability and suitability for regression tasks. Hamdan and Garibaldi [97] applied ANFIS to predict breast cancer survival, showing promising results in clinical decision support applications. Mahmoudi et al. [98] used ANFIS on microarray gene expression data for cancer classification, comparing it with SVM, KNN, and CART, and found ANFIS to be the most effective. Bora and Barman [99] demonstrated that ANFIS could effectively classify diabetes risk, achieving high AUROC scores and low classification error rates. These studies underline the adaptability of ANFIS models to various medical datasets, where careful preprocessing, relevant feature selection, and integration with hybrid techniques consistently enhance predictive performance.

As healthcare systems increasingly adopt machine learning approaches for diagnostic and prognostic purposes, models like the one proposed here can help bridge the gap between computational intelligence and human reasoning. By addressing the limitations of traditional algorithms in dealing with ambiguous or incomplete data, the neuro-fuzzy inference system enhances the reliability and robustness of decision-making systems. This aligns with the growing demand for personalized medicine, where treatment strategies are tailored to individual patient profiles.

6 Conclusion

This study demonstrates the successful application of a neuro-fuzzy inference system in predicting the risk of death in heart failure patients. The proposed model not only delivers accurate risk predictions but also provides a transparent and interpretable framework that supports clinical decision-making. The model effectively manages uncertainty, as evidenced by smooth prediction transitions across fuzzy boundaries, and its cost-effectiveness and simplicity are attributable to the use of only three key clinical inputs i.e. age, ejection fraction, and serum creatinine, without the need for complex feature engineering. The model's rule structure and positive correlation and low error values highlight its potential for straightforward implementation in healthcare settings.

However, the findings of the study are preliminary and specific to the dataset used in this study. Although the model shows promise, broader validation on diverse datasets and patient populations is necessary before generalizing the results. Future work will explore multi-center datasets and the incorporation of additional clinical variables to enhance the model's generalizability and robustness.

While the study confirms the potential of NFIS in predicting heart failure mortality, it also highlights some inherent limitations. By considering only three associated factors, its applicability is somewhat restricted, highlighting the need to incorporate additional variables and a large dataset for a more comprehensive analysis. Future studies could expand on this framework by adding more variables or applying the model to other demographic groups, further advancing the potential of neuro-fuzzy inference systems in healthcare. Moreover, a comparative study with other machine learning models, such as logistic regression, support vector machines, and random forests, can be conducted in the future to evaluate the viability of the proposed NFIS model using the same dataset. It is also important to recognize that these systems heavily rely on human expertise, which can impact their interpretability as more variables are incorporated. Therefore, to ensure the reliability and real-world applicability of these models, it is essential to involve experts from various fields, including healthcare, epidemiology, and data science, during the development process. This collaboration will help maintain the accuracy and effectiveness of neuro-fuzzy inference systems, ensuring their continued success in advancing healthcare practices.

AI Use Statement Some portions of this manuscript were enhanced with the assistance of generative AI tools. The authors carefully reviewed, edited and approved all AI-generated content. The authors take full responsibility for the accuracy and integrity of the entire chapter.

References

1. Adunlin, G., Diaby, V., Xiao, H.: Application of multicriteria decision analysis in health care: a systematic review and bibliometric analysis. Health Expect. **18**, 1894–1905 (2015). https://doi.org/10.1111/hex.12287
2. Sivakumar, G., Almehdawe, E., Kabir, G.: Developing a decision-making framework to improve healthcare service quality during a pandemic. Appl. Syst. Innov. **5**(1), 3 (2022). https://doi.org/10.3390/asi5010003
3. Hunink, M.G., Glasziou, P.P., Siegel, J.E., Weeks, J.C., Pliskin, J.S., Elstein, A.S., Weinstein, M.C.: Decision Making in Health and Medicine: Integrating Evidence and Values. Cambridge University Press, Cambridge (2001)
4. Whichello, C., Bywall, K.S., Mauer, J., Stephen, W., Cleemput, I., Pinto, C.A., van Overbeeke, E., Huys, I., de Bekker-Grob, E.W., Hermann, R., et al.: An overview of critical decision-points in the medical product lifecycle: where to include patient preference information in the decision-making process? Health Policy **124**, 1325–1332 (2020). https://doi.org/10.1016/j.healthpol.2020.07.007
5. Shahmoradi, L., Safadari, R., Jimma, W.: Knowledge management implementation and the tools utilized in healthcare for evidence-based decision making: a systematic review. Ethiop. J. Health Sci. **27**, 541–558 (2017). https://doi.org/10.4314/ejhs.v27i5.13

6. Shahid, N., Rappon, T., Berta, W.: Applications of artificial neural networks in health care organizational decision-making: a scoping review. PLoS ONE **14**, e0212356 (2019). https://doi.org/10.1371/journal.pone.0212356
7. Damman, O.C., Jani, A., de Jong, B.A., Becker, A., Metz, M.J., de Bruijne, M.C., Timmermans, D.R., Cornel, M.C., Ubbink, D.T., van der Steen, M., Gray, M., van El, C.: The use of PROMs and shared decision-making in medical encounters with patients: an opportunity to deliver value-based health care to patients. J. Eval. Clin. Pract. **26**(2), 524–540 (2020). https://doi.org/10.1111/jep.13321
8. Thokala, P., Devlin, N., Marsh, K., Baltussen, R., Boysen, M., Kalo, Z., Longrenn, T., Mussen, F., Peacock, S., Watkins, J., et al.: Multiple criteria decision analysis for health care decision making—an introduction: report 1 of the ISPOR MCDA emerging good practices task force. Value Health **19**, 1–13 (2016). https://doi.org/10.1016/j.jval.2015.12.003
9. Vis, C., Bührmann, L., Riper, H., Ossebaard, H.C.: Health technology assessment frameworks for EHealth: a systematic review. Int. J. Technol. Assess. Health Care **36**, 204–216 (2020). https://doi.org/10.1017/S026646232000015X
10. Gupta, S., Kamboj, S., Bag, S.: Role of risks in the development of responsible artificial intelligence in the digital healthcare domain. Inf. Syst. Front. **25**, 2257–2274 (2023). https://doi.org/10.1007/s10796-021-10174-0
11. Mardani, A., Zavadskas, E.K., Fujita, H., Köppen, M.: Big data-driven large-scale group decision-making under uncertainty (BiGDM-U). Appl. Intell. **52**, 13341–13344 (2022). https://doi.org/10.1007/s10489-022-04113-y
12. Orlu, G.U., Abdullah, R.B., Zaremohzzabieh, Z., Jusoh, Y.Y., Asadi, S., Qasem, Y.A.M., Nor, R.N.H., Mohd Nasir, W.M.H.b.: A systematic review of literature on sustaining decision-making in healthcare organizations amid imperfect information in the big data Era. Sustainability **15**, 15476 (2023). https://doi.org/10.3390/su152115476
13. Glaize, A., Duenas, A., Di Martinelly, C., Fagnot, I.: Healthcare decision-making applitions using multicriteria decision analysis: a scoping review. J. Multi-Criteria Decis. Anal. **26**(1–2), 62–83 (2019). https://doi.org/10.1002/mcda.1659
14. Agrawal, R.: Predictive analysis of breast cancer using machine learning techniques. Ing. Solidaria **15**(3), 1–23 (2019). https://doi.org/10.16925/2357-6014.2019.03.01
15. Omondiagbe, D.A., Veeramani, S., Sidhu, A.S.: Machine learning classification techniques for breast cancer diagnosis. IOP Conf. Ser. Mater. Sci. Eng. **495**(1), 012033 (2019). https://doi.org/10.1088/1757-899X/495/1/012033
16. Lazuwardi, B.: A logistic regression modeling on infant mortality rates in East Java Province. Indonesian J. Public Health **15**(2), 146–152 (2020). https://doi.org/10.20473/ijph.v15i2.2020.146-152
17. Ye, Y., Xiong, Y., Zhou, Q., Wu, J., Li, X., Xiao, X.: Comparison of machine learning methods and conventional logistic regressions for predicting gestational diabetes using routine clinical data: a retrospective cohort study. J. Diabetes Res. **2020**, 1–10 (2020). https://doi.org/10.1155/2020/4168340.eCollection2020
18. Rani, K.K.: Diabetes prediction using machine learning. Int. J. Sci. Res. Comput. Sci. Eng. Inf. Technol. **6**(4), 294–305 (2020). https://doi.org/10.32628/CSEIT206463
19. Soni, M., Varma, S.: Diabetes prediction using machine learning techniques. Int. J. Eng. Res. Technol. **9**(9), 921–925 (2020). https://www.scribd.com/document/848013239/Diabetes-Prediction-Using-Machine-Learning-Techniques-IJERTV9IS090496. Accessed 20 Jan 2025
20. Kishore, G.N., Rajesh, V., Reddy, A.V.A., Sumedh, K., Reddy, T.R.S.: Prediction of diabetes using machine learning classification algorithms. Int. J. Sci. Technol. Res. **9**(1), 1805–1808 (2020). https://www.ijstr.org/final-print/jan2020/Prediction-Of-Diabetes-Using-Machine-Learning-Classification-Algorithms.pdf. Accessed 20 Jan 2025
21. Khanam, J.J., Foo, S.Y.: A comparison of machine learning algorithms for diabetes prediction. ICT Express **7**(4), 432–439 (2021). https://doi.org/10.1016/j.icte.2021.02.004
22. Abaker, A.A., Saeed, F.A.: A comparative analysis of machine learning algorithms to build a predictive model for detecting diabetes complications. Informatica **45**(1), 117–125 (2021). https://doi.org/10.31449/inf.v45i1.3111

23. Llaha, O., Rista, A.: Prediction and detection of diabetes using machine learning. In: RTA-CSIT Proceeding, pp. 94–102 (2021). https://ceur-ws.org/Vol-2872/paper13.pdf
24. Butt, U.M., Letchmunan, S., Ali, M., Hassan, F.H., Baqir, A., Sherazi, H.H.R.: Machine learning based diabetes classification and prediction for healthcare applications. J. Healthc. Eng. 1–17 (2021). https://doi.org/10.1155/2021/9930985
25. Srivastava, S., Upadhyay, S.K., Chauhan, S., Alagarajan, M.: Preceding child survival status and its effect on infant and child mortality in India: an evidence from National Family Health Survey 2015–16. BMC Public Health **21**(1), 1577 (2021). https://doi.org/10.1186/s12889-021-11569-z
26. Nyoni, S.P., Nyoni, T.: Prediction of infant mortality in morocco using artificial neural networks. Int. Res. J. Innov. Eng. Technol. **5**(3), 627–631 (2021). https://doi.org/10.47001/IRJIET/2021.503108
27. Hajipour, M., Taherpour, N., Fateh, H., Yousefi, E., Etemad, K., Zolfizadeh, F., Rajabi, A., Valadbeigi, T., Mehrabi, Y.: Predictive factors of infant mortality using data mining in Iran. J. Compr. Pediatr. **12**(1), e108575 (2021). https://www.sid.ir/paper/676725/en
28. Wegbom, A.I., Bademosi, A., Edet, C.K., Green, K.I., Sapira-Ordu, L., Fagbamigbe, A.F.: Rural–urban disparities in birth interval among women of reproductive age in Nigeria. Sci. Rep. **12**(1), 17488 (2022). https://doi.org/10.1038/s41598-022-22142-y
29. Kazemi, A., Mohammadi, A.: Presenting a fuzzy expert system for diagnosis of diabetes. Int. J. Decis. Intell. **1**(1), 17–27 (2023). https://sanad.iau.ir/Journal/ijdi/Article/938763
30. Singh, S.K., Abolghasemi, V., Anisi, M.H.: Fuzzy logic with deep learning for detection of skin cancer. Appl. Sci. **13**(15), 8927 (2023). https://doi.org/10.3390/app13158927
31. Apasiya, E.A., Salifu, A.M., Agbedemnab, P.A.N.: New approaches to the prognosis and diagnosis of breast cancer using fuzzy expert systems. J. Comput. Commun. **12**(5), 151–169 (2024). https://doi.org/10.4236/jcc.2024.125011
32. Ali, M.L., Sadi, M.S., Goni, M.O.: Diagnosis of heart diseases: a fuzzy-logic-based approach. PLoS ONE **19**(2), e0293112 (2024). https://doi.org/10.1371/journal.pone.0293112
33. El-Ibrahimi, A., Daanouni, O., Alouani, Z., El Gannour, O., Saleh, S., Cherradi, B., Bouattane, O.: Fuzzy based system for coronary artery disease prediction using subtractive clustering and risk factors data. Intell. Based Med. **11**, 100208 (2025). https://doi.org/10.1016/j.ibmed.2025.100208
34. Jang, J.S.R.: ANFIS: adaptive network-based fuzzy inference systems. IEEE Trans. Syst. Man Cybern. **23**(3), 665–685 (1993). https://doi.org/10.1109/21.256541
35. Buragohain, M.: Adaptive network based fuzzy inference system (ANFIS) as a tool for system identification with special emphasis on training data minimization. Ph.D. thesis, Indian Institute of Technology Guwahati (2008). https://iitg.ac.in/chitra/notes/Thesis_Mrinal.pdf. Accessed 21 Jan 2025
36. Bushara, N.O., Abraham, A.: Using adaptive neuro-fuzzy inference system (ANFIS) to improve the long-term rainfall forecasting. J. Network Innov. Comput. **3**, 146–158 (2015). https://mirlabs.net/jnic/secured/Volume3-Issue1/Paper15.pdf. Accessed 20 Jan 2025
37. Bushara, N., Abraham, A.: Novel ensemble method for long term rainfall prediction. Int. J. Comput. Inf. Syst. Ind. Manag. Appl. **7**(1), 116–130 (2015). https://www.mirlabs.net/ijcisim/regular_papers_2015/IJCISIM_12.pdf. Accessed 20 Jan 2024
38. Chicco, D., Jurman, G.: Machine learning can predict survival of patients with heart failure from serum creatinine and ejection fraction alone. BMC Med. Inform. Decis. Mak. **20**(1), 1–16 (2020). https://doi.org/10.1186/s12911-020-1023-5
39. Meng, F., Zhang, Z., Hou, X., Qian, Z., Wang, Y., Chen, Y., Wang, Y., Zhou, Y., Chen, Z., Zhang, X., Yang, J.: Machine learning for prediction of sudden cardiac death in heart failure patients with low left ventricular ejection fraction: study protocol for a retroprospective multicentre registry in China. BMJ Open **9**(5), e023724 (2019). https://doi.org/10.1136/bmjopen-2018-023724
40. Pfeffer, M.A., Braunwald, E.: Treatment of heart failure with preserved ejection fraction: reflections on its treatment with an aldosterone antagonist. JAMA Cardiol. **1**(1), 7–8 (2016). https://doi.org/10.1001/jamacardio.2015.0356

41. Lewis, G.A., Schelbert, E.B., Williams, S.G., Cunnington, C., Ahmed, F., McDonagh, T.A., Miller, C.A.: Biological phenotypes of heart failure with preserved ejection fraction. J. Am. Coll. Cardiol. **70**(17), 2186–2200 (2017). https://doi.org/10.1016/j.jacc.2017.09.006
42. Katz, D.H., Deo, R.C., Aguilar, F.G., Selvaraj, S., Martinez, E.E., Beussink-Nelson, L., Kim, K.Y.A., Peng, J., Irvin, M.R., Tiwari, H., Rao, D.C.: Phenomapping for the identification of hypertensive patients with the myocardial substrate for heart failure with preserved ejection fraction. J. Cardiovasc. Transl. Res. **10**(3), 275–284 (2017). https://doi.org/10.1007/s12265-017-9739-z
43. Nanayakkara, S., Kaye, D.M.: Targets for heart failure with preserved ejection fraction. Clin. Pharmacol. Ther. **102**(2), 228–237 (2017). https://doi.org/10.1002/cpt.723
44. Nauta, J.F., Jin, X., Hummel, Y.M., Voors, A.A.: Markers of left ventricular systolic dysfunction when lrftventricular ejection fraction is normal. Eur. J. Heart Fail. **20**(12), 1636–1638 (2018). https://doi.org/10.1002/ejhf.1326
45. Mesquita, E.T., Grion, D.C., Kubrusly, M.C., Silva, B.B.F.F., Santos, É.A.R.: Phenotype mapping of heart failure with preserved ejection fraction. Int. J. Cardiovasc. Sci. **31**(6), 652–661 (2018). https://doi.org/10.5935/2359-4802.20180047
46. Tao, R., Zhang, S., Huang, X., Tao, M., Ma, J., Ma, S., Zhang, C., Zhang, T., Tang, F., Lu, J., Shen, C., Xie, X.: Magnetocardiography-based ischemic heart disease detection and localization using machine learning methods. Trans. Biomed. Eng. **66**(6), 1658–1667 (2018). https://doi.org/10.1109/TBME.2018.2877649
47. Mohan, S., Thirumalai, C., Srivastava, G.: Effective heart disease prediction using hybrid machine learning techniques. IEEE Access **7**, 81542–81554 (2019). https://doi.org/10.1109/ACCESS.2019.2923707
48. Spencer, R., Thabtah, F., Abdelhamid, N., Thompson, M.: Exploring feature selection and classification methods for predicting heart disease. Digital Health **6**, 1–10 (2020). https://doi.org/10.1177/2055207620914777
49. Wang, J., Liu, C., Li, L., Li, W., Yao, L., Li, H., Zhang, H.: A stacking-based model for non-invasive detection of coronary heart disease. IEEE Access **8**, 37124–37133 (2020). https://doi.org/10.1109/ACCESS.2020.2975377
50. Zhenya, Q., Zhang, Z.: A hybrid cost-sensitive ensemble for heart disease prediction. BMC Med. Inform. Decis. Mak. **21**, 1–18 (2021). https://doi.org/10.1186/s12911-021-01436-7
51. Fitriyani, N.L., Syafrudin, M., Alfian, G., Rhee, J.: HDPM: an effective heart disease prediction model for a clinical decision support system. IEEE Access **8**, 133034–133050 (2020). https://doi.org/10.1109/ACCESS.2020.3010511
52. Mienye, I.D., Sun, Y.: Improved heart disease prediction using particle swarm optimization based stacked sparse autoencoder. Electronics **10**(19), 2347 (2021). https://doi.org/10.3390/electronics10192347
53. Alarsan, F.I., Younes, M.: Analysis and classification of heart diseases using heartbeat features and machine learning algorithms. J. Big Data **6**(1), 1–15 (2019). https://doi.org/10.1186/s40537-019-0244-x
54. Ahmad, T., Munir, A., Bhatti, S.H., Aftab, M., Raza, M.A.: Survival analysis of heart failure patients: a case study. PLoS ONE **12**(7), e0181001 (2017). https://doi.org/10.1371/journal.pone.0181001
55. Weng, S.F., Reps, J., Kai, J., Garibaldi, J.M., Qureshi, N.: Can machine-learning improve cardiovascular risk prediction using routine clinical data? PLoS ONE **12**(4), e0174944 (2017). https://doi.org/10.1371/journal.pone.0174944
56. Zahid, F.M., Ramzan, S., Faisal, S., Hussain, I.: Gender based survival prediction models for heart failure patients: a case study in Pakistan. PLoS ONE **14**(2), e0210602 (2019). https://doi.org/10.1371/journal.pone.0210602
57. Jeyaganesan, J., Sathiya, A., Keerthana, S., Aiyer, A.: Diagnosis and prediction of heart disease using machine learning techniques. Elementary Educ. Online **19**(2), 1817–1827 (2020). https://ilkogretim-online.org/index.php/pub/article/view/6769. Accessed 21 Jan 2025
58. Oguntade, A.S., Ajayi, I.O.: Risk factors of heart failure among patients with hypertension attending a tertiary hospital in Ibadan, Nigeria: the RISK-HHF case-control study. PLoS ONE **16**(1), e0245734 (2021). https://doi.org/10.1371/journal.pone.0245734

59. Wang, J.: Heart failure prediction with machine learning: a comparative study. J. Phys. Conf. Ser. **2031**(1), 012068 (2021). https://doi.org/10.1088/1742-6596/2031/1/012068
60. Assiri, A.S.: Effect of gender difference in management of heart failure patients in Aseer, Saudi Arabia. Heart Views Official J. Gulf Heart Assoc. **12**(1), 18–21 (2011). https://doi.org/10.4103/1995-705X.81555
61. Ahmadi, A., Etemad, K., Khaledifar, A.: Risk factors for heart failure in a cohort of patients with newly diagnosed myocardial infarction: a matched, case-control study in Iran. Epidemiol. Health **38**, 1–7 (2016). https://doi.org/10.4178/epih.e2016019
62. Chin, M.H., Goldman, L.: Gender differences in 1-year survival and quality of life among patients admitted with congestive heart failure. Med. Care 1033–1046 (1998). https://doi.org/10.1097/00005650-199807000-00010
63. Regitz-Zagrosek, V., Oertelt-Prigione, S., Seeland, U., Hetzer, R.: Sex and gender differences in myocardial hypertrophy and heart failure. Circ. J. **74**(7), 1265–1273 (2010). https://doi.org/10.1253/circj.cj-10-0196
64. Samuel, R.S., Hausdorff, J.M., Wei, J.Y.: Congestive heart failure with preserved systolic function: is it a woman's disease? Womens Health Issues **9**(4), 219–222 (1999). https://doi.org/10.1016/s1049-3867(99)00008-0
65. Kajimoto, K., Minami, Y., Sato, N., Otsubo, S., Kasanuki, H.: Investigators of the acute decompensated heart failure syndromes (ATTEND) registry: gender differences in anemia and survival in patients hospitalized for acute decompensated heart failure with preserved or reduced ejection fraction. Am. J. Cardiol. **120**(3), 435–442 (2017). https://doi.org/10.1016/j.amjcard.2017.04.043
66. Razzolini, R., Dal Lin, C.: Gender differences in heart failure. Ital. J. Gender-Specif. Med. **1**(1), 15–20 (2015). https://doi.org/10.1723/2012.21914
67. McKeigue, P.M., Miller, G.J., Marmot, M.G.: Coronary heart disease in south Asians overseas: a review. J. Clin. Epidemiol. **42**(7), 597–609 (1989). https://doi.org/10.1016/0895-4356(89)90002-4
68. Savarese, G., Lund, L.H.: Global public health burden of heart failure. Card. Fail. Rev. **3**(1), 7–11 (2016). https://doi.org/10.15420/cfr.2016:25:2
69. Wannamethee, S.G., Shaper, A.G., Perry, I.J.: Serum creatinine concentration and risk of cardiovascular disease: a possible marker for increased risk of stroke. Stroke **28**(3), 557–563 (1997). https://doi.org/10.1161/01.str.28.3.557
70. Abid, A.R., Rafique, S., Tarin, S.M., Ahmed, R.Z., Anjum, A.H.: Age-related in-hospital mortality among patients with acute myocardial infarction. J. Coll. Phys. Surg. Pak. **14**(5), 262–266 (2004). https://pubmed.ncbi.nlm.nih.gov/15225451/. Accessed 20 Jan 2025
71. Hippisley-Cox, J., Coupland, C., Vinogradova, Y., Robson, J., Minhas, R., Sheikh, A., Brindle, P.: Predicting cardiovascular risk in England and Wales: prospective derivation and validation of QRISK2. BMJ **336**(7659), 1475–1482 (2008). https://doi.org/10.1136/bmj.39609.449676.25
72. Bui, A.L., Horwich, T.B., Fonarow, G.C.: Epidemiology and risk profile of heart failure. Nat. Rev. Cardiol. **8**(1), 30–41 (2011). https://doi.org/10.1038/nrcardio.2010.165
73. Lu, Y., Hajifathalian, K., Ezzati, M., Woodward, M., Rimm, E.B., Danaei, G.: Metabolic mediators of the effects of body-mass index, overweight, and obesity on coronary heart disease and stroke: a pooled analysis of 97 prospective cohorts with 1.8 million participants. Lancet **383**, 970–983 (2014). https://doi.org/10.1016/S0140-6736(13)61836-X
74. Dunlay, S.M., Roger, V.L.: Understanding the epidemic of heart failure: past, present, and future. Curr. Heart Fail. Rep. **11**, 404–415 (2014). https://doi.org/10.1007/s11897-014-0220-x
75. Sahoo, P.K., Jeripothula, P.: Heart failure prediction using machine learning techniques (2020). https://papers.ssrn.com/sol3/papers.cfm?abstract_id=3759562. Accessed 21 Jan 2025
76. Hosten, A.O.: BUN and creatinine. In: Clinical Methods: The History, Physical, and Laboratory Examinations, 3rd edn. (1990)
77. Shahbaz, H., Gupta, M.: Creatinine clearance (2019). https://www.ncbi.nlm.nih.gov/books/NBK544228/. Accessed 30 Nov 2024
78. Burtis, C.A., Ashwood, E.R.: Tietz Textbook of Clinical Chemistry. American Association for Clinical Chemistry (1994)

79. Verma, M., Khadapkar, R., Sahu, P.S., Das, B.R.: Comparing age-wise reference intervals for serum creatinine concentration in a "Reality check" of the recommended cut-off. Indian J. Clin. Biochem. **21**, 90–94 (2006). https://scholar.google.co.in/citations?view_op=view_citation&hl=en&user=PVfaSbUAAAAJ&citation_for_view=PVfaSbUAAAAJ:roLk4NBRz8UC. Accessed 19 Jan 2025
80. North, B.J., Sinclair, D.A.: The intersection between aging and cardiovascular disease. Circ. Res. **110**(8), 1097–1108 (2012). https://doi.org/10.1161/CIRCRESAHA.111.246876
81. Curtis, A.B., Karki, R., Hattoum, A., Sharma, U.C.: Arrhythmias in patients $\geq$ 80 years of age: pathophysiology, management, and outcomes. J. Am. Coll. Cardiol. **71**(18), 2041–2057 (2018). https://doi.org/10.1016/j.jacc.2018.03.019
82. Lang, R.M., Badano, L.P., Mor-Avi, V., Afilalo, J., Armstrong, A., Ernande, L., Flachskampf, F.A., Foster, E., Goldstein, S.A., Kuznetsova, T., Lancellotti, P., Muraru, D., Picard, M.H., Rietzschel, E.R., Rudski, L., Spencer, K.T., Tsang, W., Voigt, J.: Recommendations for cardiac chamber quantification by echocardiography in adults: an update from the American Society of Echocardiography and the European Association of Cardiovascular Imaging. J. Am. Soc. Echocardiogr. **28**(1), 1–39 (2015). https://doi.org/10.1016/j.echo.2014.10.003
83. Barasa, A., Schaufelberger, M., Lappas, G., Swedberg, K., Dellborg, M., Rosengren, A.: Heart failure in young adults: 20-year trends in hospitalization, aetiology, and case fatality in Sweden. Eur. Heart J. **35**(1), 25–32 (2014). https://doi.org/10.1093/eurheartj/eht278
84. Rosengren, A., Åberg, M., Robertson, J., Waern, M., Schaufelberger, M., Kuhn, G., Åberg, D., Schiöler, L., Toren, K.: Body weight in adolescence and long-term risk of early heart failure in adulthood among men in Sweden. Eur. Heart J. **38**(24), 1926–1933 (2017). https://doi.org/10.1093/eurheartj/ehw221
85. Lam, C.S., Solomon, S.D.: Classification of heart failure according to ejection fraction: JACC review topic of the week. J. Am. Coll. Cardiol. **77**(25), 3217–3225 (2021). https://doi.org/10.1016/j.jacc.2021.04.070
86. Tromp, J., Paniagua, S.M., Lau, E.S., Allen, N.B., Blaha, M.J., Gansevoort, R.T., Hillege, H.L., Lee, D.E., Levy, D., Vasan, R.S., Van Der Harst, P., Van Gilst, W.H., Larson, M.G., Shah, S.J., De Boer, R.A., Lam, C.S.P., Ho, J.E.: Age dependent associations of risk factors with heart failure: pooled population based cohort study. BMJ **372**, 1–9 (2021). https://doi.org/10.1136/bmj.n461
87. Shanthini, C.: A new operation on triangular fuzzy number. Malaya J. Matematik **S**(2), 4082–4085 (2020). https://doi.org/10.26637/MJM0S20/1064
88. Zhu, W., Zeng, N., Wang, N.: Sensitivity, Specificity, Accuracy, Associated Confidence Interval and ROC Analysis with Practical SAS® Implementations. Health Care and Life Sciences, NESUG (2010). www.lexjansen.com/nesug/nesug10/hl/hl07.pdf. Accessed 21 Nov 2024
89. Bhuvaneswari, A.N.G.: An intelligent approach based on principal component analysis and adaptive neuro fuzzy inference system for predicting the risk of cardiovascular diseases. In: 2013 Fifth International Conference on Advanced Computing (ICoAC). IEEE (2013). https://doi.org/10.1109/ICoAC.2013.6921957
90. Ziasabounchi, N., Askerzade, I.: ANFIS based classification model for heart disease prediction. Int. J. Eng. Comput. Sci. **14**(2), 7–12 (2014). https://citeseerx.ist.psu.edu/document?repid=rep1&type=pdf&doi=b04a43eb6f0c8affd6bfdf49f1d4da08698fc4be. Accessed 20 Jan 2025
91. Zabbah, I., Koohjani, Z., Maroosi, A., Layeghi, K.: Diagnosis of coronary artery disease using neuro-fuzzy-based method. J. Torbat Heydariyeh Univ. Med. Sci. **6**(3), 48–59 (2018). http://jms.thums.ac.ir/article-1-564-en.html. Accessed 18 Jan 2025
92. Opeyemi, O., Justice, E.O.: Development of neuro-fuzzy system for early prediction of heart attack. Int. J. Inf. Technol. Comput. Sci. **4**(9), 22–28 (2012). https://doi.org/10.5815/ijitcs.2012.09.03
93. Feng, J., Wang, Q., Li, N.: An intelligent system for heart disease prediction using adaptive neuro-fuzzy inference systems and genetic algorithm. J. Phys. Conf. Ser. **1**, 012172 (2021). https://doi.org/10.1088/1742-6596/2010/1/012172
94. Rajabzadeh, Z., Akbari Sharak, N., Esmaeili, H., Ghayour-Mobarhan, M., Shakeri, M.T., Pasdar, Z.: Assessment of the performances of adaptive neuro-fuzzy inference system and

two statistical methods for diagnosing coronary artery disease. Med. J. Islam Repub. Iran 37–46 (2023). https://doi.org/10.47176/mjiri.37.46
95. Polat, K., Güneş, S.: An expert system approach based on principal component analysis and adaptive neuro-fuzzy inference system to diagnosis of diabetes disease. Digit. Sig. Process. **17**(4), 702–710 (2007). https://doi.org/10.1016/j.dsp.2006.09.005
96. Sancar, N., Tabrizi, S.S.: Body mass index estimation by using an adaptive neuro fuzzy inference system. Procedia Comput. Sci. **108**, 2501–2506 (2017). https://doi.org/10.1016/j.procs.2017.05.137
97. Hamdan, H., Garibaldi, J.M.: Adaptive neuro-fuzzy inference system (ANFIS) in modelling breast cancer survival. In: WCCI 2010 IEEE World Congress on Computational Intelligence, pp. 573–580, CCIB, Barcelona, Spain, 18–23 June 2010. https://doi.org/10.1109/FUZZY.2010.5583997
98. Mahmoudi, S., Lahijan, B.S., Kanan, H.R. (eds.): ANFIS-based wrapper model gene selection for cancer classification on microarray gene expression data. In: 2013 13th Iranian Conference on Fuzzy Systems (IFSC). IEEE (2013). https://doi.org/10.1109/IFSC.2013.6675687
99. Bora, D., Barman, M.P.: Diabetes risk classification comparison of machine learning techniques. Neuroquantology **20**(14), 303–308 (2022). https://doi.org/10.4704/nq.2022.20.14.NQ88042

Decision-Making System Based on Fuzzy Logic for Arterial Diameter Variation Signals Evaluation

Mariela Azul Gonzalez and Leandro Zabala

Abstract A novel theoretical decision support system upon compensatory fuzzy logic is presented. The system's objective is the evaluation of cardiovascular risk through Arterial Diameter Variation signals. This system is designed as a foundational framework for the development of clinical tools that are oriented towards user requirements. A proprietary dataset, derived from pulse wave analysis of 829 individuals exhibiting both normotensive and hypertensive conditions, was employed to extract signal indices, including the Radial Augmentation Index, Diastolic Augmentation Index, and the Propagation Speed of the Arterial Pressure Wave. These indices were subjected to statistical and exploratory analysis to facilitate the development of fuzzy inference rules. Expert-defined membership functions and a rule base were constructed to enable the fuzzy characterization of the indices. Subsequently, a fuzzy inference system was proposed for the evaluation of cardiovascular risk. This decision-making system incorporates dual-implication rules and compensatory logic to address physiological variability and enhance risk assessment. The primary objective of this preliminary work is to augment clinical decision-making by formally integrating expert knowledge into the assessment of arterial health and the prediction of cardiovascular disease risk, including hypertension. This approach facilitates data interpretation under conditions of uncertainty, accounting for inter-individual variability and the inherent imprecision of arterial signal indices. Discernible patterns of arterial health were identified, providing robust tools for disease prevention and enabling clinicians to tailor evaluations to individual patient profiles. The results demonstrate the system's potential for the early detection of arterial dysfunction

M. A. Gonzalez (✉)
Laboratorio de Bioingeniería, Facultad de Ingeniería, Instituto de Investigaciones Científicas y Tecnológicas en Electrónica (ICyTE), Universidad Nacional de Mar del Plata, Mar del Plata, Argentina
e-mail: mazulgonzalez@fi.mdp.edu.ar

L. Zabala
Ingeniero Electrónico, Facultad de Ingeniería, Instituto de Investigaciones Científicas y Tecnológicas en Electrónica (ICyTE), Universidad Nacional de Mar del Plata, Mar del Plata, Argentina
e-mail: leandrozabala@mdp.edu.ar

W. Pedrycz et al. (eds.), *Decision-Making in Computational Intelligence-Based Systems*, Studies in Systems, Decision and Control 628,
https://doi.org/10.1007/978-3-032-04056-5_14

and its applicability in personalized cardiovascular risk monitoring. Validation with clinical experts affirmed the system's interpretability and potential clinical utility.

Keywords Biomedical signals · Cardiological tree assessment · Arterial diameter variation signals · Pulse wave analysis · Computational intelligence · Fuzzy logic · Compensatory fuzzy logic · Fuzzy inference system

1 Introduction

1.1 Hypertension State-of-the-Art Review

Hypertension remains a major global health concern and is one of the leading risk factors for cardiovascular disease (CVD), morbidity, and premature mortality. Accurate and early diagnosis is essential for effective prevention and management strategies, established by the World Health Organization in 2023 [1]. Early and accurate diagnosis plays a critical role in both prevention and effective treatment [2].

Traditional cardiovascular risk calculators incorporate variables such as age, sex, blood pressure, cholesterol, smoking status, and family history [3]. However, these models often overlook physiological variability and may lack generalizability across diverse populations Current blood pressure monitoring techniques, while non-invasive, remain limited in accuracy and often require frequent recalibration, particularly in long-term or ambulatory contexts [4, 5].

Cardiovascular risk assessment is inherently complex and requires advanced analytical tools to enhance precision and reduce implementation biases. These limitations underscore the need for improved diagnostic systems capable of capturing individual variability and physiological complexity [6, 7].

Moreover, conventional non-invasive monitoring techniques, though widely used, have limitations in long-term precision and fail to capture dynamic vascular behavior. Although widely used in clinical practice, such models have limitations, including insufficient incorporation of physiological biomarkers and reduced generalizability across diverse populations.

Artificial Intelligence (AI) has shown promise in biomedical signal analysis, improving diagnostic accuracy for cardiovascular diseases, as demonstrated in the work of [8]. Computational Intelligence (CI) provides new opportunities for biomedical signal processing and clinical decision support systems. Among CI methods, Fuzzy Logic (FL) offers a robust framework for managing uncertainty and physiological variability, enabling more accurate interpretation of complex biological data [9].

1.2 Signals Related to Arterial Diameter Variation as an Indicator of Cardiovascular Risk

Arterial Diameter Variation (ADV) signals, derived from radial pulse wave analysis, provide valuable information on vascular compliance, reflection patterns, and hemodynamic dynamics. These signals enable real-time, non-invasive evaluation of arterial health, making them suitable for early CVD risk detection. Indices such as the Radial Augmentation Index (RAI), Diastolic Augmentation Index (DAI), and Propagation Speed of the Arterial Pressure Wave (AVP) have shown potential in reflecting arterial stiffness and dysfunction [10].

1.3 Advanced Technologies Pertaining to Artificial Intelligence Within the Medical Field

Artificial Intelligence (AI) and computational intelligence (CI) have shown promise in enhancing biomedical signal interpretation. Among CI paradigms, fuzzy logic offers a powerful framework for managing imprecision and uncertainty inherent in biological data. In particular, compensatory fuzzy logic enables more nuanced inference by allowing trade-offs between variables—closely mimicking clinical reasoning processes [9].

1.4 SWOT (Strengths, Weaknesses, Opportunities, Threats)

SWOT (Strengths, Weaknesses, Opportunities, Threats) analysis is a strategic planning tool used to evaluate internal and external factors that affect the success of a system or technology. It is commonly used to enhance focus, guide strategic development, and identify potential risks and opportunities. In the context of AI in healthcare, SWOT analysis can be used to break down current challenges, highlight opportunities for improvement, and identify areas of vulnerability.

The integration of AI in healthcare aligns with the goals of the United Nations Sustainable Development Goals (SDGs) 2030 Agenda [11]. AI enables personalized medicine, cost reductions through process optimization, and more efficient management of preventive care. It facilitates early disease detection and resource optimization. However, AI cannot replace the empathy and interpersonal connection provided by healthcare professionals. Moreover, the healthcare sector lacks clear regulatory frameworks for AI use, raising concerns about data privacy, ethical risks, and the secure use of sensitive patient data.

Initially analyzed by [12], a Fuzzy SWOT (Strengths, Weaknesses, Opportunities, Threats) addressing the inherent uncertainty and vagueness in human assessments. By using fuzzy sets and membership functions, fuzzy logic allows for a more nuanced evaluation of factors, capturing the degree to which a factor is a strength, weakness, opportunity, or threat. This approach can lead to more robust and realistic strategic planning by accounting for ambiguous or uncertain factors. Analysis was proposed, which applies fuzzy theory to convert qualitative and quantitative information into structured evaluations. This method accommodates vagueness and leverages expert knowledge in decision-making. In healthcare, the strengths of AI include aiding physicians in diagnostics and enabling precision medicine by tailoring treatment plans to individual patient profiles. Within this framework, compensatory fuzzy logic emerges as an effective tool to manage signal variability and support clinical decisions by identifying patterns and relationships in complex datasets.

A SWOT analysis of Arterial Diameter Variation (ADV) signal analysis highlights several benefits. ADV facilitates the transition from treatment to prevention by enabling the early detection of vascular dysfunction. It allows real-time, non-invasive monitoring of arterial behavior, offering surrogate markers for arterial stiffness, which is essential in CVD risk assessment. Integrating fuzzy logic improves the interpretability of complex, multivariable data [13, 14], and collaboration with clinicians ensures practical clinical applicability.

The integration of AI into biomedical signal analysis allows for enhanced interpretation of complex and nonlinear physiological data, thereby increasing the sensitivity and specificity of cardiovascular disease detection tools. Among the AI techniques, compensatory fuzzy logic has emerged as particularly suitable for modeling systems with high degrees of uncertainty and variability, as commonly found in biological signals. Fuzzy logic is based on the gradual representation of truth, which allows for modeling uncertainty and variability in biomedical data. In the case of cardiovascular risk assessment, fuzzy predicates are used to assess whether certain indices meet criteria established by experts.

ADV signal analysis represents a promising, non-invasive approach for vascular assessment. Its strengths include clinical relevance, dynamic behavior, and integration with computational intelligence systems [15]. Nevertheless, challenges persist, including signal variability, limited interpretability, and a lack of standardization. Ethical concerns, device heterogeneity, and barriers to clinical integration must also be addressed. Combining ADV data with machine learning enables predictive modeling and personalized risk stratification [16].

Expert oversight is critical to avoid overreliance on automated systems. When developed in collaboration with clinical experts and supported by robust computational tools, ADV signal analysis offers a valuable and evolving method for monitoring arterial health and preventing disease.

This work proposes a flexible approach to represent knowledge extracted from ADV datasets. It combines strict rule-based inference with fuzzy logic to accommodate physiological variability and uncertainty, enhancing support for clinical decision-making.

1.5 System Development Planning

In this study, ADV signals were collected from 829 normotensive and hypertensive volunteers using non-invasive pulse wave analysis. Relevant indices were calculated and explored using statistical and exploratory techniques, such as histograms, correlation maps, and distributional indices. Fuzzy characterization of these indices was performed using previous findings. Finally, the steps to develop a decision support system, refined through expert collaboration is described.

Decision Making System Planification Data Flow

1. ADV Signal Collection: ADV (Arterial Diameter Variation) signals are collected from volunteers using non-invasive pulse wave analysis. These signals are the initial data input.
2. Calculation and Exploration of Indices: Relevant indices are calculated from the ADV signals. These indices are then explored using statistical and exploratory techniques. Methods mentioned include histograms, correlation maps, and distributional indices.
3. Fuzzy Characterization: Fuzzy characterization of these indices is performed using previous findings. This involves applying fuzzy logic to the calculated indices.
4. Decision Support System Refinement: The developed decision support system is refined through expert collaboration. This indicates an iterative process involving input from clinical experts.

2 Materials

The ADV indexes have shown significant potential in the early prediction and management of cardiovascular conditions, as they provide valuable information on vascular compliance and hemodynamic behavior under various physiological and pathological states.

This study presents a flexible fuzzy inference framework specifically designed to analyze Arterial Diameter Variation (ADV) signals, leveraging the capacity of fuzzy systems to model nuanced relationships between variables in cardiovascular monitoring. ADV signals were obtained from a group of healthy normotensive volunteers and hypertensive volunteers.

2.1 ADV Acquisition Process

ADV signals were captured non-invasively using 27 mm piezoelectric transducers positioned over the radial artery, providing real-time insights into arterial dynamics. The sensors were embedded in a wrist-mounted enclosure and electrically isolated

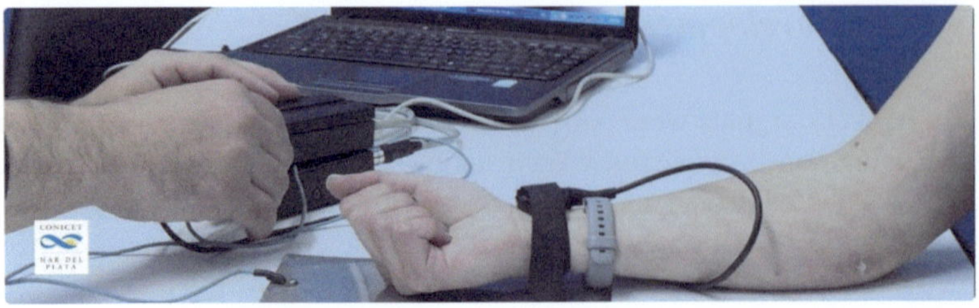

Fig. 1 ADV acquisition process. Own source

from the acquisition circuit. Figure 1 shows the entire acquisition process. The analog front-end includes a differential charge amplifier, followed by a first-order low-pass filter with a cutoff frequency of 34 Hz, resulting in an overall analog bandwidth from 0.16 to 34 Hz. The conditioned signal is then digitized using a 16-bit sigma-delta ADC. Additionally, a second-order IIR notch filter was implemented to suppress 50 Hz power line interference. An automatic gain control (AGC) algorithm ensures optimal signal amplitude across varying subjects and conditions. Statistical analysis of calculated indexes and waveform morphology is discussed in the context of system implementation.

ADV indexes statistical analysis and its interpretation is detailed for system implementation. ADV signal presents a normalized radial pulse wave contour, typically obtained via applanation tonometry or photoplethysmography, and plotted as amplitude (%) over time (ms). This waveform comprises three main features: the systolic peak (P1) resulting from direct ventricular ejection, the reflected systolic wave (P2) caused by pulse wave reflection from peripheral arterial sites, and the diastolic wave, which occurs later due to elastic recoil and wave re-entry phenomena. From this signal, several hemodynamic indices can be derived.

The signal is shown as a percentage of amplitude over time in milliseconds and exhibits three distinct characteristics: The waveform is characterized by three key components: The Systolic Wave, the Systolic Reflection, and the Diastolic Wave. The Systolic Wave corresponds to the initial sharp increase in amplitude due to the ventricular ejection of blood, corresponds to the initial sharp amplitude increase resulting from ventricular ejection, reaching the signal's apex or of the peak signal (P1).This is followed by the Systolic Reflection, a secondary peak generated by the reflection of the pulse wave at peripheral sites, which provides information about arterial stiffness and wave reflection timing (P2). The Diastolic Wave appears as a smaller peak during the declining phase of the signal, representing the diastolic component of the cardiac cycle [10].

Figure 2 shows ADV indexes illustrate a typical radial pulse wave signal acquired non-invasively, represented as a percentage of amplitude over time in milliseconds.

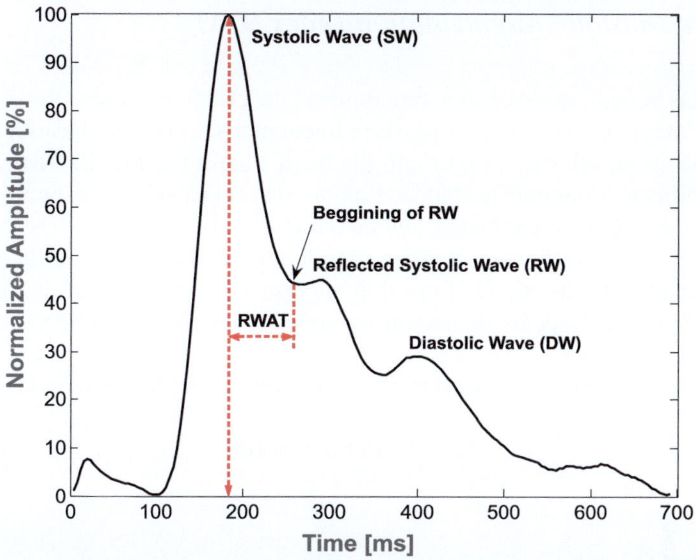

Fig. 2 Arterial diameter variation (ADV) signal utilized for index calculation (own source)

2.2 *The Radial Augmentation Index (RAI)*

One of the primary indices extracted from this waveform is the Radial Augmentation Index (RAI), calculated as the ratio of the amplitude of the reflected wave to the systolic peak. This index is a marker of arterial stiffness and is influenced by age, blood pressure, and vascular health. The temporal location and relative amplitudes of these components are crucial for the physiological and pathological interpretation of cardiovascular function.

The Radial Augmentation Index (RAI) is calculated as:

$$RAI(\%) = \frac{P1}{P2} * 100$$

where P1 is the amplitude of the first (systolic) peak and P2 is the amplitude of the reflected wave. IAR serves as an indirect marker of arterial stiffness and wave reflection intensity—higher values are associated with increased vascular resistance and aging. Measures the amplitude of fluctuations in blood pressure over a cardiac cycle. It is primarily used to assess the dynamic regulation of peripheral vascular resistance and compliance. A high RAI suggests a well-regulated cardiovascular system with preserved vascular compliance and effective sympathetic-parasympathetic modulation. A low RAI may reflect impaired vascular tone regulation or autonomic dysfunction, both of which are associated with increased cardiovascular risk. Changes in RAI may also correlate with aging, arterial stiffening, and the progression of chronic hypertension; low RAI signals dysfunction or instability in blood pressure.

2.3 The Diastolic Augmentation Index (DAI)

DAI reflects both diastolic cardiac function and the elastic properties of the vascular system. It measures the extent of pressure fluctuations during the diastolic phase of the cardiac cycle, offering insight into the heart's ability to fill effectively and the vessel's ability to accommodate blood volume. Alterations in DAI can be attributed to vascular tone, arterial compliance, and autonomic control, evaluate diastolic reflection phenomena's contribution to wave morphology, and the heart's efficient filling during diastole. Monitoring DAI can thus support early detection of diastolic heart failure and guide therapeutic interventions aimed at improving vascular and cardiac performance.

DAI refers to the ratio of the diastolic wave amplitude (Pd) to the systolic peak:

$$DAI = (PD/P1) * 100$$

2.4 Propagation Speed of the Arterial Pressure Wave (AVP)

Propagation Speed of the Arterial Pressure Wave (AVP) reflects the magnitude of oscillations in blood pressure, and captures the rate at which pulse waves propagate through the arterial tree. High AVP suggests cardiovascular instability and increased risk of hypertension. High AVP values are also indicative of increased pulse wave velocity (PWV), often a surrogate for arterial stiffness, which is a hallmark of vascular aging and early-stage hypertension [10]. An increased AVP thus suggests cardiovascular instability, reduced vascular elasticity, and heightened afterload, which may precede the onset of systolic hypertension or signal target organ damage.

Together, these indices provide a comprehensive view of arterial function and wave dynamics, and are essential for assessing cardiovascular risk, especially in normotensive and hypertensive populations. Thus, elevated AVP and reduced RAI or DAI may be employed in risk stratification and therapeutic monitoring in hypertensive patients. Collectively, these indices provide a comprehensive assessment of arterial function and wave dynamics, and are essential for evaluating cardiovascular risk, particularly in normotensive and hypertensive populations.

RAI, AVP, and DAI are linked to arterial stiffness, a key factor in hypertension. Arterial stiffness increases cardiovascular risk and is affected by prolonged hypertension, leading to vascular damage and changes in these indices. DAI assesses diastolic cardiac function and vascular reactivity, reflecting vascular tone and compliance. These indices comprehensively evaluate arterial function and wave dynamics, essential for assessing cardiovascular risk in both normotensive and hypertensive individuals. Elevated AVP with reduced RAI or DAI can be used for risk stratification and therapeutic monitoring in hypertension.

3 Methods

In this work, a Fuzzy Inference System (FIS) is proposed to assess Cardiovascular Risk. From ADV signals, indices such as the Radial Augmentation Index (RAI), Aortic Velocity Parameter (AVP), Diastolic Augmentation Index (DAI), heart rate (HR), systolic pressure (SP), diastolic pressures (DP), hypertension (HT), medicated (MED) were extracted. These variables serve as the basis for fuzzy set generation and rule construction, integrating clinical expertise into the model. The collaboration with cardiologists ensured that fuzzy rule design and inference processes reflected relevant clinical decision-making pathways. The inference mechanism was fine-tuned through iterative feedback to optimize diagnostic accuracy and generalizability.

Fuzzy logic application to ADV signal analysis represents a promising strategy for addressing inter-patient variability, improving diagnostic robustness, and assisting healthcare professionals in making informed decisions. The integration of artificial intelligence into the evaluation of biomedical signals improves the accuracy of cardiovascular disease detection. In this context, compensatory fuzzy logic is presented as an effective tool for managing variability in physiological signals and improving clinical decision-making process.

3.1 Computational Intelligence

The discovery of associations, relationships, and latent patterns in biomedical datasets is significantly enabled by computational intelligence (CI) techniques. These methods offer robust capabilities for modeling the complexity, uncertainty, and heterogeneity often found in physiological signals. Among the most prominent CI paradigms, fuzzy logic has emerged as a powerful framework for handling imprecision and ambiguity inherent in biomedical data. CI encompasses adaptive, interactive, and data-driven tools such as neural networks, support vector machines, and deep learning architectures, which are effective in the detection of nonlinear patterns in high-dimensional datasets. These are often integrated with fuzzy logic systems to enhance interpretability, an essential factor in clinical contexts where transparency in decision making is critical. Fuzzy logic provides semantic rule-based reasoning and incorporates expert knowledge via membership functions and fuzzy rules, improving the interpretability of complex models. This model is particularly applicable in clinical environments, where risk assessments frequently involve weighing multiple indicators.

Computational intelligence encompasses adaptive, informative, and interactive computational tools essential for processing and interpreting large biomedical datasets. As noted by [9], techniques such as fuzzy logic enable systems to handle uncertainty and imprecision, making them particularly useful in biomedical applications [10]. Similarly, neural networks, as demonstrated in the work of [17, 18], have been fundamental in identifying complex patterns within vast datasets.

3.2 Statistical Analysis

Exploratory and descriptive statistical analyses were performed to characterize the distribution, central tendency, and variability of physiological parameters derived from ADV signals. Based on these analyses, fuzzy predicates and inference rules are constructed to model the nonlinear relationships between physiological indices.

The dataset consisted of 829 individuals, including balanced cases of normotensive, hypertensive, women and men, spanning a wide age range from adolescence to older adulthood. Key variables analyzed included systolic pressure (SP), diastolic pressure (DP), heart rate in beats per minute (BPM), radial augmentation index (RAI), arterial velocity propagation index (AVP), and diastolic augmentation index (DAI), along with metadata such as age, sex, medication indication (MED) and hypertension status (HT).

Descriptive statistics—mean, standard deviation (SD), median, interquartile range (IQR), minimum, and maximum—were computed for each variable. Based on distributional properties, group comparisons between normotensive and hypertensive individuals were conducted.

To evaluate the linear relationships between continuous variables, Pearson's correlation coefficients were calculated. A correlation matrix is used to visualize inter dependencies among key indices and to assist in identifying redundant variables for potential dimensionality reduction prior to fuzzy rule-based modeling. Heatmaps organize correlation results data in a grid, with different colors or shades indicating different levels of correlation's magnitude. Google Colaboratory heatmaps uses Sea born Library, by default calculate Pearson Correlation Coefficient for measuring the linear relationship between two continuous variables. It assumes the normal distribution of the involved variables and is sensitive to outliers. It measures the strength of the linear relationship between two variables. If there is a strong linear relationship, the correlation coefficient is close to 1 or -1, and 0 means no linear relationship. In the case of univariate feature selection, these variables are the feature. A heat map is a way to represent data points in a data set in a visual manner. All heat maps use different colors or different shades of the same color to represent different values and to communicate the relationships that may exist between the variables plotted on the x-axis and y-axis. Usually, a darker color or shade represents a higher or greater quantity of the value being represented in the heat map. The visual nature of heat maps allows for immediate recognition of patterns, such as clusters, trends, and anomalies. This makes heatmaps an effective tool for exploratory data analysis. The changing colors across the two axes will reveal patterns in one or both variables and whether there are relationships between them [19, 20].

3.3 Fuzzy Logic and Fuzzy Inference Systems (FIS)

Computational intelligence integrates adaptive, interactive, and computational tools that are well-suited for analyzing large biomedical datasets. The discovery of associations, relationships, and latent patterns in biomedical datasets is enabled by computational intelligence (CI) techniques.

Zadeh highlighted the utility of fuzzy systems in managing imprecision, making them ideal for medical domains where signal ambiguity is common [13, 14]. Similarly, artificial neural networks (ANN) architectures, as demonstrated by [17], offer capabilities for pattern recognition in high-dimensional data. ANN facilitates the extraction of meaningful correlations from ADV signals and supports the identification of cardiovascular anomalies [10]. System rules are not only tested unidirectional (e.g., IF RAI is high THEN risk is elevated) but also exhibit bidirectional implications (e.g., IF risk is low THEN RAI is likely moderate or low), which enables robust reasoning and scenario testing. This bidirectional reasoning contributes to improved clinical interpretability and supports dynamic patient monitoring.

These rules are used to construct fuzzy sets and rules representing expert clinical knowledge. Fuzzy rules are designed and weighted in collaboration with cardiologists, and the inference engine is optimized using iterative feedback.

This work focuses on analyzing indices derived from pulse waveforms in the radial artery to identify patterns useful for diagnosing and monitoring arterial tree diseases. By combining conventional statistical analysis with fuzzy logic, this study offers a comprehensive perspective for exploring biomedical signals. Fuzzy logic handles imprecision and variability in signal data, which is essential for evaluating biomedical parameters and adapting to diverse clinical applications. Among these, fuzzy logic provides a structured method for representing imprecise or uncertain knowledge.

This study proposes a flexible framework that uses fuzzy logic to model the complexity and variability inherent in Arterial Diameter Variation (ADV) signals. The principal objective is to implement a FIS to evaluate the state of the arterial tree. The FIS predicts future risks of hypertension and related cardiovascular diseases. The use of computational intelligence tools in collaboration with medical specialists increases the data analysis usability and improves decision-making process. Fuzzy Inference Systems (FIS) are employed in various domains, including process control, recommendation systems, pattern recognition, and decision-making in complex environments.

Unlike classical logic systems, which rely on binary and well-defined values, fuzzy systems utilize fuzzy logic—rooted in the theory of fuzzy sets and membership functions—to model the reasoning process. The development of an FIS begins with the fuzzification of input and output variables, which are represented as fuzzy sets rather than crisp values. Fuzzy rules are then formulated; a fuzzy rule p is a linguistic proposition (i.e., an “if–then” statement) associated with a truth degree $\mu(p) \in [0, 1]$. This adheres to the principle of gradualism, wherein a proposition can simultaneously possess varying degrees of truth and falsehood. A simple fuzzy predicate sp is defined as one whose truth value μ(sp) can be obtained by applying a membership function to a quantitative variable. For example, the predicate "intensity is high" associates the linguistic term "high" with a fuzzy set defined over the variable "intensity," which may be measured in meters.

The rule base, typically composed of statements in the form "if A then B," establishes the logical relationships between input and output variables. Both A and B are fuzzy sets, and the system uses these rules to infer outputs from observed inputs. To enhance reliability and mutual consistency, biconditional logic ($A \leftrightarrow B$)—or double implication—can be incorporated into rule structures. A biconditional expression stipulates that both A and B must be simultaneously true or false, reinforcing mutual dependency and ensuring consistent system behavior across diverse conditions. This becomes particularly relevant in contexts where antecedent and consequent bear equal significance, as is often the case in biomedical signal interpretation. Here, biconditional rules were use to capture bidirectional dependencies between signal variations and associated physiological effects, thereby improving interpret ability and robustness.

Finally, the defuzzification process produces a crisp output value, which is utilized for decision-making or action execution. Model validation frequently employs biconditional rules to ensure mutual dependability between antecedents and consequent, which is essential in accurately modeling complex, nonlinear relationships in biomedical data.

3.4 FIS Resume Operates Through Four Key Stages

Fuzzification: Transformation of crisp input values into fuzzy sets using predefined membership functions.
Rule Evaluation: Application of fuzzy rules (if–then statements) to infer fuzzy outputs.
Aggregation: Integration of the outputs from all activated rules.
Defuzzification: Conversion of the aggregated fuzzy output into a single crisp value for decision-making.

The present study proposes the design of an FIS that formalizes reasoning based on natural language structures, using disjunctions of simple conjunctions (or vice versa). Knowledge is extracted from indices derived from pulse wave variation signals. This approach is supported by the representational completeness of fuzzy logic: all logical models can be expressed in fuzzy normal forms. Consequently, this provides a broad and flexible framework for knowledge discovery and decision making.

3.5 Compensatory Logic: Definition and Application

In fuzzy logic systems, compensatory logic facilitates integration of diverse input parameters with varying degrees of uncertainty and relevance. This enhances model robustness and adaptability, and mirrors human cognitive processes in complex reasoning tasks. Zadeh [9, 13, 14], emphasized this capability as foundational to the development of intelligent systems that can emulate expert-level decision-making under uncertainty. Compensatory logic is a decision-making model in which the disadvantages of one alternative can be offset by advantages in other criteria or attributes. It is based on the idea that decision-makers weigh multiple factors and assign weights to each, allowing a low score in one aspect to be balanced by a high score in another.

This approach is common in contexts where multiple quantifiable variables are evaluated, such as in clinical decisions, treatment selection, or risk models. For example, in the medical setting, a patient might have a high-risk factor, such as hypertension, but if they have good kidney function, are nonsmokers, and have low cholesterol, their overall cardiovascular risk may be considered acceptable. This is because positive variables "compensate" for negative ones within a weighted model, as occurs in tools like the Framingham Risk Score. Furthermore, in artificial intelligence and inference systems, such as those based on fuzzy logic, compensatory logic is used to integrate multiple inputs with varying degrees of certainty and relevance, enabling flexible and more human-like reasoning [9].

4 Preliminary Results

These exploratory analyses were essential to identify signal patterns that differentiate normotensive and hypertensive groups. Statistical descriptors such, radial augmentation index (RAI), aortic velocity parameter (AVP), and diastolic augmentation index (DAI) were computed, revealing significant differences between groups across all ages.

The descriptive statistics of the calculated indices are summarized in Table 1.

4.1 Exploratory and Descriptive Analysis

Exploratory and descriptive statistical analyses were performed to characterize the distribution, central tendency, and variability of physiological parameters derived from Arterial Diameter Variation (ADV) signals. The dataset consisted of 829 individuals, including both normotensive and hypertensive participants, spanning a wide age range from adolescence to older adulthood and similar number of women and men. All cases were ordered by Age to analyze CVR variation.

Table 1 Descriptive statistics for all indexes

	Age	SP	DP	RAI	AVP	DAI
Mean	43.97	126.4	80.97	69.61	79.78	11.81
Std.	15.08	16.86	11.51	19	2.39	5.7
Min	13	80	95	13.5	3.9	0
Max	74	200	110	116	17.8	31.7
25% percentile	50	115	75	65	56	6
50% percentile	25	120	80	78	78	7.5
75% percentile	28	125	90	84	9.2	15.4

Key variables analyzed included float numbers of systolic pressure (SP), diastolic pressure (DP), heart rate in beats per minute (BPM), radial augmentation index (RAI), Propagation Speed of the Arterial Pressure Wave (AVP), and diastolic augmentation index (DAI), along with metadata such as age, sex, med and hypertension status (HT) in a binary bases categorization. The goal of the exploratory analysis was to detect meaningful patterns and group differences that may serve as indicators of pathological alterations in arterial function. Table 1 summarizes the descriptive statistics for these variables.

Descriptive statistics—mean, standard deviation (SD), median, interquartile range (IQR), minimum, and maximum—were computed for each variable. To evaluate the linear relationships between continuous variables, Pearson's correlation coefficients were calculated. A correlation matrix (Fig. 3) was constructed to visualize interdependencies among key indices and to assist in identifying redundant variables for potential dimensionality reduction prior to fuzzy rule-based modeling.

Statistically significant differences ($p < 0.01$) were found between normotensive and hypertensive groups across several variables, including SP, DP, RAI, and AVP, indicating their potential as robust discriminative features for cardiovascular risk classification. Conversely, variables such as BPM and DAI exhibited substantial inter individual variability and weaker group separation.

These metrics revealed statistically significant differences between normotensive and hypertensive subjects across multiple indices, supporting their potential use as features in fuzzy rule generation and risk evaluation models. For instance, elevated values of RAI and AVP were consistently associated with hypertensive status, while heart rate and DAI showed high inter individual variability.

Figure 3 illustrates the correlation matrix of these parameters, indicating high linear correlations between systolic pressure, RAI, and AVP—key features used in the design of fuzzy inference rules. The correlation matrix also facilitates visualization of latent structure in the dataset and aids in reducing dimensionality for model development.

These findings form the foundation for constructing fuzzy sets that represent the cardiovascular states of patients. Through these analyses, variables most associated with hypertensive patterns were identified and then used to define fuzzy membership

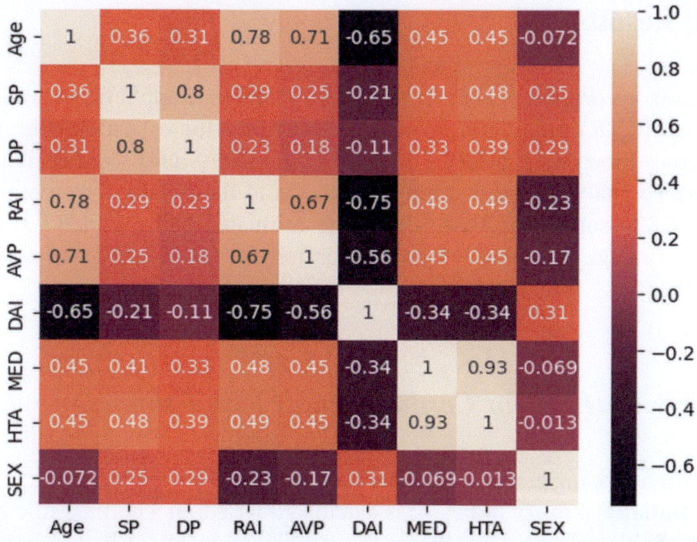

	Age	SP	DP	RAI	AVP	DAI	MED	HTA	SEX
Age	1	0.36	0.31	0.78	0.71	-0.65	0.45	0.45	-0.072
SP	0.36	1	0.8	0.29	0.25	-0.21	0.41	0.48	0.25
DP	0.31	0.8	1	0.23	0.18	-0.11	0.33	0.39	0.29
RAI	0.78	0.29	0.23	1	0.67	-0.75	0.48	0.49	-0.23
AVP	0.71	0.25	0.18	0.67	1	-0.56	0.45	0.45	-0.17
DAI	-0.65	-0.21	-0.11	-0.75	-0.56	1	-0.34	-0.34	0.31
MED	0.45	0.41	0.33	0.48	0.45	-0.34	1	0.93	-0.069
HTA	0.45	0.48	0.39	0.49	0.45	-0.34	0.93	1	-0.013
SEX	-0.072	0.25	0.29	-0.23	-0.17	0.31	-0.069	-0.013	1

Fig. 3 ADV correlation index for all signals

functions, aiding the system in distinguishing between different risk levels in the arterial tree.

These exploratory analyses were essential to identify signal patterns that were computed, revealing significant differences between groups across all ages (see Table 1).

4.2 Fuzzy Rule Extraction and Double Implications

Based on the correlations and descriptive patterns found, fuzzy predicates and inference rules were developed to capture the nonlinear relationships among variables. The preliminary following fuzzy rules were used to evaluate bidirectional reasoning:

Proposed Rules

- If **Age** is Adult **OR** Old Adult, **then RAI is High, AVP is High, and DAI is Low**.
- If **RAI is High, AVP is High, and DAI is Low**, **then Age** is Adult **OR** Old Adult
- If **RAI and AVP are High and DAI is Low**, **then** Cardiovascular Risk (CVR) is High.
- If Cardiovascular Risk (CVR) is High, then **RAI and AVP are High and DAI is Low**.

5 Final Results

In this section, we present the outcomes of fuzzy inference applied to age and cardiovascular risk (CVR) prediction using classical fuzzy logic (Zadeh operators) and Compensatory Fuzzy Logic (CFL) operators. The fuzzy rules were designed based on exploratory analysis and domain expertise, and the results were computed for a dataset of 829 subjects. These rules were evaluated using Zadeh's t-norm and Compensatory Fuzzy Logic and s-norm operators for conjunction and aggregation, respectively.

5.1 Fuzzy Rule Set for Cardiovascular Risk Estimation

A method for assessing cardiovascular risk (CVR) involves a multi-stage fuzzy logic approach. Initially, a fuzzy indicator is calculated based on a defined rule base. The evaluation of this rule base results in the classification of CVR into one of three distinct levels: Low, indicating minimal risk; Moderate, suggesting an intermediate level of concern; or High, signifying a substantial likelihood of cardiovascular events. Following this initial classification, a more refined analysis is performed. Using the preliminary CVR level, an aggregation process is initiated within a fuzzy inference system. Finally, through the application of a fuzzy decision system, all the aggregated data is consolidated into a single, definitive risk value. Defuzzification was performed using the centroid method, which computes the center of gravity of the resulting fuzzy output.

Final Base Rule Implementation (Zadeh and Compensatory)

Forward Rules

- **Rule 1**: If RAI is high, AVP is high, and DAI is low, then CVR is *High*.
- **Rule 2**: If RAI is moderate and DAI is low, then CVR is *Moderate*.
- **Rule 3**: If RAI is low, AVP is low, and DAI is high, then CVR is *Low*.

Inverse Rules (Double Implications)

- **Rule 4**: If CVR is *High*, then RAI is high, AVP is high, and DAI is low.
- **Rule 5**: If CVR is *Moderate*, then RAI is moderate and DAI is low.
- **Rule 6**: If CVR is *Low*, then RAI is low, AVP is low, and DAI is high.

Zadeh and Compensatory Logic (LDC) systems across all 829 cases using various rule configurations were tested. These rules were evaluated using Zadeh's t-norm and Compensatory Fuzzy Logic and s-norm operators for conjunction and aggregation, respectively. Fuzzy Rule Set for Cardiovascular Risk Estimation. This final value provides an assessment of the individual's overall cardiovascular risk. The use of fuzzy logic allows for the handling of uncertainties and ambiguities inherent in medical data, leading to a more accurate and clinically relevant CVR evaluation. The output grade of truth was used to quantify rule activation per subject. Compensatory base rule, we only rules were only tested forward due to its inherent definition. Final CVR risk final evaluations were determined in all cases, Fig. 4 shows all values ordered by age.

Table 2 presents statistical descriptors of the fuzzy cardiovascular risk (CVR) outputs.

Eight representative cases were selected from the cohort of 829 individuals for qualitative system testing and expert validation (see Table 3). For each individual, CVR predictions were obtained using both Zadeh and Compensatory operators. Predictions were consistent with known physiological risk profiles. Compensatory fuzzy logic, which blends conjunctive and disjunctive logic, allowing smoother transitions and partial compensation between indicators.

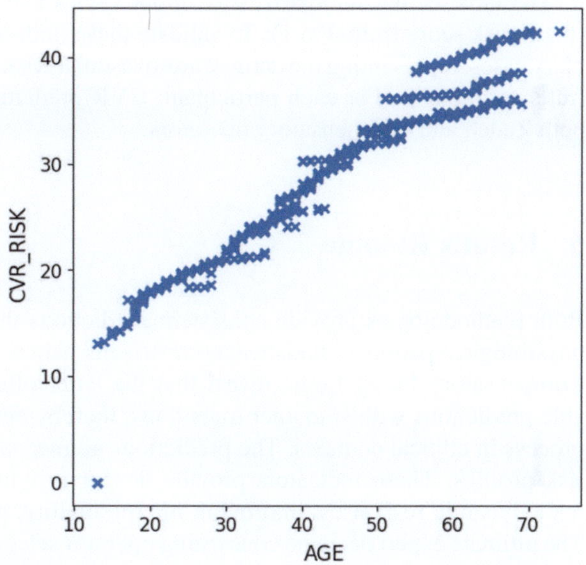

Fig. 4 CVR values for all analyzed cases

Table 2 Clinical insight from initial model testing

829 Cases	Zadeh rule	Zadeh inverse rule	LCD rule	Zadeh/CVR	LDC/CVR
Mean	0.497	0.412	0.317	0.370	0.465
Std.	0.293	0.361	0.373	0.41	0.220
Min	0	0	0	0	0
Max	0.25	0	0	1	0.94
25% percentile	0.484	0.24	0	0.462	0.548
50% percentile	0.733	0.72	0.674	0.528	0.606
75% percentile	1	0.860	1	0.58	0.706

Table 3 Clinical insight from representative selected cases

Age	SP	DP	RAI	AVP	DAI	Zadeh CVR	CFL CVR
20	90	60	50	8	18.5	0	0.25
21	120	75	52.8	5.5	24	0	0.35
44	125	95	73	6.87	9.8	0.2	0.47
46	145	98	74	7	11.1	0.34	0.49
55	120	90	90	13.17	4.5	0.6	0.71
67	130	90	85.9	11.02	6.8	0.6	0.64
70	138	88	83	7.3	5.6	0.66	0.71
74	113	59	92.5	17.8	0.6	0.7	0.76

The table below summarizes the input values and corresponding CVR outputs (fuzzy risk score from 0 to 1): To validate these findings, a selection of eight exemplary cases, representing the entire Cardiovascular Risk (CVR) spectrum, was chosen (refer to Table 3). For each participant, CVR predictions were generated utilizing both Zadeh and Compensatory operators.

6 Results Resume

Both methodologies provide consistent predictions that correspond with expected physiological patterns. Statistical comparisons between Zadeh's classical logic and Compensatory Fuzzy Logic reveal that the latter offers more consistent and reliable predictions with smoother transitions, thereby enhancing the decision-making process in clinical contexts. The predictions aligned with established physiological risk profiles. These indicators provide deeper insight into vascular behavior and hemodynamic responses, improving interpretability, precision, and clinical utility. The ultimate expert decision concerning operator selection, as well as cardiovascular risk determination, was based upon Compensatory Fuzzy Logic. Figure 4 represents the final cardiovascular risk indicator using fuzzy compensatory logic.

7 Discussion

The results confirm that Compensatory Fuzzy Logic offers some advantages over traditional Zadeh logic in clinical applications where decision-making involves complex trade-offs. Its ability to model partial compensation and gradual transitions makes it more aligned with expert reasoning and better suited for risk stratification in cardiovascular health assessments.

The inclusion of vascular indices further improves the sensitivity and utility of the system, generating more informative risk profiles. While both methodologies show consistent rule evaluation.

Compensatory Logic offers smoother transitions and more reliable predictions, enhancing clinical decision support. CFL-based models produced smoother, more nuanced output distributions, offering greater differentiation between mid-range CVR levels. Statistical evaluation across 829 individuals demonstrated that the CFL-based system consistently produced more refined output distributions, with lower variance and improved differentiation between intermediate risk states—an essential characteristic in clinical risk stratification.

Zadeh-based models tended to yield more discrete outputs, often near extreme values (0 or 1), with less granularity in intermediate risk ranges. Zadeh operator contribution for system development was to verify double implications, that was not possible using compensatory fuzzy logic. The final CVR model based on compensatory logic showed a balanced mean (0.465) and a coherent percentile spread, indicating stable and clinically useful output. Rule configurations demonstrated lower variance and more symmetrical distributions, enhancing robustness and interpretability.

Our study confirms that the application of CFL in the analysis of arterial diameter variation (ADV) enhances the accuracy and clinical relevance of cardiovascular risk (CVR) predictions by enabling more flexible and nuanced decision-making processes.

Unlike classical fuzzy inference systems (FIS) that rely strictly on conjunctive or disjunctive logic, compensatory logic incorporates a blending of both, which allows partial trade-offs between variables. This property generates smoother, more gradual transitions between truth values and better simulates the reasoning patterns of expert clinicians.

The integration of Artificial Intelligence (AI) into biomedical signal analysis offers significant potential for improving the interpretation of complex and nonlinear physiological data. In particular, compensatory fuzzy logic (CFL) emerges as a powerful methodology for modeling systems characterized by uncertainty and variability—key attributes of biological signals.

From a broader perspective, the adoption of intelligent systems in healthcare aligns with the trend toward precision medicine and data-driven diagnostics. As highlighted by [20], simplifying computational models without compromising diagnostic performance requires close collaboration with domain experts. This interdisciplinary cooperation ensures that the models are not only technically sound but also clinically relevant. Moving forward, optimizing membership functions through evolutionary algorithms such as genetic programming could further enhance the adaptability and precision of the system.

8 Conclusion

This work presents a novel decision support system (DSS) that integrates ADV signal indices into a fuzzy inference system based on compensatory logic. The model was constructed and validated in collaboration with clinical experts and incorporates fuzzy rules using double implications to enhance interpretability. By combining statistical analysis with fuzzy modeling, this system offers a reliable and adaptable tool for cardiovascular risk stratification, especially in normotensive and hypertensive populations.

We demonstrate the successful application of compensatory fuzzy logic, combined with exploratory and statistical analysis, to the evaluation of arterial diameter variation signals in the context of cardiovascular risk prediction. The developed Fuzzy Inference System (FIS) provides a clinically interpretable, adaptive, and accurate tool for assessing arterial health, particularly in asymptomatic stages of cardiovascular disease.

The use of ADV signals—obtained through non-invasive pulse wave analysis—proves to be an effective technique for assessing arterial morphology, aging, stiffness, and treatment efficacy. Moreover, the proposed FIS framework enhances decision-making capabilities, facilitating personalized cardiovascular care and early intervention strategies.

Future research is oriented on optimization of fuzzy rule bases using reinforcement learning and evolutionary computation. Ongoing collaboration among physicians, engineers, and data scientists is crucial for ensuring translational impact and clinical adoption of this promising methodology. Further projects will focus on developing a self-user interface and on real-time implementation of this system in clinical settings, integration with wearable monitoring devices.

AI Use Statement Some portions of this manuscript were enhanced with the assistance of generative AI tools. The authors carefully reviewed, edited and approved all AI-generated content. The authors take full responsibility for the accuracy and integrity of the entire chapter.

References

1. World Health Organization: Priorities for research on hypertension care delivery: web annex: results of post-webinar surveys (2024). ISBN: 978-92-4-009061-3. https://www.who.int/publications/i/item/9789240090613
2. Sánchez, R., Coca, A., Molina de Salazar, D.I., Alcocer, L., Aristizabal, D., Barbosa, E., Brandao, A.A., Díaz-Velazco, M.E., Hernández-Hernández, R., López-Rivera, J., López-Jaramillo, J., Ortellado, L., Parra-Carrillo, E., Parati, G., Peñaherrera, W.K., Ramirez, J., Valdez, O., Wyss, F., Heagerty, A., Mancia, G.: Latin American Society of Hypertension guidelines on the management of arterial hypertension and related comorbidities in Latin America. J. Hypertens. **42**(000), 1–3899 (2024). https://doi.org/10.1097/HJH.0000000000003899

3. Yadlowsky, S., Hayward, R.A., Sussman, J.B., McClelland, R.L., Min, Y.I., Basu, S.: Clinical implications of revised pooled cohort equations for estimating atherosclerotic cardiovascular disease risk. Ann. Intern. Med. **169**(1), 20–29 (2018). https://doi.org/10.7326/M17-3011
4. Laurent, S., Cockcroft, J., Van Bortel, L., Boutouyrie, P., Giannattasio, C., Hayoz, D., Pannier, B., Vlachopoulos, C., Wilkinson, I.B., Struijker-Boudier, H.: Expert consensus document on arterial stiffness: methodological issues and clinical applications. Eur. Heart J. **27**(21), 2588–2605 (2006). https://doi.org/10.1093/eurheartj/ehl254
5. Van Bortel, L.M., Laurent, S., Boutouyrie, P., Chowienczyk, P., Cruickshank, J.K., De Backer, T., Filipovsky, J., Huybrechts, S., Mattace-Raso, F., et al.: Expert consensus document on the measurement of aortic stiffness in daily practice using carotid–femoral pulse wave velocity. J. Hypertens. **30**(3), 445–448 (2012). https://doi.org/10.1097/HJH.0b013e32834fa8b0
6. Noh, S.A., Kim, H.S., Kang, S.H., et al.: History and evolution of blood pressure measurement. Clin. Hypertens. **30**, 9 (2024). https://doi.org/10.1186/s40885-024-00268-7
7. Kim, H.L.: Arterial stiffness and hypertension. Clin. Hypertens. **29**(1), 31 (2023). https://doi.org/10.1186/s40885-023-00251-9
8. Aminorroaya, A., Biswas, D., Pedroso, A.F., Khera, R.: Harnessing artificial intelligence for innovation in interventional cardiovascular care. J. Soc. Cardiovasc. Angiography Interv. **4**(3, Part B), 102562 (2025). https://doi.org/10.1016/j.jscai.2025.102562
9. Zadeh, L.: Fuzzy Sets, Fuzzy Logic and Fuzzy Systems. World Scientific, Singapore (1996). https://doi.org/10.1142/2895
10. Wang, X., Zhang, L., Li, H., Liu, S.: Simultaneous measurement of local pulse wave velocities in radial arteries using a soft sensor based on the fiber Bragg grating technique. Micromachines **15**(4), 507 (2024). https://doi.org/10.3390/mi15040507
11. United Nations: Sustainable development goals (SDGs) 2030, United Nations (n.d.). https://sdgs.un.org/goals
12. Ghazinoory, S., Abolhasani, M., Sadeghi, N.: A fuzzy SWOT analysis of the application of fuzzy logic to the decision-making process in strategic planning. Int. J. Manag. Sci. Eng. Manag. **2**(3), 183–194 (2007)
13. Zadeh, L.A.: The concept of a linguistic variable and its application to approximate reasoning—Part I. Inf. Sci. **8**(3), 199–249 (1975). https://doi.org/10.1016/0020-0255(75)90036-5
14. Zadeh, L.A.: Outline of a new approach to the analysis of complex systems and decision processes. IEEE Trans. Syst. Man Cybern. **3**(1), 28–44 (1973). https://doi.org/10.1109/TSMC.1973.5408575
15. Segers, P., Rietzschel, E.R., Chirinos, J.A.: Wave reflection: clinical applications and recent advances. Hypertension **76**(4), 1056–1064 (2020). https://doi.org/10.1161/HYPERTENSIONAHA.120.14868
16. Alber, M., Buganza Tepole, A., Cannon, W.R., De, S., Dura-Bernal, S., Garikipati, K., Karniadakis, G., Lytton, W.W., Perdikaris, P., Petzold, L., Kuhl, E.: Integrating machine learning and multiscale modeling—perspectives, challenges, and opportunities in the biological, biomedical, and behavioral sciences. NPJ Digit. Med. **2**(1), 115 (2019). https://doi.org/10.1038/s41746-019-0193-y
17. LeCun, Y., Bengio, Y., Hinton, G.: Aprendizaje profundo. Nature **521**, 436–444 (2015). https://doi.org/10.1038/nature1453
18. Castillo, O., Melin, P., Kacprzyk, J.: Hybrid Intelligent Systems: Analysis and Design. Springer (2008). https://doi.org/10.1007/978-3-540-77465-0; Shiraz, H.E., Alias, N.: Using fuzzy logic to enhance the classification and diagnosis of hypertension. J. Theor. Appl. Inf. Technol. **97**(20) (2019). https://www.jatit.org/volumes/Vol97No20/16Vol97No20.pdf
19. Bishop, C.M.: Pattern Recognition and Machine Learning. Springer (2006)
20. Russell, S.J., Norvig, P.: Artificial Intelligence: A Modern Approach, 4th edn. Pearson (2021)

Exploring Diabetes Indicators Utilizing Advanced Machine Learning Algorithms for Improved Diagnosis

S. S. Thakur, Soma Bandyopadhyay, Mahika Thakur, Aditya Singh, Aditya Kumar Singh, Shyam Sunder Singh, and Srajan Mishra

Abstract Diabetes, or Diabetes Mellitus, stands as a prevalent chronic ailment with a global reach, affecting numerous individuals across the world. It involves elevated levels of blood sugar, posing potential serious consequences for various organs in the human body. Persistently high blood sugar levels in individuals with diabetes can result in metabolic imbalances, paving the way for a cascade of complications like coronary disorder, neuropathy, retinopathy. Detecting diabetes in its early stages is crucial, as prompt intervention can avert the onset of severe complications. Many researchers in the past have delved deeply into the realm of utilizing machine learning algorithms for predicting diabetes. In this work various machine learning methods, including Logistic Regression (LR), Support Vector Classifier (SVC), and Decision Tree (DT) and Random Forest (RF), Adaptive Boosting (AB) and Gradient Boosting (GB) have been applied to identify diabetic patients using publicly available Kaggle dataset. AdaBoost achieved an impressive overall accuracy of 97.22%, while Gradient Boosting demonstrated a slightly superior performance with an overall accuracy of 97.24%.

Keywords Diabetes prediction · Logistic regression · Support vector machine · Decision tree · Random forest · Adaptive boosting · And Gradient Boosting

1 Introduction

The worldwide prevalence of diabetes is escalating at an unprecedented pace, emerging as a significant health challenge and burden in both developed and developing nations. Diabetes is a long-term condition that arises when the pancreas does not produce enough insulin or when the body is unable to use the insulin efficiently. Insulin is a hormone responsible for controlling the levels of glucose in the blood

S. S. Thakur (✉) · S. Bandyopadhyay · M. Thakur · A. Singh · A. K. Singh · S. S. Singh · S. Mishra
MCKV Institute of Engineering, Liluah, Howrah, West Bengal, India
e-mail: subroto_thakur@yahoo.com

W. Pedrycz et al. (eds.), *Decision-Making in Computational Intelligence-Based Systems*, Studies in Systems, Decision and Control 628,
https://doi.org/10.1007/978-3-032-04056-5_15

[1]. Hyperglycemia, commonly known as a common outcome is elevated blood glucose, or high blood sugar resulting in unmanaged diabetes. Over time, it results in significant harm to various bodily systems, particularly the nerves and blood vessels of eyes, kidneys and heart [2]. Indications of diabetes encompass heightened thirst, fatigue, recurrent urination, blurred vision, unaccounted-for weight loss, and protracted recovery from injuries. Furthermore, people might encounter sensations of tingling or numbness in their extremities, face recurring infections, and notice changes in their appetite. Individuals diagnosed with diabetes face an elevated risk of experiencing health issues such as heart attacks, strokes, and kidney failure. Damage to the ocular blood vessels due to diabetes can result in permanent vision loss. Nerve damage and inadequate blood flow in individuals with diabetes often lead to complications in their feet, potentially resulting in foot ulcers and, in severe cases, necessitating amputation. It has been found in research that there are mainly three type of diabetes Type 1, Type 2 and Gestations diabetes. Previously known as juvenile diabetes or insulin-dependent diabetes, Type 1 diabetes is a persistent medical condition.

It mainly occurs in childhood or at the time of adolescence, however it can also develop in adults. It is believed that genetics and environmental factors are the cause of Type 1 diabetes [3]. Type 2 diabetes are mainly developed due to excess weight, insufficient physical activity, and genetic predisposition. To identify diabetes in its early stages, it is advisable to undergo routine check-ups and blood tests. As the symptom of diabetes is mild timely identification of diabetes is crucial to mitigate the severe consequences of Type 2 diabetes. Eating healthy diet, active lifestyle, oral medicine or insulin injection as recommended by doctor can control the blood sugar level of individuals and prevent complications. Gestational diabetes manifests during pregnancy, elevating the risk of complications for both the mother and the baby during childbirth. Furthermore, women with gestational diabetes, and potentially their offspring, face an augmented likelihood of eventually developing Type 2 diabetes. The exact cause of diabetes is unknown, but researchers suggest that a combination of genetic factors and lifestyle choices plays a significant role.

Although incurable, diabetes can be managed with medication and treatment, which helps reduce the likelihood of complications such as cardiovascular disease and nerve damage [4]. Predictive analysis is a technique that utilizes machine learning algorithms, data mining, and statistical methods and it examines both current and historical data to uncover patterns and insights. This approach is used to predict future events based on the analyzed information. Predictive analytics focuses on accurately diagnosing diseases, improving patient care, optimizing resources, and enhancing clinical outcomes. Predictive analysis applied to healthcare data helps in making informed decisions. It also allows for generating accurate predictions about future health outcomes. The current method for detecting diabetes involves lab tests like fasting plasma glucose tests and oral glucose tolerance assessments and this process requires a significant amount of time [5]. In medicine, doctors and current research agree that early detection of a disease significantly improves the chances of recovery. The earlier a disease is discovered, the better the prospects for successful treatment.

As technology continues to advance, machine learning, and deep learning techniques have proven highly valuable for early disease prediction and analysis.

Diabetes Prediction and its Clinical Perspective

Diabetes mellitus is a chronic metabolic disorder characterized by high blood sugar levels over a prolonged period. It is a major global health concern, affecting millions of individuals and significantly increasing the risk of heart disease, kidney failure, vision loss, and nerve damage. According to the World Health Organization (WHO), the prevalence of diabetes has been steadily rising, especially in low- and middle-income countries. Early detection of diabetes is crucial, as it enables timely medical intervention, lifestyle modifications, and better disease management, which can significantly reduce the likelihood of complications. Traditional diagnostic procedures often rely on periodic blood tests, which may not always be accessible or affordable for all individuals.

From a clinical perspective, leveraging machine learning techniques for diabetes prediction can assist healthcare professionals in identifying at-risk individuals before the onset of symptoms and providing personalized care based on risk profiles. Additionally, reducing the burden on healthcare systems through early intervention and supporting public health initiatives by identifying vulnerable populations. By integrating predictive analytics into clinical workflows, we can move toward a more proactive and preventive model of healthcare.

This study aims to explore how machine learning algorithms can contribute to this goal by accurately predicting diabetes using patient data. These methods enhance the ability to detect and assess diseases at earlier stages [6]. In this work Decision Tree (DT), Random Forest (RF), Logistic Regression (LR), Support Vector Classifier (SVC), Adaptive Boosting (AB) and Gradient Boosting (GB) algorithm have been used to build machine learning models, that predict the likelihood of diabetes in patients by analyzing their medical background and demographic data. The rest of the paper is arranged as follows. Related work on this area is presented in Sect. 2. In Sect. 3, the proposed methodology has been explained. Results and discussion of the work are depicted and discussed in Sect. 4, while concluding remarks of the work is addressed in Sect. 5.

2 Related Work

Diagnosis of diabetes is an expanding field of research. In [7] authors explored various deep learning and classification techniques, including artificial neural networks (ANN), Decision Tree (DT), Random Forest (RF) and Support Vector machine (SVM). In [8] a model combining naïve Bayes (NB) and Support Vector machine (SVM) for diabetes prediction was proposed. They split the dataset evenly, with 50% used for training and 50% for testing. In this work Ensemble of naïve Bayes (NB) and SVM has reached an accuracy of 97.6%, significantly outperforming individual algorithms run on the dataset. For instance, naïve Bayes achieved an accuracy of

94.52%, while Support Vector machine reached 95.52%. In [9] researchers compared the accuracy of Random Forest, naïve Bayes (NB), artificial neural networks (ANN) using the Matthews correlation coefficient for evaluation. The study concentrated on glucose levels and used Support Vector Regression (SVR) to forecast diabetes, framing it as a multivariate regression problem [10]. Furthermore, an increasing number of studies have utilized ensemble methods to enhance accuracy [11].

The authors [12] proposed a soft voting classifier model by ensembling three machine learning algorithms: Random Forest (RF), Logistic Regression, and naïve Bayes (NB). Initially they conducted experiments with the Primary Immunodeficiency Diseases (PIDD) and attained an accuracy of 79.08%. Later they applied the model to the Breast Cancer Dataset. On the breast cancer dataset, it produced even higher accuracy, with correct results at 97.02%. In [13], the authors proposed a hybrid system that combines three common machine learning algorithms: Decision Tree, artificial neural networks (ANN) and Random Forest. This system was developed to improve diabetes prediction accuracy. The data for this research, sourced from Luzhou Hospital in China, comprised 68,994 healthy and diabetic patients. Principal Component Analysis (PCA) was employed to reduce the dataset's dimensionality and they have achieved an accuracy of 80%. The study [14] sought to improve the accuracy of the Decision Tree (DT) C4.5 and K-Nearest Neighbor (KNN) classification algorithms on the PIDD by applying discretization and the ensemble method Adaptive Boosting (AB) to handle numerical attributes. The improved results showed an accuracy of 83.67% for the Decision Tree and 83.18% for K-Nearest Neighbors (KNN), which are 7.36% and 9.97% more respectively with the single algorithms. A tree-based machine learning framework for the classification of PIDD was developed, utilizing Mutual Information (MI) for feature selection and Adaptive Boosting (AB) to enhance performance, with the Extra Tree (ET) algorithm as the AdaBoost classifier's base estimator yielding the highest accuracy of 90.5% [15].

The article [16] proposes an ensemble-based framework called eDiaPredict for predicting diabetes status in patients with Primary Immunodeficiency Diseases (PIDD). This framework combines various machine learning algorithms, including XGBoost, Random Forest, Support Vector Machine, Artificial Neural Networks (ANN), and Decision Tree. By leveraging ensemble modeling, eDiaPredict aims to improve the accuracy of diabetes prediction. It shows XGBoost alone achieves 92% accuracy, and when ensembled with Random Forest, the prediction accuracy improves to 95%. Two methods for feature selection, Principal Component Analysis (PCA) and Linear Discriminant Analysis (LDA), were applied to identify important features from the Primary Immunodeficiency Diseases (PIDD). These methods aimed to improve model performance by focusing on the most relevant features [17]. The article also presented a comparative analysis of different feature selection methods. Machine learning techniques, including radial basis function kernel, k-nearest neighbors (KNN), and AdaBoost, were utilized for classifying the dataset.

3 Proposed Work

This section outlines the methodologies and execution of different machine learning approaches to suggest an automated system for predicting diabetes. It details how these techniques are applied in the design process. Figure 1 depicts the block diagram of the proposed Diabetes Prediction system, along with different stages involved in this work.

The dataset was sourced from Kaggle, i.e. diabetes-prediction-dataset, which is a.csv file. It contains 1,00,000 records and contains 9 attributes, out of which 1 attribute indicates the class label. It is split into two sections: training and testing. For training purpose 75% data has been used, remaining 25% has been used for testing.

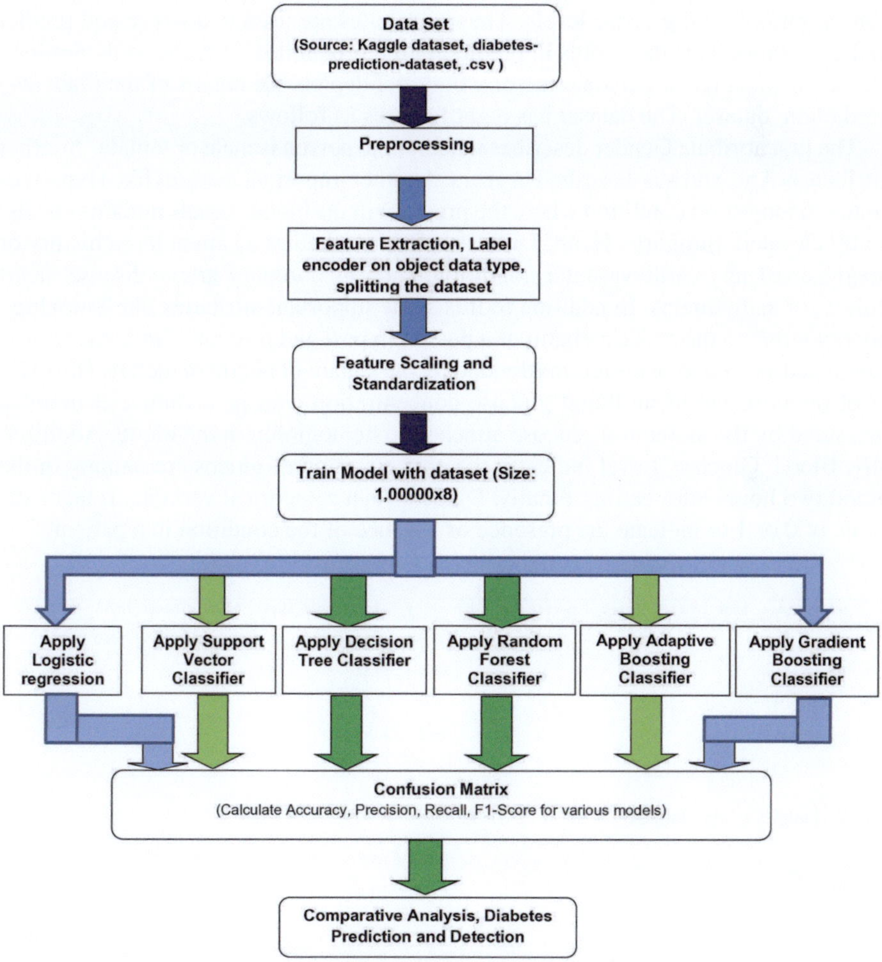

Fig. 1 Block diagram of the proposed diabetes prediction system

After that feature scaling has been applied on the dataset. Finally, various machine learning techniques like Linear Regression, Support Vector Classifier, Decision Tree, Random Forests, Adaptive Boosting and Gradient Boosting classifier have been applied on the dataset.

3.1 Dataset

The file diabetes_prediction_dataset.csv [18] includes information regarding patients and their condition of diabetes, indicating whether they are positive or negative for the condition. It includes a range of characteristics such as age, gender, body mass index (BMI), hypertension, cardiovascular ailments, smoking history, HbA1c concentrations, and glucose levels. These attributes are used to analyze and predict diabetes status. This data helps in predicting the probability of diabetes determined by various health and personal factors. Figure 2 depicts the details of the diabetes_prediction_dataset. The dataset has eight features as follows.

The first attribute Gender describes whether the person is male or female. Another attribute is Age, and it is described in years. Another important features like Hypertension, is a long-term condition where the pressure in the blood vessels remains consistently elevated. Similarly, Heart_Disease refers to whether a patient has a history or diagnosis of any cardiovascular condition, such as coronary artery disease, heart failure, or arrhythmias. In addition to this some important attributes like Smoking_history refers to the smoking habit of a person in past and present. Similarly, BMI is calculated as weight in kilograms divided by the square of height in meters. HbA1C_level refers to the mean blood glucose concentration over preceding 2–3 months, measured by the amount of glucose attached to hemoglobin in the blood. Additionally, Blood_Glucose_Level indicates the concentration of glucose remaining in the blood two hours after eating. Finally, Diabetes is a categorical variable. It takes the value of 0 or 1 to indicate the presence or absence of the condition in a patient.

	gender	age	hypertension	heart_disease	smoking_history	bmi	HbA1c_level	blood_glucose_level	diabetes
0	Female	80.0	0	1	never	25.19	6.6	140	0
1	Female	54.0	0	0	No Info	27.32	6.6	80	0
2	Male	28.0	0	0	never	27.32	5.7	158	0
3	Female	36.0	0	0	current	23.45	5.0	155	0
4	Male	76.0	1	1	current	20.14	4.8	155	0

Fig. 2 Details of the dataset

3.2 Machine Learning Classifier

The primary aim of the Machine learning based system is to diagnose and predict diabetes. This research focuses on developing a system using six classification methods: Logistic Regression, Support Vector Classifier, Decision Tree, Random Forest, Adaptive Boosting and Gradient Boosting. Machine learning methods are extensively used for diabetes prediction, yielding highly favorable results. Among these, the decision tree is a popular method in the medical field due to its strong classification capabilities. Random Forest further enhances this by generating multiple decision trees [13]. Adaptive Boosting, or AdaBoost, integrates several weak classifiers to form a robust single classifier.

Logistic Regression Classifier. Logistic Regression (LR) is a statistical method employed to predict binary outcomes. It models an S-shaped curve, which can be used to predict outcomes. Optimization revealed that the Logistic Regression model required only 150 iterations to reach convergence and considered adequate [19]. In medical diagnosis, LR can classify a patient as either positive or negative for a specific condition, such as diabetes. Logistic Regression measures the relationship between the response variable and by evaluating the likelihood using the logit function with one or more predictors. The logit of the response variable (Y) as shown in Eq. (1) is expressed as a linear amalgamation of the predictors (X), which can be articulated as follows:

$$\log it(P_j) = \log_e\left(\frac{P_j}{1-P_j}\right) = \sum_{i=0}^{K} \beta_i X_i \tag{1}$$

where P_j is defined as the probability for Y = 1 (diabetic) and $1 - P_j$ (control) is defined when Y = 0. β_i's (i = 0, 1,…, K) are the unknown regression coefficient, K is the total number of predictors (8 factors) and X_i's (i = 1,…, K) are the predictors and $X_0 = 1$. The regression coefficients are estimated using the maximum likelihood estimator (MLE), and the odds ratio (OR) is obtained by exponentiating the regression coefficients [20].

Support Vector Machine. Support Vector Machine (SVM) is a supervised learning technique that utilizes a hyperplane to distinguish between two classes by creating a decision boundary [16]. The hyperplane can be 2D or 3D. In 2D, it represents a boundary or decision surface, while in 3D, it represents a plane. When the data are linearly separable, they can be distinguished by a boundary line, defined by the equation as shown in Eq. (2):

$$y = mx + c \tag{2}$$

Here, m denotes the slope, while c is the intercept. When the data cannot be separated linearly, it is not feasible to distinguish the classes using just the x–y plane. In such cases, a new dimension, the z-axis is introduced to differentiate the classes in a three-dimensional plane. The z-axis is given by the equation as shown in Eq. (3).

$$z = x^2 + y^2 \tag{3}$$

It seeks to find the optimal hyperplane that separates data points from different classes in a high-dimensional space. By maximizing margin between nearest points of each class, known as support vectors, Support Vector Classifiers (SVC) ensures the best possible separation of the classes. Support Vector Machine (SVM) operates on the principle of the Maximum Margin Classifier, aiming to maximize the geometric margin while minimizing the error. The goal of Support Vector Classification (SVC) is to maximize the margin between the support vectors and the nearest data points from each class and the hyperplane. It manages both linear and non-linear classification problems by employing kernel functions, like polynomial or radial basis function (RBF) kernels, to map data into higher dimensions for more complex decision boundaries.

Decision Tree Classifier. A Decision Tree (DT) represents the learning process as a series of decision rules. The Decision Tree approach is utilized to model and approximate target functions with discrete values. Metrics such as Gini impurity or entropy are employed to evaluate information gain, guiding the selection of nodes at each step. These metrics assist in measuring the level of impurity or randomness within the dataset, guiding the decision-making process at each node in a tree-based model.

By minimizing impurity at each split, the model aims to enhance classification accuracy. The formula of information gain using Gini Index and using entropy respectively are shown in Eq. 4 and 5. In both (4) and (5), the variable n represents the total number of distinct class values.

$$Gini = 1 - \sum_{k=1}^{n} \left(P_{i,k}\right)^2 \tag{4}$$

$$Entropy = \sum_{i=1}^{n} -P_i \log_2 P_i \tag{5}$$

Figures 3 and 4 show the construction of Decision Tree using Gini Index and Entropy respectively. A visual representation of a Decision Tree classifier trained using the Gini Index is shown in Fig. 3. The tree displays feature-based splits that minimize impurity, showing how decisions are made to classify patients as diabetic or non-diabetic based on various thresholds.

Similarly, Fig. 4 shows the same Decision Tree structure but trained using Entropy (Information Gain) as the criterion. It illustrates how splits are chosen to maximize information gain, influencing the path the model takes in decision-making.

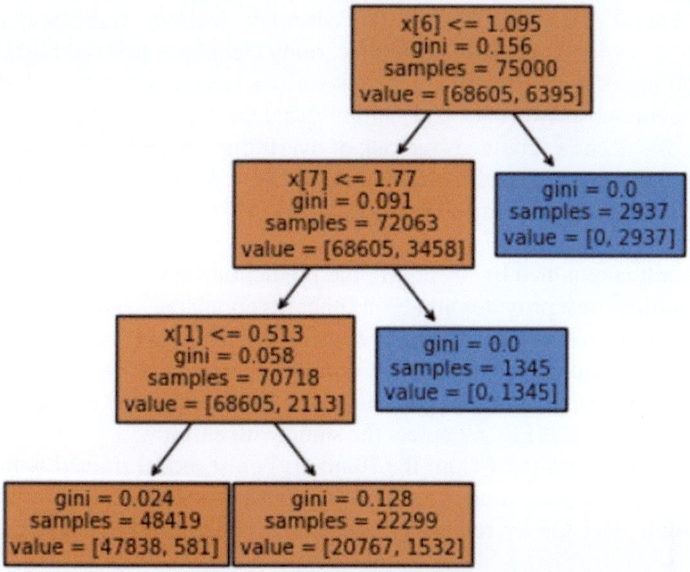

Fig. 3 Decision tree using Gini index

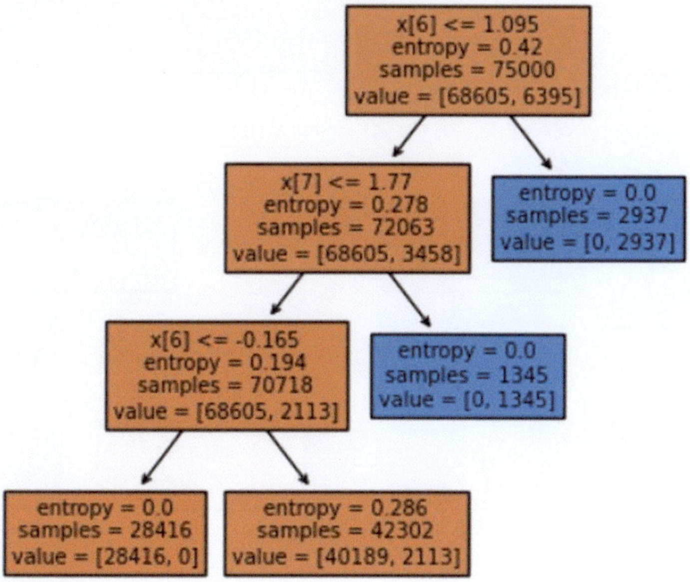

Fig. 4 Decision tree using entropy

Random Forest Classifier. Within the ensemble learning framework, Random Forest (RF) is a widely adopted machine learning technique, utilized for both classification and regression tasks. It aggregates various Decision Trees, each developed using different subsets of data and features, and combines their predictions. This method boosts accuracy, reduces the risk of overfitting, and effectively handles both continuous and categorical variables, including those with missing values. Each tree in the forest is built by training a Decision Trees (DT) model on a randomly selected portion of the dataset.

Final result is obtained by averaging the predictions made by all the trees in the forest. Random Forest provides numerous benefits compared to single decision trees, such as increased precision, minimized overfitting, and greater clarity in interpretation. It can also manage a combination of continuous and categorical features and accommodate missing values. Figure 5 shows the construction of Random Forest using Gini Index, while Fig. 6 depicts the same with entropy.

One of the decision trees from the Random Forest model trained with the Gini Index is shown in Fig. 5. It demonstrates how multiple such trees contribute to the final ensemble decision by reducing variance and improving overall classification accuracy.

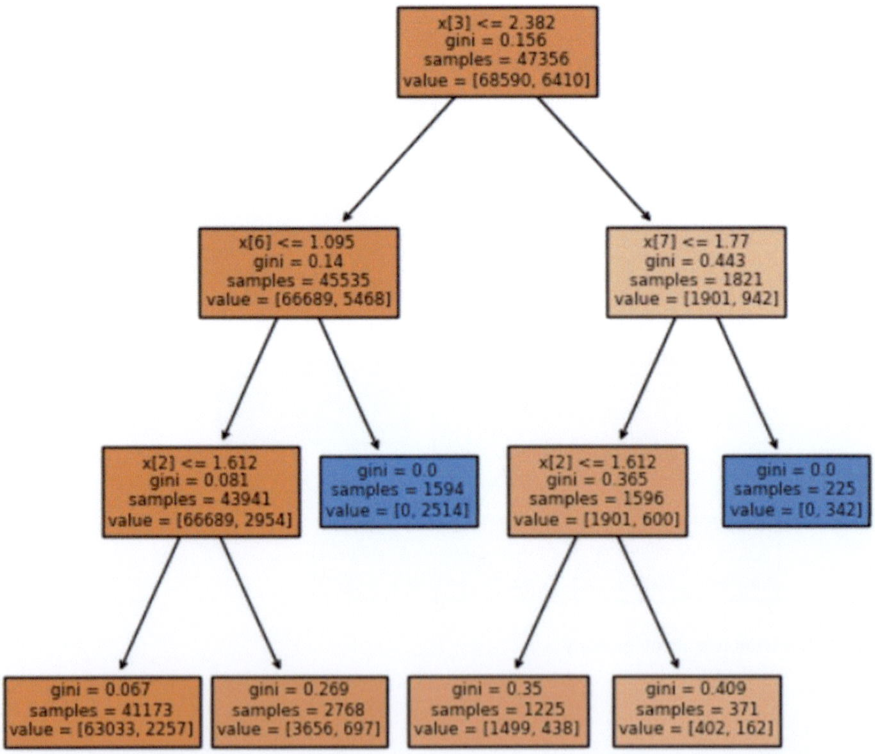

Fig. 5 Random forest using Gini index

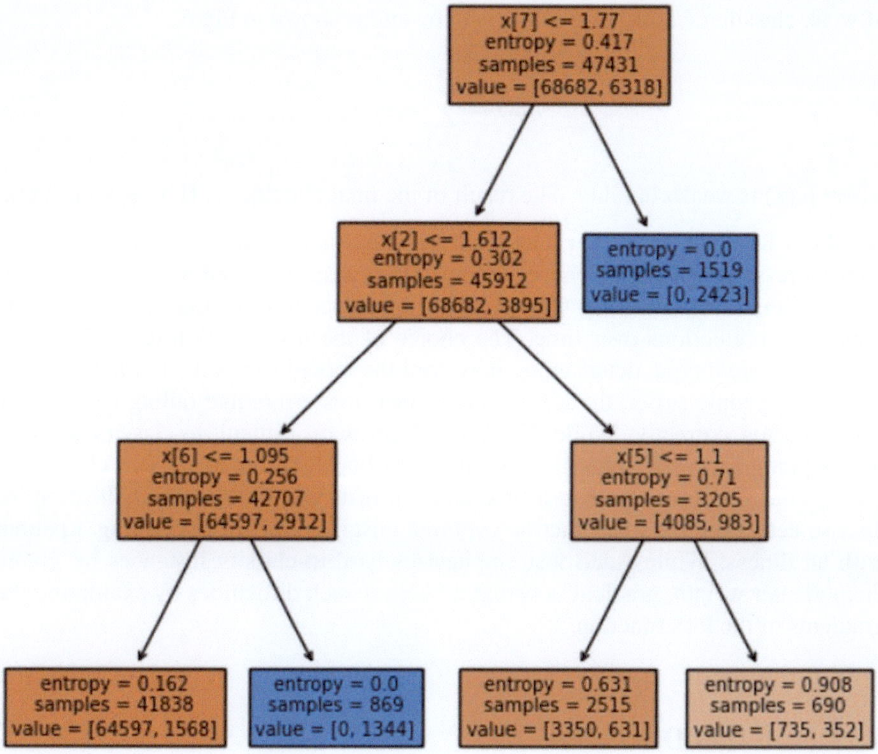

Fig. 6 Random forest using entropy

A tree from the Random Forest ensemble trained using Entropy is shown in Fig. 6. It highlights how entropy-based splitting influences feature selection and class separation within the ensemble structure.

AdaBoost Classifier. AdaBoost, which stands for Adaptive Boosting, is a method that allocates weights to weak classifiers and merges them to create a more robust classifier [21, 22]. AdaBoost is an ensemble technique that begins by training a classifier on the original dataset. It then iteratively refines the model by fitting multiple versions of the classifier to the same dataset, adjusting the focus toward instances that were previously misclassified. This framework increases the weights of misclassified instances, ensuring that they receive more attention in subsequent iterations. As a result, each new classifier focuses more on the challenging cases [23]. AdaBoost enhances accuracy by assigning weights to individual weak classifiers and iteratively training on sample data. During each iteration, it places greater emphasis on correctly classifying outliers or atypical observations [15]. The concept behind this classification method is that while a single classifier may perform well on certain portions of the dataset, it might make mistakes in other areas. Nevertheless, other weak classifiers can accurately predict the portions that were misclassified. The representation of weak classifiers' combination is given by and is shown in Eq. 6.

$$f(x) = \sum_{t=1}^{T} \alpha_t h_t(x) \tag{6}$$

where $h_t(x)$ is weak classifier. The result of the final classifier is H(x) = sign(f(x)).

Gradient Boosting Classifier. Gradient boosting is an ensemble technique used for both regression and classification tasks [24]. Its goal is to minimize a loss function by iteratively adjusting the model in the opposite direction of the gradient, improving predictions over time. The choice of the loss function depends on the specific problem type, determining how well the model performs. In binary classification of people suffer, the loss function measures errors like failing to diagnose a sick person correctly. While AdaBoost focuses on difficult-to-classify samples by assigning higher weights to them, gradient boosting identifies such challenges through the gradients of the loss function. For instance, in binary classification for disease detection, the loss function captures mistakes like misclassifying a patient with an illness. While AdaBoost emphasizes hard-to-classify instances by giving them greater weight, gradient boosting addresses such difficulties by examining the gradients of the loss function.

4 Result and Discussion

This section elucidates the findings and offers an in-depth discussion of the proposed automated diabetes prediction system. We evaluated a variety of machine learning models using classification accuracy, AUC, precision, recall, and F1 score. The formulas for these metrics are shown in Eqs. 7, 8, 9, and 10 respectively.

$$Accuracy = \frac{TP + TN}{TP + TN + FP + FN} \tag{7}$$

$$Precision = \frac{TP}{TP + FP} \tag{8}$$

$$Recall = \frac{TP}{TP + FN} \tag{9}$$

$$F1\,Score = \frac{2 \times Recall \times Precision}{Recall + Precision} \tag{10}$$

True Positive (TP) refers to instances where both the model's prediction and the actual result are positive. False Positive (FP) occurs when the model predicts a positive outcome, but the true result is negative. True Negative (TN) is used when the model predicts a negative result and the actual outcome matches this prediction.

False negative (FN) describes situations where the model predicts a negative result, but the actual outcome is positive. The comparison of different classifier is shown in Table 1.

From Table 1, the comparison of various machine learning techniques for diabetes prediction system is shown. Logistic Regression and Support Vector Classifier the accuracy obtained is 96 and 96.42%. Whereas for Decision Tree, Random Forest, AdaBoost and Gradient Boosting the accuracy obtained for classification is 97.20%, 97.06%, 97.22% and 97.24% respectively. It can be observed that for AdaBoost although the accuracy is 97.22%, the Precision value obtained is 0.97, Recall is 1.0, whereas F1 Score obtained is 0.99 for patients who are not diabetes. Similarly, for the diabetes patient, the Precision value obtained is 0.97, Recall is 0.69, whereas F1 Score obtained is 0.99 respectively. Similarly, for Gradient Boosting though the accuracy

Table 1 Comparison between different classifier

Classifier	Metrics	Class—0	Class—1
Logistic regression	Precision	0.97	0.86
	Recall	0.99	0.63
	F1 score	0.98	0.73
	Overall accuracy **96.00%**		
Support vector classifier (SVC)	Precision	0.96	0.97
	Recall	1	0.59
	F1 score	0.98	0.74
	Overall accuracy **96.42%**		
Decision tree	Precision	0.97	1
	Recall	1	0.67
	F1 score	0.98	0.80
	Overall accuracy **97.20%**		
Random forest	Precision	0.97	1
	Recall	1	0.67
	F1 score	0.98	0.80
	Overall accuracy **97.06%**		
AdaBoost	Precision	0.97	0.97
	Recall	1	0.69
	F1 score	0.99	0.81
	Overall accuracy **97.22%**		
Gradient boosting	Precision	0.97	0.99
	Recall	1.00	0.68
	F1 score	0.99	0.81
	Overall accuracy **97.24%**		

Done to emphasize the value and easy understanding

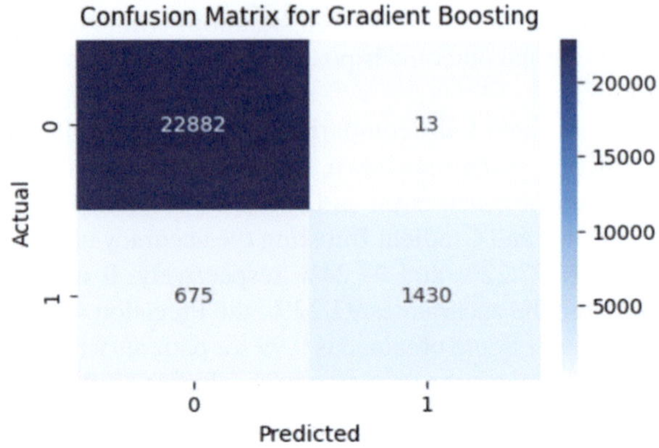

Fig. 7 Confusion matrix of gradient boosting

is 97.24%, the Precision value obtained is 0.97, Recall is 1.0, whereas F1 Score obtained is 0.99 for patients who are not diabetes. Similarly, for the diabetes patient, the Precision value obtained is 0.99, Recall is 0.68, whereas F1 Score obtained is 0.81 respectively. As the Gradient Boosting model accuracy obtained in our work, is highest among all other machine learning models, the Confusion matrix for Gradient Boosting model is evaluated and it is shown in Fig. 7.

4.1 Confusion Matrix

The Confusion Matrix (CM) provides insights which are valuable for the model's performance and highlights areas where predictive accuracy can be improved. The evaluation process produces an error matrix, commonly referred to as a Confusion Matrix (CM). In evaluating our model's performance, the CM plays a crucial role in understanding how effectively the classification model is performing. The Confusion Matrix, illustrated in Fig. 7, arranges these terms where columns denote actual values of the target variable and rows denote predicted values.

By examining these metrics, we gain a comprehensive insight into the model's performance in classification, along with its strengths and weaknesses in predicting diabetic patients from the available dataset. In this work we have shown the Confusion Matrix results for Gradient Boosting. It can be observed that our model has correctly predicted 22,882 as True positive samples and 1430 as True negative. The count of samples that were misclassified by the model as False positive which is 13, and False negative as 675 samples which can be ignored as the accuracy of the model is concerned.

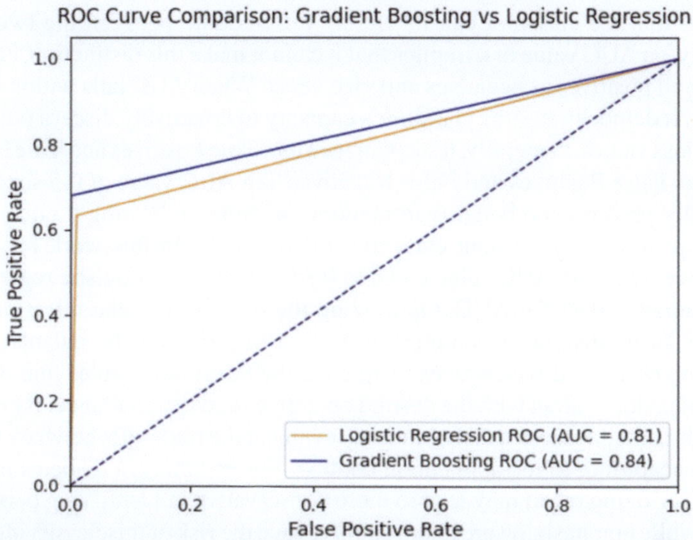

Fig. 8 ROC curve comparison: gradient boosting versus logistic regression

4.2 ROC Curve Comparison

Finally, in this work comparative analysis has been done by comparing the performance of Logistic Regression with Gradient Boosting using a receiver operating characteristic curve (ROC curve), that indicates the effectiveness of a classification model, which allows us to understand how the model makes decisions across varying levels of confidence. The receiver operating characteristic (ROC) curve is a graphical representation of how well a classification model performs. Figure 8 shows the ROC curve comparison between gradient boosting and logistic regression.

Primarily used in binary classification problems, it is a valuable evaluation metric. The ROC curve is a graph that displays the relationship between the true positive rate (TPR) and the false positive rate (FPR) at different threshold levels. Its main purpose is to effectively differentiate the 'signal' from the 'noise' in classification outcomes. In more elaborate terms, the ROC curve demonstrates the efficacy of a classification model in distinguishing between positive and negative instances across a spectrum of decision thresholds. The Area Under the Curve (AUC) is a quantitative metric extracted from the receiver operating characteristic (ROC) curve, as illustrated in Fig. 8. It assesses the classifier's proficiency in differentiating between the two classes, providing a succinct overview of the ROC curve's effectiveness. A higher AUC value signifies superior overall performance of the model in binary classification, with a value of 1 representing flawless discrimination, while a value of 0.5 signifies a lack of discrimination, akin to random guessing.

AUC values range from 0 to 1, with higher values indicating superior model performance in discerning between negative and positive classes. An AUC value of

1, signifies that the classifier can flawlessly differentiate between the two classes. Conversely, an AUC value of 0 implies that it cannot make this distinction, effectively predicting all positives as negatives and vice versa. When AUC falls within the range of 0.5 to 1, it demonstrates the classifier's capacity to effectively discern positive and negative class values. Generally, it recognizes more True Positives and True Negatives than it does False Positives and False Negatives. An AUC value of 0.5 suggests that the classifier performs no better than random guessing, indicating it either predicts randomly or assigns a constant class to all data points. In this work for Gradient Boosting we obtain an AUC value equal to 0.84, whereas for Logistic regression the value obtained is 0.81 for AUC. Optimizing the threshold in classification models allows for customization depending on distinct criteria and the balance between False Positives and False Negatives. Adjusting the threshold enables fine-tuning the model's behavior to align with the desired objectives and risk tolerance. By exploring different threshold values, stakeholders can evaluate the trade-offs between precision and recall according to the application context. For instance, in diabetes prediction system, prioritizing recall may lead to more conservative in identifying persons with diabetes, while emphasizing precision could reduce the risk of misclassifying patients with diabetes.

It can be observed from Fig. 9, that our Web-based application, i.e. Diabetes Prediction has correctly predicted whether the patient is diabetic or not based on the input data. In this work, a Streamlit application has been developed where users can choose from various Machine Learning models, such as Logistic Regression, Support Vector Classifier, Decision Tree, Random Forest, AdaBoost and Gradient Boosting etc. The application will then assess the prediction accuracy of the selected model and provide recommendations to patients.

In this work, essential libraries like pandas for data handling, matplotlib and seaborn for visualization, and sklearn for machine learning tasks are used. The dataset, sourced from Kaggle in CSV format, is loaded with pandas. Visualizations such as pair plots and heatmaps explore data distribution and correlations. Preprocessing involves encoding categorical variables and normalizing numerical features, followed by splitting the dataset into training and testing sets. Various machine learning models, including Logistic regression, Support Vector classifier, Decision tree, Random forests, Adaboost and Gradient boosting are trained using sklearn. Evaluation metrics like accuracy, precision, and ROC-AUC assess model performance. This workflow integrates data analysis and machine learning to predict the likelihood of diabetes.

5 Concluding Remarks and Limitations

Future research can focus on advancing predictive modeling techniques, exploring innovative features, and incorporating external data sources to enhance model accuracy and predictive power. Investigating novel algorithms and methodologies can address existing limitations and drive innovation in diabetes prediction. Based on

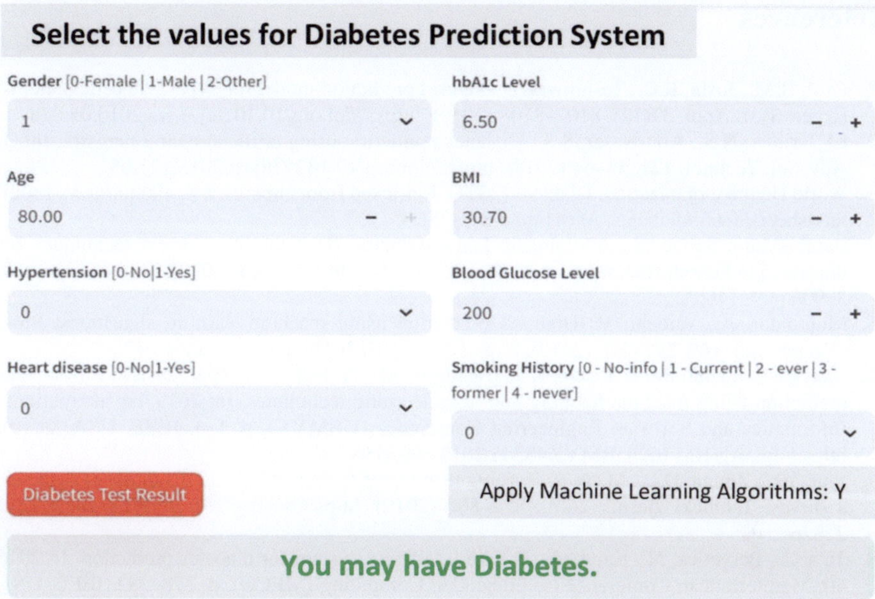

Fig. 9 Diabetes prediction system

the comparison results and the analysis of advantages and disadvantages, offer clear recommendations for the use of AdaBoost Classifier and Gradient Boosting Classifier in identifying patients with diabetes.

While the proposed diabetes prediction system demonstrates promising results using traditional machine learning models, there are several limitations such as Dataset constraints and Class imbalance problems. The study is based on the Diabetes dataset, which widely represents a specific population group. As a result, the model's applicability to other populations may be limited due to demographic and ethnic differences. Additionally, the dataset exhibits a mild imbalance between diabetic and non-diabetic cases. Although stratified sampling was applied, no advanced resampling techniques (e.g., SMOTE) were used, which could potentially impact the model's sensitivity, especially for positive diabetic cases. The model has not yet been validated on real-world clinical datasets, therefore, its ability to generalize to unseen data and practical healthcare settings remains uncertain. The impact of false positives (unnecessary testing and anxiety) and false negatives (missed diagnoses) was not quantified in detail. These outcomes carry significant weight in healthcare and requires further investigation.

References

1. Patil, B.M., Joshi, R.C., Toshniwal, D.: Hybrid prediction model for Type-2 diabetic patients. Expert Syst. Appl. **37**(12), 8102–8108 (2010). https://doi.org/10.1016/j.eswa.2010.05.078
2. El_Jerjawi, N.S., Abu-Naser, S.S.: Diabetes prediction using artificial neural network. Int. J. Adv. Sci. Technol. **121**, 55–64 (2018). https://doi.org/10.14257/ijast.2018.121.05
3. World Health Organization: Diabetes (2024). Retrieved from https://www.who.int/news-room/fact-sheets/detail/diabetes. Accessed 4 Sept 2024
4. Larabi-Marie-Sainte, S., Aburahmah, L., Almohaini, R., Saba, T.: Current techniques for diabetes prediction: review and case study. Appl. Sci. **9**(21), 4604 (2019). https://doi.org/10.3390/app9214604
5. Mujumdar, A., Vaidehi, V.: Diabetes prediction using machine learning algorithms. Proc. Comput. Sci. **165**, 292–299 (2019). https://doi.org/10.1016/j.procs.2020.01.047
6. Yahyaoui, A., Jamil, A., Rasheed, J., Yesiltepe, M.: A decision support system for diabetes prediction using machine learning and deep learning techniques. In: 2019 1st International Informatics and Software Engineering Conference (UBMYK), p. 1–4. IEEE, USA (2019). https://doi.org/10.1109/UBMYK48245.2019.8965556
7. Sun, Y.L., Zhang, D.L.: Machine learning techniques for screening and diagnosis of diabetes: a survey. Tehnički vjesnik **26**(3), 872–880 (2019). https://doi.org/10.17559/TV-20190421122826
8. Tafa, Z., Pervetica, N., Karahoda, B.: An intelligent system for diabetes prediction. In 2015 4th Mediterranean Conference on Embedded Computing (MECO), p. 378–382. IEEE, USA (2015). https://doi.org/10.1109/MECO.2015.7181948
9. Hussain, A., Naaz, S.: Prediction of diabetes mellitus: comparative study of various machine learning models. In: International Conference on Innovative Computing and Communications: Proceedings of ICICC 2020, vol. 1166, p. 103–115. Springer, Singapore (2021). https://doi.org/10.1007/978-981-15-5148-2_10
10. Georga, E.I., Protopappas, V.C., Ardigo, D., Marina, M., Zavaroni, I., Polyzos, D., Fotiadis, D.I.: Multivariate prediction of subcutaneous glucose concentration in Type 1 diabetes patients based on support vector regression. IEEE J. Biomed. Health Inform. **17**(1), 71–81 (2012). https://doi.org/10.1109/TITB.2012.2219876
11. Kavakiotis, I., Tsave, O., Salifoglou, A., Maglaveras, N., Vlahavas, I., Chouvarda, I.: Machine learning and data mining methods in diabetes research. Comput. Struct. Biotechnol. J. **15**, 104–116 (2017). https://doi.org/10.1016/j.csbj.2016.12.005
12. Kumari, S., Kumar, D., Mittal, M.: An ensemble approach for classification and prediction of diabetes mellitus using soft voting classifier. Int. J. Cogn. Comput. Eng. **2**, 40–46 (2021). https://doi.org/10.1016/j.ijcce.2021.01.001
13. Zou, Q., Qu, K., Luo, Y., Yin, D., Ju, Y., Tang, H.: Predicting diabetes mellitus with machine learning techniques. Front. Genet. **9**, 515 (2018). https://doi.org/10.3389/fgene.2018.00515
14. Majid, A.M., Utomo, W.H.: Application of discretization and adaboost method to improve accuracy of classification algorithms in predicting diabetes mellitus. ICIC Exp. Lett. **12**(12), 1177–1184 (2021). https://doi.org/10.24507/icicelb.12.12.1177
15. Hasan, S.M., Rabbi, M.F., Champa, A.I., Zaman, M.A.: (2020) An effective diabetes prediction system using machine learning techniques. In: 2020 2nd International Conference on Advanced Information and Communication Technology (ICAICT), p. 23–28. IEEE, USA (2021). https://doi.org/10.1109/ICAICT51780.2020.9333497
16. Singh, A., Dhillon, A., Kumar, N., Hossain, M.S., Muhammad, G., Kumar, M.: EDiaPredict: an ensemble-based framework for diabetes prediction. ACM Trans. Multimidia Comput. Commun. Appl. **17**(2s), 1–26 (2021). https://doi.org/10.1145/3415155
17. Choubey, D.K., Kumar, M., Shukla, V., Tripathi, S., Dhandhania, V.K.: Comparative analysis of classification methods with PCA and LDA for diabetes. Curr. Diab. Rev. **16**(8), 833–850 (2020). https://doi.org/10.2174/1573399816666200123124008
18. Mustafa, T.Z.: Diabetes prediction dataset (2023). Retrieved from https://www.kaggle.com/datasets/iammustafatz/diabetes-prediction-dataset/data. Accessed Sept 2025

19. Mitchell, T. M., & Learning, M.: The McGraw-Hill Companies. Inc., New York (1997)
20. Maniruzzaman, M., Rahman, M.J., Ahammed, B., Abedin, M.M.: Classification and prediction of diabetes disease using machine learning paradigm. Health Inf. Sci. Syst. **8**, 1–14 (2020). https://doi.org/10.1007/s13755-019-0095-z
21. Géron, A.: Hands-on machine learning with Scikit-Learn, Keras, and TensorFlow. O'Reilly Media, Inc (2022)
22. Li, D., Wang, S., Li, D., Li, D., Wang, S., Li, D.: Spatial data mining. Springer, Berlin, Heidelberg (2015). https://doi.org/10.1007/978-3-662-48538-5_2
23. Tasin, I., Nabil, T.U., Islam, S., Khan, R.: Diabetes prediction using machine learning and explainable AI techniques. Healthc. Technol. Lett. **10**(1–2), 1–10 (2023). https://doi.org/10.1049/htl2.12039
24. Bahad, P., Saxena, P.: Study of Adaboost and gradient boosting algorithms for predictive analytics. In: International Conference on Intelligent Computing and Smart Communication 2019: Proceedings of ICSC 2019, pp. 235–244. Springer, Singapore (2020). https://doi.org/10.1007/978-981-15-0633-8_22

Challenging Case Studies and Applications

Correlation Between Time Series Features and the sMAPE of the Best Methods of M4 Competition

Mario Alberto Martínez-Ponce, J. David Terán-Villanueva, Alejandro Santiago, Mirna Patricia Ponce Flores, and Eduardo Alvarez-Navarro

Abstract This study analyzes the correlation between time series characteristics and sMAPE forecasting error for the top five methods in the M4 Competition. We use two sets of financial time series with daily and monthly frequencies from the M4 Competition. Additionally, we extracted key characteristics such as trend, curvature, and seasonal strength, among others, for each series. The sMAPE error is calculated using the results provided by the competition. These variables are used in a Random Forest classifier to train ten different prediction models to identify the best forecasting method for each set of time series based on the forecasting error. The results indicate a strong correlation between the time series characteristics and the performance of the top forecasting methods in the M4 Competition, especially in the monthly time series. These series exhibit a remarkable ability to identify the most accurate forecasting method, suggesting that the characteristics of this set of series can be a determining factor in the selection process of the optimal forecasting method.

Keywords Financial time series · Random forest · Trend and seasonal strength

M. A. Martínez-Ponce (✉) · J. D. Terán-Villanueva · A. Santiago · M. P. Ponce Flores · E. Alvarez-Navarro
Universidad Autónoma de Tamaulipas, Tampico, Tamaulipas, Mexico
e-mail: alberto_dx86@hotmail.com

J. D. Terán-Villanueva
e-mail: david_teran01@yahoo.com.mx

A. Santiago
e-mail: aurelio.santiago@uat.edu.mx

M. P. Ponce Flores
e-mail: mirna_poncef@hotmail.com

E. Alvarez-Navarro
e-mail: ccalvar@docentes.uat.edu.mx

W. Pedrycz et al. (eds.), *Decision-Making in Computational Intelligence-Based Systems*, Studies in Systems, Decision and Control 628,
https://doi.org/10.1007/978-3-032-04056-5_16

1 Introduction

Forecasts are future events determined through a variable that changes over time [1]. Various forecasting methods are used to make predictions based on historical data from time series. Among the most widely recognized forecasting methods are Support Vector Machines, Autoregressive Integrated Moving Average (ARIMA), and Linear Regression, among others [2].

Given the wide range of available forecasting techniques, assessing their effectiveness in practical scenarios is essential. One of the most systematic ways to achieve this is through organized forecasting competitions, the M Competitions.

1.1 M1 Competition (1982)

In recent decades, several M Competitions have been held to identify the best forecasting methods on a provided set of time series. The M1 Competition, held in 1982, marked the beginning of a series of competitions designed to evaluate and compare the accuracy of forecasting methods in time series. Organized by Spyros Makridakis and his collaborators, this competition used a set of 1001 time series covering various domains, such as economics, demographics, and industry [3]. The main goal was to determine which of the existing methods at that time were more accurate in predicting the future behavior of the time series. Among the methods evaluated were classical techniques such as Exponential Smoothing (ES) and the ARIMA model, as well as simpler approaches like the Naive Method [3].

The results of the M1 Competition revealed that simpler methods, such as Naive and ES, often outperformed more complex techniques like ARIMA in terms of accuracy [3]. This finding surprised the forecasting community, as more sophisticated methods were expected to yield better results. However, the competition also highlighted the importance of simplicity and interpretability in forecasting methods, especially when dealing with time series that exhibit irregular patterns or noise [3]. The M1 laid the foundation for future competitions, establishing a benchmark for evaluating forecasting methods and demonstrating the need for continued research. Additionally, this event spurred the development of more robust evaluation metrics, such as the Symmetric Mean Absolute Percentage Error (sMAPE), which became a standard for measuring forecast accuracy [4].

1.2 M2 Competition (1993)

Following the success of the M1 Competition, the M2 Competition was held in 1993 to explore previous findings further and investigate new forecasting methods. Unlike M1, which used a set of synthetic data, M2 was based on 29 real-world time series, allowing for a more practical evaluation [5]. These series encompass various domains such as retail sales, industrial production, and economic data, presenting a more challenging environment for forecasting methods. In the M2 Competition, simple methods once again stood out, with ES and the Naive method achieving competitive results. However, this competition also introduced the evaluation of expert judgment-based methods, where human forecasters adjusted their predictions based on domain knowledge [5]. Although these approaches did not consistently outperform automatic methods, they demonstrated that combining quantitative and qualitative techniques could enhance accuracy in certain contexts. Furthermore, M2 reinforced the idea that simplicity and robustness are key qualities in forecasting methods, especially when dealing with real-world time series that may contain noise or non-stationary patterns [5].

The M2 Competition also served as a reminder of the importance of transparency and reproducibility in forecasting research. The organizers emphasized the need to document and share the methods used, enabling other researchers to replicate and validate the results. This approach laid the groundwork for later competitions, where fair comparisons and transparency became fundamental pillars [5]. Moreover, M2 contributed to the ongoing discussion on how to measure forecasting accuracy, a topic that remains relevant today and has been widely debated in the literature [4, 6].

1.3 M3 Competition (2000)

The M3 Competition, held in 2000, represented a significant advancement in the field of forecasting by introducing a much larger and more diverse dataset than its predecessors. With a total of 3003 time series, this competition included data from domains such as economics, finance, demographics, and industry, allowing for a more comprehensive evaluation of forecasting methods [7]. One of the most notable aspects of M3 was the inclusion of neural network-based methods, which were beginning to gain popularity at the time. Although these methods did not outperform traditional approaches like ES and ARIMA, they demonstrated promising potential for handling complex and nonlinear time series [7].

The results of the M3 Competition once again confirmed that simple methods, such as the Naive approach and ES, remained competitive in terms of accuracy. However, it was also observed that combination methods, which integrate predictions from multiple models, tended to outperform individual approaches [7]. This finding reinforced the idea that diversity in forecasting methods can enhance accuracy, especially when dealing with time series exhibiting heterogeneous patterns. Additionally, M3 emphasized the importance of rigorous evaluation and fair comparison among methods, setting a standard for future competitions [7].

M3 was also a pioneer in discussing the scalability of forecasting methods. With such a large dataset, the organizers were able to assess not only the accuracy but also the computational efficiency of the methods. This led to a greater interest in the development of algorithms that were accurate and scalable, a topic that remains highly relevant in the era of big data [8].

1.4 M4 Competition (2018)

The M4 Competition, held in 2018, marked a milestone in the history of forecasting competitions using 100,000 time series. This massive volume of data allowed for a more robust evaluation of forecasting methods, especially in terms of their ability to generalize across different domains and frequencies [9]. M4 also stood out due to the inclusion of hybrid methods that combined classical techniques such as ES with modern artificial intelligence approaches, such as Recurrent Neural Networks (RNN). The winning method, a hybrid of ES and RNN, demonstrated that combining statistical techniques with machine learning could outperform traditional approaches [10].

One of the most important findings of M4 was the superiority of the ensemble and combination methods. These approaches, which integrate predictions from multiple models, showed greater accuracy and robustness in the face of data variability [11]. Furthermore, M4 emphasized the importance of automation in the forecasting process, particularly when working with large volumes of data. Methods that incorporated automatic model selection and hyperparameter tuning techniques tended to yield better results [9].

M4 also contributed to the discussion on the interpretability of forecasting methods. While neural network-based approaches and other machine learning techniques have demonstrated high accuracy, their "black-box" nature creates challenges regarding transparency and trust. This concern has sparked a rising interest in developing methods that are both accurate and easily explainable, an area of research that continues to evolve [12].

1.5 M5 Competition (2020)

The M5 Competition, held in 2020, marked a significant shift in the focus of forecasting competitions by addressing the challenge of predicting large-scale retail sales data. Organized by Walmart, the competition involved forecasting 42,840 time series that represent sales data in various categories of products and stores [13]. Unlike previous competitions, the M5 emphasized the importance of granularity and scalability, as participants were required to predict sales at multiple hierarchical levels, including product, store, and regional levels. This added complexity highlighted the need for methods capable of handling high-dimensional data and capturing intricate dependencies between different levels of aggregation [14].

One of the key findings from the M5 Competition was the dominance of machine learning algorithms, particularly gradient-boosting methods. Four of the five best-performing methods used the LightGBM algorithm, a highly efficient implementation of gradient boosting, which demonstrated superior performance in handling the large and complex dataset [13]. The success of LightGBM underscored the growing importance of machine learning techniques in forecasting, especially when dealing with high-dimensional and hierarchical data. Additionally, the competition highlighted the value of feature engineering and the integration of external data sources, such as promotional calendars and economic indicators, in improving forecast accuracy [14].

The M5 Competition also brought attention to the challenges of evaluating forecasts in hierarchical settings. Traditional error metrics, such as sMAPE, were found to be less effective in capturing the nuances of hierarchical forecasting. As a result, new evaluation metrics, such as the Weighted Root Mean Squared Scaled Error, were introduced to better assess the accuracy of forecasts across different levels of aggregation [13]. This innovation in evaluation methodology has since influenced subsequent competitions and research in hierarchical forecasting.

1.6 M6 Competition (2022)

Finally, the M6 Competition, launched in 2022, shifted the focus from traditional time series forecasting to financial market prediction, specifically targeting the accuracy of asset price forecasts and their impact on investment decisions. This competition involved 100 assets, and participants were tasked with predicting future asset prices and providing probabilistic forecasts to guide investment strategies [15]. The M6 Competition introduced a novel approach by emphasizing the importance of collaboration and risk management in forecasting, encouraging participants to form teams and share insights to improve prediction accuracy [16].

One of the unique aspects of the M6 Competition was its focus on the practical application of forecasts in investment decision-making. Participants were required to not only predict asset prices but also optimize portfolios based on their forecasts, taking into account risk and return trade-offs. This approach highlighted the real-world implications of the accuracy of the forecast, as even small improvements in the prediction could lead to significant financial gains or losses [15]. The competition also emphasized the role of uncertainty quantification in forecasting, and participants were encouraged to provide probabilistic forecasts that captured the range of possible outcomes [12].

The M6 Competition underscored the growing importance of collaborative approaches in forecasting. By encouraging participants to work in teams and share knowledge, the competition demonstrated that collective intelligence could lead to more accurate and robust forecasts. Additionally, the M6 highlighted the need for advanced risk models and the integration of macroeconomic factors in financial forecasting, paving the way for future research in this area [12].

1.7 The Role of Competitions in Advancing Research

The International Institute of Forecasters is one of the main sponsors of these competitions, which focus on the development, dissemination, and application of forecasting knowledge. One of the key objectives of these competitions is to identify the best algorithms and techniques used, allowing for their replication across different time series datasets [9, 17]. Consequently, these competitions are highly relevant as they generate time series and forecasting methods that can serve as valuable references for various research efforts.

The M4 competition was selected for this study because it represents the most extensive evaluation of forecasting methods to date, covering a wide variety of domains and incorporating both statistical and machine learning-based approaches. The competition's results provide a well-documented benchmark that allows for a deeper analysis of how time series features influence forecasting accuracy. By focusing on the top five methods from M4, this research aims to uncover patterns that explain variations in sMAPE, contributing to a better understanding of method selection based on time series characteristics.

This research examines the relationship between time series features and the sMAPE of forecasts generated by the top five methods from the M4 Competition, aiming to identify the most suitable forecasting method. Section 1 introduces the research context, objectives, and significance of analyzing forecasting accuracy, with a focus on the M4 Competition's leading methods. Section 2 reviews the evolution of forecasting competitions (M1 to M6), highlighting key findings and methodological advancements that shape the study's foundation.

Section 3 presents the theoretical framework, covering time series fundamentals, traditional and machine learning forecasting methods, and accuracy metrics like sMAPE, while emphasizing the role of Random Forest (RF) in feature analysis. Section 4 details the methodology, including dataset extraction from M4, feature decomposition (trend, seasonality, curvature), correlation analysis, and RF model development optimized with Hyperband. Section 5 compares sMAPE errors and feature correlations across daily and monthly frequencies, revealing stronger predictive relationships in monthly series. Finally, Sect. 6 concludes by proposing data-driven method selection based on feature relevance, supported by references and tables throughout the analysis.

2 State of the Art

Recent years have witnessed significant progress in machine learning approaches for time series forecasting. As comprehensively reviewed in [18], contemporary methods encompass both linear techniques like penalized regressions and ensemble models, as well as nonlinear approaches including deep neural networks and tree-based algorithms. This study particularly emphasizes how hybrid architectures combining multiple methodologies have demonstrated superior predictive performance across various domains. The work systematically evaluates model assessment protocols and their application in economic and financial forecasting scenarios where high-frequency data analysis is crucial.

The application of deep learning to financial time series analysis has undergone a significant transformation in recent years. As documented in [19], modern approaches have shifted from traditional statistical methods to sophisticated neural architectures, particularly for handling the nonlinear and chaotic nature of financial markets. This comprehensive survey reveals that long short-term memory (LSTM) networks and hybrid models have emerged as particularly effective solutions, demonstrating superior capability in capturing complex temporal dependencies characteristic of financial data. The study further identifies key challenges in model interpretability and computational efficiency that continue to motivate current research directions.

Recent advances in feature engineering for time series analysis have demonstrated significant improvements in both predictive performance and model interpretability. The multivariate time series classification with fusion features framework proposed in [20] represents a breakthrough in multivariate time series classification through its innovative fusion of frequency-domain, temporal, and spatial features. By combining continuous wavelet transforms with attention mechanisms and graph neural networks, this approach achieves superior classification accuracy while maintaining interpretability through feature visualization capabilities. The methodology's effectiveness across diverse UEA datasets highlights the value of multi-domain feature integration for complex temporal patterns.

The multivariate time series for finance model presented in [21] further advances multivariate financial forecasting through its innovative multi-feature extraction framework. By combining LSTM networks with attention mechanisms for temporal patterns, dynamic periodic graphs for recurring trends, and Convolutional Neural Networks (CNN)-based spatial correlation analysis, this approach demonstrates how comprehensive feature extraction can enhance prediction accuracy in complex financial time series. The model's superior performance on benchmark datasets underscores the value of integrated feature analysis in financial applications.

The financial sector has witnessed transformative innovations in time series forecasting through advanced neural architectures. Cheng et al. [22] introduces a multimodal graph neural network that revolutionizes quantitative analysis by integrating heterogeneous data sources, including market prices, news sentiment, and economic events, into a unified knowledge graph framework. The model's dual attention mechanism not only improves prediction accuracy but also enhances interpretability, enabling traders to quantify the relative impact of different information modalities. This represents a significant advancement over traditional univariate approaches, particularly for high-frequency trading scenarios. Building on these foundations, [16] introduces an innovative BiLSTM-Graph Convolutional Network architecture that combines the temporal processing strengths of bidirectional LSTM with the relational learning capabilities of graph convolutional networks. This hybrid approach addresses a critical limitation of conventional models by simultaneously modeling both sequential patterns and inter-asset dependencies, particularly valuable in commodity markets. Empirical results demonstrate consistent performance improvements across multiple error metrics, with the integrated architecture reducing forecasting errors by 15–20% compared to standalone models.

Expanding this direction, [23] presents a novel ensemble architecture specifically designed for financial time series forecasting. The proposed framework synergistically combines CNN for spatial feature extraction, LSTM networks for temporal pattern recognition, and auto-regressive moving average components for modeling linear dependencies. Experimental validation demonstrates that this integrated approach achieves superior stability and accuracy compared to individual model components, particularly in volatile market conditions. The study provides empirical evidence that such hybrid architectures can effectively capture both short-term fluctuations and long-term trends in complex financial datasets.

To address latency-sensitive scenarios, [24] proposes a novel dual-LSTM framework that significantly enhances prediction stability in volatile market conditions. The sequential training mechanism, where each forecast informs subsequent predictions, represents a paradigm shift from traditional batch-processing approaches. This adaptive architecture demonstrates particular efficacy in cryptocurrency markets, where it achieves 30% greater accuracy than ARIMA models while maintaining computational efficiency.

Beyond architectural improvements to LSTM networks, recent work has focused on enhancing both predictive performance and interpretability in financial forecasting. The hybrid framework developed in [25] represents this trend by combining RNNs and attention mechanisms with feature attribution techniques. This approach successfully addresses the dual challenges of capturing financial market complexities while maintaining model transparency, demonstrating that sophisticated deep learning architectures can be made interpretable without sacrificing accuracy.

Further advancing LSTM architectures, [26] develops an adaptive LSTM-Batch Normalization network incorporating batch normalization and specialized loss functions. This innovation specifically targets the challenge of abrupt market shifts that often degrade model performance. The voting mechanism among parallel networks and inherited training strategy yields a 25% improvement in forecasting sharp price movements, as validated across multiple international market indices. These architectural refinements highlight the growing sophistication of neural approaches in addressing domain-specific forecasting challenges.

For practical forecasting applications, [27] introduces a sophisticated hybrid architecture combining temporal convolutional networks with feature selection techniques. The study's three-stage approach, employing random forests for dimensionality reduction, wavelet transforms for signal decomposition, and a temporal convolutional network for final prediction, demonstrates remarkable effectiveness in urban water demand forecasting. With a mean absolute percentage error of 1.22%, this work establishes a new benchmark for municipal resource planning, while its feature selection methodology provides valuable insights into the most influential demand drivers.

Complementing this perspective, [28] provides a meta-analysis of automated machine learning (AutoML) frameworks for temporal prediction tasks. The authors trace the methodological evolution from classical linear models to sophisticated deep learning architectures, while identifying persistent challenges in forecasting accuracy and computational efficiency. Their analysis reveals that AutoML solutions are increasingly critical for domains requiring real-time predictions, such as financial markets and meteorological modeling, where they help bridge the gap between theoretical models and operational requirements.

Agricultural forecasting presents unique challenges due to extreme volatility from climatic and market factors. Ray et al. [29] addresses this through an ARIMA-LSTM-RF hybrid that combines econometric modeling with machine learning. The model's generalized autoregressive conditional heteroskedasticity component successfully captures volatility clustering in commodity prices, while the RF-based feature selection identifies key predictive indicators from diverse agrometeorological datasets. Field tests demonstrate consistent performance across crop types, providing actionable insights for both farmers and commodity traders.

Complementing this work, [30] demonstrates the effectiveness of CNN-LSTM hybrids for intraday stock price forecasting using five-minute resolution data from India's national stock exchange. The study's systematic comparison of multiple architectural variants provides empirical evidence that convolutional layers effectively extract local price patterns, which LSTMs then contextualize temporally, achieving superior RMSE compared to conventional technical analysis methods. This approach offers practical advantages for algorithmic trading systems requiring both precision and computational efficiency.

The random forest algorithm has established itself as a fundamental tool in machine learning applications for time series analysis. As comprehensively examined in [31], this ensemble method operates by constructing multiple decision trees through bootstrap aggregation, significantly enhancing prediction stability and accuracy. The algorithm's inherent advantages include robust handling of missing values, effective management of class imbalance, and intrinsic feature selection capabilities, making it particularly suitable for complex temporal datasets. Furthermore, its parallelizable architecture enables efficient processing of high-dimensional data, a critical requirement in modern time series forecasting and classification tasks. The interpretability challenge in time series models has been addressed through significant improvements to the KernelSHAP algorithm, as presented in [32]. The enhanced version specifically addresses temporal dependencies that traditional feature importance methods often overlook. By optimizing computational efficiency and reliability in time-series contexts, this work enables more accurate identification of critical forecasting features while maintaining model transparency, a crucial requirement for operational decision-making systems.

Complementing these approaches, [33] advances correlation analysis through Dynamic Time Warping (DTW), offering superior performance for nonlinear time series compared to conventional methods. The DTW-based approach proves particularly valuable in environmental applications where temporal misalignments and scaling variations are common. Experimental results demonstrate its effectiveness in climate studies, providing more accurate similarity assessments for complex temporal patterns in remote sensing data.

Building upon these foundational strengths, [34] demonstrates the practical efficacy of random forests in agricultural remote sensing applications. The study's innovative integration of spectral, vegetation, and texture features for cotton classification achieves a remarkable 92.12% accuracy at the county scale. Notably, the random forest-based feature selection mechanism successfully identified gray-level co-ocurrence matrix texture features as the second most discriminative predictors, highlighting the algorithm's capability to uncover non-obvious patterns in multivariate temporal data. This work establishes random forest as a superior alternative to both support vector machines and artificial neural networks for precision agriculture applications, particularly when dealing with multi-temporal spectral data.

Recent advances in anomaly detection for multivariate time series have revealed significant challenges that require further investigation. As detailed in [35], while deep learning approaches like LSTM networks and autoencoders have shown effectiveness in identifying various anomaly types (point, contextual, and collective), they often lack explanatory capabilities. The study emphasizes that nearly 78% of current anomaly detection systems fail to provide interpretable explanations for flagged anomalies, severely limiting their practical utility in critical domains such as industrial monitoring and financial fraud detection. This interpretability gap presents a crucial research direction for developing more transparent anomaly detection frameworks.

Finally, the challenge of data scarcity in time series analysis is addressed through transfer learning techniques, as explored in [36]. Their comprehensive experiments across seismology, healthcare, and finance domains demonstrate that feature transferability depends heavily on the similarity of temporal patterns between source and target domains. Notably, the research found that models pretrained on seismic data showed 23% better performance when applied to financial volatility prediction compared to random initialization, suggesting that certain temporal characteristics may be domain-agnostic. However, the study also identifies key limitations in current transfer learning methods, particularly their sensitivity to hyperparameter choices and the minimum required dataset size for effective fine-tuning.

3 Theoretical Framework

3.1 Time Series and Forecasting

A time series is a sequence of data points collected over time, such as stock prices or monthly sales figures. Forecasting involves predicting future values based on past data. Financial time series are particularly challenging because they often exhibit non-linear and non-stationary patterns, meaning their behavior changes unpredictably over time [37]. To address this, researchers use a variety of methods, from simple statistical techniques to advanced machine learning models [1].

3.2 *Forecasting Methods*

Forecasting methods fall into two main categories: traditional statistical models and machine learning approaches. Traditional methods include ES, which weights recent observations more heavily [3], and ARIMA, a more sophisticated technique that models trends and seasonality [7]. These approaches are widely used for their interpretability and reliability with structured time series data.

Machine learning can be implemented using two main types of models: hybrid and non-hybrid. Non-hybrid models, such as RNN and Long LSTM networks, are specialized in understanding sequential data, like time series or text [19]. On the other hand, hybrid models, like the ES-RNN, combine traditional statistical techniques with artificial intelligence methods to leverage the strengths of both. This combination enables hybrid models to achieve higher accuracy, especially when dealing with highly variable or complex datasets [10]. By addressing the limitations of standalone approaches, hybrid methods are particularly effective in real-world applications that involve unpredictable or high-dimensional data.

3.3 *Measuring Forecast Accuracy*

sMAPE is a widely utilized metric for assessing the accuracy of forecasts. It quantifies the percentage discrepancy between forecasted and actual values, wherein lower sMAPE values denote superior performance [4]. The depiction of sMAPE is provided in Eq. 1.

3.4 *Time Series Features*

Time series analysis generally identifies important features vital for making predictions. The trend indicates the general movement of the data over time, whether it is ascending or descending. Seasonality involves consistent and repetitive patterns, like the increase in monthly sales during holidays. Curvature, on the other hand, addresses irregular variations that cannot be attributed to trend or seasonality alone [21]. Combined, these characteristics offer essential insights into the inherent patterns within time series data, aiding analysts in choosing the most suitable forecasting approach for any specific dataset.

3.5 *Random Forest for Forecasting*

RF constitutes a robust machine learning methodology characterized by the construction of multiple decision trees, whose aggregated predictions render it particularly effective for the forecasting of time series. Among its principal advantages is the capacity to process complex non-linear data relationships and to discern the most influential features that affect forecast accuracy [38]. Recent developments, such as the Hyperband optimization for parameter tuning, have significantly improved RF efficacy in the analysis of time series data [39], thereby enhancing its value as a tool for forecasters managing intricate temporal patterns.

4 Methodology

The proposed methodology for this research is shown in Fig. 1. It begins with the extraction of two subsets of financial time series (daily and monthly) from those used in the M4 Competition, labeled as $s_1, s_2, ..., s_n$. Additionally, forecasts provided by the competition for methods $m_1, m_2, ..., m_5$ are extracted. Next, the sMAPE error is calculated for the forecasts generated by the top five methods. Furthermore, various characteristics of $s_1, s_2, ..., s_n$ are obtained for both time series subsets. Finally, predictive models referred to as $RFm_1, RFm_2, ..., RFm_5$, are developed using RF to explore the possible correlation between the characteristics of the series and the error of the sMAPE.

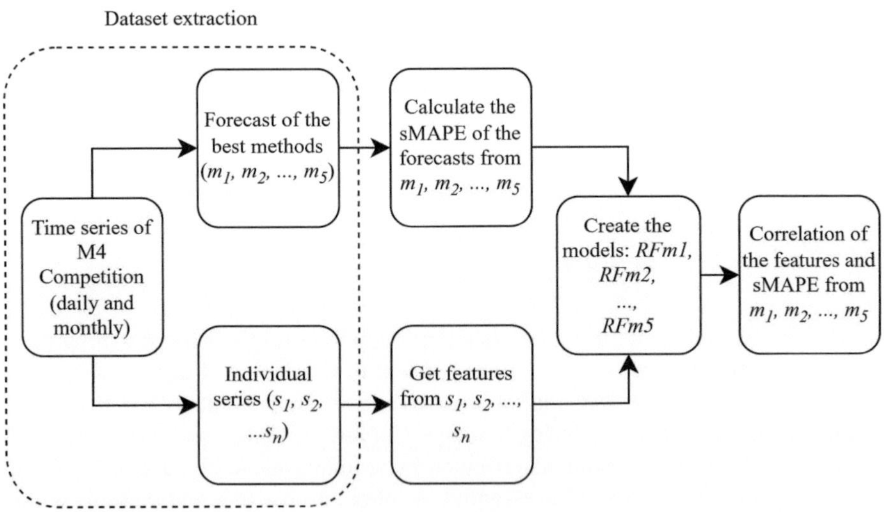

Fig. 1 Methodology

RF stands out as a highly effective machine learning approach for classification and forecasting tasks, particularly when handling complex data sets with high-dimensionality and non-linear relationships [40]. Its ensemble nature, which integrates multiple decision trees trained on random subsets of data, enhances its robustness and reduces overfitting, making it a reliable choice for pattern recognition and predictive modeling [41]. Additionally, recent advancements in RF techniques, such as refining tree selection based on accuracy and correlation metrics, have enhanced their predictive capabilities over standard implementations [40, 41]. These improvements make RF particularly suitable for examining the relationship between time series characteristics and forecasting errors, as it efficiently captures complex dependencies that simpler models might overlook.

Regarding the methodology, in the next section, the dataset content is presented.

4.1 Dataset Extraction

The complete dataset of the M4 Competition is detailed in Table 1. This research focuses on daily and monthly frequency financial time series. Financial time series are used due to their non-linear, non-stationary random patterns [37]. As shown in the table, the daily financial series includes a total of 1559 observations, while the monthly series comprises 10,987 observations.

4.2 Forecasting Methods in the M4 Competition

This research uses the forecasts from the five top-performing methods in the M4 Competition to analyze their correlation with various characteristics of the financial time series used in that competition. Table 2 presents the names of these methods along with their final classifications.

Table 1 Complete dataset [42]

Frequency	Demographic	Financial	Industry	Macro	Micro	Other	Total
Yearly	1088	6519	3716	3903	6538	1236	23,000
Quarterly	1858	5305	4637	5315	6020	865	24,000
Monthly	5728	10,987	10,017	10,016	10,975	277	48,000
Weekly	24	164	6	41	112	12	359
Daily	10	1559	422	127	1476	633	4227
Hourly	0	0	0	0	0	414	414
Total	8708	24,534	18,798	19,402	25,121	3437	100,000

Table 2 Best methods of M4 competition [9, 43]

Rank	Acronym	Method
1	m_1	Exponential smoothing—Recurrent neural Network (ES-RNN) hybrid models
2	m_2	M4metalearning (FFORMA)
3	m_3	Weighted ensemble of statistical models
4	m_4	Combination based forecasting method
5	m_5	GROEC

4.3 Error Calculation and Time Series Features

Error metrics are used to compare the performance of different datasets [1]. The $sMAPE$ error is widely used to assess the accuracy of forecasting models. The formula is expressed as follows:

$$sMAPE = \frac{2}{n}\sum_{i=1}^{n}\frac{|y_i - \hat{y}_i|}{|y_i| + |\hat{y}_i|} \times 100 \tag{1}$$

where n is the total number of observations, y_i is the actual value of the i-th observation, and $\hat{y}_i$ represents the forecasted value for that observation.

Table 3 presents the average $sMAPE$ error obtained from the results of methods m_1, m_2, ..., m_5, broken down by daily and monthly frequency. It can be observed that the monthly frequency showed higher average errors compared to the daily frequency, indicating the increased difficulty of this time series subset.

The decomposition of a time series is essential for understanding its behavior. Figure 2 illustrates the feature extraction process from time series, resulting in three main features: trend, seasonality, and curvature. The trend reflects the overall direction of the series over time, while seasonality represents the periodic patterns that repeat at regular intervals, and curvature shows the variations in the series that are not explained by trend and seasonality. However, in the case of daily time series, this approach is not applicable because the data may exhibit multiple patterns within the data set [1].

Table 3 sMAPE of the best methods

Frequency	m_1	m_2	m_3	m_4	m_5
Daily	3.5648	3.5941	3.1000	3.4654	3.3722
Monthly	13.7317	13.5570	13.6035	13.6481	13.7525

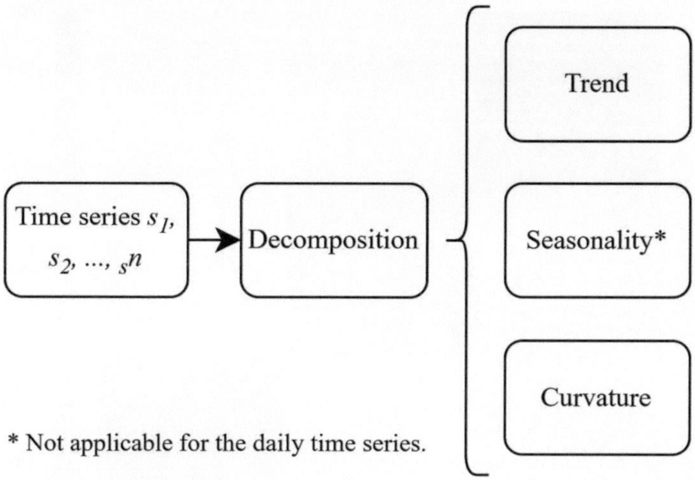

Fig. 2 Time series decomposition

4.4 *Correlation Analysis*

Correlation matrices are essential for assessing the relationship between the features and the accuracy of m_1, m_2, ..., m_5. Figures 3 and 4 present the correlation matrices between the features of the time series and the sMAPE error of m_1 for both daily and monthly frequencies.

In Fig. 3, it can be seen that the correlation between the features and the sMAPE error is relatively low for the subset of the daily time series. This suggests that the extracted features may not significantly influence the forecasting accuracy in this case.

However, in Fig. 4, it can be seen that the correlation coefficient of all the features with the sMAPE error is higher than that of the daily series. Furthermore, it is noteworthy that the trend feature stands out with the highest correlation value, indicating a stronger relationship.

The correlation analysis presented in this section highlights the relationship between time series features and forecasting accuracy, particularly in terms of their impact on sMAPE error across different frequencies. These insights serve as the foundation for developing predictive models, where a RF classifier is employed to further explore these dependencies. By leveraging the most influential features identified in the correlation matrices, the predictive models aim to enhance the accuracy of forecasting method selection, ensuring a more data-driven approach in determining the optimal technique for different time series characteristics.

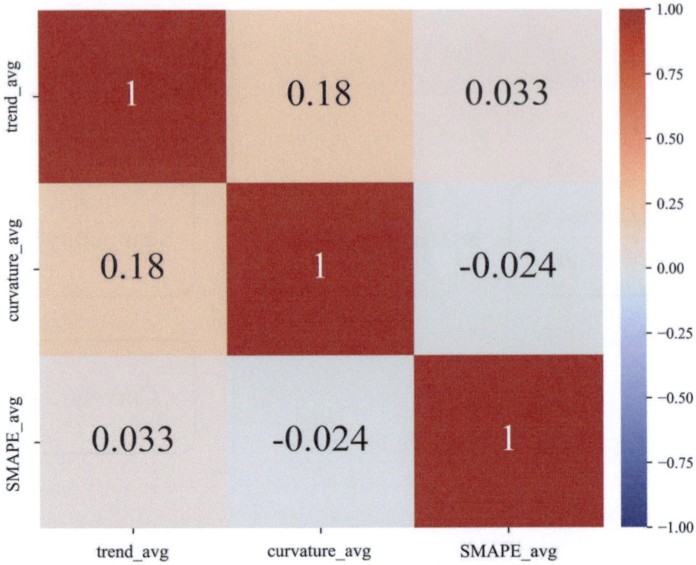

Fig. 3 Daily correlation matrix of m_1

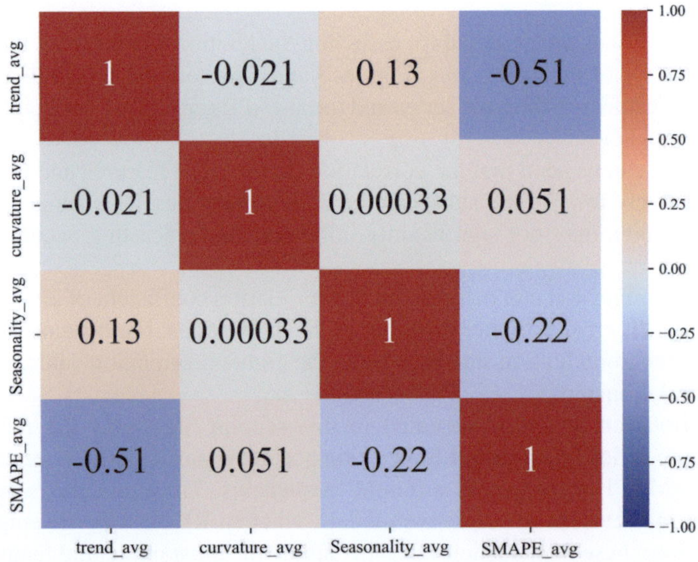

Fig. 4 Monthly correlation matrix of m_1

4.5 Development of Predictive Models

RF is a powerful ensemble learning method that has gained popularity due to its robustness, scalability, and ability to handle high-dimensional data. Introduced by Breiman [38], RF operates by constructing multiple decision trees during training and outputting the mode of the classes (for classification) or the mean prediction (for regression) of the individual trees. This ensemble approach reduces the risk of overfitting and improves generalization, making RF particularly effective for complex datasets with non-linear relationships [38]. In the context of time series forecasting, RF has been successfully applied to capture intricate patterns and interactions between features, especially when combined with feature engineering and hyperparameter optimization techniques.

Figure 5 illustrates the input process of the features and the sMAPE error, as well as the data output in the RF classifier applied to methods $m_1, m_2, ..., m_5$, resulting in the generation of models RFm_1, RFm_2, ..., RFm_5. Finally, the correlation coefficient between the sMAPE error of the models and the forecasting methods is calculated.

In addition, Table 3 shows the parameters configured for the RF classifier. To optimize the hyperparameters of the RF model, we employed Hyperband, a state-of-the-art algorithm for hyperparameter optimization. Hyperband is designed to efficiently allocate resources to the most promising configurations by iteratively evaluating multiple hyperparameter combinations and discarding underperforming ones early in the process [39]. This approach significantly reduces the computational cost compared to traditional methods like grid search or random search, while still achieving competitive performance.

Table 4 shows the key parameters tuned using Hyperband include the "Number of trees", which specifies the number of trees in the RF forest, "Min samples split", which represents the minimum number of samples required to split an internal node, and "Random state", which sets the seed for the random number generator to ensure reproducibility. By leveraging Hyperband, we were able to efficiently explore a wide

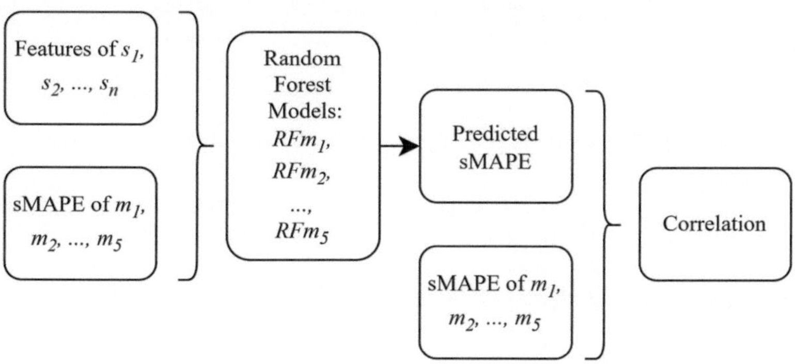

Fig. 5 Random forest process

Table 4 Random forest parameters

Parameter	Value
Number of trees	[100–300]
Min samples split	[2–5]
Random state	1

range of hyperparameter configurations and identify the optimal settings for RF models, thereby improving their predictive accuracy and robustness.

5 Results

In Tables 5 and 6, the correlations and sMAPE error for the daily and monthly series models are shown, respectively. The correlation of the sMAPE error is low overall for all models, with values ranging from 0.0349 to 0.0801 in Table 5. Additionally, the sMAPE error values are high, ranging from 54.1333 to 66.0571. This suggests that, for the daily series, the relationship between the features and the error is considerably low. On the other hand, Table 6 shows that the correlations are higher, with values from 0.5285 to 0.05826. Also, the sMAPE values are lower, ranging from 40.9428 to

Table 5 Results of random forest regression on daily financial time series

Daily		
Model	Correlation	sMAPE
RFm_1	0.0462	54.1333
RFm_2	0.0801	55.0275
RFm_3	0.0357	66.0571
RFm_4	0.0480	55.4135
RFm_5	0.0349	55.8368

Table 6 Results of random forest regression on monthly financial time series

Monthly		
Model	Correlation	sMAPE
RFm_1	0.5826	40.9428
RFm_2	0.5313	43.8302
RFm_3	0.5398	42.9459
RFm_4	0.5306	43.7933
RFm_5	0.5285	43.9104

43.9104. This suggests that for the monthly frequency series, the extracted features are more relevant in predicting the sMAPE error.

6 Conclusions

In the analysis conducted, significant differences were identified in the predictive capability of the extracted features for financial time series of different frequencies. The correlation between the features and the sMAPE error in the daily time series is relatively low. This suggests that, in such a series, the extracted features have a limited impact on the accuracy of the forecasting models. Additionally, the sMAPE error values are higher, indicating that accurately predicting in this subset of series may be complicated due to the variable and noisy nature of daily time series.

On the other hand, the correlation between the features and the sMAPE error in the monthly time series is stronger. This suggests that, in series with this frequency, the extracted features are more effective in capturing significant patterns that help identify the most suitable forecasting method. Furthermore, the sMAPE error values are lower compared to the models for the daily time series.

The research suggests that by understanding the relationship, the process of selecting the most accurate forecasting method could be simplified, especially in monthly time series, where a correlation between the features and the sMAPE error of the method is observed.

AI Use Statement: Some portions of this manuscript were enhanced with the assistance of generative AI tools. The authors carefully reviewed, edited and approved all AI-generated content. The authors take full responsibility for the accuracy and integrity of the entire chapter.

References

1. Hyndman, R.J., Athanasopoulos, G.: Forecasting: principles and Practice, 3rd edn. OTexts, Monash University, Australia (2024). https://otexts.com/fpp3/
2. Kumar, J., Goomer, R., Singh, A.K.: Long short term memory recurrent neural network (lstm-rnn) based workload forecasting model for cloud datacenters. Procedia Comput. Sci. **125**, 676–682 (2018), The 6th International Conference on Smart Computing and Communications. https://doi.org/10.1016/j.procs.2017.12.087. https://www.sciencedirect.com/science/article/pii/S1877050917328557
3. Makridakis, S., Andersen, A., Carbone, R., Fildes, R., Hibon, M., Lewandowski, R., Newton, J., Parzen, E., Winkler, R.: The accuracy of extrapolation (time series) methods: results of a forecasting competition. J. Forecasting **1**(2), 111–153 (1982). https://doi.org/10.1002/for.3980010202. https://onlinelibrary.wiley.com/doi/abs/10.1002/for.3980010202
4. Hyndman, R.J., Koehler, A.B.: Another look at measures of forecast accuracy. Int. J. Forecast. **22**(4), 679–688 (2006). https://doi.org/10.1016/j.ijforecast.2006.03.001. https://www.sciencedirect.com/science/article/pii/S0169207006000239

5. Makridakis, S., Chatfield, C., Hibon, M., Lawrence, M., Mills, T., Ord, K., Simmons, L.F.: The m2-competition: a real-time judgmentally based forecasting study. Int. J. Forecast. **9**(1), 5–22 (1993). https://doi.org/10.1016/0169-2070(93)90044-N. https://www.sciencedirect.com/science/article/pii/016920709390044N
6. Armstrong, J.S.: Principles of Forecasting: a Handbook for Researchers and Practitioners. Springer Science and Business Media (2001). https://www.springer.com/gp/book/9780792374015
7. Makridakis, S., Hibon, M.: The m3-competition: results, conclusions and implications. Int. J. Forecast. **16**(4), 451–476 (2000), The M3- Competition. https://doi.org/10.1016/S0169-2070(00)00057-1. https://www.sciencedirect.com/science/article/pii/S0169207000000571
8. Fildes, R., Hibon, M., Makridakis, S., Meade, N.: Generalising about univariate forecasting methods: further empirical evidence. Int. J. Forecast. **14**(3), 339–358 (1998). https://doi.org/10.1016/S0169-2070(98)00009-0. https://www.sciencedirect.com/science/article/pii/S0169207098000090
9. Makridakis, S., Spiliotis, E., Assimakopoulos, V.: The m4 competition: results, findings, conclusion and way forward. Int. J. Forecast. **34**(4), 802–808 (2018). https://doi.org/10.1016/j.ijforecast.2018.06.001. https://www.sciencedirect.com/science/article/pii/S0169207018300785
10. Fry, C., Brundage, M.: The m4 forecasting competition 'a practitioner's view. Int. J. Forecast. **36**(1), 156–160 (2020). https://doi.org/10.1016/j.ijforecast.2019.02.013. https://www.sciencedirect.com/science/article/pii/S0169207019301189. M4 Competition
11. Makridakis, S., Spiliotis, E., Assimakopoulos, V.: Statistical and machine learning forecasting methods: concerns and ways forward. PLOS ONE **13**(3), 1–26 (2018). https://doi.org/10.1371/journal.pone.0194889. https://doi.org/10.1371/journal.pone.0194889
12. Talagala, T.S., Hyndman, R.J., Athanasopoulos, G.: Meta-learning how to forecast time series. J. Forecasting **42**(6), 1476–1501 (2023). https://doi.org/10.1002/for.2963. https://onlinelibrary.wiley.com/doi/abs/10.1002/for.2963
13. Makridakis, S., Spiliotis, E., Assimakopoulos, V.: M5 accuracy competition: results, findings, and conclusions. Int. J. Forecast. **38**(4), 1346–1364 (2022). https://doi.org/10.1016/j.ijforecast.2021.11.013. https://www.sciencedirect.com/science/article/pii/S0169207021001874. Special Issue: M5 competition
14. Theodorou, E., Wang, S., Kang, Y., Spiliotis, E., Makridakis, S., Assimakopoulos, V.: Exploring the representativeness of the m5 competition data. Int. J. Forecast. **38**(4), 1500–1506 (2022), Special Issue: M5 competition. https://doi.org/10.1016/j.ijforecast.2021.07.006. https://www.sciencedirect.com/science/article/pii/S0169207021001175
15. Makridakis, S., Spiliotis, E., Hollyman, R., Petropoulos, F., Swanson, N., Gaba, A.: The m6 forecasting competition: bridging the gap between forecasting and investment decisions (2023). https://arxiv.org/abs/2310.13357
16. Lazcano, A., Herrera, P.J., Monge, M.: A combined model based on recurrent neural networks and graph convolutional networks for financial time series forecasting. Mathematics **11**(1) (2023). https://doi.org/10.3390/math11010224. https://www.mdpi.com/2227-7390/11/1/224
17. Petropoulos, F., Makridakis, S., Assimakopoulos, V., Nikolopoulos, K.: 'horses for courses' in demand forecasting. Eur. J. Oper. Res. **237**(1), 152–163 (2014). https://doi.org/10.1016/j.ejor.2014.02.036. https://www.sciencedirect.com/science/article/pii/S0377221714001714
18. Masini, R.P., Medeiros, M.C., Mendes, E.F.: Machine learning advances for time series forecasting. JJ. Econ. Surv. **37**(1), 76–111 (2023). https://doi.org/10.1111/joes.12429. https://onlinelibrary.wiley.com/doi/abs/10.1111/joes.12429
19. Tang, Y., Song, Z., Zhu, Y., Yuan, H., Hou, M., Ji, J., Tang, C., Li, J.: A survey on machine learning models for financial time series forecasting. Neurocomputing **512**, 363–380 (2022). https://doi.org/10.1016/j.neucom.2022.09.003. https://www.sciencedirect.com/science/article/pii/S092523122201089X
20. Du, M., Wei, Y., Hu, Y., Zheng, X., Ji, C.: Multivariate time series classification based on fusion features. Expert. Syst. Appl. **248**, 123452 (2024). 10.1016/j.eswa.2024.123452. https://www.sciencedirect.com/science/article/pii/S0957417424003178

21. Chen, J., Guan, A., Du, J., Ayush, A.: Multivariate time series prediction with multi-feature analysis. Expert. Syst. Appl. **268**, 126302 (2025). https://doi.org/10.1016/j.eswa.2024.126302. https://www.sciencedirect.com/science/article/pii/S0957417424031695
22. Cheng, D., Yang, F., Xiang, S., Liu, J.: Financial time series forecasting with multi-modality graph neural network. Pattern Recogn. **121**, 108218 (2022). https://doi.org/10.1016/j.patcog.2021.108218. https://www.sciencedirect.com/science/article/pii/S003132032100399X
23. He, K., Yang, Q., Ji, L., Pan, J., Zou, Y.: Financial time series forecasting with the deep learning ensemble model. Mathematics **11**(4) (2023). https://doi.org/10.3390/math11041054
24. Gajamannage, K., Park, Y., Jayathilake, D.I.: Real-time forecasting of time series in financial markets using sequentially trained dual-lstms. Expert. Syst. Appl. **223**, 119879 (2023). https://doi.org/10.1016/j.eswa.2023.119879. https://www.sciencedirect.com/science/article/pii/S0957417423003809
25. Liu, S., Wu, K., Jiang, C., Huang, B., Ma, D.: Financial time-series forecasting: towards synergizing performance and interpretability within a hybrid machine learning approach. arXiv preprint (2023). https://doi.org/10.48550/arXiv.2401.00534. https://arxiv.org/abs/2401.00534
26. Fang, Z., Ma, X., Pan, H., Yang, G., Arce, G.R.: Movement forecasting of financial time series based on adaptive lstm-bn network. Expert. Syst. Appl. **213**, 119207 (2023). https://doi.org/10.1016/j.eswa.2022.119207. https://www.sciencedirect.com/science/article/pii/S0957417422022254
27. Guo, J., Sun, H., Du, B.: Multivariable time series forecasting for urban water demand based on temporal convolutional network combining random forest feature selection and discrete wavelet transform. Water Resour. Manag. **36**(9), 3385–3400 (2022). https://doi.org/10.1007/s11269-022-03207-z
28. Alsharef, A., Aggarwal, K., Sonia, Kumar, M., Mishra, A.: Review of ML and AutoML solutions to forecast time-series data. Arch. Comput. Methods Eng. **29**(7), 5297–5311 (2022). https://doi.org/10.1007/s11831-022-09765-0
29. Ray, S., Lama, A., Mishra, P., Biswas, T., Sankar Das, S., Gurung, B.: An arima-lstm model for predicting volatile agricultural price series with random forest technique image 1. Appl. Soft Comput. **149**, 110939 (2023). https://doi.org/10.1016/j.asoc.2023.110939. https://www.sciencedirect.com/science/article/pii/S1568494623009572
30. Mehtab, S., Sen, J.: Analysis and forecasting of financial time series using CNN and LSTM-based deep learning models. In: Sahoo, J.P., Tripathy, A.K., Mohanty, M., Li, K.C., Nayak, A.K. (eds.) Advances in Distributed Computing and Machine Learning, pp. 405–423. Springer Singapore, Singapore (2022). https://doi.org/10.1007/978-981-16-4807-6_39
31. Salman, K., Steiti: Random forest algorithm overview. Babylon. J. Mach. Learn. **2024**, 69–79 (2024). https://doi.org/10.58496/BJML/2024/007. https://journals.mesopotamian.press/index.php/BJML/article/view/417
32. Villani, M., Lockhart, J., Magazzeni, D.: Feature importance for time series data: improving KernelSHAP. arXiv preprint (2022). https://doi.org/10.48550/arXiv.2210.02176. https://arxiv.org/abs/2210.02176
33. Liu, Y., Guo, H., Zhang, L., Liang, D., Zhu, Q., Liu, X., Lv, Z., Dou, X., Gou, Y.: Research on correlation analysis method of time series features based on dynamic time warping algorithm. IEEE Geosci. Remote. Sens. Lett. **20**, 1–5 (2023). https://doi.org/10.1109/LGRS.2023.3285788
34. Fei, H., Fan, Z., Wang, C., Zhang, N., Wang, T., Chen, R., Bai, T.: Cotton classification method at the county scale based on multi-features and random forest feature selection algorithm and classifier. Remote Sens. **14**(4) (2022). https://doi.org/10.3390/rs14040829
35. Li, G., Jung, J.J.: Deep learning for anomaly detection in multivariate time series: Approaches, applications, and challenges. Inf. Fusion **91**, 93–102 (2023). https://doi.org/10.1016/j.inffus.2022.10.008. https://www.sciencedirect.com/science/article/pii/S1566253522001774
36. Otović, E., Njirjak, M., Jozinović, D., Mauša, G., Michelini, A., Štajduhar, I.: Intra-domain and cross-domain transfer learning for time series data—how transferable are the features? Knowl Based Syst **239**, 107976 (2022). https://doi.org/10.1016/j.knosys.2021.107976. https://www.sciencedirect.com/science/article/pii/S0950705121010984

37. Cao, J., Li, Z., Li, J.: Financial time series forecasting model based on Ceemdan and LSTM. Physica A: Stat. Mech. Its Appl. **519**, 127–139 (2019). https://doi.org/10.1016/j.physa.2018.11.061. https://www.sciencedirect.com/science/article/pii/S0378437118314985
38. Breiman, L.: Random forests. Mach. Learn. **45**(1), 5–32 (2001). https://doi.org/10.1023/A:1010933404324
39. Li, L., Jamieson, K., DeSalvo, G., Rostamizadeh, A., Talwalkar, A.: Hyperband: a novel bandit-based approach to hyperparameter optimization. J. Mach. Learn. Res. **18**(1), 6765–6816 (2018). https://www.jmlr.org/papers/volume18/16-558/16-558.pdf
40. Random forest algorithm overview. Babylon. J. Mach. Learn. **2024**, 69–79 (2024). https://doi.org/10.58496/BJML/2024/007. https://journals.mesopotamian.press/index.php/BJML/article/view/417
41. Sun, Z., Wang, G., Li, P., Wang, H., Zhang, M., Liang, X.: An improved random forest based on the classification accuracy and correlation measurement of decision trees. Expert Syst. Appl. **237**, 121549 (2024). https://doi.org/10.1016/j.eswa.2023.121549. https://www.sciencedirect.com/science/article/pii/S0957417423020511
42. Makridakis, S., Spiliotis, E., Assimakopoulos, V.: The m4 competition: 100,000 time series and 61 forecasting methods. Int. J. Forecast. **36**(1), 54–74 (2020), M4 Competition. https://doi.org/10.1016/j.ijforecast.2019.04.014. https://www.sciencedirect.com/science/article/pii/S0169207019301128
43. Jaganathan, S., Prakash, P.: A combination-based forecasting method for the m4-competition. Int. J. Forecast. **36**(1), 98–104 (2020). https://doi.org/10.1016/j.ijforecast.2019.03.030. https://www.sciencedirect.com/science/article/pii/S0169207019301542. M4 Competition

Analysis and Improvement of Runway Operations in an Airport in Central Mexico: A Simulation Case Study

David-Leonardo Padilla-Georgge, Emmy Getsel Sánchez-Córdova, José-Abel Bañuelos-Terés, Santiago-Omar Caballero-Morales, and Diana Sánchez-Partida

Abstract Discrete-event simulation has become an essential strategic tool for analyzing, assessing, and enhancing operational efficiency across service and manufacturing sectors. Airport operations present particularly challenging scenarios for simulation due to the inherent complexity, interdependencies among various processes, and stringent constraints. This study focuses on modeling and optimizing takeoff and landing operations at Bajío International Airport (BJX) in León, Guanajuato, Mexico. Real-time runway usage data from 368 flights was collected using the global aircraft tracking service Flightradar24. Using PROMODEL software, the current state of runway occupancy and utilization at BJX was analyzed. The simulation identified delays and inefficient use of runways as taxiways, negatively impacting air traffic operations. Based on this information, an improved taxiway configuration was proposed and validated through additional simulations. The results confirm that introducing direct taxiways connecting runway thresholds significantly reduces runway occupancy times, improving overall operational efficiency.

Keywords Discrete-event simulation · Optimization · Airport operations · Runway efficiency · Congestion management

D.-L. Padilla-Georgge · E. G. Sánchez-Córdova · J.-A. Bañuelos-Terés ·
S.-O. Caballero-Morales · D. Sánchez-Partida (✉)
Department of Logistics and Supply Chain Management, UPAEP University, Puebla, Puebla, Mexico
e-mail: diana.sanchez@upaep.mx

E. G. Sánchez-Córdova
Department of Physics, Mathematics and Engineering, University of Sonora, Navojoa, Sonora, Mexico

W. Pedrycz et al. (eds.), *Decision-Making in Computational Intelligence-Based Systems*, Studies in Systems, Decision and Control 628,
https://doi.org/10.1007/978-3-032-04056-5_18

1 Introduction

Airport operations constitute complex systems characterized by a network of interrelated activities, ranging from passenger arrival to aircraft departure; air transport is a relevant indicator of territorial structuring and organization, from mobility to accessibility, with great dynamism. The efficient management of these processes significantly influences airport facilities' overall performance, customer satisfaction, and economic viability [1]. Among these, Bajío International Airport (BJX) is a vital transportation hub, strategically located in Mexico's central Bajío region—a dynamic area notable for substantial industrial and economic activities.

Airport infrastructure typically comprises runways, taxiways, aircraft parking stands, boarding bridges, passenger and cargo terminals, and integrated ground transportation hubs [2]. Efficient runway operations, including takeoff and landing, are crucial in maintaining flight punctuality and overall passenger satisfaction [3]. Baggage handling demands precise synchronization with flight schedules, accounting for weight constraints and cargo hold capacities. Similarly, aircraft refueling and servicing require meticulous timing to minimize ground times and enhance turnaround efficiency [4].

The growing complexity of these operations underscores the importance of simulation techniques as analytical tools. Simulation allows stakeholders to explore hypothetical scenarios, identify operational bottlenecks, and make informed strategic decisions [5]. Recent literature highlights discrete-event simulation as an effective method for airport capacity assessment, strategic resource allocation, and operational optimization. Moreover, simulation models facilitate proactive management, enabling airport authorities to anticipate operational disruptions, mitigate delays, and enhance passenger experiences [6, 7].

In recent years, BJX has witnessed significant growth in passenger and cargo volumes, driven largely by the expanding industrial activity in Guanajuato and the broader Bajío region. Despite this positive growth, the airport faces challenges including runway congestion, prolonged occupancy times, and inefficiencies in taxiway management. These issues affect operational efficiency and limit BJX's ability to accommodate future demand increases.

Addressing these operational constraints, this research applies discrete-event simulation techniques using PROMODEL software [8]. Real-time operational data from Flightradar24 was utilized to construct and calibrate simulation models that accurately reflect current runway usage patterns. The primary objective was to evaluate the impact of existing taxiway configurations on runway occupancy and propose an optimized layout to mitigate identified delays.

This paper provides a comprehensive analytical assessment of runway and taxiway operations, proposing tangible infrastructure improvements and evaluating their operational impact through robust simulation methods. It contextualizes the results within broader operational management practices in airport systems, considering alternatives and trade-offs. Ultimately, this research provides valuable insights into infrastructure planning and operational efficiency applicable to BJX and airports experiencing similar growth-driven operational pressures.

2 Airport Runway Operations

An essential element of air traffic management (ATM) at BJX, which directly impacts its design and operational capacity, is aircraft turnaround time (TAT). According to the International Air Transport Association (IATA), TAT refers to the duration of the turnaround process, from the aircraft reaching its parking position (Actual In Block Time, AIBT) to departing from the position (Actual Off Block Time, AOBT) [9]. This process involves specific actions such as positioning and removing aircraft wheel chocks, passenger boarding and disembarkation, baggage handling, refueling, catering, and various maintenance checks.

Gate occupancy time (GOT) is closely related to TAT, as prolonged GOT translates directly into lost commercial revenue. Airports typically charge airlines fees based on when aircraft occupy gates for boarding and disembarkation, making minimizing GOT a critical financial and strategic imperative [10]. Efficient management of passenger boarding and disembarkation significantly impacts GOT, as confirmed by numerous field studies highlighting operational bottlenecks during these critical phases. Each flight involves diverse ground assistance activities, varying wildly in execution, frequency, and service requirements. Thus, each turnaround cycle becomes unique, demanding meticulous coordination among multiple ground-handling teams. Airlines focus their efforts on the highly efficient use of airport infrastructure to make their operations financially viable [11].

The complexity and variability of ground operations need robust scheduling and resource allocation practices. For instance, coordinating ground handling vehicles, personnel availability, baggage transfer logistics, and refueling schedules requires precise timing to avoid delays. Unpredictable weather conditions, mechanical issues, and passenger behavior influence ground operations, further complicating efficient scheduling and operational flow.

Taxiways are critical airport infrastructure components designed to enhance operational efficiency by enabling aircraft's smooth and safe movement between runways, gates, and parking stands. High-traffic airports typically benefit from parallel taxiways, facilitating simultaneous aircraft operations, reducing operational delays, and enhancing throughput. An innovative example is Munich Airport's (MUC) implementation of a taxiway roundabout system designed to manage complex traffic flows efficiently and minimize local bottlenecks [12]. Such designs demonstrate significant potential for improving aircraft ground operations.

Advanced technological tools and platforms, including Flightradar24 [13] and FlightAware [14], have become integral in managing airport runway operations by providing comprehensive real-time, historical, and predictive flight data. These platforms facilitate detailed operational analysis, allowing airport managers to optimize resource allocation, identify inefficiencies, and streamline operations. Data obtained from these systems is typically managed through sophisticated relational databases and analyzed using statistical modeling techniques, enabling systematic performance evaluations and targeted operational improvements [15].

Aircraft response time remains a critical operational metric, significantly influencing overall airport efficiency. Delays in aircraft turnaround processes can quickly escalate into broader operational disruptions, affecting subsequent flights, passenger satisfaction, and airline schedules. Hence, understanding and optimizing response times through continuous monitoring, analysis, and strategic interventions is essential to maintain punctuality, efficiency, and overall operational effectiveness in the various commercial aviation [16].

3 Case Study Description

Bajío International Airport (BJX) is located in Silao, in the state of Guanajuato, Mexico, near the city of Leon. Strategically located in the Bajío region, it is key in connecting the regional economy with national and international markets. It serves passenger and cargo operations and is a key gateway to the economic zones surrounding Leon, Silao, and the state capital.

As illustrated in Fig. 1, BJX is equipped with a single runway, designated 13-31, which is utilized for both takeoff and landing operations. Notably, the same runway also functions as a taxiway. This dual-purpose use significantly limits operational flexibility, often resulting in delays and decreased efficiency. The taxiway infrastructure at BJX is composed of three designated taxiways—

A (Alfa), B (Bravo), and C (Charly)—which guide aircraft to and from the main runway and aprons.

The commercial apron includes nine aircraft stands labeled 1 through 9, designed to support boarding, deboarding, refueling, and ground servicing operations. Additionally, the General Aviation Apron has been adapted to accommodate two positions suitable for regional jets such as the Embraer 190, thus providing flexibility in stand allocation during peak periods or irregular operations.

The operational bottlenecks at BJX are intensified by the need to manage takeoff and landing slots efficiently while handling ground services such as baggage loading/unloading and aircraft refueling. These processes are critical in minimizing turnaround time and ensuring on-time departures and arrivals. According to several authors [17], delays in these areas can cascade throughout the flight network, leading to increased fuel consumption, extended crew and ground staff working hours, and possible compensation for delayed passengers. It is further emphasized that inefficient gate management and tarmac operations contribute directly to operational costs and customer dissatisfaction [18].

Historical data highlights the airport's increasing significance within the national aviation system. As illustrated in Fig. 2, BJX has consistently grown in passenger traffic, rising from 479,783 passengers in 1992 to over 3 million in 2024. During this

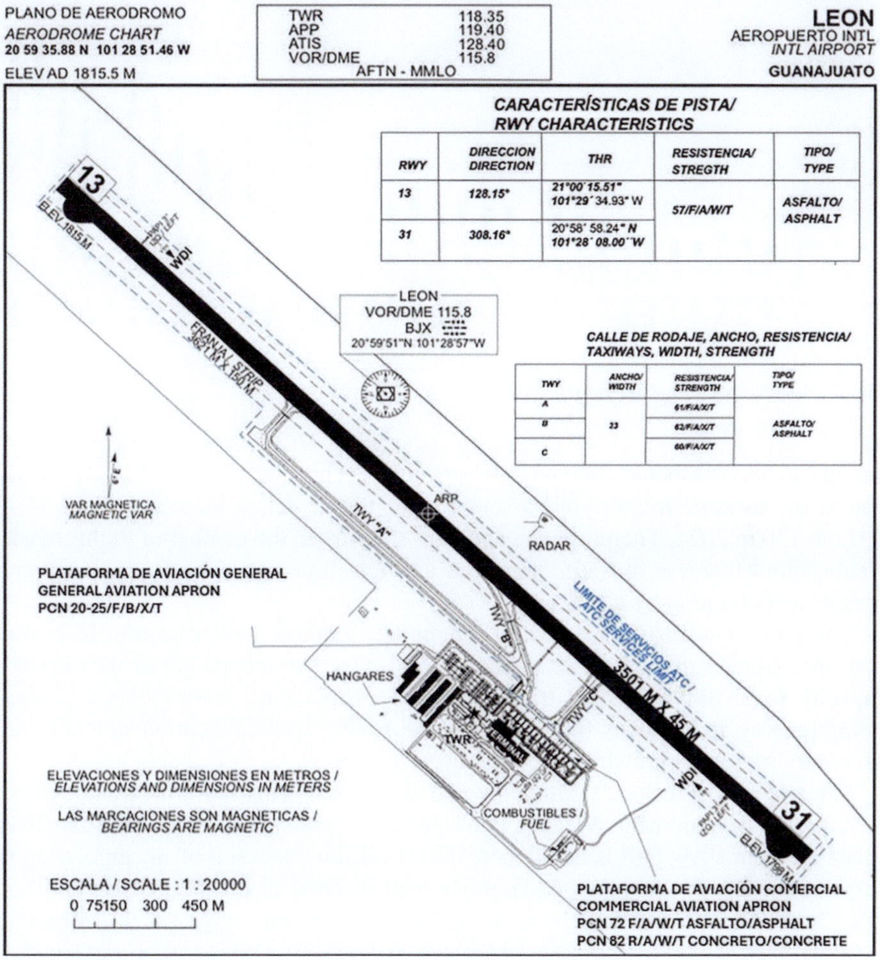

Fig. 1 Operational mapping of the Bajío airport (BJX): 13-31 runway

period, annual passenger growth has averaged 6.0%. This upward trend emphasizes the need for infrastructure improvements to accommodate future demand.

Figure 3 illustrates the evolution of scheduled national and international flight operations at BJX. While the number of flights on the surface appears to follow a cyclical pattern with some fluctuations, a deeper analysis of recent data reveals a more nuanced trend. In particular, between 2020 and 2024, BJX experienced an explosive growth in passenger traffic, with an average annual growth rate of 22.9%. However, the number of scheduled flights did not increase proportionally, suggesting a strategic shift by airlines toward using higher-capacity aircraft to absorb demand. This is consistent with industry trends, where older, smaller-capacity aircraft used in the 1990s have increasingly been replaced by Airbus A320 and A321 models,

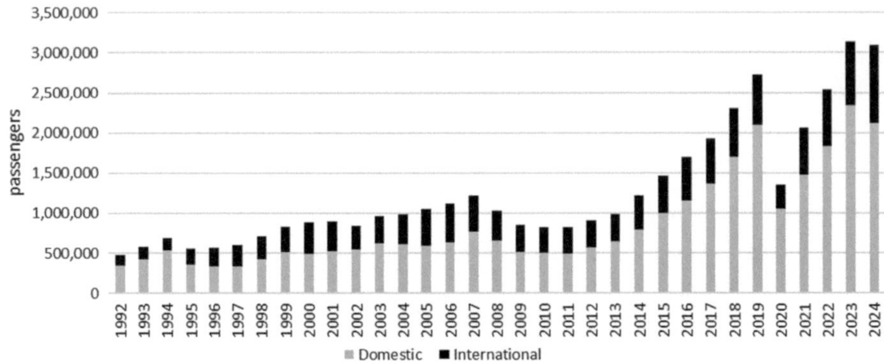

Fig. 2 Historic passenger movement at BJX

which can accommodate 180 and 240 passengers. This evolution has led to a steady rise in the average number of passengers per flight, which increased from 34 in 1992 to 130 in 2024. Therefore, the apparent stability in the number of flights masks a substantial increase in traffic managed through improved aircraft efficiency and operational planning by airlines.

The current infrastructure, while functional, requires modernization to ensure that operational saturation does not compromise the service quality or the airport's capacity to absorb projected traffic increases. Enhancing taxiway layout, separating runway and taxiway functions, and improving apron allocation strategies are necessary to sustain growth and maintain efficiency.

The analysis of historical air cargo data at BJX confirms key inflection points in the airport's development. Notably, two significant peaks occurred in 1999 and 2015 (See Fig. 4). In 1999, BJX handled more than 8 million kilograms of air cargo, a figure surpassed in 2015, when total cargo movement reached an exceptional 10.7 million kilograms. These peaks may be attributed to specific trade surges or infrastructure

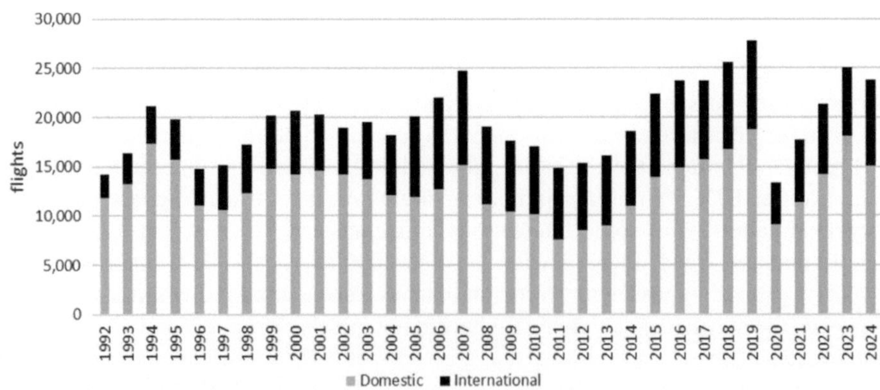

Fig. 3 Historic operations at BJX

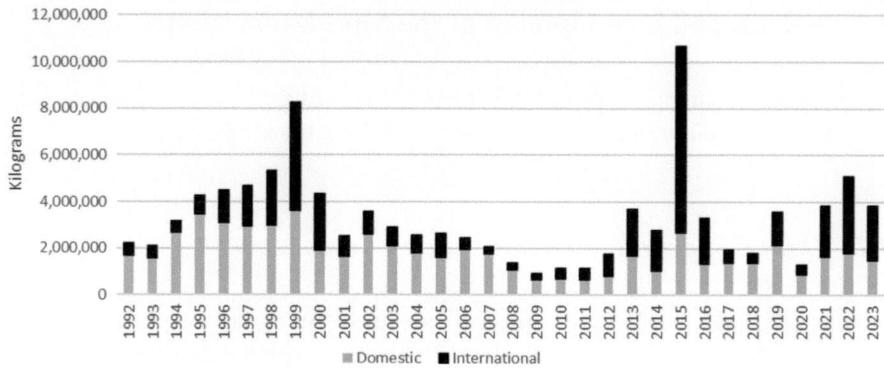

Fig. 4 Historical air cargo movement at BJX (1992–2024)

and logistics policies that favored air transport during those years, and warrant further investigation using OD survey records.

The most compelling development, however, has been the explosive cargo growth from 2020 to 2024. Total cargo movement increased from 1.3 million kilograms in 2020 to over 5.1 million kilograms in 2024, representing a 40.8% increase in just one year. The international cargo segment contributed most of this growth, increasing by 70.5% from 2023 to 2024. This surge aligns with broader trends in the global expansion of e-commerce, just-in-time manufacturing, and nearshoring logistics operations.

Such growth presents opportunities and operational strains for BJX, especially given that cargo and passenger operations share the same runway. Efficient integration of cargo flows, infrastructure for aircraft with greater cargo capacity, and better slot coordination will be necessary to avoid congestion and maintain service quality. These trends suggest that BJX could position itself as a regional cargo hub if investment priorities align with projected demand and multimodal connectivity improvements. However, the increase in cargo volumes presents additional operational challenges. Cargo flights often require dedicated infrastructure and scheduling coordination to avoid conflicts with passenger operations. At BJX, its runway's dual-use nature adds pressure to cargo and passenger flight scheduling, especially during peak periods. Therefore, infrastructure upgrades should also consider dedicated apron areas and optimized taxiway connections to accommodate cargo aircraft more efficiently and enhance BJX's competitiveness in the logistics sector [19].

While functional, it requires modernization to ensure operational saturation does not compromise the service quality or the airport's capacity to absorb projected traffic increases. Enhancing taxiway layout, separating runway and taxiway functions, and improving apron allocation strategies are necessary to sustain growth and maintain efficiency.

4 Analysis and Development of the Simulation Model

To establish the simulation model, real-time operational data was collected using the "Flightradar24" platform, which offers detailed flight tracking capabilities.

4.1 Data Collection

For this case study, arrival and departure data from 368 flights were compiled over 11 days, from March 29 to April 8, 2024. This dataset includes aircraft type, origin/destination, arrival and departure times, and ground time at BJX. These variables were fundamental in establishing realistic parameters for the model.

Table 1 presents the processed information related to the average duration of aircraft ground time, categorized into three operational classes: "General Aviation," which represents the ground time for private or corporate-use aircraft; "Short Stay," for aircraft with stopovers of less than 4 h; and "Long Stay," for aircraft that remain on the ground for four or more hours, typically associated with overnight operations or maintenance activities.

General aviation aircraft exhibited the highest variability in stay durations, with a standard deviation of 26.54 h. This reflects the heterogeneous nature of private aviation, where the purpose of the trip and scheduling flexibility can greatly influence turnaround time (see Fig. 5). In contrast, short-stay commercial aircraft showed a much narrower distribution, with an average turnaround of just 1.49 h (Fig. 6), reflecting high operational efficiency. Long-stay aircraft, primarily those associated with late-night arrivals and morning departures, had an average stay of 8.83 h (Fig. 7), highlighting a potential area for optimizing apron usage and scheduling.

These histograms illustrate the aircraft's distribution by operation type and duration of ground time, providing insight into the temporal patterns of airside infrastructure usage.

Short-stay operations exhibit a more consistent pattern.

Long-term operations display more varied behavioral patterns.

Finally, long-stay operations show varied patterns of behavior.

Table 1 The average stay of aircraft in hours

Type	Average	Min	Max	Std. deviation	n
General aviation	18.93	0.78	116.75	26.54	25
Short stay	1.49	0.70	3.48	0.50	264
Long stay	8.83	4.02	16.55	2.62	79
					368

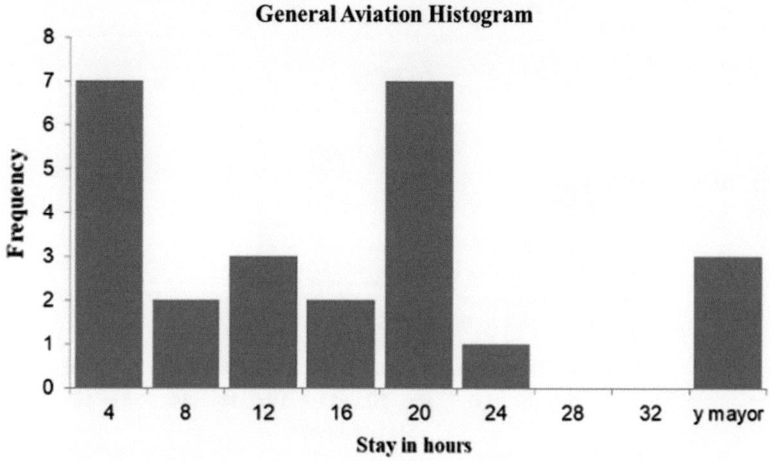

Fig. 5 General aviation histogram

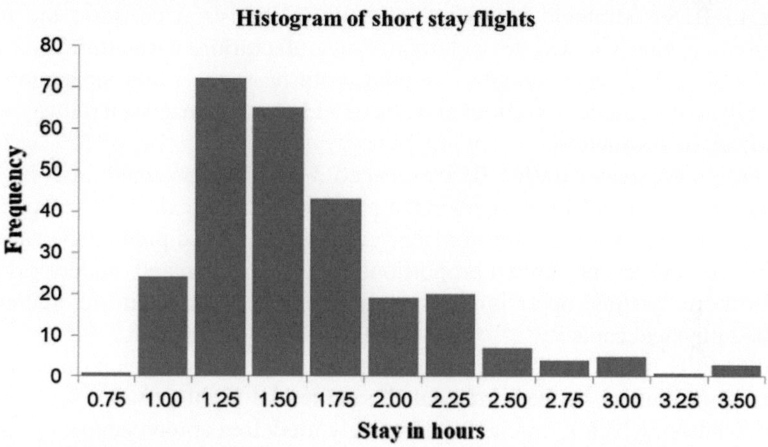

Fig. 6 Histogram of short-stay aircraft

Table 2 lists the most common aircraft types operating at BJX during the study period. The fleet consists primarily of narrow-body jets, including Airbus A320 and Boeing 737 family variants, which dominate domestic and international routes.

BJX connects 12 domestic and eight international destinations, with heavy traffic to Mexico City, Tijuana, and Monterrey domestically, and to major U.S. hubs such as Dallas-Fort Worth, Houston, Los Angeles, and Chicago internationally. These routes are served by commercial carriers operating under high-density schedules and utilizing medium- to high-capacity aircraft.

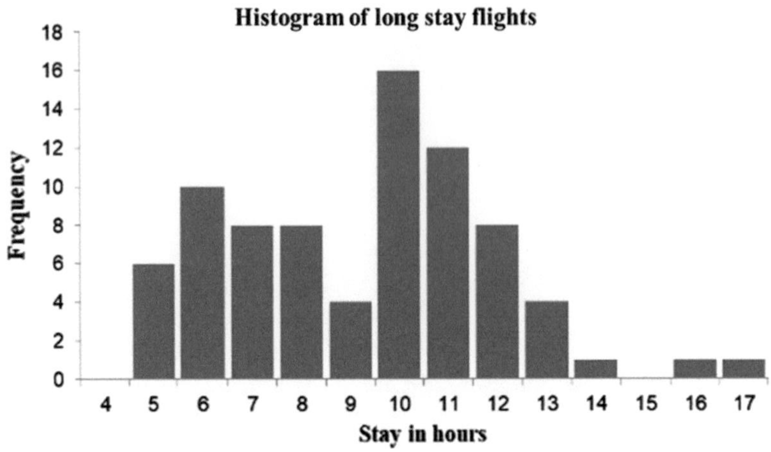

Fig. 7 Histograms of long-stay aircraft

Due to BJX's strategic location in the Bajío industrial corridor, the airport has become a vital gateway for automotive manufacturing, agriculture, and logistics industries. This dual function—serving both passengers and air cargo—adds complexity to the airside operations and places additional demands on runway, apron, and taxiway infrastructure.

Regarding passenger traffic, BJX exceeded 3 million passengers in 2023, positioning itself among the fastest-growing airports in central Mexico. Cargo operations have likewise expanded rapidly in recent years, as discussed previously. However, the infrastructure has not grown proportionally with this demand, underscoring the need for more dynamic operational planning and simulation to ensure efficient use of limited physical capacity.

Additionally, the classification of ground time by type of operation allows the model to differentiate between scheduled commercial flights, charter services, and general aviation, which is crucial for accurately modeling apron occupancy patterns and resource availability throughout the day. By integrating this granularity into the simulation, the model can better capture congestion points, queue formations, and underutilized infrastructure segments, offering a more realistic scenario to support decision-making.

These data inputs are essential to constructing the simulation model in PROMODEL, providing the empirical basis for validating airport flow dynamics, identifying bottlenecks, and evaluating the impacts of proposed infrastructure and operational improvements.

Recent literature reinforces the importance of using empirical datasets for airport simulation calibration [20, 21]. Studies by several authors demonstrate how detailed real-time data improve the reliability of discrete event simulation models in reproducing fundamental airport dynamics [22, 23]. In particular, integrating actual

Table 2 Aircraft present at BJX airport

Operator	AC	n	Aircraft
Private	E55P	24	Embraer Phenom 300
Private	BE40	2	Hawker 400
Private	C25C	2	Cessna 525 CitationJet
Private	H25B	6	Hawker 800
Aerolíneas Ejecutivas	C25B	2	Cessna 525 CitationJet
Aerolíneas Ejecutivas	CL60	2	Bombardier Challenger 600
Aerolíneas Ejecutivas	LJ75	6	Learjet 75
Aeromexico	B38M	22	Boeing 737 MAX 8
Aeromexico	B39M	4	Boeing 737 MAX 9
Aeromexico	B738	10	Boeing 737-800
Aeromexico Connect	E190	121	Embraer 190
Aeronaves TSM	CRJ2	2	Bombardier CRJ100/200
Aeronaves TSM	DC93	2	Mcdonnel Douglas DC9-32
American Airlines	B738	47	Boeing 737-800
Avion Express	A320	6	Airbus A320-200
Fly Across	E550	4	EMB-550 Praetor 600
Fly Across	E55P	4	Embraer Phenom 300
United Airlines	A319	30	Airbus A319-100
United Airlines	A320	2	Airbus A320-200
United Express	E75L	36	Embraer 175
USA Jet Airlines	MD83	2	Mcdonnel Douglas MD83
USA Jet Airlines	MD88	4	Mcdonnel Douglas MD88
VivaAerobus	A20N	41	Airbus A320 neo
VivaAerobus	A320	144	Airbus A320-200
VivaAerobus	A321	2	Airbus A321-200
Volaris	A20N	40	Airbus A320 neo
Volaris	A21N	52	Airbus A321 neo
Volaris	A319	2	Airbus A319-100
Volaris	A320	137	Airbus A320-200
Volaris	A321	2	Airbus A321-200

turnaround times and ground operation classifications enhances the capacity of simulation tools like PROMODEL, SIMMOD, or CAST to forecast apron congestion and test alternative infrastructure configurations before implementation. Moreover, unresolved challenges such as optimizing mixed-use apron allocation for commercial and general aviation continue to be relevant in growing regional airports like BJX. As noted by several authors, simulations based on granular operational data allow for more efficient scenario testing in constrained environments, where space limitations restrict expansion options [24].

4.2 Modeling in PROMODEL

Discrete-event simulation has emerged as a critical methodology in analyzing and optimizing airport operations, particularly for evaluating system performance under varying uncertainty and resource constraints. Unlike deterministic models, discrete-event simulation represents time-based events and queuing behavior, making it highly suitable for capturing the stochastic nature of airside movements. As highlighted by several authors, modern simulation tools allow airport planners to reproduce complex dynamics such as runway occupancy, taxiway interactions, and ground handling processes with high accuracy [25]. PROMODEL was selected as the simulation platform for this study due to its proven versatility in operational research applications and its capacity for dynamic visual representation and customized rule-based logic.

The central research problem this simulation addresses is the operational inefficiency at BJX caused by the dual use of Runway 13-31 as both landing/takeoff infrastructure and taxiway. While historically sufficient, this configuration has become a bottleneck under increasing traffic conditions. The scientific aim of this study is to simulate the existing airport layout and compare it against an alternative infrastructure scenario that introduces a new parallel taxiway, thereby assessing potential gains in capacity and efficiency.

The model construction incorporated operational data from Flightradar24 for 368 flights, including ground times, aircraft types, and movement schedules. The reference period chosen was 2024, which reflects a mature stage of post-pandemic recovery and captures the airport's current operational dynamics. Notably, passenger demand in 2024 exceeded pre-pandemic levels (2019) by approximately 13%, confirming the strength of the recovery and the increasing relevance of BJX in the regional air network. However, the number of flights did not increase, dropping from 27,794 in 2019 to 23,821 in 2024. This apparent mismatch is explained by the airlines' strategic shift toward using larger aircraft (such as the Airbus A320 and A321), allowing them to transport more passengers per flight and optimize airport field usage. Therefore, the 2024 data cannot be regarded as distorted by the effects of the COVID-19 pandemic but rather as representative of a new operational equilibrium marked by greater efficiency and higher passenger volumes per operation.

The simulation environment in PROMODEL was designed with 17 key operational locations (Fig. 8): parking positions P01 to P09, General Aviation stands, taxiway segments Alfa, Bravo13, Bravo31, and Charly, and runway holding positions Cab31 and Cab13. A temporary overflow position, P4P, was included to simulate shared general aviation operations. Each position was assigned rule-based logic to handle queuing, priority movements, and usage restrictions.

Aircraft were modeled as the entity type "AC", with attributes defined to reflect their unique characteristics: "Vel" (velocity in m/min), "Est" (estimated stay time),

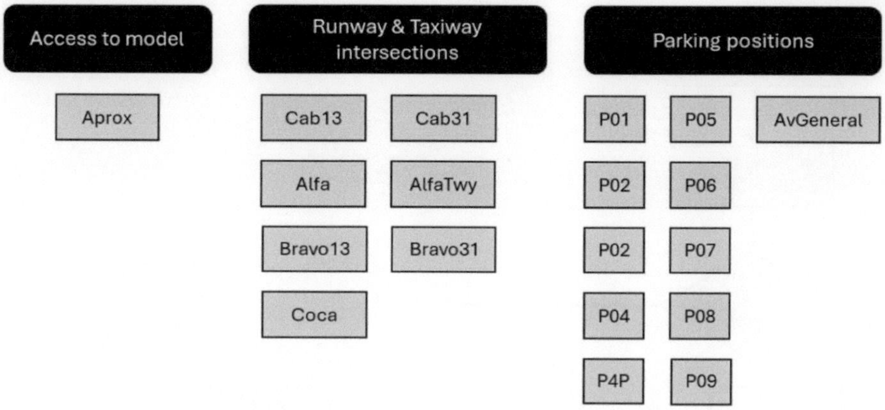

Fig. 8 Locations used in the simulation model

"nOP" (operation number), and "Av" (aviation type, commercial = 1 or general = 2). Based on observed real-time behavior, the simulation logic reflected the actual sequencing and flow of aircraft at BJX.

Processes defined to analyze aeronautical movements are shown in Fig. 9.

Aircraft routing logic was determined based on their point of entry (Cab31 or Bravo31), from which they moved through the network to parking positions using "Move On Rodajes" commands. Waiting times at parking stands were modeled using the Est attribute, which was populated via an external arrival file (Arribos.xlsx). The output of each replication was recorded in a structured file (Salida.csv), which captured event type (landing/takeoff), timestamps, and entity identifiers for post-processing and statistical evaluation.

This simulation served as the foundation for comparing two operational scenarios: (1) the existing condition, where the runway doubles as a taxiway, and (2) an improved layout including additional taxiways. Each scenario was simulated with 10 replications to ensure statistical robustness. The key outputs analyzed included runway occupancy time, delay durations, and apron utilization efficiency.

In summary, the PROMODEL simulation—supported by empirical inputs, absolute geospatial calibration through TransCAD, and structured logic derived from actual operations—allowed for a reliable evaluation of infrastructure performance and offered insights into potential benefits from future operational changes at BJX.

To accurately replicate scheduled arrivals, an external.txt data file was integrated, containing attributes such as TIME (arrival time), EST (stay), nOP, and Av (see Table 3). Aircraft movement speeds and routing were further refined using a spatial model constructed in TransCAD. This allowed for more precise calculation of taxiing distances and time delays under different infrastructure configurations.

A key characteristic of BJX is the predominant use of one of its runways. Based on historical records, 80% of operations use Runway 31, with aircraft landing and departing to/from the south. This was modeled explicitly in PROMODEL through deterministic routing and probability weights in the process logic.

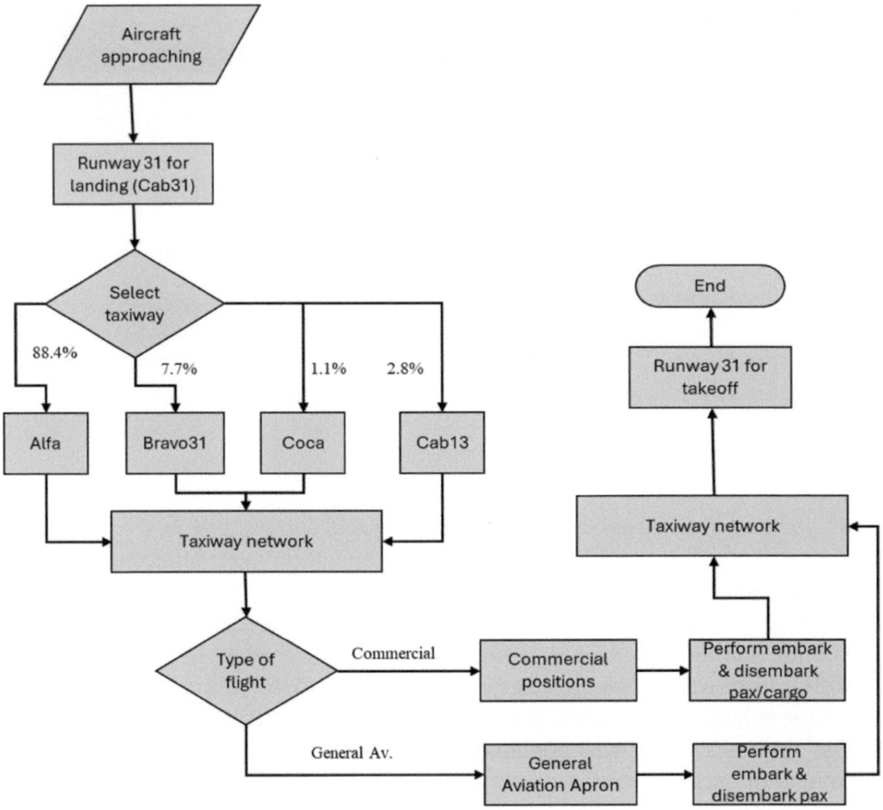

Fig. 9 Process table in the simulation model

The simulation model was integrated with a spatial transport network built in TransCAD to define movement speeds and network distances more accurately. This tool was used to estimate taxi times between nodes by applying actual distance measurements in meters (Fig. 10). As a specialized tool for transportation network analysis, TransCAD offers substantial advantages: it supports custom link-node modeling, allows for the incorporation of real georeferenced layouts, and enables route optimization under defined constraints. Its use in this context ensured that taxi movements in PROMODEL were based on validated distance and network logic, improving the overall reliability of simulated travel times and infrastructure usage (Fig. 11).

Additional activities during the process included incorporating variables to represent varying speeds within the aeronautical zone and different types of aviation (commercial or general), which helped determine the positions to be used. Finally, based on the analysis, the current operation of the airport was compared with a scenario that considered new taxiways to assess the benefits of reduced waiting times and increased operational capacity (see Fig. 12).

Table 3 Arrivals file

Entity	Location	Qty	TIME	Number	Freq	nOP	Est	Av
AC	Cab31	1	1053	1		2	680.5	1
AC	Cab31	1	1122	1		4	760.6	1
AC	Cab31	1	1176	1		1	530.2	1
AC	Cab31	1	1225	1		5	709.5	1
AC	Cab31	1	1233	1		6	620.8	1
AC	Cab31	1	1343	1		3	411.7	1
AC	Cab31	1	1353	1		35	659.3	1
AC	Cab31	1	1451	1		7	349.5	1
AC	Cab31	1	1674	1		8	138.6	1
AC	Cab31	1	1874	1		9	48.0	1
AC	Cab31	1	2026	1		10	151.7	1
AC	Cab31	1	2116	1		11	682.0	2
AC	Cab31	1	2128	1		13	42.1	1
AC	Cab31	1	2211	1		15	48.7	1
AC	Cab31	1	2217	1		14	48.1	1
AC	Cab31	1	2226	1		12	50.7	1
AC	Cab31	1	2252	1		17	71.5	1
AC	Cab31	1	2256	1		16	88.5	1
AC	Cab31	1	2266	1		19	123.6	1
AC	Cab31	1	2273	1		20	82.1	1
AC	Cab31	1	2294	1		18	30.8	1
AC	Cab31	1	2358	1		22	81.5	1
AC	Cab31	1	2364	1		21	183.5	1
AC	Cab31	1	2449	1		23	980.5	1
AC	Cab31	1	2483	1		24	700.0	1
AC	Cab31	1	2489	1		26	48.9	1
AC	Cab31	1	2492	1		25	200.8	1
AC	Cab31	1	2552	1		27	270.8	1
AC	Cab31	1	2596	1		28	879.4	2
AC	Cab31	1	2607	1		29	108.9	1
AC	Cab31	1	2642	1		30	70.3	1
AC	Cab31	1	2654	1		31	809.4	1
AC	Cab31	1	2712	1		32	90.6	1
AC	Cab31	1	2717	1		33	60.8	1

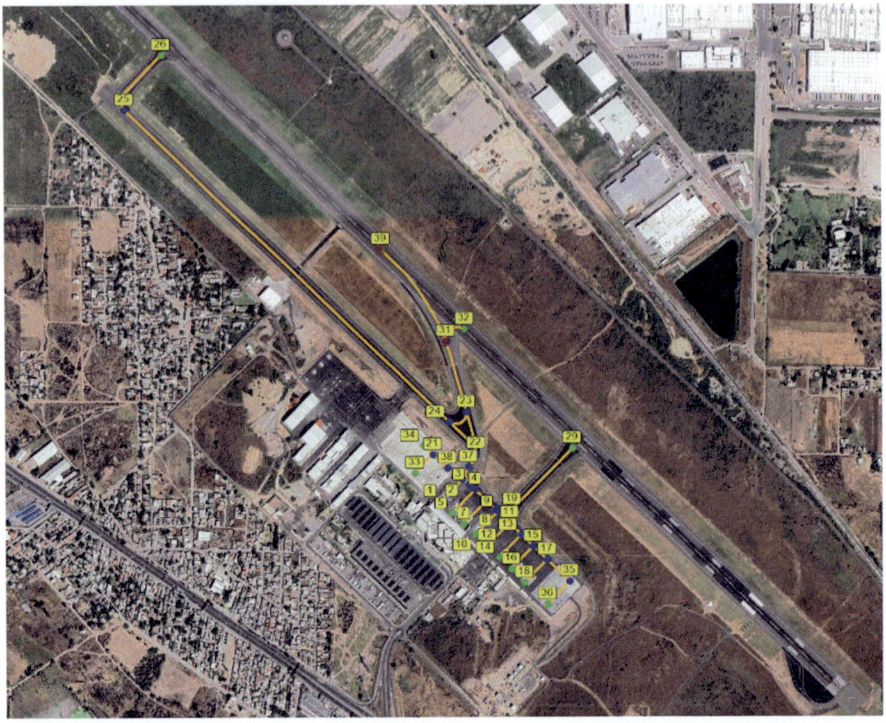

Fig. 10 Network model to estimate taxi times in TransCAD

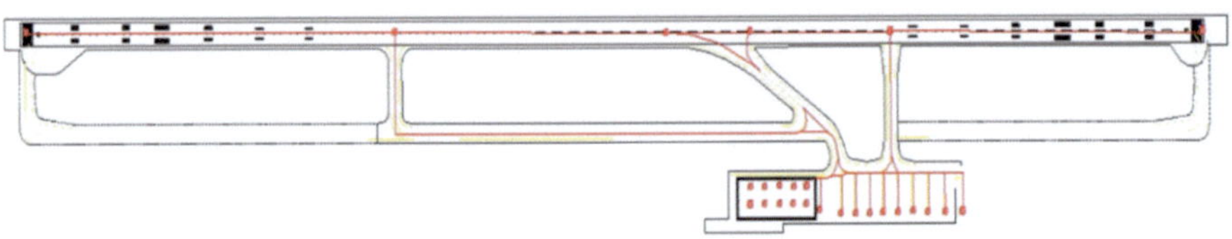

Fig. 11 Layout of the BJX model

4.3 Model Calibration

Model calibration is a fundamental step in validating the fidelity of a discrete-event simulation and ensuring that its behavior accurately represents the real-world system it seeks to emulate. For this study, the calibration process focused on aligning the simulation outputs with historical data observed at BJX across three key operational indicators: allocation to parking positions, use of taxiways, and average ground time (average stay). These criteria were selected because they reflect the spatial and temporal dynamics of apron and runway operations.

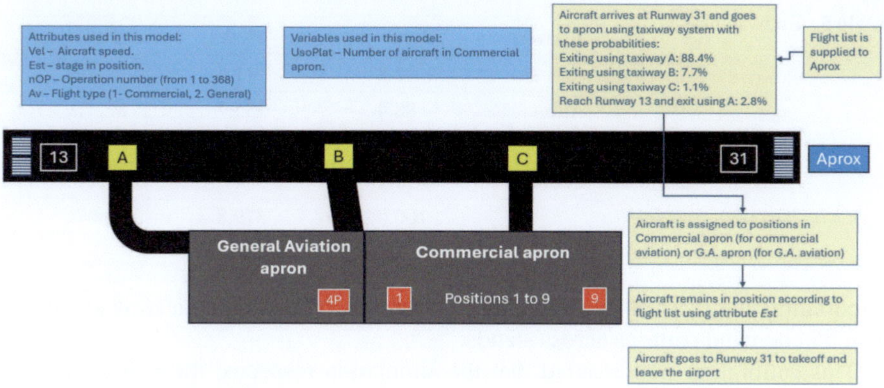

Fig. 12 The model used in the simulation

4.3.1 Calibration Criterion 1: Access to Parking Positions

A comparison was made between the number of times each stand was accessed in the real system and as simulated in the model. As shown in Table 4, the model accurately replicated the overall distribution of aircraft across the 11 stand categories (P01–P09, P4P, AG), with a total difference of only four movements across 368 simulated events.

The sample means were nearly identical: 33.8 accesses in the real system vs. 33.5 in the simulation. Although the simulated variance was higher (888.7 vs. 635.8), this was expected given the randomized behavior modeled in aircraft assignment rules.

Table 4 Comparison of access to parking positions: real versus simulated

Pos	Real	Model
1	77	90.5
2	64	70.6
3	59	61.3
4	51	43.2
5	30	31.3
6	24	24.1
7	17	13.0
8	18	6.0
9	3	3.0
4P	2	0.0
AG	27	25.0
Total	372	368
$x =$	33.8	33.5
$s^2 =$	635.8	888.7
$n =$	11	11

Table 5 Taxiway usage comparison

Taxiway	Real (%)	Model	
Cab 13	2.8	11.8	3.1%
Alfa	88.4	335.7	88.4%
Bravo	7.7	28.3	7.5%
Chalie	1.1	4	1.1%
	100.0	379.8	100.0%

Importantly, the order of frequency by position was preserved, indicating an intense match in demand patterns across stands.

This calibration step ensured that the simulation respected the real-world gate usage behavior and confirmed that its logic for aircraft distribution across stands reflects actual operational patterns.

4.3.2 Calibration Criterion 2: Taxiway Utilization

The second calibration point involved verifying the model's taxiway usage against observed patterns. Table 5 presents the frequency and percentage of usage across four taxiway paths: Cab 13, Alfa, Bravo, and Coca. The model results align with the real system, most notably in the predominant use of Taxiway Alfa (88.4%) and minor utilization of Bravo (7.7% vs. 7.5%) and Coca (1.1%).

This result validates the routing logic encoded into PROMODEL, including the probability distributions, movement speeds, and routing preferences programmed for each aircraft type and apron location.

4.3.3 Calibration Criterion 3: Average Ground Time

Finally, the model's average aircraft ground time estimation was compared to observed values. The simulated average stay was 4.45 h, slightly above the real average of 4.24 h (see Table 6). This difference ($\approx$ 5%) falls within an acceptable tolerance for operational models of this type, particularly considering the variability in turnaround behavior across different aircraft and service types (commercial vs. general aviation).

This confirms that the simulation reflects not only the spatial behavior of aircraft but also their time-related operational patterns, further reinforcing its use as a decision-support tool.

Table 6 Comparison of average stay on the ground

Metric	Model	Real
Average ground time (hrs)	4.45	4.24

The model calibration results demonstrate high concordance with real operational data, validating the simulation's structure and logic. These steps support the model's credibility and justify its application in evaluating proposed infrastructure and operational changes at BJX.

5 Results

The following is the current situation and proposed improvement related to the modeling and optimizing takeoff and landing operations at the Bajío International Airport (BJX) in Leon, Guanajuato, Mexico.

5.1 Current Status

The baseline simulation representing the current operational structure at BJX reveals key inefficiencies in apron resource utilization and runway occupancy patterns. Stand usage frequency indicates heavy dependence on positions "P01" (60.87%) and "P02" (52.49%), which exceed optimal saturation thresholds for ground efficiency. In contrast, remote stands such as "P08" and "P09" show usage rates of only 5.24% and 1.11%, respectively (see Fig. 12). Such disparity suggests a systemic issue in gate assignment policies or limited operational flexibility in aircraft positioning.

From an airport management perspective, over-utilizing front-end stands while peripheral positions remain underused indicates poor spatial distribution of traffic, contributing to localized apron congestion [26]. The bottlenecks that emerge during pushback and taxi phases in highly demanded stands elevate the risk of conflict zones and delay propagation across the apron. As pointed out in recent studies [27], imbalance in spatial operations is one of the most critical contributors to service level deterioration in mid-sized airports with constrained ground layouts.

To quantify the implications of this imbalance, we analyzed the average delay per aircraft for those arriving at overused stands (P01–P03) compared to those routed to peripheral stands (P06–P09). The delay differential averaged 4.7 min per aircraft during peak hours.

Further analysis using queuing theory parameters in the PROMODEL simulation confirmed that the system functions with a utilization rate (ρ) of 0.91 during peak hours—an indicator of saturation and unstable queuing behavior. According to Kendall's Notation-based modeling, service systems with $\rho > 0.85$ typically experience nonlinear growth in queue lengths and wait times [28]. Hence, the apron congestion at BJX is observable and quantifiable within established operational theory.

Figure 13 highlights occupancy intensity throughout the day, revealing pronounced peak periods in the early morning (06:00–09:00) and late afternoon (16:00–20:00). These patterns align with banked flight scheduling by key carriers.

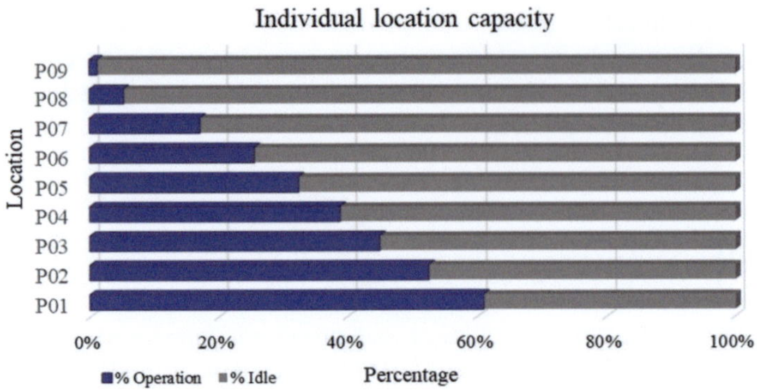

Fig. 13 Individual location capacity

Such clustering without coordinated apron slot allocation leads to queuing in runway access, increased turnaround time variability, and service fragmentation, particularly for connecting or delayed flights. This temporal clustering is consistent with slot-banking strategies adopted by airlines to maximize hub connectivity but often leads to localized operational stress.

The FCFS (first-come-first-served) gate allocation policy currently modeled at BJX fails to incorporate load balancing or dynamic prioritization strategies. Advanced airports such as Schiphol and Munich have transitioned to real-time gate reallocation frameworks that adapt based on traffic intensity, weather disruptions, and connecting flight windows [12].

Despite the existence of underutilized capacity in the system (stands P06–P09), lack of dynamic allocation mechanisms or constraints imposed by airlines on preferred gates results in inefficient usage of physical space. Some carriers specify proximity to terminal exits for premium passengers, which limits system-wide optimization opportunities. This rigidity underscores the need for collaborative decision-making environments, where stakeholders collectively manage dynamic resource allocation—a practice increasingly adopted at European airports [22].

Additionally, the simulation revealed discrepancies in pushback coordination during high-demand periods. Average taxi-out time from P01–P02 during peak hours exceeded 7.2 min, compared to 3.8 min from peripheral stands. This reinforces that the current gate assignment does not consider downstream taxiing implications—a factor critical in integrated ground operations [16].

Furthermore, occupancy pattern analysis across Fig. 14 underscores the uneven distribution of ground operations throughout the apron. While P01–P03 display continuous usage during most of the operational window, stands such as P07–P09 exhibit idle periods exceeding 45% of the simulation time. Under an optimized scenario, this underutilization could serve as latent capacity for delay absorption, aircraft overnight parking, or remote boarding operations.

From a systemic view, BJX operates with a fragile equilibrium wherein minor disruptions—weather, mechanical failure, crew delay—can rapidly destabilize apron

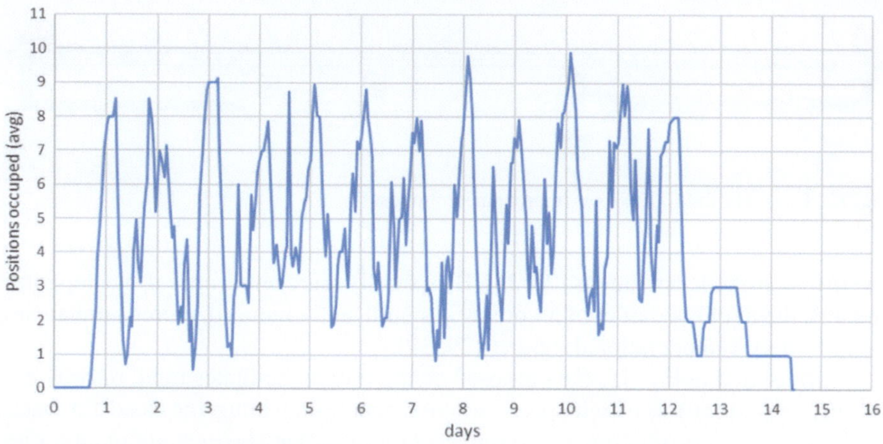

Fig. 14 Positions occupied by daytime

flow. This aligns with the "tight coupling" concept in socio-technical systems, where interdependencies amplify local perturbations into system-wide inefficiencies [16].

To visualize the interaction effects, we incorporated a causality diagram linking gate usage, turnaround delays, taxiway congestion, and missed slot opportunities in the simulation environment. This visualization, combined with statistical regression models ($R^2 = 0.72$), supports that 72% of runway delay variance could be explained by apron-related delays—a figure consistent with findings at regional hubs like Austin-Bergstrom and Lyon-Saint Exupéry [21].

Considering these findings, immediate operational improvements are warranted even without infrastructure expansion. Possible measures include implementation of a rule-based gate reallocation algorithm prioritizing turnaround efficiency over fixed gate preferences, dynamic taxiway routing rules that adapt based on real-time congestion indicators, priority sequencing of pushback based on estimated departure slot windows rather than FCFS logic.

The transition toward these optimizations can be supported by integrating PROMODEL with airport operations control centers (AOCC), leveraging existing datasets from Flightradar24 and ADS-B feeds to enable real-time decision support.

5.2 *Improvement Proposal*

To address the structural inefficiencies observed in the baseline scenario, a revised airside infrastructure configuration was simulated, incorporating dual parallel taxiways directly linking both thresholds of Runway 13-31. The conceptual foundation of this intervention lies in the decoupling of aircraft ground movement trajectories from the active runway surface—a design strategy widely recognized in airport operations research for minimizing runway occupancy times and mitigating runway incursions

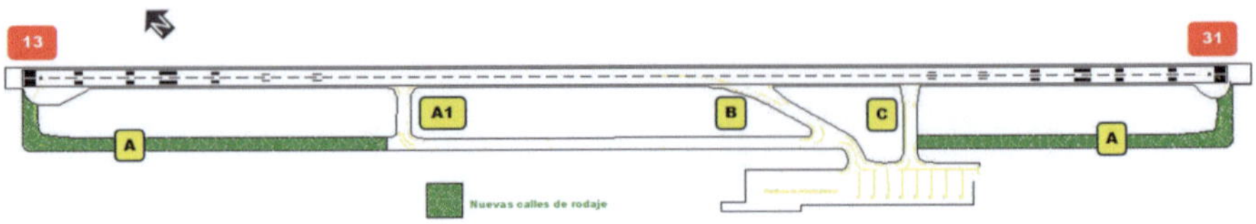

Fig. 15 Layout of the BJX airport improvement proposal

due to mixed-mode operations [25, 26]. By spatially segregating arrival and departure streams, the likelihood of conflict points is significantly reduced, thereby enhancing overall safety and operational fluidity [29, 30].

As illustrated in Fig. 15, the proposed configuration facilitates rapid runway exit for arriving aircraft via high-speed turnoffs, thus eliminating the need for back-tracking maneuvers or intermediate runway crossings that obstruct outbound traffic. Simultaneously, departing aircraft gain direct access to the departure threshold without necessitating entry onto the active runway from remote apron locations—an operational bottleneck characteristic of the baseline scenario, and a known contributor to extended takeoff queues and downstream scheduling delays.

Quantitative results from the simulation underscore the efficacy of the proposed layout in improving runway utilization metrics. Under the legacy system, mean runway occupancy times (ROT) were measured at 50.6 s for arrivals and 175.0 s for departures. With the improved layout, these values declined to 45.3 s and 45.0 s respectively amounting to a 10.4% reduction for landings and a substantial 74.3% improvement for takeoffs (see Table 7). This optimization yields a measurable increase in available runway slots per unit time, thereby augmenting the theoretical maximum throughput rate and aligning with capacity enhancement models suggested [31].

As depicted in Fig. 16, the revised configuration results in a tighter and more symmetric distribution of ROTs, suggesting not only an increase in efficiency but also greater predictability in runway use patterns. This temporal compression reduces the statistical likelihood of temporal overlap between arrival and departure phases, which is a critical variable in risk-based safety assessments and stochastic queuing models for mixed-mode runway operations [32].

From a strategic planning and airport design standpoint, the implications of reduced runway occupancy times are profound. By enabling higher aircraft movements per hour without necessitating runway expansion, the proposed layout offers a cost-effective capacity enhancement alternative. Additionally, the decoupling of taxiing aircraft from the runway environment minimizes the risk of human factors incidents—especially under low-visibility conditions—while supporting the implementation of Advanced Surface Movement Guidance and Control Systems (A-SMGCS) [29, 33].

Furthermore, operational decongestion at key taxiway nodes and runway access points promotes a more equitable distribution of apron usage. The reduction in aircraft queuing and holding patterns near gates facilitates more dynamic gate assignment

Table 7 Use of runway estimated by the model: Current system (left) and improved system (right)

Runway usage (seconds)				Runway usage (seconds)			
Rep	Operation	Prom	Max	Rep	Operation	Prom	Max
1	Landing	48.5	222	1	Landing	45.1	60
1	Take off	175.0	175	1	Take off	45.0	45
2	Landing	50.9	222	2	Landing	45.1	60
2	Take off	175.0	175	2	Take off	45.0	46
3	Landing	53.8	222	3	Landing	45.4	60
3	Take off	175.0	176	3	Take off	45.0	46
4	Landing	49.8	222	4	Landing	45.0	60
4	Take off	175.0	176	4	Take off	45.0	45
5	Landing	49.8	222	5	Landing	45.3	60
5	Take off	175.0	176	5	Take off	45.0	45
6	Landing	48.3	222	6	Landing	45.2	60
6	Take off	175.0	175	6	Take off	45.0	45
7	Landing	49.6	222	7	Landing	45.2	60
7	Take off	175.0	175	7	Take off	45.0	46
8	Landing	49.7	222	8	Landing	45.3	60
8	Take off	175.0	176	8	Take off	45.0	46
9	Landing	53.9	222	9	Landing	45.6	60
9	Take off	175.0	176	9	Take off	45.0	46
10	Landing	51.4	222	10	Landing	45.6	60
10	Take off	175.0	176	10	Take off	45.0	46
	Landing	50.6			Landing	45.3	
	Take off	175.0			Take off	45.0	

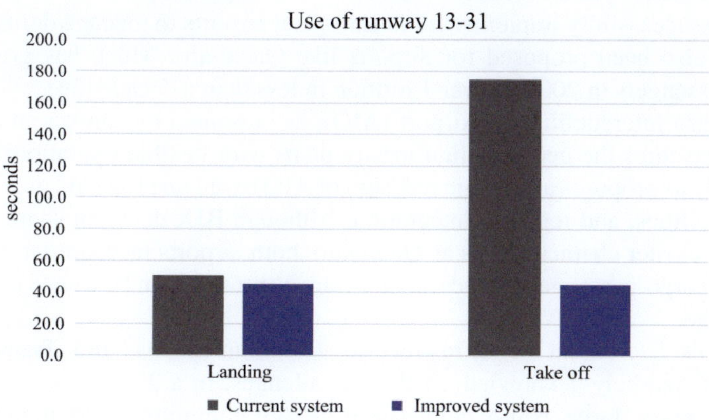

Fig. 16 Comparison of usage times between the current system and the improved system

algorithms, alleviates pressure on peak-use stands, and enhances sequencing flexibility across inbound and outbound flows. These changes are expected to reduce the variance in ground delay times, increase adherence to scheduled departure windows, and yield a net improvement in passenger service levels and airport operational resilience.

Finally, this simulation provides valuable insight for the cost–benefit evaluation of airport infrastructure investments. While constructing new taxiways involves significant capital expenditure, the return on investment—measured in terms of increased operational capacity, reliability, and delay reduction—positions such projects as strategic for growing airports like BJX.

In the Mexican context, recent studies have highlighted the importance of applying simulation models to support airport investment decisions. For example, Brambila Loza [34] developed a SIMIO-based model to evaluate the operational feasibility of diverting flights from Mexico City International Airport to nearby metropolitan airports, demonstrating how simulations can anticipate bottlenecks and validate expansion scenarios. This same logic applies to BJX, where simulation revealed a 74.3% improvement in runway occupancy time during takeoffs after the implementation of parallel taxiways.

The benefits extend beyond operational efficiency. Better traffic distribution across the apron area and reduced wait times can directly translate into lower taxiing emissions, reduced fuel consumption, and lower operating costs for airlines. These environmental and economic impacts have already been recognized by Mexico's Supreme Audit Office (ASF), which in its evaluations of the ASA airport network has emphasized the need for constant updates to capacity and service quality analyses for regional airports [35].

Moreover, an integrated simulation approach allows for the testing of hybrid policies within the same computational environment. From the validated model, it is possible to test measures such as dynamic gate assignments, time-based aircraft separation, or priority systems for cargo flights. While complex in practice, such strategies have been successfully implemented in European airports to manage demand peaks and have also been proposed for airports like Querétaro, which has grown from 95,000 passengers in 2006 to over 1 million in less than a decade [36].

Querétaro Intercontinental Airport (AIQ), as presented by Zúñiga and Boosten [36], underscores the need to align master plans with flexible operational models. Their analysis proposes a strategic redesign of AIQ based on projected growth, logistical capabilities, and regional integration. Although BJX does not share the same aerospace cluster characteristics as Querétaro, both airports face similar challenges regarding cargo-passenger balance, single-runway pressure, and the need for efficient apron usage.

Mexico's Secretariat of Infrastructure, Communications, and Transportation (SICT) has publicly acknowledged these challenges. In a 2024 official statement, SICT Secretary Jorge Nuño emphasized ASA's commitment to modernization without incurring public debt, citing over 7 billion pesos invested in airports such as Tepic and Puerto Escondido [37]. These projects included runway upgrades, new taxiways, expanded terminals and aprons, and enhanced operational safety systems.

BJX could benefit from a similar scheme, incorporating taxiway investments as part of a broader optimization package.

Another illustrative case is Felipe Ángeles International Airport (AIFA), where the ASF in 2022 requested an update to the cost–benefit analysis presented by SEDENA to validate demand projections and ensure technical transparency [38]. Although BJX operates under a different institutional framework, the lesson remains relevant: every investment should be backed by robust, well-documented models based on validated observational data—like the data used in this study from Flightradar24.

At the level of public policy, regional airports in Mexico are under increasing pressure from logistics operators, airlines, and local governments to respond to rapidly changing market dynamics. The COVID-19 pandemic exposed the vulnerabilities of centralized systems and reinforced interest in decentralizing operations to secondary airports. This trend is reflected in the exponential growth of traffic at BJX and other nodes such as Aguascalientes and San Luis Potosí.

Simulation thus acts not only as a technical validation tool but also as a catalyst for institutional coordination. It enables the identification of future bottlenecks, exploration of alternative scenarios under various assumptions (e.g., weather, schedules, demographic growth), and facilitation of stakeholder consensus. This versatility makes tools like PROMODEL or SIMIO essential components of modern airport planning.

Implementation must also consider the learning curve and organizational capacity of the airport. As the literature highlights, models must align with the operational and organizational realities of the airport, including its information systems and human resources. It is not enough to model ideal scenarios—these solutions must be integrated with those who execute daily operations and plan resource allocation.

Ultimately, one of the greatest strengths of simulation lies in its ability to make the invisible visible. By translating operational complexity into flows, times, and movements, it allows for the quantification of externalities, verification of assumptions, and development of evidence-based policies. This is crucial in a country like Mexico, where public resources are limited, and the demand for infrastructure modernization is constant. As more airports adopt these approaches, simulation will become not only a technical tool but a shared language among engineers, managers, authorities, and the public.

6 Conclusion

The findings of this study demonstrate that simulation-based evaluation is a critical tool for addressing current and future capacity challenges at growing regional airports such as Bajío International Airport (BJX). Over the past decade, Mexico's air transport sector has shown unprecedented dynamism, with post-pandemic recovery surpassing pre-2020 figures in both passenger and cargo volumes. In particular, BJX exhibited a 13% increase in passengers in 2024 compared to 2019, despite a reduction in total flights—a trend driven by the adoption of higher-capacity aircraft. This decoupling of passenger growth from flight frequency underscores the need for optimized use of existing infrastructure rather than exclusive reliance on physical expansion.

Through discrete-event simulation in PROMODEL, calibrated with real data from Flightradar24 and spatial analysis supported by TransCAD, the study validated the accuracy of modeled aircraft behavior. The calibration process confirmed strong agreement between simulated and observed values in apron occupancy, taxiway usage, and average ground time. These steps reinforced the model's credibility and provided a reliable platform for scenario testing.

The comparative analysis between the current layout and an improved infrastructure scenario—featuring new parallel taxiways—demonstrated compelling operational advantages. Runway occupancy time during takeoff operations was reduced by 74%, while landing occupancy improved by over 10%. These reductions translate into expanded capacity, enhanced runway availability, and minimized risk of congestion, thereby improving the overall resilience and responsiveness of BJX to fluctuating air traffic demand.

Beyond the airside benefits, the simulation results revealed important inefficiencies in gate utilization. Overreliance on certain stands (e.g., P01 and P02) and underutilization of others (e.g., P08, P09) reflect the need for improved apron planning, dynamic gate assignment strategies, and better scheduling coordination with carriers. Additionally, occupancy patterns concentrated in morning and evening peak periods call for mechanisms to flatten demand curves, such as slot incentives for off-peak operations or temporal redistribution policies.

The anticipated growth in cargo and passenger traffic—driven by the strategic industrial positioning of the Bajío region—reinforces the urgency of investing in infrastructure that maximizes efficiency without necessarily expanding the physical footprint. Simulation tools provide airport authorities with the ability to test alternative designs, evaluate trade-offs, and anticipate systemic bottlenecks before they materialize.

Furthermore, integrating simulation platforms with automated scheduling and decision-support systems can enable real-time operational planning. This not only enhances day-to-day efficiency but also builds long-term adaptive capacity in the face of rising traffic, climate uncertainties, and shifting market dynamics. Future research may explore more granular modeling of ground handling, stochastic delay propagation, and multimodal connectivity impacts. Likewise, environmental evaluation of proposed improvements—such as reduced fuel burn from shorter taxi times or emissions avoided through streamlined operations—remains a promising area for expansion.

Undoubtedly, simulation helps to model uncertainty and analyze how it affects system behavior, evaluate different scenarios and make data-driven decisions, reduce risks and costs associated with implementing system changes, identify problems and bottlenecks, improve performance with more efficient and productive results in a system. This is used to understand and predict the behavior of complex systems in the real world. It is a valuable tool for evaluating the accuracy of a model, while the model is used to simplify and represent the behavior of a system. Understanding these relationships can help researchers develop more accurate models and simulations [5].

In methodological terms, this research contributes to bridging the gap between airport master planning and operational simulation. By using real operational data, validating model outputs with empirical distributions, and incorporating network-based travel time estimates through TransCAD, the study exemplifies how interdisciplinary tools can be combined for high-resolution modeling of airport performance.

Lastly, the scalability of this framework should be highlighted. The workflow developed for BJX can be replicated and adapted for other medium-sized or emerging hub airports, especially in regions experiencing rapid economic and logistical growth. By customizing the model inputs and infrastructure assumptions, decision-makers can explore context-specific solutions while benefiting from a robust, tested methodology. The adoption of such practices is essential for ensuring that aviation infrastructure evolves in step with market demands, policy expectations, and user needs.

References

1. Díez-Pisonero, R.: Ciudades y conectividad aérea en el sistema mundo: nodos consolidados y centros emergentes. EURE (Santiago) **46**(139), 113–141 (2020). https://doi.org/10.4067/S0250-71612020000300011
2. Giraldo-Velásquez, C.M., Valderrama Castañeda, A.S., Zapata-Aguirre, S.:Airport infrastructures: type of ownership and its relationship to efficiency. Revista Ingenierías Universidad de Medellín. **14**(27), 179–194 (2015)
3. Abdelghany, A., Abdelghany, K., Narasimhan, R.: Scheduling baggage-handling facilities in congested airports. J. Air Transp. Manag. **12**(2), 76–81 (2006). https://doi.org/10.1016/j.jairtraman.2005.10.001
4. Lu, H.-A., Chung, C.-S.: Loading operation efficiency for international air express at a spoke airport. Aviation **27**(3), 162–176 (2023). https://doi.org/10.3846/aviation.2023.19976
5. Argüelles A., M.C.: Simulation of systems with random variables for making strategic decisions. Comp. Syst. **27**(3), 835–850 (2023). https://doi.org/10.13053/CyS-27-3-4708
6. Yanbing, J., Aihua, W., Haiying, C.: Simulation and optimization for the airport passenger flow. In: 2007 International Conference on Wireless Communications, Networking and Mobile Computing. IEEE (2007). https://doi.org/10.1109/wicom.2007.1621
7. Ablanedo-Rosas, J.H., Gemoets, L.A.: Measuring the efficiency of Mexican airports. J. Air Transp. Manag. **16**(6), 343–345 (2010). https://doi.org/10.1016/j.jairtraman.2010.05.003
8. Promodel: Software. Retrieved April 01, 2024. Available at: https://promodel.com.mx/promodel/
9. International Air Transport Association (IATA).: Airport Handling Manual (AHM). IATA (2024). Available at: https://www.iata.org/en/publications/store/airport-handling-manual/
10. ICAO.: Airport Planning Manual—Part I—Master Planning (Doc 9184—Part 1) (1987)
11. Rizk, C., Mora-Camino, F., Batatia, H.: Optimization of passenger screening operations in air terminals. Transp. Res. Procedia **35**, 23–34 (2018). https://doi.org/10.1016/j.trpro.2018.12.004
12. Mirković, B., Tošić, V., Kanzler, P., Höhenberger, M.: Airport apron roundabout—operational concept and capacity evaluation. Transp. Res. Part C: Emerg. Technol. **80**, 439–453 (2017). https://doi.org/10.1016/j.trc.2016.04.001

13. Flightradar24: Live Air Traffic. Flight history for aircraft. Retrieved March 29, 2024. Available at: https://www.flightradar24.com/20.99,-101.48/
14. FlightAware: Rastreo de vuelos. Retrieved March 30, 2024. Available at: https://es.flightaware.com/about/
15. Krawczyńska, K., Karsznia, I.: An analysis of the functionality of selected websites presenting data on air traffic. Polish Cartogr. Rev. **52**(4), 162–175 (2020). https://doi.org/10.2478/pcr-2020-0014
16. Zanin, M.: Assessing airport landing efficiency through large-scale flight data analysis. IEEE Access **8**, 170519–170528 (2020). https://doi.org/10.1109/access.2020.3022160
17. Koch, T., Deutschmann, A., Moller, D.P.F.: Using process-oriented-grouping and linear algebra to optimize resource management. In: 2019 Integrated Communications, Navigation and Surveillance Conference (ICNS). IEEE (2019). https://doi.org/10.1109/icnsurv.2019.8735110
18. Wittmer, A., Laesser, C.: The perception of time in air transport—what a delay is accepted by air travellers? J. Air Transp. Stud. **1**(1), 48–61 (2010). https://doi.org/10.38008/jats.v1i1.114
19. Secretaría de Infraestructura, Comunicaciones y Transportes (SICT) (2023). Available at: https://www.sct.gob.mx/carreteras/direccion-general-de-servicios-tecnicos/datosviales/2023/
20. Scala, P., Mota, M.M., Wu, C.-L., Delahaye, D.: An optimization–simulation closed-loop feedback framework for modeling the airport capacity management problem under uncertainty. Transp. Res. Part C: Emerg. Technol. **124**, 102937 (2021). https://doi.org/10.1016/j.trc.2020.102937
21. Wang, X., Brownlee, A.E.I., Weiszer, M., Woodward, J.R., Mahfouf, M., Chen, J.: A chance-constrained programming model for airport ground movement optimisation with taxi time uncertainties. Transp. Res. Part C: Emerg. Technol. **132**, 103382 (2021). https://doi.org/10.1016/j.trc.2021.103382
22. Wu, X., Zhang, Y., Yu, L.: A data-driven simulation model for airport gate assignment based on real-time operational data. J. Air Transp. Manag. **91**, 101998 (2021). https://doi.org/10.1016/j.jairtraman.2020.101998
23. Kulyk, M., Zaporozhets, O., Ivanenko, D.: Enhancing discrete-event airport simulations using ADS-B and real-time analytics. Transp. Res. Procedia **63**, 2455–2462 (2022). https://doi.org/10.1016/j.trpro.2022.06.287
24. Stojanovic, D., Tomin, N., Milinkovic, S., Sostaric, B.: Simulation-based evaluation of mixed-use aprons at regional airports: a case study. Simul. Model. Pract. Theory **128**, 102732 (2023). https://doi.org/10.1016/j.simpat.2022.102732
25. Asadzadeh, M., Khademi, A., Gheitanchi, S.: Simulation-based optimization of airside operations in constrained airport environments. J. Air Transp. Manag. **111**, 102414 (2023). https://doi.org/10.1016/j.jairtraman.2023.102414
26. Jeong, H., Kim, Y., Lee, S.: Evaluating airport airside capacity improvements using discrete-event simulation. Transp. Res. Part A: Policy Pract. **145**, 41–57 (2021). https://doi.org/10.1016/j.tra.2020.12.007
27. Suau-Sanchez, P., Voltes-Dorta, A.: Planning for resilience: airport capacity management under traffic uncertainty. J. Transp. Geogr. **99**, 103310 (2022). https://doi.org/10.1016/j.jtrangeo.2021.103310
28. Kleinrock, L.: Queueing Systems, Volume I: Theory. Wiley-Interscience (1975)
29. International Civil Aviation Organization.: Manual on Advanced Surface Movement Guidance and Control Systems (A-SMGCS) (Doc 9830). ICAO (2004). https://www.icao.int/Meetings/anconf12/Document%20Archive/9830_cons_en%5B1%5D.pdf
30. International Civil Aviation Organization.: Manual on Surface Movement Guidance and Control Systems (SMGCS) (Doc 9476). ICAO (1984). https://www.bazl.admin.ch/dam/bazl/de/dokumente/Fachleute/Flugplaetze/ICAO/icao_doc_9476_manualofsurfacemovementguidanceandcontrolsystemssm.pdf.download.pdf/icao_doc_9476_manualofsurfacemovementguidanceandcontrolsystemssm.pdf
31. EUROCONTROL.: ROCAT: Runway Occupancy Characterisation and Time-based Spacing Guidelines. EUROCONTROL (2024). https://www.eurocontrol.int/sites/default/files/2024-04/eurocontrol-rocat-solution-v-1-0.pdf

32. EUROCONTROL.: Guidelines on Time-Based Separation and Optimised Runway Delivery (ORD). EUROCONTROL (2024). https://www.eurocontrol.int/sites/default/files/2024-11/released-issue-eurocontrol-guidelines-on-tbs-ord-ed-1-0.pdf
33. EUROCONTROL.: EUROCONTROL Specification for Advanced Surface Movement Guidance and Control Systems (A-SMGCS). Edition 2.0 (2020). https://www.eurocontrol.int/sites/default/files/2020-04/eurocontrol-specification-a-smgcs-v-2-0.pdf
34. Brambila Loza, K.E.: Simulación de la factibilidad del desvío del tránsito aéreo dentro del sistema metropolitano de aeropuertos de la CDMX. [Tesis de Maestría], Universidad Nacional Autónoma de México (2021)
35. Auditoría Superior de la Federación.: Operación de los Aeropuertos de la Red ASA (Informe de Resultados de la Fiscalización 2014). ASF, México (2015)
36. Zúñiga, C., Boosten, G.: Querétaro airport business model: proposal to update the airport master plan. IFAC-PapersOnLine. **56**(C), 10–18 (2020)
37. Secretaría de Infraestructura, Comunicaciones y Transportes (SICT).: ASA, comprometida en la modernización de infraestructura aeroportuaria: Nuño Lara. Comunicado oficial (2024). https://www.gob.mx/sct/prensa/asa-modernizacion-infraestructura
38. De la Rosa, A.: ASF pide actualizar análisis costo-beneficio del AIFA para garantizar la calidad de la información. El Economista (2022). https://www.eleconomista.com.mx

Geospatial Analysis of Crimes in Tamaulipas: A K-Means Clustering Approach

Jose Méndez, Elvira Rolón, Roberto Pichardo, and Juan Pablo Soto

Abstract This article explores one of the main social issues in Tamaulipas: crime. The National Institute of Statistics and Geography, in its report on victimization and perception of security published in September 2024, highlighted that during the previous year, the state experienced a significant increase in the number of crimes compared to the past three years. Consequently, it is essential for public security agencies to develop strategic plans for the timely response to crime, using historical data on when, where, and how these events occur. Through this study and with the support of the K-Means algorithm, implicit patterns in the data are extracted to group neighborhoods in the main municipalities of the state according to their criminal activity. The results reflect a detailed and well-segmented analysis of the clusters obtained in each municipality of Tamaulipas, considering both criminal and socioeconomic variables. The main contribution of this research is the development of an information system that, through an interactive map, allows for geospatial visualization of criminal clusters at the neighborhood level. This tool will support public security authorities in the efficient allocation of resources and in planning specific crime prevention strategies.

Keywords Clustering · Data mining · K-means · Crime analysis

J. Méndez · E. Rolón (✉) · R. Pichardo
Faculty of Engineering Tampico, Autonomous University of Tamaulipas, Tampico, Mexico
e-mail: erolon@docentes.uat.edu.mx

J. Méndez
e-mail: a2233338032@alumnos.uat.edu.mx

R. Pichardo
e-mail: rpichardo@docentes.uat.edu.mx

J. P. Soto
Interdisciplinary Faculty of Exact and Natural Sciences, University of Sonora, Hermosillo, Sonora, Mexico
e-mail: juanpablo.soto@unison.mx

W. Pedrycz et al. (eds.), *Decision-Making in Computational Intelligence-Based Systems*, Studies in Systems, Decision and Control 628,
https://doi.org/10.1007/978-3-032-04056-5_19

1 Introduction

Insecurity in Latin America has increased in the last two decades, making it the most violent region in the world [1, 2]. In Mexico is recognized as the main social problem, affecting the country's economic and social environment, since it discourages investment and economic activity [3]. In September 2024, the National Institute of Statistics and Geography (INEGI), published the results of the National Survey of Victimization and Perception of Public Safety (ENVIPE), which highlights that during the year 2023 at least 20.2% of households had at least one victim of crime, a situation that represents a statistically significant change with respect to the three previous years that were below this figure. The ENVIPE measures crimes that directly affect victims or households, such as robbery, fraud, extortion and threats, and estimates that, during the year under study, the crime rate was 25,129 per 100,000 inhabitants, the cost of insecurity and crime represented an amount of 5 billion pesos, which are distributed mainly in direct economic losses due to the fact and in expenses for prevention measures, as for the perception of citizens, 59.6% of the elderly population felt unsafe in the place where they live [4]. Through these exercises, the federal government provides an overview of the crime situation in each state, enabling them to allocate public spending to address the problem. However, as mentioned in [5], allocating a larger or smaller budget in this area does not guarantee satisfactory results. This underscores the necessity of developing innovative models focused on crime prevention and reduction [6]. Nevertheless, the exponential increase in data, its diversification, and its large volume exceed human capacity for efficient analysis, paving the way for tools that can perform this task automatically or semi-automatically [7].

Therefore, this study addresses these challenges by applying a data-driven approach focused on the spatial analysis of criminal activity at the neighborhood level, specifically within municipalities in Tamaulipas, Mexico.

The main contributions of this work are as follows:

- Proposal of a clustering-based framework using the K-Means algorithm to segment neighborhoods according to the frequency of criminal events recorded over a seven-year period.
- Comprehensive characterization of the obtained cluster profiles, analyzing the incidence of different types of crimes and their spatial patterns to identify priority intervention areas.
- Implementation of a geospatial information system that facilitates the visualization and prioritization of critical zones for preventive intervention.
- Assessment of cluster cohesion and separation through internal validation metrics, namely the Silhouette Index and the Davies-Bouldin Index.

The remainder of this article is structured as follows:

Section 2 reviews related works in crime analysis and clustering applications. Section 3 describes the proposed methodology, including the data collection, preprocessing, and modeling procedures. Section 4 presents the experimental results and provides an interpretation of the crime cluster profiles. Finally, Sect. 5 concludes the study and outlines directions for future research.

2 Related Works

In the literature, numerous studies have explored the segmentation of regions based on their criminal activity, leveraging clustering techniques. For example, [8] analyzed a dataset from the Republic of Indonesia using the K-Medoids algorithm, resulting in two clusters that divide provinces according to the number of reported crimes. Similarly, the authors in [9, 10] applied the K-Means algorithm to segment areas within Indonesia. The first study incorporated social and economic indicators such as poverty and unemployment, identifying three distinct groups that exhibit strong correlations with these socioeconomic factors. The second study focused specifically on Central Java, using where the process includes data preprocessing, descriptive analysis and determination of the optimal number of clusters using methods such as Elbow and Silhouette, such as incorporating population data by region and developing an information system to map the safest and least vulnerable sectors.

In the study conducted by [11], the authors report that, while several algorithms are available for crime prediction, the K-Means algorithm delivers superior results. Their analysis predicts regions with higher crime rates in India and identifies age groups with a greater or lesser tendency to commit crimes. To determine the optimal number of clusters, they used the elbow method and implemented the project in Python. In the same region, [12, 13] explored serious property fraud (e.g., fraud, breach of trust) in various states. The first study revealed patterns and trends related to property loss, while the second focused on crimes against women, enabling the creation of crime zoning maps.

In Latin America, the application of clustering techniques for crime analysis is relatively limited. Nonetheless, authors such as [14] have contributed to the field with a study that characterizes crimes in Colombia using machine learning. Their findings employ the Predictive Apriori algorithm to identify relationships between categorical attributes, as well as the K-Means algorithm to segment groups based on attributes such as age, municipality, and type of crime.

In this context, the present research seeks to address one of society's most pressing issues through the processes of data mining and machine learning algorithms. In the U.S., [15] applied K-Means for crime mapping in Chicago, integrating interactive visualizations via NetBeans and Google Maps for actionable insights.

Studies like [16] and [17] expand methodological diversity by incorporating data mining algorithms. While [16] uses KNN and decision trees to predict fraudulent activities, [17] combines clustering with Random Forest Regression, demonstrating the latter's potential to forecast high-risk crime types through interactive tools like Tkinter GUI.

Finally, [18] focused on crimes against women in India, employing Linear Regression and Random Forest Classifier within a Flask-based web application. This study achieved a higher accuracy (83%) using Linear Regression and emphasized scalable state-wise analysis and visualization of crime hotspots.

The similarities among all studies analyzed is that commonly employ clustering techniques (e.g., K-Means) for segmentation and utilize data preprocessing to refine datasets. Many explore tools to optimize cluster determination, like the Elbow method, and emphasize interactive visualizations for practical implementation.

Some differences are that while some studies prioritize socioeconomic correlations ([9, 10]), others target specific crime types like property fraud [12], crimes against women ([13, 18]), or criminal demographics ([11, 14]). Methodological variations, such as combining clustering with Random Forest Regression [17] or KNN [16], demonstrate diverse approaches tailored to regional contexts and datasets. Advanced visualization methods and machine learning frameworks also distinguish studies like [15] and [18] by their user-focused interfaces.

Finally, it is worth mentioning that none of the works described above corresponds to the analysis of crimes in Mexico. This is clearly evidenced in [19] where the authors examine over 150 articles to explore the various machine learning and deep learning algorithms applied to predict crime. As shown in the systematic review of the literature, there is no study or proposal for crime analysis in Mexico.

The K-Means algorithm has been widely adopted in crime analysis studies due to its algorithmic simplicity, computational efficiency, and scalability for handling large datasets [9–11, 15]. Nevertheless, it is important to acknowledge that this method presents inherent limitations, such as the assumption that clusters are spherical and of similar size, which may not accurately represent the real-world spatial distribution of criminal phenomena [11, 20]. Despite these constraints, multiple studies have demonstrated that, through appropriate data preprocessing including normalization, outlier management, and the selection of relevant variables, K-Means can generate meaningful groupings that effectively characterize geographic areas based on their crime profiles [9, 10, 14, 15].

In particular, its ease of interpretation through centroids facilitates the identification of actionable patterns for public security agencies, which require clear and practical results for strategic planning and efficient resource allocation [15, 17]. Considering the nature of the data employed in this study, accumulated crime frequencies combined with socioeconomic attributes at the neighborhood level, and the need to produce segmentations that are easily applicable in operational public security contexts, K-Means emerges as an effective and suitable tool for the classification of urban areas in the state of Tamaulipas, Mexico. Additionally, internal validation of the clustering results using metrics such as the Silhouette Coefficient [21] and the Davies-Bouldin Index [20] further reinforces the robustness of the findings.

In this context, the present research seeks to address one of society's most pressing issues through the processes of data mining and machine learning algorithms, for the analysis of crimes in the state of Tamaulipas, Mexico, considering that of the techniques used so far, K-means is the one that generates the best results when classifying areas according to crime rates.

3 Methodology

This project was developed using the Python programming language, alongside specialized libraries for data mining and machine learning, such as pandas and sklearn. To enable interactive data exploration and create dynamic visualizations, the Dash library from Plotly was utilized, while GeoPandas provided robust tools for managing and analyzing geospatial data.

The methodology followed in this research, as outlined in Fig. 1, is based on the KDD process, ensuring a systematic approach to extracting meaningful insights.

3.1 Data Collection

In Mexico, the Executive Secretariat of the National Public Security System (SESNSP) is the government body responsible for collecting and recording the occurrence of crimes documented through preliminary inquiries or investigation files reported to the prosecutorial agencies tasked with delivering justice. Each month, the SESNSP publishes crime incidence figures for statistical purposes.

At the state level, in Tamaulipas, the State Information Center, an entity under the SESNSP, manages and consolidates crime data to develop strategies for crime prevention, deterrence, and containment. This organization provides access to a platform called the Comprehensive Platform for Analysis, Information, Georeferencing, and Security Studies, from which the dataset used in this project was obtained.

The Government of Mexico also operates a governmental body known as the National Population Council (CONAPO), responsible for planning the country's demographic dynamics. Its mission is to ensure that the government's economic and social development aligns with the population's needs and the demographic changes occurring over time.

In October 2021, CONAPO released its latest report on socioeconomic indicators,[1] providing data at various levels of geographic disaggregation. Among these indicators, the Marginalization Index stands out. This index serves as a tool for classifying different geographic areas of the country based on the overall level of deprivation faced by their population. It evaluates factors such as limited access to education, healthcare services, adequate housing, basic goods, or sufficient income.

The datasets for this project were obtained from the two government agencies described above. The first provides crime data, while the second contributes socioeconomic variables at the neighborhood level, including population, population without basic education, the marginalization index, and digital cartography for the municipalities of Tamaulipas.

The crime dataset contains 48,768 records documenting crimes committed between 2017 and 2023. It comprises seven variables, Table 1: the year, month,

[1] https://www.gob.mx/conapo/documentos/indices-de-marginacion-2020-284372.

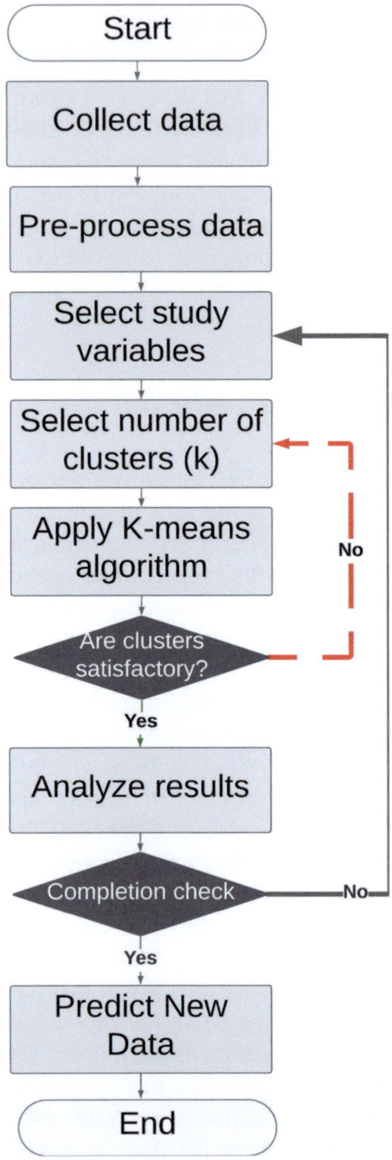

Fig. 1 Methodology

and day of the incident; the location (municipality and neighborhood); the number of occurrences; and the specific type of crime.

The dataset covers only nine municipalities, primarily from the border region, including Nuevo Laredo, Reynosa, Rio Bravo, Matamoros, and Valle Hermoso. From the central region, it includes Ciudad Victoria and San Fernando, while the southern region is represented by El Mante, Altamira, Ciudad Madero, and Tampico. The

Table 1 Crime dataset variables

Variable	Data type
Year	Numeric, discrete
Month	Numeric, discrete
Day	Numeric, discrete
Municipality	Categorical
Neighborhood	Categorical
Quantity	Numeric, discrete
Crime	Categorical

Table 2 **a** Example of crime dataset (sample records) **b** Example of socioeconomic dataset (sample records)

(a)

Year	Month	Day	Municipality	Neighborhood	Quantity	Crime type
2021	5	14	Reynosa	Las Fuentes	1	Intentional homicide
2022	3	22	Victoria	Zona Centro	2	Business robbery
2023	9	7	Matamoros	Solidaridad	1	Kidnapping

(b)

Municipality	Neighborhood	Population	Marginalization index	Without basic education
Reynosa	Las Fuentes	146.57	50	36.96
Victoria	Zona Centro	150.12	1856	26.36
Matamoros	Solidaridad	152.35	132	11.09

crimes recorded in the dataset include burglary, vehicle theft, robbery, theft from businesses, homicide, and kidnapping.

To enhance the understanding of the datasets used in this study, examples of sample records are presented below. Table 2a shows a subset of the crime dataset, illustrating the structure and types of information collected regarding criminal incidents. Similarly, Table 2b presents a sample of the socioeconomic dataset, displaying key attributes such as population size, marginalization index, and educational attainment levels. These examples provide a clearer view of the raw data used in the preprocessing and modeling stages of the analysis.

3.2 *Preprocessing*

As the first step, the crime dataset was preprocessed. There was no missing data, so it was not necessary to handle null values. Only the attributes for Municipality, Colony, Quantity, and Crime were selected. The names of neighborhoods in the central zone were standardized due to discrepancies with similar names, such as Sector Centro and

Sector Zona Centro. The digital cartography provided by CONAPO only includes the neighborhood name field, excluding the type of settlement where it is located (e.g., Housing Development, Residential Unit, Village, Locality). Therefore, a new attribute called 'category' was added to the crime dataset through the colony column, which extracts the area type from the colony name. Finally, to complete the matching of attributes between both datasets, spatial characters were removed, and neighborhood names were changed from lowercase to uppercase. In Table 3, an example of the preprocessed dataset is presented.

For the purposes of this project, the analysis of criminal behavior is performed at the neighborhood level for the municipality under study. The clustering technique aims to group sectors that share a similar frequency of crime. Therefore, the crime dataset first filters the data based on the municipality chosen by the user, then creates a pivot table that groups crime occurrences by neighborhood and category.

If the incorporation of socioeconomic variables is considered in the analysis, it is essential to consider that the data provided by CONAPO covers up to the year 2021, while crime data extends to 2023. This time gap, along with the constant phenomenon of demographic changes, such as the creation or renaming of neighborhoods, directly impacts the size of the dataset. As a result, the number of records available for applying clustering techniques may be reduced.

If the user selects an analysis with socioeconomic variables, since the CONAPO dataset contains information from across the nation, a filter is first applied to select only data corresponding to Tamaulipas. Then, it is filtered by the municipality under study, and the columns of interest (which contain the socioeconomic variables) are selected, with each column being clearly and identifiably renamed. Finally, special characters are removed. In Table 4, socioeconomic variables are added.

Since the numerical scale of the data differs for each variable, and because this situation affects the performance of machine learning algorithms, it is necessary to transform the values into a common range without altering their meaning. Therefore, MinMax normalization, Eq. 1, from the sklearn library was applied, which adjusts the values to a range from 0 to 1.

$$V_i' = \frac{V_i - \min_A}{\max_A - \min_A}\left(new\max_A - new\min_A\right) + new\min_A \tag{1}$$

3.3 Modeling

The modeling process involves several stages, from variable selection to result analysis, with the goal of constructing a predictive model based on data. Below are the key steps of this process.

Table 3 Sample of the preprocessed crime dataset

#	Neighborhood	Category	Intentional homicide	Burglary (home burglary)	Business robbery	Street robbery	Carjacking (vehicle robbery with violence)	Motor vehicle theft (without violence)	Kidnapping
0	2 DE MAYO	COLONIA	5	5	3	1	5	13	0
1	20 DE NOVIEMBRE	COLONIA	1	4	3	0	0	0	0
2	ABEDUL	FRACCIONAMIENTO	0	6	1	0	2	2	1
3	ACAPULQUITO	COLONIA	0	2	0	4	4	4	0
4	ADOLFO LOPEZ MATEOS	AMPLIACIÓN	1	2	2	0	0	2	0

Table 4 Sample of the crime dataset merged with the CONAPO dataset

#	Neighborhood	Category	Intentional homicide	Burglary (home burglary)	Business robbery	Street robbery	Carjacking (vehicle robbery with violence)	Motor vehicle theft (without violence)	Kidnapping	Marginalization index	Population	Without basic education
0	2 DE MAYO	COLONIA	5	5	3	1	5	13	0	148.51	21	29.78
1	20 DE NOVIEMBRE	COLONIA	1	4	3	0	0	0	0	145.68	257	31.22
2	ABEDUL	FRACCIONAMIENTO	0	6	1	0	2	2	1	150.59	747	21.03
3	ACAPULQUITO	COLONIA	0	2	0	4	4	4	0	145.02	178	23.25
4	ADOLFO LOPEZ MATEOS	AMPLIACIÓN	1	2	2	0	0	2	0	145.02	644	33.81

3.3.1 Variable Selection

Proper selection of variables is crucial to ensure that the model is as efficient and representative as possible. In this project, the Python Pandas library is used, which provides exploratory data analysis (EDA) tools to evaluate and preprocess the variables. A key step is identifying correlations between variables, which can be done using a Pearson correlation matrix. This matrix helps to understand how the variables relate to each other, showing whether there is a positive, negative, or no correlation at all.

To visualize these relationships, a heatmap based on the correlation matrix is created, where darker areas indicate strong correlations, and lighter areas indicate weak or no correlations. This visualization helps to identify the most important variables, those with the highest correlation with others. From here, the user can select the features they wish to investigate further, choosing the most relevant combinations for subsequent analysis.

3.3.2 Clustering Algorithm: K-Means

Among clustering techniques, K-Means stands out as the most popular method, as noted by several authors [7, 22, 23]. Its primary concept involves defining the number of clusters to be created (parameter k). Initially, k elements are selected randomly to represent the centers or means of the clusters. Subsequently, each data point is assigned to the nearest cluster center based on the Euclidean distance. Once all points are assigned, the average position (centroid) of the points within each cluster is calculated to establish a new cluster center. These steps are iterated until the assignments of data points and cluster centers stabilize, remaining unchanged across iterations, (See Algorithm 1).

Algorithm 1: K-means clustering algorithm

Input:
- k: Number of clusters
- D: Dataset containing n data points

Output:
- Set of k clusters

Steps:
1. Randomly select k initial centroids from the dataset D
2. Repeat until convergence:
   a. Assign each data point to the nearest centroid, based on a distance metric (e.g., Euclidean distance)
   b. Update each centroid by calculating the mean of all data points assigned to its cluster
3. Return the final set of clusters and their centroids

Once the relevant numerical variables were selected and the data had been preprocessed, the K-Means algorithm was applied to the dataset. The process involved creating a data frame containing the scaled and selected variables.

The algorithm then grouped the observations by assigning a cluster label to each one, adding a new column labeled "cluster" to the dataset. This column indicated the specific group to which each observation belonged.

The resulting clustered dataset enabled the visualization of neighborhood groupings and facilitated the analysis of common crime characteristics within each identified cluster.

3.3.3 Distance Metric

The similarity between elements is typically determined by their spatial proximity, often measured using a distance function. Among these, the Euclidean distance (Eq. 2) is particularly notable, in this context, the quality of a cluster is commonly represented by the maximum distance between any two elements within the group.

$$d(x, y) = \sqrt{\sum_{i=1}^{n} (X_i - Y_i)^2} \tag{2}$$

3.3.4 Cluster Number Selection

To determine the optimal number of clusters, several methods can be employed; among them, the elbow method is one of the simplest and most widely used [22]. This method identifies the point at which increasing the number of clusters no longer yields substantial improvements in clustering quality. In this project, the K-Means model was implemented using the scikit-learn library, and the inertia metric was calculated, which measures the within-cluster sum of squared distances between each data point and its assigned centroid.

The interface developed includes a function that automatically runs the K-Means algorithm for values of k ranging from 1 to 10. For each k, the corresponding inertia value is calculated and stored, allowing the construction of an interactive plot to visualize how inertia varies as the number of clusters increases. This plot was created using the Plotly library and is integrated into the dashboard to facilitate the selection of the optimal k.

Additionally, the system incorporates a secondary function that computes the first and second derivatives of the inertia curve. This analytical approach provides a more objective mechanism to detect the elbow point, where the inertia reduction stabilizes. By combining visual interpretation with derivative-based analysis, the selection process becomes both more robust and data-driven, supporting informed decision-making for determining the ideal number of clusters.

3.3.5 Result Analysis

Result analysis is one of the most critical phases of the modeling process. In this project, this analysis was conducted using digital maps provided by CONAPO in shapefile format, which are suitable for geospatial processing as they contain geometries, polygons, and geographic names. To represent the spatial results of the clustering, the GeoPandas library was used to load the GeoJSON file corresponding to the municipality being studied. Once loaded, a base map is generated where all geographical zones are projected.

The data resulting from the clustering process, which includes the "cluster" label for each observation, is merged with the geospatial data, enabling each neighborhood to be associated with its respective cluster. This allows the representation of clusters on a choropleth map, where each region is colored according to its cluster assignment. Plotly's tools facilitate the creation of these interactive maps, supporting intuitive exploration of spatial patterns.

The integration of geospatial clustering with thematic maps provides public safety authorities with a clear visual representation of neighborhood groupings and their associated crime patterns. This dynamic mapping approach contributes significantly to the design of more effective, evidence-based intervention strategies, allowing decision-makers to allocate resources efficiently and implement targeted prevention efforts in the most vulnerable areas.

3.3.6 Prediction

Finally, the system allows for predictions on new data. The interface dynamically generates text input components related to the study variables selected initially, allowing the user to enter new values. Once the values are entered, the model automatically assigns the new data points to one of the previously defined clusters. This step enables the system to classify new observations into the identified groups, which is useful, for example, for assigning new crime data to areas or groups with similar characteristics.

This prediction capability facilitates real-time decision-making, allowing public safety authorities not only to understand current dynamics but also to predict future crime patterns in specific areas and act accordingly.

4 Results

In order to have an information system that visually and interactively displays the analysis of crime data through data mining and machine learning, the Plotly framework was used. This framework enables the creation of such applications through an HTML interface with components like buttons, charts, and maps, (See Fig. 2).

The main interface has the following key features:

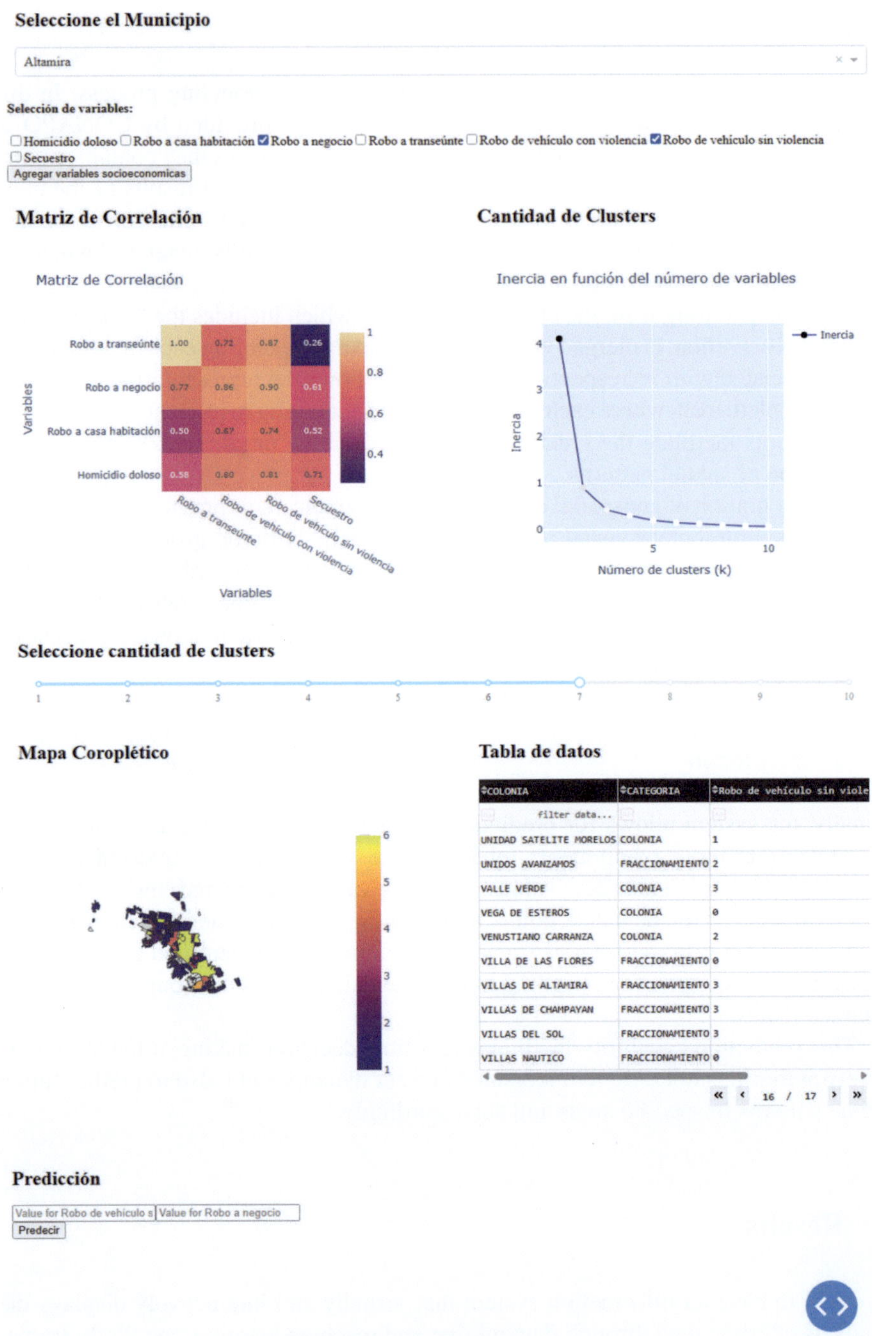

Fig. 2 Main interface for crime analysis with K-means

1. Dropdown: To select the municipality under study.
2. Checklist: Contains the names of the attributes used by the machine learning algorithm for clustering.
3. Button: To add socioeconomic variables to the previous checklist.
4. Correlation Chart: Displays a heatmap of Pearson correlations for each variable.
5. Inertia Chart: Dynamically shows the calculation of the elbow method to determine the ideal number of clusters based on the selected variables.
6. Slider and Button: To set the number of clusters or the parameter k for the clustering technique.
7. Choropleth Map and Data Table: Displays the distribution of clusters and summarizes the data in a table for analysis.
8. Prediction Section: Through a textbox and a button, allows predicting the assignment of new data to the already created clusters.

It is important to note that the system interface is presented in Spanish, considering that the primary target users are public security agencies and local authorities in Mexico.

For the purposes of this study, the clustering algorithm was applied exclusively to those variables that exhibited high correlation within each region, as identified through the Pearson correlation matrix displayed by the system interface. This selection process ensured that only the most relevant and interrelated crime types were included, thereby enhancing the robustness and interpretability of the clustering results. Table 5 provides the correlation coefficients for each region concerning the combination of crime variables.

The elbow method was applied to each of the datasets from the previous table, yielding the results presented in Table 6, which lists the inertia values and the parameter k representing the optimal number of clusters. Subsequently, two complementary metrics were used to evaluate the clustering quality: the Davies–Bouldin index [20] and the silhouette coefficient [21]. The Davies–Bouldin index assesses within-cluster dispersion and inter-cluster separation; lower values indicate compact and well-separated clusters. In contrast, the silhouette coefficient evaluates cluster cohesion and separation, with values ranging from -1 to 1, where a score close to 1 suggests that the observations are well clustered.

On the other hand, when socioeconomic variables are integrated into the analysis, the correlation coefficients are as presented in Table 7.

Table 8 describes the inertia and elbow selection for municipalities concerning crimes and socioeconomic variables.

The system's interface is designed to provide an interactive and dynamic experience that allows the user to analyze the distribution of clusters through a thematic map, thus facilitating the visualization and understanding of how different geographic areas are grouped based on the selected characteristics. This tool enables the user to effectively explore the spatial relationships of the clusters, providing a visual representation that highlights areas with similar patterns.

In addition to the map, the system displays a data table that provides detailed information about each neighborhood, including the frequency of crimes and/or

Table 5 Correlation analysis with crime variables

Municipality	Combination of variables	Correlation coefficient
Altamira	Business robbery and motor vehicle theft (without violence)	0.895
Ciudad Madero	Home burglary and motor vehicle theft (without violence)	0.845
El Mante	Business robbery and motor vehicle theft (without violence)	0.985
Matamoros	Street robbery and motor vehicle theft (without violence)	0.931
Nuevo Laredo	Street robbery and motor vehicle theft (without violence)	0.877
Reynosa	Kidnapping and motor vehicle carjacking (vehicle robbery with violence)	0.874
Rio Bravo	Street robbery and business robbery	0.923
San Fernando	Carjacking (vehicle robbery with violence) and intentional homicide	0.980
Tampico	Business robbery and motor vehicle theft (without violence)	0.883
Valle Hermoso	Carjacking (vehicle robbery with violence) and intentional homicide	0.971
Victoria	Vehicle theft (without violence) and street robbery	0.960

the corresponding socioeconomic variables for each area. This table also includes the cluster to which each neighborhood belongs, allowing the user to perform a comparative analysis between different zones and better understand the underlying factors affecting each group.

Figure 3 shows a representative example of an analysis conducted with the system, providing a clear view of the most vulnerable areas or those with similar patterns. The combination of the thematic map and the data table facilitates the identification of trends, patterns, and relationships between the analyzed variables, providing a powerful tool for data-driven decision-making.

The following section provides a set of tables (Tables 9, 10, 11, 12, 13, 14, 15, 16, 17, 18, 19, 20, 21, 22, 23, 24, 25, 26, 27, 28, 29, 30) describing the clusters obtained through the clustering analysis. First, the results based solely on crime-related variables are presented, followed by the results that include socioeconomic variables to enhance the interpretation of the patterns found.

As a core part of this analysis, narrative cluster profiles were developed for each municipality. These profiles offer clear interpretations of each cluster's distinctive traits, including levels of criminal activity, the distribution of settlement types, and the number of neighborhoods within each group.

Each profile was supported by validation metrics such as the Davies-Bouldin index and the silhouette score, ensuring the internal consistency and separability of the resulting clusters. This narrative and quantitative approach provides a valuable

Table 6 Inertia values and evaluation metric whit crime variables

Municipality	Inertia values	Parameter k (inertia difference)	Davies-Bouldin index	Silhouette coefficient
Altamira	4.102, 0.902, 0.42, 0.296, 0.193, 0.14, 0.114, 0.089, 0.067, 0.059	7	0.398	0.701
Ciudad Madero	5.712, 2.427, 1.015, 0.776, 0.638, 0.583, 0.353, 0.279, 0.241, 0.214	6	0.731	0.461
El Mante	2.169, 0.237, 0.084, 0.039, 0.031, 0.013, 0.011, 0.008, 0.005, 0.004	5	0.344	0.631
Matamoros	4.076, 0.974, 0.459, 0.428, 0.282, 0.21, 0.148, 0.116, 0.093, 0.069	4	0.651	0.627
Nuevo Laredo	3.046, 1.185, 0.497, 0.343, 0.236, 0.152, 0.129, 0.101, 0.087, 0.065	9	0.492	0.655
Reynosa	3.739, 0.839, 0.452, 0.362, 0.264, 0.195, 0.133, 0.099, 0.084, 0.069	4	0.664	0.722
Rio Bravo	2.237, 0.296, 0.141, 0.071, 0.025, 0.012, 0.004, 0.003, 0.001, 0.001	8	0.189	0.92
San Fernando	1.989, 0.123, 0.051, 0.034, 0.019, 0.015, 0.012, 0.009, 0.005, 0.004	8	0.410	0.602
Tampico	4.068, 1.382, 0.622, 0.442, 0.391, 0.312, 0.247, 0.184, 0.144, 0.12	5	0.729	0.558

(continued)

Table 6 (continued)

Municipality	Inertia values	Parameter k (inertia difference)	Davies-Bouldin index	Silhouette coefficient
Valle Hermoso	4.51, 1.863, 1.059, 0.581, 0.373, 0.326, 0.24, 0.208, 0.122, 0.114	8	0.186	0.894
Victoria	3.182, 0.318, 0.2, 0.12, 0.091, 0.065, 0.042, 0.034, 0.028, 0.024	8	0.506	0.583

Table 7 Correlation analysis with socioeconomic variables

Municipality	Combination of variables	Correlation coefficient
Altamira	Home burglary and population	0.787
Ciudad Madero	Home burglary and population	0.833
El Mante	Motor vehicle theft (without violence) and population	0.880
Matamoros	Intentional homicide and population	0.386
Nuevo Laredo	Home burglary and population	0.687
Reynosa	Home burglary and population	0.691
Rio Bravo	Home burglary and population	0.821
San Fernando	Home burglary and population	0.676
Tampico	Home burglary and population	0.711
Valle Hermoso	Motor vehicle theft (without violence) and population	0.935
Victoria	Home burglary and population	0.698

tool for both academic interpretation and public policy design focused on crime prevention.

In the municipality of Altamira, clustering analysis revealed a predominance of neighborhoods with very low crime incidence. Cluster 1, encompassing the largest number of neighborhoods, is characterized by minimal business robberies and motor vehicle thefts, suggesting relatively peaceful areas requiring only basic preventive strategies. In contrast, Clusters 2 and 7 each identified a single neighborhood with extremely high crime levels, highlighting critical zones that demand urgent and specialized interventions. The findings underscore the spatial asymmetry of crime distribution in Altamira, with most neighborhoods exhibiting low criminal activity and a few areas facing severe vulnerability.

In the municipality of Altamira, the clustering analysis based on socioeconomic variables identified a predominance of low-crime residential areas. Cluster 1, the

Table 8 Inertia values and evaluation metric whit socioeconomic variables

Municipality	Inertia values	Parameter k (inertia difference)	Davies-Bouldin index	Silhouette Coefficient
Altamira	6.13, 2.715, 1.555, 1.22, 1.001, 0.844, 0.662, 0.497, 0.463, 0.435	6	0.760	0.438
Ciudad Madero	6.099, 2.328, 1.004, 0.627, 0.55, 0.423, 0.396, 0.346, 0.307, 0.267	5	0.768	0.514
El Mante	2.234, 0.426, 0.14, 0.075, 0.055, 0.047, 0.026, 0.024, 0.02, 0.018	6	0.598	0.487
Matamoros	5.976, 3.741, 2.002, 1.32, 0.955, 0.779, 0.653, 0.512, 0.454, 0.367	9	0.509	0.526
Nuevo Laredo	7.283, 2.652, 2.025, 1.451, 1.193, 1.122, 0.935, 0.502, 0.428, 0.401	7	0.731	0.523
Reynosa	9.942, 3.798, 2.502, 2, 1.569, 1.328, 1.059, 0.869, 0.776, 0.66	6	0.856	0.60
Rio Bravo	5.784, 1.881, 1.293, 0.845, 0.679, 0.609, 0.524, 0.371, 0.32, 0.259	7	0.769	0.413
San Fernando	2.835, 0.976, 0.694, 0.428, 0.277, 0.223, 0.093, 0.071, 0.05, 0.04	6	0.508	0.434
Tampico	7.983, 3.215, 1.807, 1.325, 1.198, 1.019, 0.712, 0.555, 0.482, 0.434	6	0.805	0.422

(continued)

Table 8 (continued)

Municipality	Inertia values	Parameter k (inertia difference)	Davies-Bouldin index	Silhouette Coefficient
Valle Hermoso	2.2, 0.433, 0.099, 0.069, 0.04, 0.036, 0.018, 0.016, 0.006, 0.003	6	0.371	0.478
Victoria	6.436, 3.303, 1.936, 1.263, 1.01, 0.798, 0.555, 0.483, 0.402, 0.366	6	0.621	0.508

⇕COLONIA	⇕CATEGORIA	⇕Secuestro	⇕Robo de vehículo con violencia	⇕Cluster
filter data...				=5
ALMAGUER	COLONIA	5	108	5
BENITO JUAREZ	COLONIA	3	79	5
JOSE DE ESCANDON - PETROLERA	COLONIA	2	102	5
LAS FUENTES	COLONIA	1	143	5
LAS FUENTES SECCION LOMAS	COLONIA	2	130	5
LOMAS DEL REAL DE JARACHINA NORTE	FRACCIONAMIENTO	5	71	5
REYNOSA	FRACCIONAMIENTO	2	127	5
RODRIGUEZ	COLONIA	3	105	5
ZONA CENTRO	COLONIA	5	83	5

Fig. 3 Choropleth map and data table for analysis the distribution of clusters

largest group, includes predominantly small to medium-sized residential neighborhoods with low home burglary incidence. In contrast, Clusters 2 and 4 revealed critical hotspots: Cluster 2 groups highly populated areas with very high burglary rates, while Cluster 4 corresponds to a single neighborhood with extreme burglary concentration associated with its maximum population density. These findings highlight a spatial imbalance, where most neighborhoods maintain low residential crime

Table 9 Cluster profile—Altamira municipality (crime variables)

Cluster	Key characteristics	Interpretation
Cluster 1: Low-incidence neighborhoods	• Very low levels of business robberies (between 0 and 7 cases, averaging approximately 1) • Sparse motor vehicle thefts (between 0 and 7 cases, averaging around 1) • Great diversity in neighborhood types (residential areas, planned developments, expansions) • Largest cluster, comprising 140 neighborhoods	These neighborhoods represent relatively peaceful areas with low criminal activity. Suitable for basic preventive strategies and light monitoring
Cluster 2: Critical area with extremely high crime incidence	• Very high levels of business robberies (81 cases) • Extremely high incidence of motor vehicle thefts (84 cases) • Represented by a single specific neighborhood	A single extremely critical area urgently requiring specialized and continuous interventions due to extreme crime concentration
Cluster 3: High incidence concentrated in business robbery	• Significant business robberies (8 to 21 cases, average around 14) • Low motor vehicle thefts (between 0 and 4 cases, average approximately 2) • Predominantly residential neighborhoods (10 neighborhoods)	Neighborhoods primarily vulnerable in commercial aspects, suitable for targeted commercial surveillance and focused preventive strategies
Cluster 4: Areas with medium–high incidence of both crimes	• Moderate to high incidence of business robberies (15 to 25 cases, average 19.7) • Moderate motor vehicle thefts (10 to 19 cases, average 13.2) • Represented by few neighborhoods (4 neighborhoods), predominantly planned residential areas ("fraccionamientos")	Mixed-use areas requiring balanced surveillance and comprehensive preventive strategies due to considerable levels of both crime types
Cluster 5: Critical areas with high mixed-crime incidence	• High business robberies (32 to 36 cases, average 34) • High motor vehicle thefts (18 to 21 cases, average 19.5) • Limited number of specific neighborhoods (2 neighborhoods)	Neighborhoods severely impacted by multiple crime categories, demanding priority interventions, enhanced surveillance, and immediate targeted actions

(continued)

Table 9 (continued)

Cluster	Key characteristics	Interpretation
Cluster 6: Moderate-high vehicle robbery incidence	• Moderate business robberies (up to 7 cases, average 3.4) • Higher motor vehicle thefts (9 to 19 cases, averaging around 13) • Predominantly typical residential neighborhoods (7 neighborhoods)	Areas specifically vulnerable to motor vehicle theft. Preventive strategies and targeted patrols focusing specifically on vehicle crime are necessary
Cluster 7: Extremely critical area in both crimes	• Very high business robberies (102 cases) • High motor vehicle thefts (77 cases) • Represented by a single specific neighborhood	Top-priority specific area due to extreme crime severity, requiring immediate attention, intensive and specialized surveillance

risks, while a few densely populated sectors exhibit acute vulnerabilities requiring urgent and specialized interventions.

In the municipality of Ciudad Madero, the clustering analysis revealed a predominance of neighborhoods with low to moderate crime incidence. Cluster 4, the largest group, includes areas with minimal home burglaries and motor vehicle thefts, suitable for maintaining basic preventive surveillance. In contrast, Clusters 2 and 5 identified critical zones with extremely high robbery concentrations in both homes and vehicles, each represented by only two neighborhoods, demanding urgent and continuous interventions. Overall, the findings highlight a spatial disparity, with most neighborhoods experiencing low crime rates and a few sectors requiring prioritized security actions.

In the municipality of Ciudad Madero, the clustering analysis based on socioeconomic variables revealed a predominance of small, low-crime neighborhoods. Cluster 1, the largest group, comprises sparsely populated areas with minimal home burglary incidence, suitable for basic preventive surveillance. In contrast, Cluster 3 identified a critical group of densely populated neighborhoods with extremely high burglary rates, requiring urgent and continuous interventions. The results highlight a clear spatial imbalance, with most neighborhoods maintaining low residential crime risk, while a few highly populated sectors present acute vulnerabilities.

In the municipality of El Mante, clustering analysis revealed that most neighborhoods exhibit very low crime levels. Cluster 4, the largest group, includes neighborhoods with minimal business robberies and motor vehicle thefts, suitable for light preventive surveillance. In contrast, Cluster 2 identified an extremely critical zone, represented by a single neighborhood with exceptionally high rates of both crimes, demanding urgent and specialized interventions. The findings highlight a marked spatial asymmetry, with the majority of areas remaining peaceful and only a few sectors requiring intensive security measures.

Table 10 Cluster profile—Altamira municipality (socioeconomic variables)

Cluster	Key characteristics	Interpretation
Cluster 1: Residential areas with low incidence of home burglary	• Low incidence of home burglaries (0 to 4 cases, average approximately 2) • Relatively small to medium-sized population (between 44 and 1773 residents, average around 685) • High number of neighborhoods in this cluster (86 neighborhoods) • Predominantly residential neighborhoods	These represent quiet residential areas with low domestic crime rates, suitable for basic preventive surveillance
Cluster 2: Highly populated areas with very high incidence of home burglary	• High incidence of home burglaries (15 to 31 cases, average 25.3) • High population density (between 2669 and 8584 residents, average about 5452) • Only 6 neighborhoods, predominantly planned residential areas ("fraccionamientos")	Critical areas with large populations experiencing high domestic crime pressure. They urgently require constant monitoring, specialized operations, and strengthened community security
Cluster 3: Moderately populated neighborhoods with medium–high burglary incidence	• Medium to high incidence of home burglaries (10 to 18 cases, average 12.6) • Moderate to high population (870 to 3417 residents, average approximately 2536) • Predominantly planned residential areas ("fraccionamientos," 13 neighborhoods)	These neighborhoods present significant vulnerabilities regarding residential security, suitable for reinforced surveillance and specific preventive community programs
Cluster 4: Neighborhood with maximum population concentration and critical burglary incidence	• Highest incidence of home burglaries (44 cases) • Very high population concentration (11,216 residents) • Represented by a single large planned residential neighborhood ("fraccionamiento")	Specific critical hotspot due to extreme burglary incidence, likely related to its large size and population density. It demands specialized interventions and continuous intensive surveillance

(continued)

Table 10 (continued)

Cluster	Key characteristics	Interpretation
Cluster 5: Large neighborhoods with moderate burglary incidence	• Moderate incidence of home burglaries (1 to 10 cases, average approximately 5.5) • High population (2178 to 4465 residents, average about 3188) • Predominantly typical residential neighborhoods ("colonias")	Areas presenting moderate residential security risks and considerable population size; preventive surveillance and active community involvement can mitigate escalating risks
Cluster 6: Small neighborhoods with medium burglary incidence	• Consistent moderate home burglary incidence (5 to 10 cases, average approximately 7) • Small-sized neighborhoods (21 to 1409 residents, average around 780) • Predominantly typical residential neighborhoods ("colonias," 19 neighborhoods)	Small neighborhoods requiring focused preventive attention due to vulnerability to residential crime, although the general crime pressure is not excessively high

In the municipality of El Mante, clustering analysis based on socioeconomic variables revealed that most neighborhoods are small to medium-sized areas with low vehicle crime incidence. Cluster 4, the largest group, comprises quiet neighborhoods with minimal criminal activity, requiring only basic surveillance. In contrast, Cluster 2 identified a single, highly populated neighborhood with extreme motor vehicle theft rates, demanding urgent and continuous interventions. The findings reflect a clear spatial imbalance, with the majority of neighborhoods remaining low-risk, while isolated areas face severe security challenges.

In the municipality of Matamoros, clustering analysis revealed that most neighborhoods exhibit very low crime incidence. Cluster 3, the largest group, comprises peaceful areas with minimal street robberies and motor vehicle thefts, suitable for light monitoring. In contrast, Cluster 2 identified a critical zone composed of only two neighborhoods with extremely high crime levels, requiring urgent, intensive, and coordinated security interventions. The results highlight a pronounced spatial asymmetry, with the majority of neighborhoods maintaining low crime rates and a few sectors facing severe criminal concentration.

In the municipality of Matamoros, clustering analysis based on socioeconomic variables revealed that most neighborhoods are small, peaceful areas with no recorded homicides. Cluster 2, the largest group, comprises low-population neighborhoods suitable for early prevention programs. In contrast, several critical clusters, such as Clusters 3, 7, and 8, identified specific neighborhoods with concentrated lethal violence, demanding urgent interventions and intensive monitoring. The findings

Table 11 Cluster profile—Ciudad Madero municipality (crime variables)

Cluster	Key characteristics	Interpretation
Cluster 1: Moderate incidence of home burglary and motor vehicle theft	• Home burglaries range between 5 and 16 cases, averaging approximately 8.5 • Motor vehicle thefts range between 0 and 7 cases, average of 3.5 • Comprises 18 neighborhoods, predominantly typical residential areas	These neighborhoods exhibit moderate levels of both home burglary and motor vehicle theft. Regular surveillance and targeted preventive measures are recommended
Cluster 2: Critical zones with high robbery incidence	• Very high home burglary incidence (46 to 61 cases, average 53.5) • Extremely high incidence of motor vehicle thefts (49 to 56 cases, average 52.5) • Represented by only 2 neighborhoods (urban expansions)	These areas are extremely critical, requiring urgent and specialized continuous monitoring due to their high crime concentrations in both homes and vehicles
Cluster 3: High incidence concentrated in both types of robberies	• High home burglary incidence (7 to 27 cases, average 16.1) • High motor vehicle thefts incidence (9 to 32 cases, average 16) • Represented by 17 typical residential neighborhoods	Neighborhoods with significant vulnerabilities in both residential and vehicle security, necessitating prioritized active surveillance and more intense preventive measures
Cluster 4: Neighborhoods with generally low crime incidence	• Low home burglaries (0 to 5 cases, average 1.68) • Very low motor vehicle thefts (0 to 4 cases, average 1.19) • Largest cluster with 41 neighborhoods, predominantly typical residential areas	Peaceful neighborhoods with low criminal incidence, suitable for maintaining preventive surveillance
Cluster 5: Critical neighborhoods with high combined incidence	• High home burglaries (44 to 48 cases, average 46) • High motor vehicle thefts (24 to 29 cases, average 26.5) • Represented by only 2 residential neighborhoods	Critical zones with significant robbery incidence in both categories, demanding immediate intervention, continuous monitoring, and strengthened preventive strategies
Cluster 6: Moderate incidence primarily focused on motor vehicle theft	• Moderate home burglaries (0 to 8 cases, average 4.45) • Higher motor vehicle thefts incidence (5 to 14 cases, average 7.82) • Comprises 11 neighborhoods	Neighborhoods particularly vulnerable to vehicle-related crimes. Focused vehicle surveillance and targeted parking area patrols are recommended

Table 12 Cluster profile—Ciudad Madero municipality (socioeconomic variables)

Cluster	Key characteristics	Interpretation
Cluster 1: Small neighborhoods with low domestic burglary incidence	• Low incidence of home burglaries (0 to 7 cases, average approximately 2.1) • Small population (27 to 1200 residents, average around 561) • Largest cluster with 35 neighborhoods	Quiet and sparsely populated areas with low residential crime risk, suitable for basic preventive surveillance
Cluster 2: Large neighborhoods with medium–high burglary incidence	• Medium to high incidence of home burglaries (13 to 27 cases, average 20.8) • High population density (3984 to 6366 residents, average approximately 5296) • Only 5 neighborhoods, predominantly typical residential areas	Large residential areas with significant domestic security issues, needing enhanced attention, constant surveillance, and specific preventive programs
Cluster 3: Densely populated neighborhoods with high burglary incidence	• Very high incidence of home burglaries (46 to 61 cases, average 51.7) • High population density (7522 to 8967 residents, average approximately 8048) • Represented by only 3 neighborhoods	Highly vulnerable neighborhoods due to combined high population density and extreme crime incidence. Urgent intensive monitoring and permanent specific strategies are required
Cluster 4: Medium-sized neighborhoods with moderate burglary incidence	• Moderate home burglary incidence (4 to 21 cases, average 8.8) • Medium population size (1267 to 2437 residents, average around 1770) • Comprising 19 typical residential neighborhoods	Medium-sized areas with moderate risks that would benefit significantly from regular surveillance
Cluster 5: Medium to large neighborhoods with variable burglary incidence	• Variable home burglary incidence (0 to 18 cases, average approximately 11.6) • Considerably high population (2980 to 4917 residents, average around 3884) • Comprising 12 neighborhoods	Neighborhoods with high variability in domestic crime incidence; differentiated surveillance and tailored strategies are recommended according to each neighborhood's specific risk levels

highlight a clear spatial disparity, with the majority of neighborhoods maintaining low violence levels while a few areas present severe security risks.

In the municipality of Nuevo Laredo, clustering analysis revealed that most neighborhoods exhibit low crime incidence. Cluster 1, the largest group, includes peaceful areas with minimal motor vehicle thefts and no street robberies, requiring only light

Table 13 Cluster profile—El Mante municipality (crime variables)

Cluster	Key characteristics	Interpretation
Cluster 1: Moderate incidence primarily vehicular	• Moderate business robberies (0–6 cases, average 2.1) • Moderate motor vehicle thefts (1–10 cases, average 4.8) • Represented by 33 neighborhoods	Neighborhoods with moderate robbery incidence, especially vehicle-related. Regular surveillance with particular emphasis on vehicle crime prevention is recommended
Cluster 2: Extremely critical zone in both crime types	• Very high business robberies (77 cases) • Extremely high motor vehicle thefts (162 cases) • Represented by a single neighborhood	Extremely critical area requiring urgent, specialized, and continuous intervention due to very high crime concentration
Cluster 3: Critical neighborhood with high vehicle crime incidence	Cluster 3: Critical neighborhood with high vehicle crime incidence	Specific critical area needing specialized attention primarily focused on intensive vehicle surveillance
Cluster 4: Broad low-incidence area	• Extremely low business robberies (0–2 cases, average 0.15) • Very low motor vehicle thefts (0–3 cases, average 0.56) • Largest cluster comprising 84 neighborhoods	Areas representing tranquility, ideal for light preventive surveillance
Cluster 5: Specific area with moderate incidence in both crimes	• Moderate business robberies (7 cases) • Moderate motor vehicle thefts (24 cases) • Represented by a single neighborhood	Specific neighborhood with moderate crime incidence, requiring targeted surveillance particularly focusing on vehicle theft prevention

preventive surveillance. In contrast, several critical clusters, such as Clusters 2, 6, 7, and 9, identified specific neighborhoods with extremely high motor vehicle thefts and, in some cases, high street robberies, demanding urgent and strategic interventions. The findings reflect a clear spatial asymmetry, with most neighborhoods maintaining low criminal activity while a few isolated sectors face severe crime concentrations.

In the municipality of Nuevo Laredo, clustering analysis based on socioeconomic variables revealed that most neighborhoods are small, low-crime areas. Cluster 2, the largest group, includes low-population neighborhoods with minimal home burglary incidence, suitable for preventive maintenance and community engagement. In contrast, Clusters 6 and 7 identified critical areas with very high population densities and elevated burglary rates, requiring immediate and sustained interventions. The results highlight a clear spatial disparity, with most neighborhoods maintaining low risks and a few sectors under significant criminal pressure.

Table 14 Cluster profile—El Mante municipality (socioeconomic variables)

Cluster	Key characteristics	Interpretation
Cluster 1: Medium-sized neighborhoods with low vehicle crime incidence	• Low motor vehicle thefts (0–9 cases, average approx. 2.7) • Medium population size (403–894 residents, average approx. 629) • Composed of 21 neighborhoods	Areas with low vehicle crime incidence, suitable for basic preventive surveillance
Cluster 2: Highly populated neighborhood with extreme vehicle crime incidence	• Extremely high motor vehicle thefts (162 cases) • Very high population (10,706 residents) • Represented by a single neighborhood	Critical neighborhood with extremely high population density and vehicle crime incidence; requires urgent special interventions and continuous intensive surveillance
Cluster 3: Large neighborhoods with moderate-high vehicle crime incidence	• Moderate-high motor vehicle thefts (6–24 cases, average approx. 12.7) • Considerably high population (2949–3655 residents, average 3207) • Composed of only 3 neighborhoods	Large residential areas significantly vulnerable to vehicle crime, recommended for specialized surveillance and targeted preventive programs
Cluster 4: Small neighborhoods with minimal vehicle crime incidence	• Very low motor vehicle thefts (0–6 cases, average approx. 1) • Small population neighborhoods (32–385 residents, average approx. 214) • Represented by 33 typical neighborhoods	Quiet, small residential areas ideal for minimal surveillance and basic community-based preventive strategies
Cluster 5: Medium-sized neighborhoods with moderate vehicle crime incidence	• Moderate motor vehicle thefts (2–8 cases, average approx. 4.5) • Medium population size (1024–1635 residents, average approx. 1259) • Represented by 11 neighborhoods	Residential neighborhoods with moderate vehicle risk, requiring focused surveillance and specific preventive actions to prevent crime escalation
Cluster 6: Medium-large neighborhoods with moderate vehicle crime incidence	• Moderate motor vehicle thefts (2–10 cases, average approx. 4.9) • Medium-large population size (1748–2416 residents, average approx. 2054) • Composed of 7 typical neighborhoods	Areas with moderate vehicle crime incidence, recommended to maintain regular surveillance and specific attention to vulnerable points

Table 15 Cluster profile—Matamoros municipality (crime variables)

Cluster	Key characteristics	Interpretation
Cluster 1: Neighborhoods with moderate incidence in both crimes	• Street robberies range from 0 to 7 cases (average 1.8) • Motor vehicle thefts range from 2 to 17 cases (average 7.2) • Composed of 59 neighborhoods	Neighborhoods with moderate crime risk in both Street Robberies and motor vehicle thefts. Require constant surveillance and balanced preventive strategies
Cluster 2: Critical zone with extremely high incidence	• Very high street robberies (36 to 55 cases, average 45.5) • Extremely high motor vehicle thefts (121 to 129 cases, average 125) • Represented by only 2 neighborhoods	Critical area with high crime concentration. Needs immediate intervention, intensive surveillance, and coordinated public safety actions
Cluster 3: Large number of neighborhoods with low incidence	• Very low street robberies (0 to 3 cases, average 0.24) • Very low motor vehicle theft (0 to 5 cases, average 1.26) • Largest cluster with 285 neighborhoods	Peaceful, low-risk areas. Suitable for light monitoring
Cluster 4: High mixed incidence, especially vehicle-related	• Street robberies range from 5 to 16 cases (average 9) • Motor vehicle thefts range from 17 to 62 cases (average 28.4) • Composed of 8 residential neighborhoods	Neighborhoods with mixed criminal issues, particularly vehicle-related. Require active surveillance, targeted patrols, and coordinated preventive actions

In the municipality of Reynosa, clustering analysis revealed that most neighborhoods exhibit low crime incidence. Cluster 1, the largest group, includes areas with minimal kidnappings and carjackings, suitable for preventive maintenance and light surveillance. In contrast, Cluster 2 identified a critical zone with extremely high levels of both crimes, demanding urgent police intervention and sustained institutional support. The findings highlight a sharp spatial disparity, where the majority of neighborhoods remain low-risk while a few areas face extreme criminal concentration.

In the municipality of Reynosa, clustering analysis based on socioeconomic variables revealed that most neighborhoods are small, low-crime areas. Cluster 2, the largest group, includes low-population neighborhoods with minimal home burglary incidence, suitable for basic preventive monitoring. In contrast, Clusters 4 and 6 identified critical sectors with very high burglary rates, demanding direct attention, intensive surveillance, and strategic crime prevention measures. The findings highlight a clear spatial imbalance, with the majority of neighborhoods remaining relatively safe while a few densely populated sectors face significant security risks.

Table 16 Cluster profile—Matamoros municipality (socioeconomic variables)

Cluster	Key characteristics	Interpretation
Cluster 1: Medium-sized neighborhoods with high homicide rates	• Homicides between 3 and 8 cases (average 4.4) • Population from 844 to 4617 residents (average 2729) • Includes 11 residential neighborhoods	Neighborhoods with elevated levels of lethal violence. Require reinforced surveillance and targeted crime containment programs
Cluster 2: Small neighborhoods with zero recorded homicides	• No homicides recorded • Small population (26 to 1623 residents, average 625) • Largest cluster with 176 neighborhoods	Peaceful areas with no violent incidents. Suitable for peacekeeping programs and early prevention strategies
Cluster 3: Single critical neighborhood with concentrated violence	• 20 homicides • Small population (428 residents) • Represented by one neighborhood	Critical area with concentrated lethal violence. Requires immediate and specialized intervention
Cluster 4: Large residential areas with low to moderate homicide rates	• Homicides between 0 and 2 (average 1) • High population (3822 to 6397 residents, average 5023) • 24 planned residential areas	Densely populated zones with low to moderate violence. Preventive surveillance and social cohesion programs are recommended
Cluster 5: Medium-sized residential areas with low homicide rates	• Homicides between 0 and 2 (average 0.55) • Medium population (1747 to 3696 residents, average 2536) • 29 planned residential areas	Areas with low violence but in need of monitoring to prevent spikes
Cluster 6: Small to medium neighborhoods with low homicide rates	• Homicides between 1 and 3 (average 1.3) • Population from 21 to 2054 (average 848) • 64 typical neighborhoods	Zones with low but scattered violence. Suitable for a mix of surveillance and community education
Cluster 7: Critical high-density area with lethal violence	• 7 homicides • Very high population (18,282 residents) • One planned residential area	Critical area due to the combination of high population and violence. Priority attention and intensive surveillance are required
Cluster 8: High-risk neighborhood with extreme homicide rate	• 21 homicides • High population (8003 residents) • Represented by one neighborhood	Critical area requiring police intervention, urgent social programs, and continuous monitoring
Cluster 9: Very large residential areas with elevated homicide	• Homicides between 2 and 5 (average 3.3) • Population from 7866 to 11,409 residents (average 9357) • 3 planned residential areas	Densely populated neighborhoods with notable violence. Need for frequent patrolling and containment programs

Table 17 Cluster profile—Nuevo Laredo municipality (crime variables)

Cluster	Key characteristics	Interpretation
Cluster 1: Numerous neighborhoods with low crime incidence	• No street robberies (0 cases) • Low motor vehicle thefts (0 to 14 cases, average 3.0) • Largest cluster with 104 neighborhoods	Peaceful, low-risk areas. Require only light preventive surveillance
Cluster 2: Critical neighborhood with extreme crime incidence	• Very high street robberies (18 cases) • Extremely high motor vehicle thefts (221 cases) • Represented by one neighborhood	Top-priority area with severe crime concentration. Needs intensive surveillance and immediate strategic intervention
Cluster 3: High crime incidence in specific neighborhoods	• High street robberies (5–6 cases) • Moderate to high motor vehicle thefts (11–32 cases, average 21.5) • Composed of 2 neighborhoods	Neighborhoods with significant risks in both crime types. Require specific intervention and reinforced patrolling
Cluster 4: High vehicle crime with moderate pedestrian risk	• Street robberies from 2 to 4 (average 2.8) • High motor vehicle thefts (36 to 48, average 42.5) • Composed of 4 neighborhoods	Areas especially vulnerable to vehicle crime with moderate pedestrian risk. Recommend combined surveillance and focused operations
Cluster 5: Neighborhoods with mixed moderate incidence	• Street robberies (2 to 3) • Motor vehicle theft (2 to 26, average 13.5) • Includes 6 neighborhoods	Intermediate crime level zones. Require continuous monitoring and preventive reinforcement based on peak times and locations
Cluster 6: Specific neighborhood with high vehicle and pedestrian crime	• 5 street robberies • 115 motor vehicle thefts • One neighborhood	Critical area with strong vehicle crime vulnerability. Priority for strategic surveillance
Cluster 7: Critical neighborhood with combined high crime	• 8 street robberies • 52 motor vehicle thefts • One neighborhood	Neighborhood with significant threats in both crime categories. Needs intensive patrol
Cluster 8: Neighborhoods with low street robberies and moderate vehicle crime	• Low street robberies (0–1, average 0.93) • Light to moderate motor vehicle thefts (2–25, average 12.5) • Includes 27 neighborhoods	Areas with low pedestrian risk but moderate vehicle vulnerability. Recommend focused monitoring in parking and transit areas
Cluster 9: High-risk neighborhood for vehicle crime only	• No street robberies • 55 motor vehicle thefts • One neighborhood	Critical zone focused on vehicle crime

Table 18 Cluster profile—Nuevo Laredo municipality (socioeconomic variables)

Cluster	Key characteristics	Interpretation
Cluster 1: Medium neighborhoods with moderate home burglary incidence	• Home burglaries between 6 and 14 cases (average 9) • Medium population (469 to 4290 residents, average 2032) • Includes 14 neighborhoods	Residential areas with intermediate burglary levels. Require regular surveillance and localized preventive strategies
Cluster 2: Small neighborhoods with low home burglary incidence	• Home burglaries between 0 and 5 cases (average 1.28) • Low population (48 to 2581 residents, average 875) • Largest cluster with 79 neighborhoods	Peaceful zones with low risk. Ideal for preventive maintenance and community participation programs
Cluster 3: Densely populated neighborhoods with moderate-high incidence	• Home burglaries between 6 and 14 cases (average 10.3) • Very high population (12,493 to 15,625 residents, average 13,595) • Represented by 3 neighborhoods	Dense areas with notable crime. Require continuous surveillance and strong institutional presence to prevent escalation
Cluster 4: Large neighborhoods with high home burglary incidence	• Home burglaries between 11 and 19 cases (average 16.25) • High population (3964 to 10,922 residents, average 7421) • Comprises 8 residential areas	High-activity areas for household crime
Cluster 5: Medium neighborhoods with low-moderate incidence	• Home burglaries between 0 and 7 cases (average 3.1) • Medium population (2157 to 6849 residents, average 3888) • 27 neighborhoods	Zones with moderate risks requiring active monitoring to prevent crime increases
Cluster 6: Critical neighborhoods with high population and elevated incidence	• Home burglaries between 24 and 28 cases (average 26) • Very high population (18,389 to 23,143 residents, average 20,766) • 2 neighborhoods	Critical areas with significant population and crime concentration. Immediate and sustained intervention is necessary
Cluster 7: Highly populated neighborhoods with critical burglary levels	• Home burglaries between 26 and 39 cases (average 31.7) • High population (2871 to 11,967 residents, average 8775) • Composed of 3 neighborhoods	High-priority zones under strong criminal pressure. Require focused attention

In the municipality of Río Bravo, clustering analysis revealed that most neighborhoods exhibit no recorded crime. Cluster 1, the largest group, includes quiet residential areas ideal for passive surveillance strategies. In contrast, Clusters 2 and 6 identified critical sectors with concentrated business and street robberies, requiring immediate intervention and increased patrolling. The findings highlight a strong spatial

Table 19 Cluster profile—Reynosa municipality (crime variables)

Cluster	Key characteristics	Interpretation
Cluster 1: Large number of neighborhoods with low kidnapping and carjacking	• Kidnappings between 0 and 1 cases (average 0.09) • Carjacking between 0 and 52 cases (average 6.95) • Largest cluster with 287 neighborhoods	Low-risk areas, suitable for preventive maintenance and light community-based surveillance
Cluster 2: Critical zone with extremely high combined incidence	• Very high kidnappings (20 to 26 cases, average 23) • Extremely high carjackings (503 to 710 cases, average 606.5) • Represented by only 2 neighborhoods	Critical zone with extreme violence. Requires urgent police intervention, intensive monitoring, and sustained institutional support
Cluster 3: Medium–high risk neighborhoods in both crimes	• Kidnappings between 0 and 5 cases (average 1.82) • Carjackings between 2 and 108 cases (average 44.7) • Composed of 38 neighborhoods	Areas with considerable risk, requiring dual strategies of preventive patrolling and community action to reduce high-impact crimes
Cluster 4: High carjackings with moderate kidnapping incidence	• Kidnappings between 0 and 2 cases (average 0.83) • Carjackings between 101 and 248 cases (average 156.2) • Includes 6 neighborhoods	Areas heavily affected by violent vehicle theft. Priority should be given to road operations, video surveillance, and rapid response strategies

asymmetry, with the majority of neighborhoods maintaining peaceful conditions while a few specific areas demand focused security efforts.

In the municipality of Río Bravo, clustering analysis based on socioeconomic variables revealed that most neighborhoods are small, low-crime areas. Cluster 1, the largest group, includes low-density neighborhoods with minimal home burglary incidence, ideal for maintaining basic preventive surveillance. In contrast, Clusters 5 and 7 identified critical planned residential areas with very high burglary rates, requiring continuous patrolling, technology-based monitoring, and institutional support. The findings highlight a strong spatial asymmetry, with most neighborhoods remaining peaceful while a few larger sectors face significant residential crime risks.

In the municipality of San Fernando, clustering analysis revealed that most neighborhoods exhibit low or no crime incidence. Cluster 7, comprising peaceful areas with no recorded carjackings or homicides, represents the largest group and is suitable for minimal preventive action. In contrast, Cluster 2 identified a critical zone with extremely high levels of violence, requiring immediate intervention and high-intensity surveillance. The results emphasize a marked spatial asymmetry, where the majority of neighborhoods remain low-risk, while a few sectors face severe security challenges.

In the municipality of San Fernando, clustering analysis based on socioeconomic variables revealed that most neighborhoods are small to medium-sized areas with low

Table 20 Cluster profile—Reynosa municipality (socioeconomic variables)

Cluster	Key characteristics	Interpretation
Cluster 1: Large planned neighborhoods with high home burglary	• Home burglaries between 28 and 43 cases (average 35.7) • Very high population (8727 to 18,637 residents, average 11,985) • Includes 9 planned residential areas	Densely populated zones with considerable home burglary rates. Intensive monitoring and focused crime prevention are recommended
Cluster 2: Small neighborhoods with low home burglary	• Home burglaries between 0 and 10 cases (average 2.78) • Low population (38 to 3785 residents, average 781) • Largest cluster with 217 neighborhoods	Peaceful neighborhoods suitable for basic preventive monitoring
Cluster 3: Medium neighborhoods with moderate home burglary	• Home burglaries between 9 and 29 cases (average 15.8) • Population between 478 and 5268 (average 2474) • Composed of 46 neighborhoods	Moderate-risk areas needing continuous monitoring and strategic preventive patrolling
Cluster 4: Very large, high-risk planned neighborhoods	• Very high home burglaries (65 to 85 cases, average 75) • Very high population (10,359 to 29,832, average 20,095) • Only 2 neighborhoods	Critical areas due to size and crime level. Should receive direct attention, permanent surveillance, and deterrent technology
Cluster 5: Large neighborhoods with intermediate incidence	• Home burglaries between 0 and 25 cases (average 11.9) • High population (5151 to 13,101, average 7126) • Composed of 13 neighborhoods	Populous zones under moderate pressure. Require proportional intervention
Cluster 6: Medium-sized neighborhoods with high burglary rates	• Home burglaries between 35 and 66 cases (average 51.1) • Population between 78 and 9110 (average 5119) • 10 neighborhoods	Neighborhoods with frequent burglaries, some with low population. Focused attention and intensive surveillance are necessary

burglary incidence. Cluster 1, representing quiet mid-sized neighborhoods, is suitable for light surveillance and community prevention efforts. In contrast, Clusters 5 and 2 identified critical areas with high burglary rates, requiring immediate, integrated, and continuous interventions. The findings highlight a spatial disparity, with most neighborhoods maintaining low crime risk while a few concentrated sectors demand urgent attention.

In the municipality of Tampico, clustering analysis revealed that most neighborhoods exhibit low crime incidence. Cluster 3, the largest group, includes peaceful areas with minimal business robberies and motor vehicle thefts, suitable for light

Table 21 Cluster profile—Río Bravo municipality (crime variables)

Cluster	Key characteristics	Interpretation
Cluster 1: Neighborhoods with no recorded crime	• No business or street robberies • Largest cluster with 84 neighborhoods (all categorized as "colonia")	Completely quiet zones. Ideal for passive surveillance strategies
Cluster 2: Critical neighborhood with high combined incidence	• 42 business robberies • 7 street robberies • One single neighborhood	High-priority area with concentrated crime. Requires immediate intervention, increased patrols, and comprehensive security efforts
Cluster 3: Specific neighborhood with moderate-high crime	• 6 business robberies • 1 street robbery • One neighborhood	Moderate-high impact area. Preventive monitoring and assessment of commercial and pedestrian areas are recommended
Cluster 4: Planned residential areas with pedestrian crime	• Very low business robberies (0–1) • 1 street robbery in each neighborhood • Includes 3 neighborhoods	Slight but consistent risk for pedestrians
Cluster 5: Medium-risk neighborhoods for business crime	• Business robberies range from 3 to 6 (average 4.75) • No street robberies • Composed of 4 neighborhoods	Moderate risk for commercial targets. Focus surveillance on shops and economic areas
Cluster 6: High-activity neighborhood with combined incidents	• 14 business robberies • 1 street robbery • One neighborhood	Commercially significant area with crime presence. Needs consistent monitoring
Cluster 7: Neighborhoods with low business crime	• 1 to 2 business robberies (average 1.08) • No street robberies • 13 neighborhoods	Low-risk neighborhoods where preventive presence can deter future incidents
Cluster 8: Isolated neighborhood with low but dual crime	• 2 business robberies • 1 street robbery • One neighborhood	Mild but varied crime presence. Apply deterrence strategies

surveillance and community-based prevention. In contrast, Cluster 2 identified a critical zone with extremely high levels of both crimes, requiring immediate intervention, technological monitoring, and permanent patrol presence. The findings highlight a pronounced spatial asymmetry, with the majority of neighborhoods maintaining low crime rates while a few sectors concentrate significant criminal activity.

In the municipality of Tampico, clustering analysis based on socioeconomic variables revealed that most neighborhoods are small, low-crime areas. Cluster 4, the

Table 22 Cluster profile—Río Bravo municipality (socioeconomic variables)

Cluster	Key characteristics	Interpretation
Cluster 1: Small neighborhoods with low burglary incidence	• Home burglaries between 0 and 4 cases (average 1.17) • Low population (22 to 514 residents, average 207) • Largest cluster with 35 neighborhoods	Quiet, low-density areas with minimal crime
Cluster 2: Large neighborhoods with high burglary incidence	• Home burglaries between 13 and 20 cases (average 15.5) • High population (2389 to 3981 residents, average 3071) • Includes 4 neighborhoods	Residential zones with significant household crime. Require regular patrols and proximity policing efforts
Cluster 3: Medium-sized neighborhoods with intermediate risk	• Home burglaries between 6 and 20 cases (average 10.23) • Population between 573 and 2128 (average 1211) • Includes 13 neighborhoods	Mid-sized neighborhoods with moderate risk. Need focused prevention to avoid escalation
Cluster 4: Large planned neighborhoods with high crime	• Home burglaries between 13 and 25 cases (average 19.7) • High population (3693 to 5219 residents, average 4456) • Includes 3 planned areas	Zones with significant residential crime. Require constant surveillance
Cluster 5: Critical planned neighborhood with extreme incidence	• 40 home burglaries • Population: 3973 • One planned residential area	High-priority area needing continuous patrol, tech-based monitoring, and institutional support
Cluster 6: Small to medium neighborhoods with low-moderate incidence	• Home burglaries between 0 and 6 cases (average 3.54) • Population between 367 and 1936 (average 825) • 28 neighborhoods	Areas with moderate risk. Recommended for periodic patrols
Cluster 7: Very large planned neighborhood with high burglary	• 33 home burglaries • Population: 6403 • One planned residential area	Critical area due to size and crime level. Prioritize intensive intervention and institutional coordination

largest group, includes peaceful neighborhoods with minimal home burglary incidence, suitable for basic preventive strategies. In contrast, Clusters 2 and 3 identified large, high-pressure residential areas with significant burglary rates, requiring proactive policing, structural interventions, and continuous monitoring. The findings highlight a spatial disparity, with the majority of neighborhoods remaining safe while a few large sectors present elevated security risks.

In the municipality of Valle Hermoso, clustering analysis revealed that most neighborhoods are crime-free. Cluster 1, the largest group, includes peaceful areas suitable for basic preventive monitoring. In contrast, Cluster 2 identified a critical zone with extremely high levels of violence, requiring immediate intervention, intensive patrols,

Table 23 Cluster profile—San Fernando municipality (crime variables)

Cluster	Key characteristics	Interpretation
Cluster 1: Low-crime neighborhoods with minimal combined incidents	• 1 Carjacking • 1 homicide • 5 neighborhoods	Very low-risk areas needing only basic monitoring
Cluster 2: Critical neighborhood with extreme crime	• 57 Carjackings • 30 homicides • One neighborhood	Top-priority zone with severe violence. Requires immediate intervention and high-intensity surveillance
Cluster 3: Neighborhoods with high combined incidence	• Carjackings (5 to 9), average 7.3 • 4 homicides • 3 neighborhoods	High-crime areas. Require intensive patrol and crime containment strategies
Cluster 4: Neighborhoods with low combined incidence	• Carjackings (1 to 3), average 1.36 • No homicides • 14 neighborhoods	Low-priority zones with minor crime. Light surveillance is recommended
Cluster 5: Medium-crime neighborhoods	• Carjackings (2 to 4), average 2.8 • Homicides (1 to 2), average 1.1 • 10 neighborhoods	Moderate-risk neighborhoods. Regular monitoring and mixed prevention actions are needed
Cluster 6: Specific critical neighborhood	• 2 Carjackings • 3 homicides • One neighborhood	Localized but critical area. Needs tactical intervention and focused security presence
Cluster 7: Crime-free neighborhoods	• No Carjacking • No homicides • 12 neighborhoods	Peaceful areas. Ideal for minimal preventive action
Cluster 8: Neighborhoods with homicides but no vehicle crime	• No Carjacking • 1 to 2 homicides (average 1.2) • 10 neighborhoods	Lethal violence without vehicle crime. Focus on social programs and community prevention

and institutional presence. The results emphasize a pronounced spatial asymmetry, where most neighborhoods maintain low risk while a few sectors face severe and concentrated criminal threats.

In the municipality of Valle Hermoso, clustering analysis based on socioeconomic variables revealed that most neighborhoods are small, peaceful areas with no or minimal motor vehicle theft. Clusters 4 and 5, the largest groups, consist of crime-free neighborhoods suitable for early prevention and light surveillance. In contrast, Cluster 2 identified a critical, densely populated neighborhood with significant crime levels, requiring ongoing patrols and targeted interventions. The findings highlight a spatial contrast, where most areas maintain security while isolated zones demand prioritized attention.

In the municipality of Victoria, clustering analysis revealed that most neighborhoods exhibit low crime levels. Cluster 6, the largest group, includes predominantly

Table 24 Cluster profile—San Fernando municipality (socioeconomic variables)

Cluster	Key characteristics	Interpretation
Cluster 1: Medium-sized neighborhoods with low burglary	• Home burglaries between 0 and 5 (average 2) • Population between 895 and 1311 (average 1182) • 6 neighborhoods	Quiet mid-sized areas with low crime. Suitable for light surveillance and community prevention programs
Cluster 2: Medium-sized neighborhoods with critical incidence	• Home burglaries between 16 and 17 (average 16.5) • Population between 569 and 1819 (average 1194) • 2 neighborhoods	Small-to-medium areas with high crime. Require priority attention and continuous monitoring
Cluster 3: Small neighborhoods with low burglary	• Home burglaries between 0 and 3 (average 1.25) • Population between 111 and 802 (average 389) • 8 neighborhoods	Very low-population and low-risk areas. Ideal for passive surveillance
Cluster 4: Large neighborhood with high incidence	• 14 home burglaries • Population: 3945 • One neighborhood	High-risk area due to size and crime. Requires specialized patrols and institutional presence
Cluster 5: Critical neighborhood with highest burglary	• 30 home burglaries • Population: 2891 • One neighborhood	Top-priority neighborhood for intervention due to extreme incidence. Needs integrated response and immediate action
Cluster 6: Large neighborhoods with intermediate-high incidence	• Home burglaries between 8 and 13 (average 10) • Population between 2075 and 2679 (average 2435) • 3 neighborhoods	Wide areas with relevant crime levels. Recommend constant monitoring and household-focused preventive strategies

gated communities with very low motor vehicle theft and no street robberies, suitable for light preventive surveillance. In contrast, Clusters 2 and 3 identified critical areas with extremely high crime concentrations, requiring immediate intervention, strategic operations, and continuous monitoring. The findings highlight a spatial asymmetry, where the majority of neighborhoods maintain low risk while a few sectors face severe security challenges.

In the municipality of Victoria, clustering analysis based on socioeconomic variables revealed that most neighborhoods are small, low-risk areas. Cluster 1, the largest group, includes small`gated communities with minimal home burglary incidence, suitable for light preventive surveillance and community-building efforts. In contrast, Clusters 3 and 6 identified critical zones with extremely high burglary rates, requiring immediate institutional intervention, intensive surveillance, and structured crime prevention measures. The findings highlight a pronounced spatial contrast, where most neighborhoods remain safe while a few sectors exhibit severe residential crime concentration.

Table 25 Cluster profile—Tampico municipality (crime variables)

Cluster	Key characteristics	Interpretation
Cluster 1: Neighborhoods with medium-level crime in both categories	• Business robberies: 0 to 35 cases (average 14.4) • Motor vehicle thefts: 1 to 20 cases (average 11.3) • 34 neighborhoods	Moderate-risk areas needing frequent patrols and dual preventive surveillance
Cluster 2: Critical zone with highest crime incidence	• Extremely high business robberies (130 to 186, average 158) • Very high motor vehicle thefts (85 to 105, average 95) • Only 2 neighborhoods, both categorized as "colonia"	Top-priority zone. Requires immediate intervention, monitoring technology, and permanent patrol presence
Cluster 3: Largest group of low-crime neighborhoods	• Business robberies: 0 to 19 (average 2.7) • Motor vehicle thefts: 0 to 9 (average 2.3) • 92 neighborhoods	Low-risk, peaceful neighborhoods. Ideal for light surveillance and community-based prevention
Cluster 4: High-incidence areas for both business and vehicle crime	• Business robberies: 9 to 31 (average 18.3) • Motor vehicle thefts: 27 to 40 (average 33.5) • 4 neighborhoods	Priority areas due to combined commercial and vehicle crime. Require joint operations and ongoing monitoring
Cluster 5: High-incidence neighborhoods in business and motor vehicle theft	• Business robberies: 29 to 61 (average 44.3) • Motor vehicle thefts: 16 to 43 (average 24.2) • 6 neighborhoods	Significant crime hotspots. Prioritize with specialized patrols and focus on commercial zones

This summary consolidates the findings obtained through the application of the K-Means clustering algorithm across eleven municipalities in the state of Tamaulipas, Mexico, considering two dimensions: variables directly related to crime, and socioeconomic variables linked to neighborhood contexts. The aim was to identify spatial and contextual patterns to support the prioritization of areas for intervention and the design of targeted public safety strategies. In summary, the following section outlines the key findings resulting from the cluster characterization process conducted through the analysis.

Crime Variables

- Most Frequent Pattern: Most municipalities show a large number of neighborhoods with low or no crime incidence, typically grouped into a single low-risk cluster.
- Critical Zones: Every municipality has at least one cluster with extreme incidence in one or more crimes, concentrating a high number of criminal events. These areas are ideal candidates for urgent intervention.

Table 26 Cluster profile—Tampico municipality (socioeconomic variables)

Cluster	Key characteristics	Interpretation
Cluster 1: Medium neighborhoods with low-moderate burglary	• Home burglaries between 3 and 8 (average 4.3) • Population from 85 to 2149 (average 821) • 32 neighborhoods (all categorized as "colonia")	Moderate-risk areas requiring regular monitoring
Cluster 2: Large neighborhoods with high burglary rates	• Home burglaries between 10 and 16 (average 13) • Very high population (13,063 to 15,253, average 14,158) • 2 neighborhoods	High-priority zones due to their size and residential crime level. Require proactive policing and control strategies
Cluster 3: Very large neighborhoods with critical incidence	• Home burglaries between 20 and 37 (average 25.3) • Population from 4553 to 11,785 (average 7692) • 8 neighborhoods	High-pressure neighborhoods requiring structural intervention and increased patrols
Cluster 4: Small neighborhoods with very low incidence	• Home burglaries between 0 and 2 (average 0.94) • Population from 74 to 2707 (average 605) • Largest cluster with 53 neighborhoods	Peaceful areas with minimal crime. Suitable for basic preventive strategies
Cluster 5: Medium neighborhoods with moderate-high incidence	• Home burglaries between 9 and 18 (average 12.3) • Population from 466 to 5028 (average 2285) • 16 neighborhoods	Intermediate-high risk areas. Need consistent monitoring
Cluster 6: Large neighborhoods with intermediate burglary	• Home burglaries between 2 and 10 (average 6.8) • Population from 2330 to 5247 (average 3662) • 21 neighborhoods	Densely populated zones with significant incidence. Recommend ongoing surveillance

- Crime-specific trends by Municipality:
    (a) Reynosa and Victoria stand out for having clusters with high levels of violent crimes or carjackings.
    (b) Matamoros and Tampico exhibited clusters with high levels of business and motor vehicle theft, suggesting pressure in commercial and transit areas.
    (c) San Fernando and Valle Hermoso reflect the presence of homicides even in low-population neighborhoods.

Socioeconomic Variables

- Common Pattern: The population variable clearly correlates with the level of crime incidence, especially in the most critical clusters.

Table 27 Cluster profile—Valle Hermoso municipality (crime variables)

Cluster	Key characteristics	Interpretation
Cluster 1: Crime-free neighborhoods	• 0 Carjacking • 0 homicides • Largest cluster with 36 neighborhoods	Peaceful areas. Suitable for basic preventive monitoring
Cluster 2: Critical neighborhood with extreme violence	• 24 Carjackings • 20 homicides • One neighborhood	Top-priority area with severe violence. Requires immediate intervention, intensive patrols, and institutional presence
Cluster 3: High-crime neighborhoods	• 4 Carjackings • 2 to 3 homicides (average 2.5) • 2 neighborhoods	High-risk zones. Need ongoing surveillance and multi-level crime containment strategies
Cluster 4: Neighborhood with lethal violence and low vehicle crime	• 1 Carjacking • 3 homicides • One neighborhood	Area with lethal crime. Suggest deploying targeted surveillance and dissuasive operations
Cluster 5: Neighborhoods with isolated homicide and no vehicle crime	• 0 Carjackings • 1 homicide • 4 neighborhoods	Low-crime zones with presence of lethal violence. Require community-based prevention focus
Cluster 6: Neighborhoods with vehicle crime but no homicides	• 1 Carjacking • 0 homicides • 4 neighborhoods	Presence of property crime. Recommend focused patrols and transit zone monitoring
Cluster 7: Neighborhood with elevated combined crime	• 6 Carjackings • 2 homicides • One neighborhood	High-impact area across multiple crime categories. Prioritize for continuous surveillance and rapid response
Cluster 8: Neighborhood with moderate crime combination	• 3 Carjackings • 1 homicide • One neighborhood	Significant crime presence. Requires systematic monitoring and preventive action

- Critical Neighborhoods: Clusters with high population density tend to concentrate more home burglaries, as observed in Victoria, Reynosa, and Matamoros.
- Small and Peaceful Neighborhoods: These are grouped into numerous clusters with minimal incidence and are ideal for early prevention strategies and community cohesion programs.

Compared to previous studies, such as those by [9] and [10], which applied K-Means clustering to segment crime data in Indonesia, the present study similarly leverages clustering to identify spatial crime patterns. However, this research introduces a distinctive approach by integrating both crime incidence and socioeconomic factors at the neighborhood level, providing a more comprehensive profile of risk areas. Additionally, while prior works often focus on broader geographic units such

Table 28 Cluster profile—Valle Hermoso municipality (socioeconomic variables)

Cluster	Key characteristics	Interpretation
Cluster 1: Small neighborhoods with low vehicle crime	• 1 motor vehicle theft per neighborhood • Population between 225 and 911 (average 419) • 11 neighborhoods	Low-density, low-crime zones. Ideal for light surveillance
Cluster 2: Critical high-population, high-incidence neighborhood	• 16 motor vehicle thefts • Population: 6399 • One neighborhood	High-priority zone due to density and crime volume. Needs ongoing patrols
Cluster 3: Large neighborhood with significant crime	• 5 motor vehicle thefts • Population: 2803 • One neighborhood	Pressure zone with considerable crime. Requires frequent monitoring and institutional support
Cluster 4: Very small, crime-free neighborhoods	• 0 motor vehicle thefts • Population between 70 and 393 (average 231.5) • 16 neighborhoods	Peaceful zones with no crime records. Suitable for early prevention and community cohesion strategies
Cluster 5: Medium-sized neighborhoods with no crime	• 0 motor vehicle thefts • Population between 421 and 1092 (average 609) • 12 neighborhoods	Crime-free but more populated areas. Require preventive surveillance to avoid future offenses
Cluster 6: Single moderate-risk neighborhood	• 2 motor vehicle thefts • Population: 2648 • One neighborhood	Moderate crime presence. Needs targeted patrols and specific deterrence strategies

as provinces or cities, this study operates at a finer granularity by analyzing neighborhoods within municipalities. This granularity, combined with internal validation through metrics such as the Silhouette Coefficient and the Davies–Bouldin Index, strengthens the robustness and practical applicability of the findings for public safety strategy design in Mexico.

5 Conclusions and Future Work

The results of this research reflect a comprehensive characterization of the cluster profiles generated for the municipalities of Tamaulipas, Mexico, considering both criminal and socioeconomic variables. Based on the characterization of these profiles, recurrent patterns were identified that provide a better understanding of the spatial distribution of crime and the associated contextual conditions.

Regarding criminal variables, it was observed that most municipalities exhibit a large number of neighborhoods with low or no crime incidence, generally grouped into a single low-risk cluster. Nevertheless, critical zones were identified in all municipalities, represented by clusters with extreme incidence in one or more crimes,

Table 29 Cluster profile—Victoria municipality (crime variables)

Cluster	Key characteristics	Interpretation
Cluster 1: Medium-risk neighborhoods with moderate vehicle crime	• No street robberies • Motor vehicle thefts from 6 to 16 (average 8.5) • 39 neighborhoods	Moderate vehicle crime risk. Require deterrent patrols in parking areas
Cluster 2: Critical neighborhood with extreme crime levels	• 39 street robberies • 285 motor vehicle thefts • One neighborhood	Top-priority area. Requires immediate intervention, deterrent tech, and continuous patrolling
Cluster 3: High-risk neighborhood with combined incidence	• 29 street robberies • 197 motor vehicle thefts • One neighborhood	High-priority zone. Needs strategic operations and strong security presence
Cluster 4: Moderate-risk neighborhoods	• 0 to 2 street robberies (average 0.58) • 18 to 44 motor vehicle thefts (average 24.1) • 12 neighborhoods	Moderate vehicle risk, light pedestrian impact. Focused monitoring recommended
Cluster 5: Low combined crime neighborhoods	• 1 to 2 street robberies (average 1.15) • 0 to 15 motor vehicle thefts (average 6.1) • 47 neighborhoods	Generally low-risk areas. Moderate monitoring encouraged
Cluster 6: Low-crime gated communities	• No street robberies • 0 to 1 motor vehicle thefts (average 0.35) • 137 neighborhoods (mainly gated)	These zones present minimal risk and are appropriate for light preventive surveillance
Cluster 7: Small neighborhoods with high crime levels	• 3 to 6 street robberies (average 3.8) • 15 to 38 motor vehicle thefts (average 24.6) • 5 neighborhoods	Critical small zones. Require prioritized attention and heavy patrols
Cluster 8: Low vehicle crime, no pedestrian crime	• No street robberies • 2 to 5 motor vehicle thefts (average 3.1) • 99 neighborhoods	Low-risk zones. Recommend strategic deterrence in specific time frames and areas

concentrating a significant proportion of criminal events. These areas constitute priority candidates for more intensive public safety interventions.

The specific analysis by municipality revealed that Reynosa and Victoria stand out due to the presence of clusters with high levels of violent crimes or vehicle robberies with violence. In Matamoros and Tampico, clusters with a high incidence of business and vehicle theft were identified, suggesting significant pressure in commercial and transit areas. In contrast, municipalities such as San Fernando and Valle Hermoso reflect the occurrence of homicides, even in neighborhoods with low population density.

Table 30 Cluster profile—Victoria municipality (socioeconomic variables)

Cluster	Key characteristics	Interpretation
Cluster 1: Small gated neighborhoods with low burglary	• Home burglaries between 0 and 20 (average 3.2) • Population from 24 to 783 (average 297) • Largest cluster with 165 neighborhoods	Very low-risk zones with small population. Best suited for light preventive surveillance and community-building
Cluster 2: Large neighborhoods with high burglary	• Home burglaries between 17 and 66 (average 31.5) • Population from 3557 to 7625 (average 4853) • 11 typical neighborhoods	Areas with significant household crime. Require high-priority response and consistent patrolling
Cluster 3: Extreme-risk neighborhood	• 182 home burglaries • Population: 9740 • One single neighborhood	Severe crime concentration. Requires immediate institutional intervention and structured crime prevention measures
Cluster 4: Medium-sized neighborhoods with moderate-high burglary	• Home burglaries between 1 and 46 (average 16.1) • Population from 1467 to 3307 (average 2349) • 36 neighborhoods	Moderate-sized areas with significant crime. Need regular monitoring
Cluster 5: Small neighborhoods with intermediate burglary	• Home burglaries between 0 and 35 (average 8.8) • Population from 661 to 1865 (average 1129) • 76 neighborhoods	Intermediate-risk areas. Recommended for frequent follow-up and directed prevention programs
Cluster 6: Medium-sized gated communities with critical burglary	• Home burglaries between 60 and 85 (average 72.5) • Population from 127 to 1914 (average 1020) •2 gated communities	High-priority zones due to elevated burglary levels. Need intensive surveillance and focused crime control strategies

Regarding socioeconomic variables, a consistent pattern was found wherein the population variable directly correlates with the level of crime incidence, particularly in the most critical clusters. High-density population clusters tend to concentrate a greater number of home burglaries, as observed in Victoria, Reynosa, and Matamoros. Similarly, numerous clusters composed of small and low-crime neighborhoods were identified, representing ideal areas for the implementation of early prevention strategies and the strengthening of community cohesion.

One of the main contributions of this research is the development of an information system that, through an interactive dynamic map, enables the geospatial visualization of criminal clusters at the neighborhood level. This tool aims to support public security authorities in the more efficient allocation of resources and the planning of specific crime prevention strategies, whether through patrol distribution, the implementation of preventive measures, or the execution of targeted operations. It is recommended to continuously update the database to provide real-time and up-to-date visualizations of criminal information, allowing for better adaptation to changes in crime patterns and strengthening the capacity to respond to crime in Tamaulipas.

Limitations

Despite the robust segmentation achieved, this study presents certain limitations. The analysis is based on historical crime data aggregated over a multi-year period (2017–2023), which may not fully capture recent dynamic changes in criminal behavior. Additionally, the current approach focuses exclusively on spatial patterns, without incorporating temporal evolution or predictive modeling components.

Future Work

As future work, it is proposed to develop a predictive model based on supervised classification algorithms. A binary classification approach will be explored to estimate the probability of crime occurrence within specific time windows of 1, 7, and 15 days. The selection of the most appropriate algorithm will be determined through a comparative evaluation of different techniques, considering the nature of the available data and the operational needs of public security authorities. Incorporating temporal forecasting capabilities would enhance the practical utility of the system, allowing for proactive resource allocation and strategic crime prevention planning.

AI Use Statement
Some portions of this manuscript were enhanced with the assistance of generative AI tools. The authors carefully reviewed, edited and approved all AI-generated content. The authors take full responsibility for the accuracy and integrity of the entire chapter.

References

1. Vilalta, C.: Violence in Latin America: an overview of research and issues. Annu. Rev. Sociol. **46**, 693–706 (2020). https://doi.org/10.1146/ANNUREV-SOC-073018-022657
2. Jaitman, L.: Frontiers in the economics of crime: lessons for Latin America and the Caribbean. Lat. Am. Econ. Rev. **28**, 19 (2019). https://doi.org/10.1186/s40503-019-0081-5
3. Rodríguez, J.C., Torres, C.E.T.: Los orígenes socioeconómicos del delito en México: análisis 2015 con regresión múltiple. Revista Logos Ciencia & Tecnología. **14**, 8–27 (2022). https://doi.org/10.22335/RLCT.V14I2.1589
4. Instituto Nacional de Estadística y Geografía.: Encuesta Nacional de Victimización y Percepción sobre Seguridad Pública (ENVIPE) 2024 (2024)
5. Orona, C.H.F., Leyva, R.E.S., Sidón, G.M.: Análisis de la contribución del gasto público de seguridad en la mitigación de la delincuencia de las entidades federativas de México (2022). https://www.gestionypoliticapublica.cide.edu/ojscide/index.php/gypp/article/view/1012
6. Montenegro Bagatella, J.C.: Inteligencia para la seguridad pública en las entidades federativas de México (2020). https://dialnet.unirioja.es/servlet/articulo?codigo=7716120&info=resumen&idioma=ENG
7. Hernández Orallo, J., Ramírez Quintana, M.J., Ferri Ramírez, C.: Introducción a la minería de datos (2004)

8. Azis, N., Ali, Y., Subekti, R., Junianto, P., Diner, L., Suhendra, S., Aisyah, S., Windarto, A.P.: Mapping study using the unsupervised learning clustering approach. IOP Conf. Ser. Mater. Sci. Eng. **1088**, 12005 (2021). https://doi.org/10.1088/1757-899X/1088/1/012005
9. Vulandari, R., Laksito, W., Aditama, D.: Application of k-means clustering in mapping of central java crime area. Indonesian J. Appl. Statis. **3**, 38 (2020). https://doi.org/10.13057/ijas.v3i1.40984
10. Wahidah, Z., Utari, D.T.: Implementation k-means algorithm to group provinces by factors influenced criminal act in Indonesia in 2019. Enthusiastic: Int. J. Appl. Stat. Data Sci. **2**, 37–46 (2022). https://doi.org/10.20885/enthusiastic.vol2.iss1.art5
11. Jha, G., Ahuja, L., Rana, A.: Criminal behaviour analysis and segmentation using k-means clustering. In: 2020 8th International Conference on Reliability, Infocom Technologies and Optimization (Trends and Future Directions) (ICRITO), pp. 1356–1360 (2020)
12. Krishnendu, S.G., Lakshmi, P.P., Nitha, L.: Crime analysis and prediction using optimized k-means algorithm. In: 2020 Fourth International Conference on Computing Methodologies and Communication (ICCMC). pp. 915–918 (2020)
13. Singh, R., Reddy, R., Kapoor, V., Churi, P.: K-means clustering analysis of crimes on indian women. J. Cybersecur. Inf. Manage., 5–25 (2020). https://doi.org/10.54216/JCIM.040101
14. Chanchí Golondrino, G.E., Ospina Alarcón, M.A., Muñoz Sanabria, L.F.: Cybercrime characterization in the department of Cundinamarca during 2021 through exploratory analysis and machine learning. Ingeniería y Competitividad. **25** (2022). https://doi.org/10.25100/iyc.v25i1.11760
15. Ali, W., Alalloush, H., Manasa, K.N.: Crime analysis and prediction using k-means clustering technique. EPRA Int. J. Econ. Bus. Rev. **5**, 277–280 (2020). https://doi.org/10.36713/epra2016
16. Jadhav Payal, U.: Crime detection system using data mining. Int. J. Comp. Sci. Mobile Comput. **10**, 82–85 (2021). https://doi.org/10.47760/ijcsmc.2021.v10i02.013
17. Umasare, S., Phirke, S., Thakur, S., Kulkarni, V.: Crime rate analysis and prediction using k-means (2022)
18. Ritisha, D., Pooja, R., Jasmine, K.S.: Crime data analysis and prediction. Grenze Int. J. Eng. Technol. (2022)
19. Mandalapu, V., Elluri, L., Vyas, P., Roy, N.: Crime prediction using machine learning and deep learning: a systematic review and future directions. IEEE Access. **11**, 60153–60170 (2023). https://doi.org/10.1109/ACCESS.2023.3286344
20. Davies, D.L., Bouldin, D.W.: A cluster separation measure. IEEE Trans Pattern Anal Mach Intell. PAMI-1, 224–227 (1979). https://doi.org/10.1109/TPAMI.1979.4766909
21. Rousseeuw, P.J.: Silhouettes: a graphical aid to the interpretation and validation of cluster analysis. J. Comput. Appl. Math. **20**, 53–65 (1987). https://doi.org/10.1016/0377-0427(87)90125-7
22. García, J., Molina, J.M., Berlanga, A., Patri-Cio, M.A., Bustamante, Á.L., Padilla, W.R.: Ciencia de datos. Técnicas analíticas y aprendizaje estadístico (2018)
23. Han, J., Kamber, M., Pei, J.: Data mining. In: Concepts and Techniques, 3rd ed. (The Morgan Kaufmann Series in Data Management Systems). Morgan Kaufmann Publishers (2011)

Technological Surveillance Applied to Decision-Making in a Municipal Clinical Engineering Service

Mariela Azul Gonzalez, Adriana Pilar Antonelli, and Antonio Morcela

Abstract Technological Surveillance is a strategic information search and analysis methodology aimed at transforming dispersed data into actionable knowledge, thereby supporting low-risk and evidence-based decision-making processes. In a rapidly evolving and uncertain technological landscape, staying ahead of emerging trends and identifying innovation opportunities enables organizations to maintain competitive advantages and foster adaptive transformation. Innovation-oriented change enhances system responsiveness to environmental challenges, promoting inclusive benefits for all stakeholders. In healthcare systems, decision-making platforms increasingly rely on intelligent mechanisms that integrate technological surveillance tools to promote innovation. One such tool is a custom-built search engine, specifically implemented for the systematic updating of medical imaging equipment and the identification of emerging resources for Clinical Engineering services. This search engine facilitates the detection of both tangible (e.g., diagnostic devices) and intangible (e.g., software tools, training programs) innovations by analyzing technological trends relevant to biomedical services. The system serves as a catalyst for continuous technological renewal, informing procurement strategies, guiding educational curricula for engineering staff, and aligning Clinical Engineering priorities with global innovation trajectories. Alert tuning to provide optimal results remains a major challenge, as user needs and requirements are expert-dependent

M. A. Gonzalez (✉)
Laboratorio de Bioingeniería, Facultad de Ingeniería, Instituto de Investigaciones Científicas y Tecnológicas en Electrónica (ICyTE), Universidad Nacional de Mar del Plata, Mar del Plata, Argentina
e-mail: mazulgonzalez@fi.mdp.edu.ar

A. P. Antonelli
División de Ingeniería Clínica (U.E. 11-00-0-0-6-01), Centro de Estudios Médicos Ambulatorios, CEMA, Mar del Plata, Argentina
e-mail: aantonelli@mardelplata.gob.ar

A. Morcela
Departamento de Ingeniería Industrial, Universidad Nacional de Mar del Plata, Mar del Plata, Argentina
e-mail: omorcela2000@gmail.com

W. Pedrycz et al. (eds.), *Decision-Making in Computational Intelligence-Based Systems*, Studies in Systems, Decision and Control 628,
https://doi.org/10.1007/978-3-032-04056-5_20

and context-specific. As an innovative aspect, the engine incorporates Fuzzy Logic predicates to manage uncertainty in the classification and prioritization of technological alerts. These fuzzy predicates enable nuanced evaluations of alert relevance, urgency, and contextual impact, particularly useful during periods of reduced operational load or budget constraints. The alerts are dynamically adapted based on expert-defined requirements and integrated into public procurement processes, enhancing transparency and strategic alignment. By embedding fuzzy inference mechanisms into the surveillance architecture, the system transforms raw data into semantically enriched knowledge, strengthening the Clinical Engineering service's capacity for proactive and intelligent decision-making. Additionally, the system features an alert rescoring technique that further reduces the number of alerts. The quantification and classification of alerts are performed using Fuzzy Logic predicates. The system integrates fuzzy logic to classify and prioritize technological alerts, supporting decision-making in public procurement processes. Each expert's requirements are used to adapt alerts via Fuzzy Logic. This search engine helps the service stay up-to-date with tangible and intangible new resources and supports the decision-making process during periods of reduced workload.

Keywords Technological surveillance · Fuzzy logic · Medical image management · Decision-making · Clinical engineering service

1 Introduction

Healthcare systems worldwide are undergoing a significant transformation as they attempt to respond to complex challenges involving constrained financial resources, a shortage of qualified personnel, and the rapid pace of technological advancement. In this context, Technological Surveillance (TS) has emerged as a strategic tool for converting raw data into actionable knowledge, enhancing decision-making processes, and guiding investment in innovative technologies [1]. This practice plays a critical role in supporting Clinical Engineering (CE) services, which are responsible for ensuring the effective procurement, maintenance, and integration of health technologies within public healthcare institutions. The long-term benefits of developing and adopting innovative products and services can enhance both the patient and professional experience, adding value to healthcare delivery and improving the quality of care provided [2]. In this context, Technology Surveillance (TS) emerges as a strategic tool to convert data into actionable knowledge, improving decision-making processes and guiding investment in innovative technologies [3]. A TS search engine develops and implements a search tool that allows the Clinical Engineering Service to efficiently monitor new technologies and medical equipment, facilitating informed decisions in acquisition and maintenance processes.

Digital Transformation (DT) requires substantial investments in R&D&I and close collaboration between public administrations, industry, and academia [4–6]. Institutional transformation begins with an assessment of the maturity level of the Clinical

Engineering service's DT, aimed at identifying improvement opportunities. DT is not only about moving from analog to digital, but also about the full, integrated, intelligent, and optimal connection of information. DT efficiently integrates information, facilitating smart, connected, and optimal systems were discussed in the World Economic Forum [7, 8]. The hard and soft capacities of the actors within the organization, as well as the implications of the actions and inactions of the stakeholders involved with the institution, are analyzed for tool planification.

Technological Surveillance (TS) can be defined according to the UNE 166006:2011 standard as a key tool for R&D systems. It is defined as "the organized, selective, and systematic process of gathering external and internal information related to science and technology, selecting, analyzing, disseminating, and communicating it." TS is a key tool in R&D&I systems, allowing the identification of relevant innovations inside and outside the organization [1].

The incorporation of technological innovation in hospital processes has become imperative to improve the quality of care, optimize decision-making, and reduce operating costs. However, these benefits can only be achieved through an efficient technological monitoring (TM) strategy that allows anticipating changes in the environment and detecting opportunities for improvement in real time. Monitoring and updating technological resources have a positive impact both on the preventive and corrective maintenance of equipment and on the acquisition of new technologies that respond to the needs of patients and health professionals.

Strategic intelligence, in combination with TS and competitive intelligence (CI), enhances the capacity of institutions to generate innovation, detect emerging threats, and respond to trends with speed and precision [3].

In a context of accelerated digital transformation, clinical engineering services have the responsibility not only to maintain the optimal functioning of biomedical devices but also to anticipate innovations that improve operational efficiency and reduce risks during public bidding processes. This rationale is aligned with the global objectives of digital transformation promoted by institutions such as the World Economic Forum [7, 8], which highlights the importance of interoperability, the adoption of disruptive technologies, and the continuous improvement of public services through data-based solutions.

There are several high-quality academic tools and systems that utilize fuzzy logic predicates to manage alerts in technological surveillance contexts. Fuzzy logic stands out for its tolerance to uncertainty, high explainability, and suitability for contexts with unstructured or scarce data, such as municipal services. These features make it a particularly useful option compared to methods like machine learning, which require historical data, or traditional expert systems, which lack dynamic adaptability. Mohaidat et al. [9] reinforce Fuzzy logic tolerance for uncertainty, high explainability, suitability, and ability to handle ambiguity for contexts with unstructured or scarce data, such as municipal services. These features make it a particularly useful option compared to methods like machine learning, providing more nuanced and adaptable alert mechanisms. Fuzzy systems are designed to handle uncertainty and imprecision inherent in complex monitoring environments. In some cases, Fuzzy predicates become complex with the addition of more semantic behaviors. Expert tuning and hardware requirements are a significant barrier to overcome. In other cases, it is difficult to maintain performance when adapting to new environments. In the following lines, we will mention the strategies most similar to the present system, as presented by [10, 11]. Table 1 summarizes DS and TS current approaches.

In particular, there are other similar DS and TS systems at work. References [16–18] proposed FuzMet in [18], a novel IDS alert management system which employs several metrics and a fuzzy-logic-based approach for scoring and prioritizing alerts. It utilizes Mamdani-style fuzzy logic to evaluate and rank these alerts based on multiple fuzzy input variables such as alert frequency, time proximity, and source credibility. FuzMet allows for an alert management engine that employs fuzzy logic to prioritize alerts generated by intrusion detection systems (IDS). By evaluating the severity and relevance of alerts, FuzMet aims to reduce false positives and assist security administrators in focusing on critical threats. IDS provide results that are inaccurate and difficult to manage. Performance may degrade in highly dynamic

Table 1 Comparative evaluation of artificial intelligence approaches for decision support in clinical engineering

Criterion	Fuzzy logic (FL)	Supervised machine learning (ML)	Expert systems (ES)
Requires labeled data	No	Yes	No
Tolerance to uncertainty	High—handles vague or incomplete data [12, 13]	Moderate—depends on model type	Moderate—limited flexibility [12]
Explainability	High—linguistic rules are interpretable [12]	Low—black-box models [13]	High—rule-based and traceable
Requires prior training	No	Yes—requires large datasets	No—but requires manual rule design
Scalability and automation	Medium—requires manual tuning of rules	High—suitable for large datasets [13]	Low—manual maintenance limits scalability
Ease of implementation	High—suitable for low-resource settings [12, 13]	Medium—requires technical expertise [13]	Medium—needs expert collaboration
Real-time adaptability	High—rules and weights can be adjusted on the fly [12]	Moderate—retraining required	Low—static rule base
Useful without historical data	Yes—works in data-scarce environments [12]	No—needs historical data	Yes—based on expert knowledge
Healthcare applications	Alert prioritization, maintenance classification [14, 15]	Disease prediction, image analysis [13, 14]	Rule-based diagnostics and alert systems [15]

environments where attack patterns change rapidly. The most significant weakness is the requirement for continuous tuning of membership functions.

I-SAFE is a real-time surveillance system that utilizes fuzzy decision-making at the edge to identify suspicious activities [16, 17]. It integrates lightweight deep learning models with fuzzy logic to mimic human decision processes, enabling instant alerting with minimal communication delay. It combines lightweight CNNs (Convolutional Neural Networks) with a fuzzy inference system. The application of fuzzy logic in analyzing large-scale surveillance video data is used to handle the ambiguity and uncertainty in video analytics, improving the accuracy of surveillance systems in detecting and interpreting events. These tools demonstrate the effective integration of fuzzy logic in surveillance technologies, enhancing the management and prioritization of alerts in various contexts. I-SAFE is limited by edge device hardware capabilities and requires careful modeling of fuzzy rules to avoid false negatives. Table 2 presents comparison results.

A SWOT (Strengths, Weaknesses, Opportunities, and Threats) analysis was conducted to evaluate the implementation of the Technological Surveillance (TS) system in a Clinical Engineering Service (CI). This rationale is aligned with the global objectives of digital transformation promoted by institutions such as the World Economic Forum [6, 7], which highlights the importance of interoperability, the adoption of disruptive technologies, and the continuous improvement of public services through data-based solutions [19, 20]. System strengths are: integration with machine learning for adaptive alert prioritization; real-time monitoring of trends, fostering innovation adoption; enhancement with natural language processing for unstructured data; expansion to other biomedical domains beyond imaging equipment; and reducing the cognitive burden on experts through automated alert classification [21]. Expert system tuning is focused on avoiding the risk of data overload and irrelevant alert generation, as well as resistance to adoption from clinical staff unfamiliar with AI systems. The flowchart shown in Fig. 1 summarizes the nine sequential phases involved in the creation and implementation of the TS system, from requirement definition to testing and validation. This represents the development workflow of the technological surveillance (TS) search engine. The visual model enhances understanding of the methodological structure outlined in Sect. 2.

Table 2 Comparative evaluation of similar decision system implementations

Comparative summary				
System	Input type	Fuzzy logic role	Unique feature	Domain
Fuzz Met	Alert logs	Prioritization	Alert overload mitigation	IDS
I-SAFE	Video frames	Suspicion detection	Edge deployment	Edge surveillance

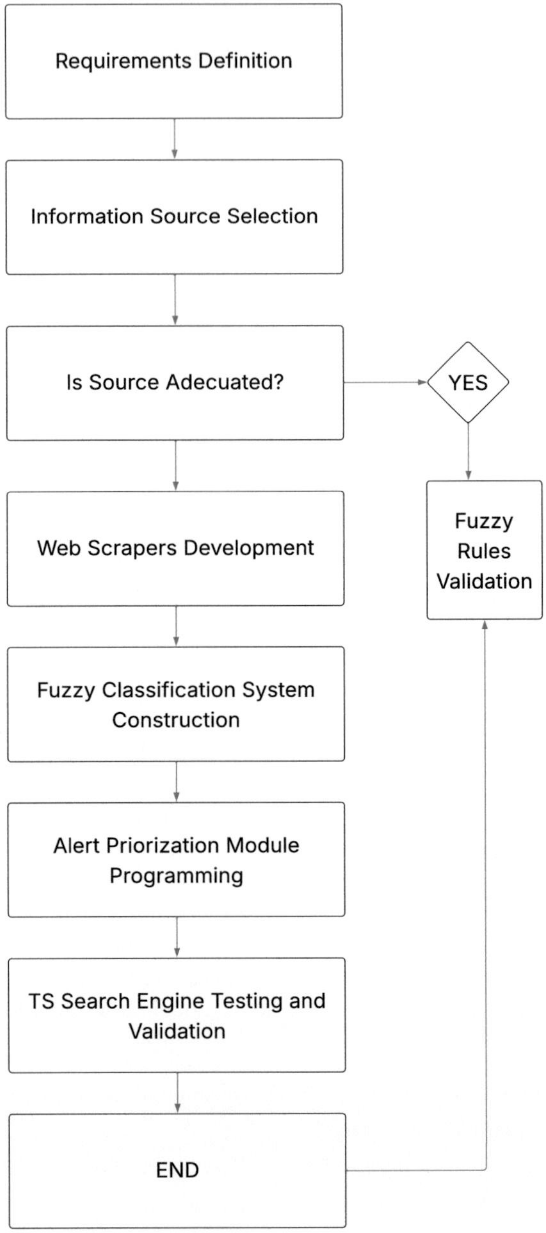

Fig. 1 Development workflow of the technological surveillance (TS) search engine

2 Materials and Methods

The process of transforming data into information and information into knowledge involves integrating digital tools, software platforms, and continuous training, enabling smart decision-making to optimize the organization's critical tangible and intangible resources. A decision system for the development of the TS search engine for the Clinical Engineering service enables efficient monitoring of new equipment and technologies, facilitating well-informed procurement decisions during public tenders. The system ensures that the service can continuously improve and integrate innovative equipment, ultimately benefiting public healthcare in the municipality.

- **Digital Transformation Assessment**

An institutional diagnosis was performed to evaluate the CI service's digital maturity, involving an analysis of current capabilities and identification of critical areas for improvement. A series of meetings with the Clinical Engineering team were conducted to elicit key questions needed for the search engine's development [22]. A diagnosis was conducted, considering the latest recommendations in the area of innovation [7, 8, 19, 20]. These questions focused on training, decision-making processes, service interconnection, and equipment maintenance. Based on these discussions, we identified relevant keywords and built search queries for specialized search engines and databases.

- **Strategic Intelligence Framework**

The model uses Technological Surveillance (TS), Competitive Intelligence (CI), and Strategic Planning (SP) as essential pillars for innovation processes [3]. These components form the foundation of Strategic Intelligence, allowing organizations to:

- Enhance their ability to innovate.
- Identify and analyze trends, threats, and opportunities.
- Continuously update strategic plans based on environmental changes.

The TS engine is embedded within a broader Strategic Intelligence (SI) model composed of three core components:

- **Technological Surveillance (TS)**: Defined by [1] as the systematic and selective process of collecting, analyzing, and disseminating relevant external and internal information to support innovation. TS is "the organized, selective, and systematic process of collecting information from external sources and the organization itself to analyze, disseminate, and communicate it."
- **Competitive Intelligence (CI)**: Refers to the ethical analysis of information, systematic collection of information about competitors, business environment stakeholders, to support decision-making and inform in strategic decisions [1].
- **Strategic Planning (SP)**: The process of goal-setting and action planning across short-, medium-, and long-term horizons based on internal and external assessments. Focuses on setting goals and determining actions to achieve them within a

specific time frame. It considers short-, medium-, and long-term objectives based on institutional analysis.

These elements collectively enhance an organization's ability to innovate, respond to environmental shifts, and align its strategy with dynamic technological trends.

- **Computational intelligence**

Computational intelligence is a set of computational methodologies and approaches inspired by nature for addressing complex real-world problems for which traditional approaches, i.e., first-principles modeling or explicit statistical modeling, are ineffective or unfeasible. Traditional models often fail to handle uncertainty, noise, and the presence of a constantly changing context. Computational intelligence provides solutions to these and other wicked problems and inverse problems [21].

The Computational Intelligence (CI) area of knowledge encompasses the theory, design, application, and development of algorithms inspired by biology and linguistics, with an emphasis on Artificial Neural Networks, Fuzzy Logic, Evolutionary Algorithms, Collective Intelligence Algorithms, and Hybrid Systems [23, 24].

- **Fuzzy Logic**

Fuzzy Logic (FL) emerged as an extension of classical or Boolean logic. In the latter, set membership is defined as binary, i.e., either 0 or 1. In FL, instead of using binary set membership, it is possible to assign a degree of membership between 0 and 1, enabling a more gradual assessment of the degree of set membership, closer to natural language. For example, a person who is two meters tall is clearly tall if the value for a short person has previously been taken and set at one meter. Both values are contextualized to individuals and referred to a standard linear metric measure, and can be smoothed and graded using fuzzy logic.

This theory was formulated in 1965 by the engineer and mathematician Lotfi A. Zadeh [23, 24]. Fuzzy logic is better suited to the real world we live in, and can even understand and work with our natural language expressions, such as "it's very hot," "it's not very high," "my heart rate is a little fast," etc.

Fuzzy set theory also defines the operations of union, intersection, difference, negation, or complement, and other operations on sets on which this logic is based. For each fuzzy set, there is an associated membership function for its elements, which indicates to what extent the element is part of that fuzzy set.

The use of FL allows us to reach a specific conclusion based on input information, which may be indefinite, vague, or inaccurate. It is an appropriate tool for formalizing linguistic expressions, such as those used in the design process of assessment instruments.

- **Expert System with Fuzzy Logic Integration**

To support alert classification and prioritization, an expert system based on FL was integrated into the TS engine. FL provides a means to handle ambiguity and incomplete information—a frequent challenge in healthcare innovation contexts. FL has proven to be particularly useful in situations where decisions are based on imprecise or incomplete information, such as in the case of health alerts. By using fuzzy predicates, each alert can be classified according to levels of relevance or urgency, allowing experts to prioritize interventions and recommendations more accurately. This is especially useful in medical contexts where multiple alerts need to be managed simultaneously [4–6].

FL systems implementation has also proven to be effective by combining fuzzy logic and neural networks to optimize decision-making in complex scenarios. A relevant example is their application in the prediction of ICU admissions during the COVID-19 pandemic, where these systems achieved competitive results compared to other classical classification techniques [15]. Therefore, by integrating this technique, we are able to assign priority levels (e.g., "high," "medium," or "low") based on fuzzy criteria such as relevance, frequency of occurrence, and content novelty. Furthermore, it reduces the cognitive load on experts, making it easier to identify alerts that deserve immediate attention.

- **Compensatory Logic Concept and Foundations**

Compensatory logic refers to a decision-making framework in which a disadvantage in one criterion can be offset by advantages in others. It is commonly used in multicriteria decision-making models that require simultaneous evaluation of various quantifiable and interdependent factors. The key premise is that no single attribute is a definitive exclusion criterion; instead, the decision is based on an aggregate evaluation of all attributes, typically through weighted sums or similar aggregation functions.

This model contrasts with non-compensatory logic, where failing to meet a single criterion can lead to the rejection of an alternative, regardless of its performance on other attributes.

Moreover, compensatory logic is foundational for intelligent decision-support systems, such as fuzzy inference systems, which are capable of incorporating uncertainty and human-like reasoning into clinical decision-making. These models enable personalized assessments that go beyond traditional population-based algorithms [21].

- **Development of the TS Search Engine**

The search engine objectives were established, defining the type of news and required functionalities, such as search forms, filters, and result storage. This system supports the early identification of relevant innovations in medical technology and facilitates their inclusion in public tendering processes. It contributes to continuous improvements in equipment modernization, benefiting the delivery of public healthcare services.

The roadmap for developing the medical equipment news search engine for the Clinical Engineering service is outlined below. In summary, first, define the requirements and scope. Then, relevant data is collected from reliable sources, and the necessary processing is performed. Once the engine is developed, extensive testing is performed, adjustments are made as necessary and used. It is an ongoing process of constant improvements to maintain the relevance of the search results. The engine is monitored for news in the area, and feedback is collected from users to make continuous improvements to the search engine [25].

The TS engine was then tailored to meet these needs by integrating interoperable components and a flexible architecture capable of evolving with emerging demands. The initial development phase involved structured interviews and collaborative meetings with CE professionals to define search priorities and key information needs. These included equipment maintenance issues, training requirements, interoperability challenges, and procurement timelines. Then, a decision support system was designed to operationalize technological surveillance for Clinical Engineering (CE).

The development of the TS search engine for the Clinical Engineering service followed a structured roadmap involving multiple iterative phases: from defining requirements to expert validation. This section presents the detailed outcomes from each development stage, the construction and application of fuzzy predicates, and the validation process with domain experts.

The development workflow included the following key steps:

- **Requirements Definition and Scope Delimitation**: Clear goals were established, focusing on real-time surveillance of medical imaging, biomedical signal acquisition, and clinical engineering tools relevant to public healthcare settings.
- **Data Collection**: News and technological updates were sourced from validated platforms, including manufacturers' websites, academic databases, Google Alerts, Biznar, and Espacenet.
- **Data Pre-processing and Indexing**: Text normalization, duplicate removal, and metadata tagging were conducted. Indexing allowed fast retrieval based on keywords and Boolean operators. Choosing the appropriate Technology Surveillance tools to design the search engine that meets the user's requirements. This may include a search form, search filters, relevant search results, and an associated page where each result is stored.
- **Technological Tool Integration**: Functional modules were implemented for search customization, alert storage, and classification interfaces.
- **Evaluation and Deployment**: Functional tests and debugging ensured system stability before final deployment in a user-accessible environment. Thorough testing of the search engine was conducted to ensure its functionality and performance, with adjustments made as necessary to improve accuracy and user experience.
- **Monitoring and Feedback Loop**: Alerts were continuously monitored, categorized, and rated by experts. Feedback was systematically incorporated into iterative improvements. The performance of the search engine was monitored to fix bugs, optimize performance, and add new functionality as needed.

3 Preliminary Results

- **Validation Process**

To validate the functional performance and decision-making relevance of the Technological Surveillance (TS) search engine, an expert-based evaluation was carried out involving four potential end users from diverse roles in the clinical engineering ecosystem:

Expert 1: Medical imaging technician
Expert 2: Electronic engineer in biomedical maintenance
Expert 3: Head of maintenance department at a public healthcare facility
Expert 4: Academic bioengineering expert.

Each expert independently received a curated list of 29 technological alerts, collected by the TS engine over a defined time window (December 4–13, 2024). The alerts covered updates in medical imaging, biomedical devices, and clinical equipment. The evaluation applied a Likert-type scale from 0 to 5, where 0 indicated that the alert was not relevant and 5 indicated that it was highly relevant. The evaluation was based on four specific criteria:

Maintainability of the device
Cost implications
Potential efficiency improvements
Contextual relevance of the news to local clinical engineering.

- **Data Processing and Fuzzy Validation**

The expert scores were aggregated and processed using the fuzzy logic model implemented in the system. This enabled the identification of priority alerts and low-impact content: 4 alerts (13.8% of the sample) were consistently rated as highly relevant, receiving scores above 4 in all four dimensions. 3 alerts (10.3% of the sample) were considered low relevance, with average scores under 2 across all evaluators. This process allowed us to confirm that the fuzzy rule engine aligns well with expert judgment, and it provided actionable insight for refining the system: High-relevance alerts were typically associated with innovations in imaging and signal-processing technologies with known integration potential. Low-relevance alerts included generic news or technologies with limited maintainability or no foreseeable impact on current municipal equipment workflows.

- **Outcome and Implications**

This validation confirms that the TS engine is capable of filtering out non-essential alerts, highlighting updates that are relevant for decision-making, and supporting the prioritization of procurement and maintenance activities. Furthermore, the insights obtained during this validation cycle enable the refinement of the search logic tree, the improvement of fuzzy predicates, and the enhancement of the system's accuracy in future iterations. Expert feedback also indicates that this process should be repeated periodically to ensure the system remains aligned with evolving technological landscapes.

Alerts were stored in structured spreadsheets with columns for date, source, type, and quantified predicates (T: training; S: software; H: hardware). Fuzzy inference, using Zadeh and compensatory operators, yielded a priority score for each alert (0–1 scale).

- **Preliminary Search Results and Source Evaluation**

Medical imaging equipment (X-rays, magnetic resonance, ultrasound) that present high technological innovation or have been recently launched are prioritized. This ensures that the clinical engineering service quickly identifies technologies that can improve diagnostic accuracy and reduce care time.

- **Search queries for recollected needs of the IC department equipment**

Equipment such as X-Ray machines, electrocardiographs, or blood pressure monitors must be compatible with hospital systems. If the equipment update frequency is high but its interoperability is limited, the alert is classified with low priority to avoid the adoption of devices that hinder integration. Hospital equipment such as stretchers or dental chairs that require frequent maintenance but do not significantly impact daily operations receive minimum priority. This optimizes the use of clinical engineering service resources, avoiding unnecessary repairs.

Medical Imaging Devices: ("Medical imaging devices") AND (X-ray OR MRI OR Ultrasound).

Biomedical Signal Acquisition Devices: ("Biomedical signal acquisition") AND (ECG OR stethoscope OR oximeter).

Clinical Engineering Devices: ("Hospital equipment") AND (wheelchairs OR stretchers OR dental chairs).

- **Search Tools Used**

Biznar: Retrieved 88 categorized alerts. Main topics included healthcare (34 alerts), manufacturers (30), innovations (12), and medical equipment (12). Although rich in results, expert feedback indicated difficulty in relevance assessment due to information overload.

Google Alerts: Delivered 54 well-structured alerts within a four-week period. Key categories: diagnostic imaging innovation (14), AI in healthcare (12), system interoperability (5), and cybersecurity (3). Experts found these alerts easy to interpret and immediately actionable. Google Alerts is a free tool from Google that allows users to receive email notifications about specific topics, such as news, research, or any content related to keywords of interest. It identified alerts over a four-week period, covering AI applications in healthcare, cybersecurity, and diagnostic equipment.

Espacenet (Patent Database): Returned 344 patents from the past five years, primarily in imaging and biomedical signal technologies (WIPO class G16H).

These results supported long-term procurement planning and preventive maintenance strategies.

To prioritize query results, fuzzy logic predicates were constructed using linguistic rules that reflect expert heuristics. Each fuzzy rule incorporates variables such as novelty, technological advancement, interoperability, maintenance cost, and clinical relevance. Rule definitions are based on CI expert requirements previously defined in the diagnosis performed.

- **Fuzzy Sentences in the Search Engine**

Medical imaging diagnostic devices

Fuzzy sentence:

"IF (alert relevance) IS high AND (frequency of novelty) IS recent THEN (priority) IS very high".

The FL result allows the system to evaluate the priority of each alert based on two factors: the relevance of the technology (for example, whether it is a state-of-the-art MRI device) and the frequency with which new versions are reported on the market. If both criteria are high, the assigned priority will be very high, suggesting that the service should immediately evaluate the inclusion of the device.

Fuzzy rule: IF (novelty) IS high AND (technology level) IS advanced THEN (priority) IS very high.

TS Objective: Prioritize emerging diagnostic technologies to improve service accuracy and responsiveness.

"IF (equipment novelty) IS moderate AND (compatibility with other systems) IS low THEN (priority) IS low".

In cases where compatibility with pre-existing systems is limited (for example, an oximeter that does not communicate with HL7 hospital software), the fuzzy system can assign a low priority, avoiding recommendations that imply high costs without clear integration.

"IF (technology) IS advanced AND (novelty) IS recent THEN (priority) IS very high".

The highest priority is assigned to advanced equipment with artificial intelligence that has been launched recently. Fuzzy logic allows treating novelty and technology not as binary values, but as degrees of belonging (for example, novelty = 0.8, advanced technology = 0.9), which ensures a flexible evaluation.

"IF (maintenance cost) IS high AND (clinical impact) IS low THEN (priority) IS minimum".

Alerts generated by the engine are evaluated using fuzzy predicates based on criteria such as novelty, relevance, and frequency. This fuzzy inference mechanism enables the assignment of qualitative values (e.g., high, medium, low) to each alert, assisting experts in quickly identifying those that warrant immediate attention. For example: The integration of FL reduces cognitive overload, supports transparent documentation of alert evaluations, and aligns procurement strategies with innovation readiness.

- **Results Validation**

The TS search engine was tested using Biznar, Google Alerts, and patent databases such as Espacenet. Biznar provided alerts, categorized by healthcare technologies, brands, services, and innovations. However, the results were hard for experts to analyze due to information overload. Google Alerts identified alerts within four weeks, covering AI applications in healthcare, cybersecurity, and diagnostic equipment, resulting in an easy assessment by the experts.

The experts' scores were aggregated and processed using the fuzzy logic model implemented in the system, allowing for the identification of priority alerts and low-impact content. The experts included a Medical imaging technician, Biomedical electronics engineer, Public hospital maintenance manager, and an Academic bioengineering specialist. Using a 0–5 Likert scale, experts rated each alert based on: Device maintainability, Cost implications, Efficiency potential, and Contextual relevance to local systems.

The fuzzy rule system aligned well with expert evaluations, demonstrating its effectiveness in filtering and classifying alerts. This validation confirmed that:

- Fuzzy logic improves prioritization accuracy.
- The system reduces expert cognitive load.
- The TS engine supports strategic procurement and technology updates.

Feedback led to refinement of fuzzy rules, enhancement of alert filtering mechanisms, and optimization of result presentation. The system will continue evolving through periodic expert consultations, ensuring alignment with technological advances and contextual clinical needs. Alert information is collected, saving valuable information such as Date, Source, and Type. Then, four predicates are qualified by the expert, resulting in a total quantification value for each alert, using Zadeh and compensatory fuzzy logic operators. Expert Total Qualification (0–1) LDC Expert Total Qualification Zadeh (0–1).

Patent searches resulted in 344 documents in the last 5 years, focusing on imaging devices and healthcare innovations. The TS search engine was implemented using various search and evaluation tools, including Biznar, Google Alerts, and patent databases such as Espacenet. Below are the results obtained in each phase of the analysis:

Biznar Alert Results: relevant grouped into several categories, highlighting: Healthcare: 34 alerts. Brands and manufacturers: 30 alerts. Technological innovations: 12 alerts. Medical equipment: 12 alerts.

The visualization of results by category allowed us to identify emerging trends in medical devices, as well as new areas of innovation in biomedical equipment.
Google Alerts results: During a four-week test period, 54 relevant alerts were received, organized into the following themes: AI implementations in health: 12 alerts, Connectivity and interoperability systems: 5 alerts, Computer security: 3 alerts, Innovation in diagnostic imaging: 14 alerts. These results demonstrated that Google Alerts is an efficient tool to keep up to date with trends and new

technologies applied in the health environment. This tool was selected for future optimizations of the TS.

Patent Results in Espacenet the search in patent databases yielded 344 relevant results, with a specific focus on: Diagnostic imaging devices: G16H Classification (WIPO). New technologies for the acquisition of biomedical signals. The results allowed us to identify patented innovations that could be incorporated in future public tenders and optimize preventive maintenance of the service.

- **Validation Process and Results**

To validate the functional performance and decision-making relevance of the Technological Surveillance (TS) search engine, an expert-based evaluation was carried out involving four potential end users from diverse roles in the clinical engineering ecosystem:

The expert scores were aggregated and processed using the fuzzy logic model implemented in the system. This enabled the identification of priority alerts and low-impact content:

- 4 alerts (13.8% of the sample) were consistently rated as highly relevant, receiving scores above 4 in all four dimensions.
- 3 alerts (10.3% of the sample) were considered low relevance, with average scores under 2 across all evaluators.

This process allowed us to confirm that the fuzzy rule engine aligns well with expert judgment, and it provided actionable insight for refining the system: High-relevance alerts were typically associated with innovations in imaging and signal-processing technologies with known integration potential.

Low-relevance alerts included generic news or technologies with limited maintainability or no foreseeable impact on current municipal equipment workflows.

This validation confirms that the TS engine is capable of filtering out non-essential alerts, highlighting updates that are relevant for decision-making, and supporting the prioritization of procurement and maintenance activities. Furthermore, the insights obtained during this validation cycle enable the refinement of the search logic tree, the improvement of fuzzy predicates, and the enhancement of the system's accuracy in future iterations. Expert feedback also indicates that this process should be repeated periodically to ensure the system remains aligned with evolving technological landscapes and the specific needs of the context.

The implementation of the Technological Surveillance (TS) search engine within the Clinical Engineering (CE) service enables efficient monitoring of emerging medical technologies. This facilitates well-informed procurement decisions during public tenders and supports the continuous improvement of health services. By integrating multiple data sources—such as Biznar, Google Alerts, and Espacenet—the system offers a structured approach to identify trends, optimize acquisition processes, and prioritize relevant innovation.

4 Discussion

Digital Transformation (DT) is critical for modernizing public health infrastructure. It demands a comprehensive institutional assessment, sustained investment in R&D&I, and robust collaboration among government agencies, academia, and industry stakeholders [7, 8]. Beyond mere digitization, DT entails the integration of advanced digital platforms, continuous workforce training, and the development of intelligent systems capable of adapting to the evolving health technology landscape.

Today's healthcare systems face significant challenges in efficiently managing human, financial, and technological resources. Investing in innovative technologies and services not only enhances patient outcomes but also improves the working conditions and effectiveness of healthcare professionals [19, 20].

The transformation of data into actionable knowledge requires the integration of digital tools, software platforms, and ongoing professional training. This process facilitates informed decision-making, enabling the optimization of both tangible and intangible organizational resources. By digitizing data collection systems and enhancing existing platforms, institutions can foster a culture of knowledge-driven decision-making supported by skilled personnel.

The TS engine plays a pivotal role in this context by converting raw data into structured knowledge. It alleviates information overload commonly experienced by Clinical Engineering (CE) professionals and supports strategic decision-making. Through real-time alerts and periodic updates, the system ensures continuous access to emerging medical technologies, contributing to the development of intelligent and sustainable healthcare systems.

A core innovation of the TS engine is the incorporation of fuzzy logic to prioritize alerts under conditions of uncertainty. Unlike traditional rule-based systems, fuzzy logic manages imprecise, incomplete, or ambiguous data more effectively. This facilitates the dynamic allocation of scarce resources, reduces unnecessary expenditures, and improves the efficiency of maintenance and acquisition processes. Fuzzy logic has demonstrated its versatility in managing complexity across domains such as surveillance, cybersecurity, and big data analysis [9, 21]. By adopting similar methodologies, the TS engine offers CE services an adaptive, intelligent decision-support tool aligned with the latest advances in artificial intelligence.

Furthermore, fuzzy predicates allow experts to manage alerts based on approximate criteria, enabling the prioritization of critical tasks and real-time adjustments. This results in more precise resource allocation, minimized operational risks, and optimized technology management [23, 24].

The implementation of the TS engine has demonstrated its capacity to support procurement decisions, particularly in public tendering processes. By integrating data from multiple sources—including Biznar, Google Alerts, and Espacenet—the system identifies trends, informs strategic choices, and ensures the adoption of relevant innovations.

The results validate the effectiveness of the TS engine in facilitating the continuous modernization of the municipality's Clinical Engineering service. The periodic update mechanism ensures access to the latest medical technologies, enhances institutional foresight, and improves public resource utilization. Ultimately, the combination of intelligent digital tools and targeted staff training supports the successful adoption of innovation in healthcare environments.

Comparatively, while machine learning approaches could offer predictive capabilities, fuzzy logic was chosen for its interpretability, flexibility, and reduced data dependency, critical for public sector environments where training datasets are often limited.

Continuous updating through TS ensures alignment with the dynamic innovation landscape, contributing to better preventive maintenance planning and strategic procurement.

5 Conclusion

Healthcare systems worldwide are undergoing profound transformations driven by financial constraints, workforce shortages, and rapid technological advancement. In this evolving landscape, Technological Surveillance (TS) emerges as a strategic tool that enables evidence-based decision-making in Clinical Engineering (CE), [1, 3]).

The TS search engine developed for a Clinical Engineering service exemplifies the application of strategic intelligence in public healthcare [22]. By facilitating the identification of emerging technologies, improving procurement planning, and supporting continuous professional development, the system contributes to the modernization of public health services. Automated, periodic monitoring ensures that institutions remain aligned with global innovation cycles.

Digital Transformation (DT) represents a systemic shift requiring alignment between institutional strategies and digital infrastructure. The integration of fuzzy logic into TS frameworks enhances the decision-making capabilities of CE professionals by allowing real-time prioritization of alerts under uncertain conditions. This is essential for optimizing both tangible assets (e.g., medical devices) and intangible ones (e.g., technical knowledge, protocols) [3, 6].

This initiative demonstrates that combining TS with fuzzy logic improves the agility, responsiveness, and sustainability of Clinical Engineering operations. The collaboration between academic institutions and public agencies fosters a dynamic environment for knowledge transfer, innovation, and resilience in the face of changing healthcare demands. The TS engine, designed to enhance decision-making in municipal CE services by incorporating expert knowledge, compensatory logic, and real-world data, offers a robust, adaptable, and user-friendly tool for prioritizing biomedical technology investment. The TS engine promotes strategic planning, improves budget efficiency, and enables evidence-informed decisions aligned with clinical and operational needs. It represents a significant step forward in the digital transformation of public health technology management.

The implemented TS engine has proven useful in identifying innovative equipment, training opportunities, and technological trends. It supports informed decision-making, particularly in periods of reduced workload, and strengthens CE's procurement processes. By integrating diverse information sources and continuously updating its databases, the engine contributes to optimizing maintenance and acquisition workflows.

The combination of digital tools and staff training also ensures the successful adoption of innovations in the healthcare setting and strengthens the adoption of new technologies. Expert feedback also indicates that this process should be repeated periodically to ensure the system remains aligned with evolving technological landscapes and the specific needs of the context.

Future work should focus on real-world validation involving clinical personnel and expanding the TS engine's scope to incorporate predictive analytics and machine learning components, further enhancing its decision-support capabilities.

AI Use Statement
Some portions of this manuscript were enhanced with the assistance of generative AI tools. The authors carefully reviewed, edited and approved all AI-generated content. The authors take full responsibility for the accuracy and integrity of the entire chapter.

References

1. AENOR: Spanish Standard UNE 166006:2011 EX: Technological surveillance and competitive intelligence systems. Spanish Association for Standardization (2011)
2. Aguirre, J.: Technological surveillance and competitive intelligence: tools for innovation management. Díaz de Santos Publishing House (2015)
3. Aguirre, J.: Strategic intelligence: a system for managing innovation. Rev. Estud. Gerenciales **31**(134), 100–110 (2015). https://doi.org/10.1016/j.estger.2014.07.001
4. Tecnogestión Group: Digital transformation in the healthcare sector. https://www.grupotecnogestion.com.ar/transformacion-digital-en-el-sector-salud/
5. Tecnogestión Group: Digital transformation in public health. https://www.collinsdictionary.com/dictionary/spanish-english/pendiente
6. Stoumpos, A.I., Kitsios, F., Talias, M.A.: Digital transformation in healthcare: technology acceptance and its applications. Int. J. Environ. Res. Public Health **20**(4), 3407 (2023). https://doi.org/10.3390/ijerph20043407
7. World Economic Forum.: Digital enterprise narrative (2016a). https://www.weforum.org/reports/digital-transformation-of-industries/
8. World Economic Forum.: Digital transformation of industries: in collaboration with Accenture (2016b). https://reports.weforum.org/digital-transformation
9. Mohaidat, M., Rashidi Fathabadi, F., Alkhamaiseh, K., Grantner, J., Shebrain, S.A., Abdel-Qader, I.: Towards the development of an automated assessment system for the fundamentals of laparoscopic surgery tests. Acta Polytechnica Hungarica. **21**(10), 37–56 (2024). https://doi.org/10.12700/APH.21.10.2024.10.3
10. Çelen, M.M.: Improved fuzzy logic based edge detection method on clinical images. ITU Graduate Thesis (2023). https://polen.itu.edu.tr/items/2665de90-2620-47c7-acab-882e684e6c13

11. Bhasme, M., Ulhe, P., Gokhe, S.: AI in agriculture: the future of robotic farming and automation. In: IEEE Conference on Artificial Intelligence in Healthcare (2024). https://ieeexplore.ieee.org/document/10842885
12. Mohaidat, M., Fathabadi, F.R.: A systematic approach to the development of an automated assessment system for laparoscopic surgery fundamentals. IEEE Trans. Artif. Intell. Healthcare (2024). https://ieeexplore.ieee.org/document/10619060
13. Hajiarbabi, M.: Glaucoma detection expert system using deep learning and certainty theory. IEEE SoutheastCon, 2025. https://ieeexplore.ieee.org/document/10971709
14. Santos, J., Torres-Machi, C., Morillas, S., Cerezo, V.: A fuzzy logic expert system for selecting optimal and sustainable life cycle maintenance and rehabilitation strategies for road pavements. Int. J. Pavement Eng. **23**(2), 425–437 (2020). https://doi.org/10.1080/10298436.2020.1751161
15. Sadat Asl, M.G., Namdari, A., Sadoghi, M., Karami, M.: Fuzzy-based clinical decision support system for prioritizing health alerts: a COVID-19 case study. IEEE J. Biomed. Health Inform. **27**(4), 1034–1042 (2023). https://doi.org/10.1109/JBHI.2023.3261789
16. Nikouei, S.Y., Chen, Y., Aved, A., Blasch, E., Faughnan, T.R.: I-SAFE: Instant Suspicious Activity identiFication at the Edge using fuzzy decision making. arXiv preprint (2019). https://arxiv.org/abs/1909.05776
17. Nikouei, S.Y., Yu, W., Yang, K., Lou, Y., Chen, C., Song, H.: Real-time traffic monitoring using fuzzy logic and edge computing. IEEE Internet Things J. **6**(3), 5697–5705 (2019). https://doi.org/10.1109/JIOT.2019.2896272
18. Alsubhi, K., Boutaba, R.: FuzMet: A fuzzy-logic based alert prioritization engine for intrusion detection systems. Int. J. Netw. Manag. **21**(5), 374–388 (2011). https://doi.org/10.1002/nem.804
19. An American Health Organization: from the evolution of health information systems to digital transformation. https://iris.paho.org/handle/10665.2/53801
20. PAHO.: Digital transformation of the health sector. Pan American Health Organization. https://www.paho.org/en/topics/digital-transformation-health-sector
21. Holland, J.H.: Adaptation in natural and artificial systems. University of Michigan Press (1975)
22. Municipality of General Pueyrredón: Outpatient Medical Specialties Center (CEMA). https://www.mardelplata.gob.ar/cema
23. Zhang, Y., Wang, S.: Fuzzy logic in surveillance big video data analysis. In: Proceedings of the 2021 ACM Symposium on Applied Computing, pp. 123–130 (2021). https://doi.org/10.1145/3444693
24. Zhang, Y., Wang, Y.: Fuzzy logic-based data analysis for cybersecurity: a review. Expert Syst. Appl. **178**, 114989 (2021). https://doi.org/10.1016/j.eswa.2021.114989
25. Pavlicevic, J.S., Guagliano, M., Tornillo, J., Servetto, D.: Survey of open source technologies applicable to teaching processes in engineering courses through the use of technological tools

Enhancing Clinical Decision-Making: An Integrated Approach Using ISPS and KMoS-SSA Framework

Oscar Rafael Contreras Estrada, Karla Olmos Sánchez, Jorge Rodas-Osollo, and Ariel Rubén Galván Astorga

Abstract Clinical psychologists navigate complex decision-making processes shaped by diverse mental health conditions, individualized patient needs, and evolving treatment options. These complexities create significant challenges, such as limited resources and the demand for personalized treatment plans. In response, this chapter introduces the *Integrated System for Psychologist Support* (*ISPS*), a comprehensive system designed to enhance informed decision-making in clinical psychology. The *ISPS* has been conceptualized using the *Knowledge Management on a Strategy through Soft Systemic Analysis* (*KMoS-SSA*) *methodological framework*, which leverages knowledge management theories and soft systems thinking to achieve desirable, effective, and feasible solutions within the evolution knowledge cycle. The *ISPS* incorporates a *Gamified Application Module* to collect patient data and a *Machine Learning Module* to categorize patient information into profiles that support clinical decision-making. Real-world applicability was pursued from the outset: practicing clinicians participated in co-design workshops, walkthrough evaluations, and iterative refinements that ensured the prototype fits seamlessly into routine assessment workflows while respecting stringent data-privacy regulations. In parallel, privacy-preserving synthetic data were incorporated to broaden the system's training base without compromising confidentiality. This project is anchored in KMoS-SSA's knowledge-evolution cycle, which captures expert insight, models domain interactions, validate proposed solutions, and continually updates them as new evidence.

O. R. Contreras Estrada (✉) · K. Olmos Sánchez · J. Rodas-Osollo · A. R. Galván Astorga
Department of Electrical and Computer Engineering, Universidad Autónoma de Ciudad Juárez, Ciudad Juárez, México
e-mail: al234517@alumnos.uacj.mx

K. Olmos Sánchez
e-mail: kolmos@uacj.mx

J. Rodas-Osollo
e-mail: jorge.rodas@uacj.mx

A. R. Galván Astorga
e-mail: al234510@alumnos.uacj.mx

W. Pedrycz et al. (eds.), *Decision-Making in Computational Intelligence-Based Systems*, Studies in Systems, Decision and Control 628,
https://doi.org/10.1007/978-3-032-04056-5_21

This iterative mechanism secures the long-term relevance and adaptability of ISPS as clinical contexts evolve.

Keywords Decision-making support system · Mental healthcare · Clinical psychology · Gamified psychometric assessments · Machine learning

1 Introduction

The global mental health crisis has significantly increased the Mental-health services worldwide are facing an unprecedented convergence of pressures: rising prevalence of anxiety and mood disorders, widening socio-economic stressors, and a chronic shortage of specialized personnel. The World Health Organization warns that unmet demand is projected to grow fastest in regions with limited public resources and fragmented primary-care infrastructures, conditions that typify much of Latin America and, specifically, northern Mexico, where this study is situated [1]. Within this context, clinical psychologists must interpret symptom constellations that often overlap or mask one another, weigh comorbidities, and design personalized interventions, all while managing high caseloads and strict confidentiality requirements. Conventional assessment workflows, lengthy anamneses, paper-based psychometric batteries, and manual synthesis of findings are increasingly mis-matched to that reality. As Kitayama et al. observe, even incremental gains in automation or patient engagement can translate into meaningful reductions in waiting lists and therapeutic drop-out rates [2].

Digital transformation offers promising, but uneven, relief. A growing body of work explores e-mental-health tools, ranging from self-guided cognitive-behavioral apps to clinician dashboards that aggregate symptom scores in real time [3, 4]. Yet three limitations persist. First, many solutions digitize questionnaires verbatim, overlooking human–computer-interaction principles that could mitigate test anxiety and sustain attention. Second, analytic engines often operate as opaque "black boxes," producing recommendations without exposing the reasoning needed for clinical accountability. Third, few systems provide a structured methodology to ensure long-term relevance: when diagnostic taxonomies evolve or local practice patterns shift, the tool itself stagnates.

To confront these gaps, we propose the Integrated System for Psychologists Support (ISPS), a modular, clinician-centered platform whose development is anchored in the Knowledge Management on a Strategy through Soft Systemic Analysis (KMoS-SSA) framework. ISPS weaves together two complementary components. The Gamified Application Module delivers validated instruments, principally the 16 Personality Factors (16PF-5) questionnaire and the Beck inventories, through an interactive visual-novel interface inspired by serious-game design. Dialogue-based narration, subtle progress cues, and micro-interactions transform what is traditionally a static questionnaire into a more tolerable, even engaging, experience; preliminary usability walkthroughs with Mexican practitioners confirmed the format reduces respondent fatigue while preserving psychometric integrity. The Machine-Learning Module then converts these raw responses into clinically meaningful profiles via a C4.5 decision-tree classifier and complements them with experience-grounded recommendations generated through Case-Based Reasoning (CBR). Crucially, both modules exchange data through a privacy-preserving pipeline that supports the offline inclusion of synthetic cases, widening the training base without exposing personally identifiable information.

KMoS-SSA provides the methodological spine that holds these technical parts together. Rooted in soft-systems thinking and knowledge-management theory, the framework unfolds as an iterative knowledge-evolution cycle: (i) eliciting domain insight from clinicians; (ii) structuring that knowledge into artefacts, (iii) modelling and validating candidate solutions; and (iv) reflecting on outcomes to refine or re-align the design. Each pass through the cycle equips ISPS with updated context, new treatment guidelines, emergent symptom patterns, or feedback from the Gamified Module, thus guarding against the obsolescence that plagues many static digital tools.

The present chapter makes four primary contributions. First, it documents how KMoS-SSA can translate systemic thinking into a concrete, clinician-usable technology. Second, it articulates the design rationale behind ISPS's modular architecture, explaining how choices of language, data flow, and interface elements were driven by both ethical and practical constraints in Mexican outpatient contexts. Third, it summarizes insights gleaned from expert walkthroughs, highlighting usability, perceived clinical utility, and areas requiring enhancement, thereby demonstrating real-world alignment without claiming a full-scale patient deployment. Finally, it identifies avenues for scalability, including cloud-based data synchronization, longitudinal validation studies, and extended support for additional psychometric batteries.

The remainder of this chapter is organized as follows. Section 2 surveys foundational psychological-assessment approaches and positions ISPS within current digital-health literature. Section 3 details the KMoS-SSA methodology as applied to the mental-health domain, emphasizing its knowledge-evolution properties. Section 4 presents the ISPS architecture, referring readers to the Gamified Application Module and the Machine-Learning Module for implementation specifics. Section 5 reports preliminary results and expert feedback, while Sect. 6 concludes with implications and future work.

2 Foundations of Psychological Therapies

Although traditional assessment protocols remain a cornerstone in clinical psychology, they exhibit well-documented limitations in terms of scalability, adaptability, and integration with digital environments. Typically, clinicians rely on standardized batteries of tests, lengthy interviews, and manual documentation processes to evaluate patients. While effective in controlled settings, these approaches struggle to keep pace with the increasing demand and complexity of mental health cases.

Freedman [5] highlights how conventional methods can be time-consuming, often slowing the decision-making process in high-pressure clinical contexts. Psychologists may spend hours conducting assessments, which restricts their capacity to accommodate new or returning patients [6]. Additionally, Pade [7] observes that although digital tools have emerged, many psychological assessments remain paper-based or static, missing the opportunity to leverage interactive interfaces and real-time data analysis.

Another core challenge is that mental health conditions often present overlapping symptoms and comorbidities. This overlap increases diagnostic ambiguity and delays intervention planning. Furthermore, when assessment frameworks are updated manually, inconsistencies may arise across providers or institutions, leading to variability in patient care. These issues underscore the need for standardized, adaptable, and knowledge-driven systems that evolve alongside clinical evidence.

3 Methodology

The increasing complexity in clinical psychology, particularly in the design of personalized treatments, requires a structured approach capable of managing expert knowledge and the dynamic nature of mental health care. To address these demands, this chapter adopts the Knowledge Management on a Strategy through Soft Systemic Analysis (KMoS-SSA) framework, which supports the conceptualization and design of an adaptable solutions in complex domains.

KMoS-SSA is a methodological framework grounded in knowledge management principles and soft systems thinking. It enables the identification of systemic structures, the organization of domain knowledge, and the generation of conceptual models based on expert elicitation. Through a structured sequence of stages, including knowledge elicitation, domain structuring, model generation, validation, and refinement, KMoS-SSA facilitates the design of solutions that are not only technically feasible but also aligned with real-world constraints and expert perspectives. Previous applications of KMoS-SSA and its predecessor Knowledge Management on a Strategy through Requirements Engineering (KMoS-RE) have been documented in diverse domains [9, 10].

To better illustrate the methodological logic and iterative nature of the process, the evolutionary cycle of knowledge within the KMoS-SSA framework is shown in Fig. 1. This cycle integrates the real-world context with systemic thinking, and depicts how knowledge is elicited, internalized, modeled, validated, and continually updated through reflection and abduction.

In this project, KMoS-SSA was applied to conceptualize the Integrated System for Psychologists Support (ISPS). The process involved interviews with clinical experts to identify key needs and challenges, followed by the construction of enriched domain representations using systemic tools such as CATWOE, conceptual modeling, and root definitions. These artifacts were used to conceptualize and structure the technological components of ISPS and to ensure alignment with real clinical workflows.

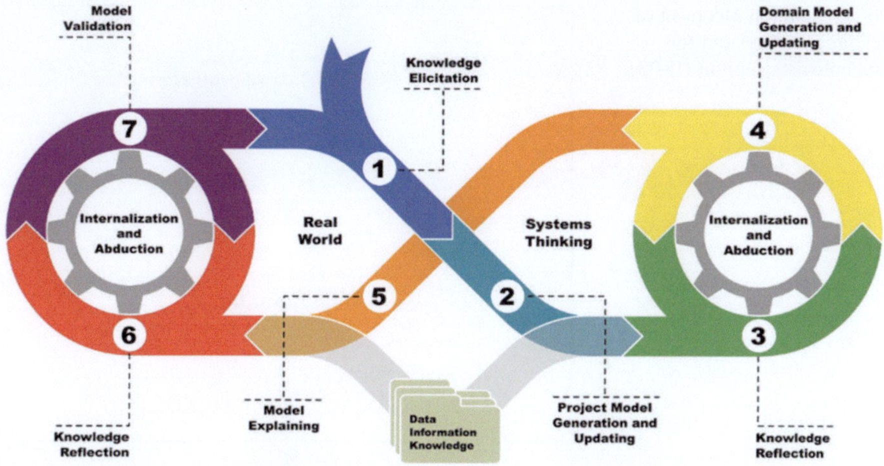

Fig. 1 Evolutionary cycle of knowledge in KMoS-SSA

The framework also guided the identification of relevant knowledge flows between modules and the integration of data processing techniques, including machine learning and synthetic data generation.

4 System's Architecture

The Integrated System for Psychologists Support (ISPS) was initially conceptualized as a modular platform aimed at enhancing clinical decision-making in psychology through the combination of gamified data collection and basic analytical mechanisms. The prototype focused on validating the feasibility of this dual-module structure, composed of: (1) the Gamified Application Module and (2) the Machine Learning Module.

Although the system did not include real-time data synchronization or cloud-based storage at this point, the architecture was designed with modularity in mind, allowing for future expansion into more scalable environments. The core objective of the prototype was to test the technical and clinical feasibility of combining gamified assessment with AI-assisted decision support in a psychology context.

Figure 2 presents the conceptual blueprint of the Integrated System for Psychologists Support (ISPS).

The Gamified Application Module, developed using the Ren'Py engine, delivers psychometric assessments in an interactive and engaging format. Standardized instruments, such as the 16PF-5 and Beck Inventories, are presented to the user through a visual-novel interface. Patient responses are collected and stored in structured local files, which serve as the input for subsequent processing.

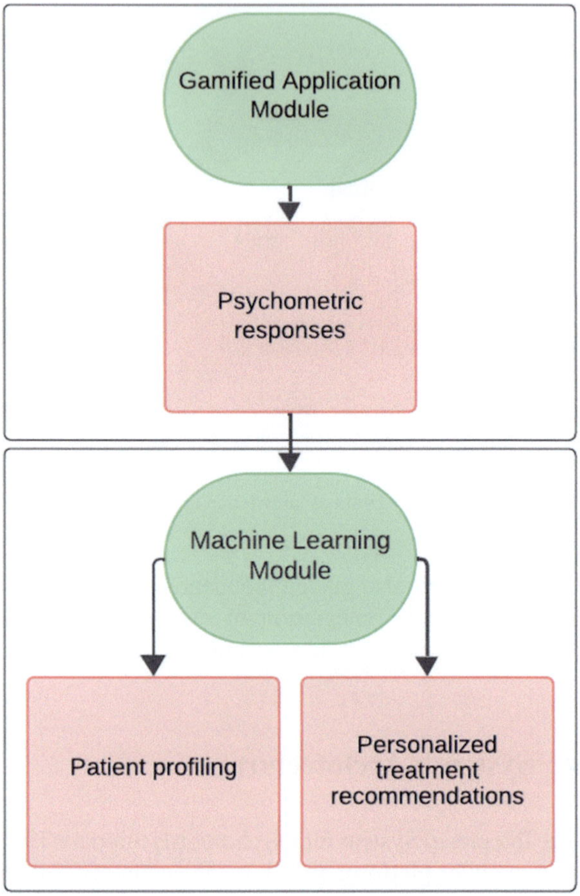

Fig. 2 Solution blueprint of the integrated system for psychologists support (ISPS)

The Machine Learning Module, implemented in Python, receives these structured inputs and applies the C4.5 decision-tree algorithm to generate patient profiles. These profiles categorize individuals based on key psychological factors such as symptom severity and personality traits. The classification results offer clinicians a clearer overview of patient conditions and support more informed decision-making.

This blueprint reflects ISPS's early integration of gamified assessment delivery and intelligent profiling techniques, laying the foundation for data-driven clinical support in psychology.

4.1 Component Description and Technical Integration

4.1.1 Data Collection via Gamified Application Module

The Gamified Application Module serves as the primary interface of the ISPS prototype, responsible for administering psychometric assessments through an interactive and visually engaging experience. Built using Ren'Py, a Python-based engine for interactive storytelling, the module was adapted to present psychological instruments in a structured and user-friendly format.

The assessments included standardized instruments such as the 16 Personality Factors (16PF-5) and the Beck: Beck Depression Inventory (BDI), Beck Anxiety Inventory (BAI) and Beck Hopelessness Scale (BHS). These were integrated into the application through scripted scenes and visual elements that simulate a game-like experience. The structure was designed to encourage patient participation while maintaining the integrity of the psychometric tools.

To reduce cognitive fatigue and promote sustained engagement, brief interactive elements, such as click-based interactions, transitions, and visual cues, were interspersed throughout the assessment flow. These elements were not designed for diagnosis, but rather to support a more tolerable and less monotonous experience for users.

Patient responses were stored in structured text or CSV files on the local device. Each record included the selected response for each item, the question identifier, and basic metadata such as the time of response. These data files served as the primary input for the Machine Learning Module, which processed them externally.

Core Functionalities of the Module: assessment delivery (Ren'Py), psychometric items are dynamically presented and navigated through Ren'Py's scripting logic, allowing conditional flows, feedback, and visual structuring, engagement features, the application uses visual progress indicators, dialogue-based narration, and light interactivity to enhance the overall experience and structured data output, responses are stored in local files using a consistent schema, ensuring compatibility with analytical tools.

By transforming traditional assessment formats into an interactive digital experience, the Gamified Application Module provides a practical and user-centered approach to collecting clinically relevant data while enhancing user engagement.

4.1.2 Patient Profiling and Recommendation Using Machine Learning Module

The Machine Learning Module of ISPS is responsible for transforming structured psychometric data into clinically meaningful classifications through a patient profiling process. This component was developed using Python and implements the C4.5 decision-tree algorithm, which classifies patients based on psychological risk factors and symptom severity.

Patient responses collected from the Gamified Application Module are formatted into structured CSV files. These are processed by the Machine Learning Module to generate labels that help clinicians interpret the patient's current psychological profile.

Key Components of the C4.5 Classification Process:

Input Features: The algorithm uses responses from standardized psychometric instruments, including the 16PF-5 and Beck Inventories (BDI, BAI, BHS), as input variables.
Classification Targets: Patients are categorized into psychological risk levels (e.g., low, moderate, high) based on symptom combinations and personality profiles. These categories assist in prioritizing clinical attention and planning treatment paths.

C4.5 Decision Tree for Patient Profiling

The first stage of analysis in the Machine Learning Module is the classification of patient profiles using the C4.5 decision-tree algorithm. Originally developed by Quinlan, C4.5 builds interpretable decision trees based on entropy measures and information gain. This algorithm was selected for its robustness in handling both continuous and categorical data, its capacity to deal with missing values, and its ability to prune overfitted branches to maintain model generalization.

Implementation Details:

Input Features: Psychometric data from the 16PF-5 personality test and Beck Inventories (BDI, BAI, BHS) are used as input attributes. These values are standardized and encoded to fit the model input schema.
Target Labels: Patients are classified into psychological risk categories (e.g., mild, moderate, severe) based on symptom intensity and composite personality profiles. Labels are defined in collaboration with clinical experts.
Training and Validation: The decision tree is trained using a combination of synthetic datasets, generated to address class imbalance and data scarcity, and real annotated cases when available. The model undergoes iterative validation using accuracy, precision, recall, and F1-score as performance metrics.

Role in ISPS:

Once trained, the C4.5 model receives structured patient data and returns a classification output in the form of a profile label and a list of decision paths (rules). These outputs serve two functions:

1. Support for Clinical Interpretation: The decision paths show clinicians which features (e.g., high anxiety + low self-reliance) influenced the classification.
2. Pre-processing for CBR: The classification results are used as filters and weights in the Case-Based Reasoning Module, improving the retrieval of similar cases and the relevance of suggested treatments.

This component enhances diagnostic clarity, improves the speed of decision-making, and ensures consistent application of classification criteria across diverse patient profiles.

Case-Based Reasoning for Treatment Recommendation

The Case-Based Reasoning (CBR) component of the Machine Learning Module is designed to generate personalized treatment recommendations by reusing knowledge derived from previously solved clinical cases. This method aligns with the principle that similar problems often share similar solutions, making it especially well-suited for domains like psychology, where variability among patients is high and interventions are context dependent.

CBR in ISPS is structured around a classic four-phase cycle: Retrieval, Reuse, Revision, and Retention. Each of these phases is supported by custom logic implemented in Python and integrates seamlessly with the patient profiling results obtained from the C4.5 decision tree.

1. Retrieval

In this phase, the system queries the case base, a database of past patient cases that includes input attributes (e.g., psychometric scores, profile labels) and the treatments that were applied. A similarity engine computes distance metrics between the new patient and existing cases using a weighted combination of: 16PF-5 personality factor scores, Beck inventory scores (BDI, BAI, BHS), C4.5-generated profile label and social support attributes (when available).

Cases with the highest similarity scores are retrieved for further evaluation.

2. Reuse

Once a set of similar cases is identified, the associated treatments from those cases are analyzed. The system selects the most recurrent or successful intervention strategies, considering outcome evaluations from the historical case and patient context.

A proposed recommendation is generated, tailored to the current patient's data.

3. Revision

The proposed recommendation is passed to a revision module, which allows clinical experts to review and adjust the suggestion based on additional data not considered in the automated process. The system can also perform automatic revision by re-evaluating cases if new patient information becomes available during treatment.

4. Retention

If the clinician validates the recommendation and the treatment proceeds, the full case, along with eventual outcomes, is stored back in the case base. This ensures that ISPS continuously evolves by learning from new experiences, expanding its ability to handle complex and novel patient presentations.

Integration with ISPS the CBR module is invoked through the Machine Learning Module's, triggered after the C4.5 classification phase. Retrieved and revised treatment plans are structured and returned as JSON objects, making them compatible with future integration into clinician dashboards or patient reports.

Benefits and Role in Decision Support:

The CBR engine augments the system's intelligence by embedding a learning-by-experience mechanism that adapts over time. Its integration provides personalized recommendations, grounded in real-world evidence, adaptability, since each new case refines the system's knowledge base and transparency, as clinicians can trace how and why a particular recommendation was made.

Together with C4.5 classification, CBR forms the decision-making backbone of ISPS, bridging statistical analysis and experiential reasoning to offer comprehensive support in clinical psychology.

4.1.3 Technologies and Communication Architecture

The ISPS prototype was developed using a lightweight, modular architecture that supported local execution and offline interaction between components. The choice of technologies emphasized accessibility, ease of testing, and adaptability to future enhancements (Table 1).

Communication Flow

The ISPS prototype followed a file-based communication model between components:

Data Collection: The Gamified Application Module saved patient responses locally in structured CSV files immediately after assessment completion.

Data Access: The Machine Learning Module read these local files as input, processed them using the C4.5 algorithm, and generated output files containing classification results.

Result Review: Output files could be visualized or interpreted directly through logs or simple interfaces during validation sessions.

Table 1 Technological stack

Component	Technology	Purpose
Gamified application	Ren'Py	Delivery of interactive psychometric assessments
Data storage	CSV	Structured storage of patient responses and classification results
Machine learning module	Python 3	Implementation of the C4.5 algorithm and processing of structured data
Supporting libraries	Scikit-learn, Pandas	Data handling, model training, and decision-tree generation

Deployment and Execution

Both modules were executed on a single machine during the prototype stage. The system was designed for modularity, allowing each component to be tested independently and adjusted without affecting the other.

This architecture provided a robust foundation for functional validation, rapid iteration, and alignment with real clinical needs. Its simplicity ensured that development remained focused on core functionalities, data acquisition, classification, and result interpretation, without requiring a full production environment.

5 Results

The initial version of the Integrated System for Psychologists Support (ISPS) was evaluated to verify the technical feasibility and usability of its two foundational components: the Gamified Application Module and the Machine Learning Module. The evaluation process focused on user experience, data collection efficiency, and the ability of the classification engine to provide consistent and interpretable profiling outputs and recommendations for decision-making.

5.1 Gamified Application Module Evaluation

The Gamified Application Module was tested with a small group of participants to assess usability and engagement during the completion of psychometric assessments. Informal feedback from both users and a participating clinical psychologist was collected. Figure 3 illustrates the main interface of the gamified module, which was evaluated during this process.

Key Observations

- User Engagement: Participants responded positively to the interactive format of the assessments, highlighting the visual flow, progress indicators, and storytelling elements as helpful for maintaining focus.
- Usability: The module was described as intuitive and easy to navigate. The scripted interaction allowed users to complete assessments without confusion, even if they were unfamiliar with psychological instruments.
- Professional Feedback: A clinical psychologist involved in the testing process expressed satisfaction with the clarity of the assessment sequence, the preservation of the psychometric structure, and the potential for broader patient engagement.

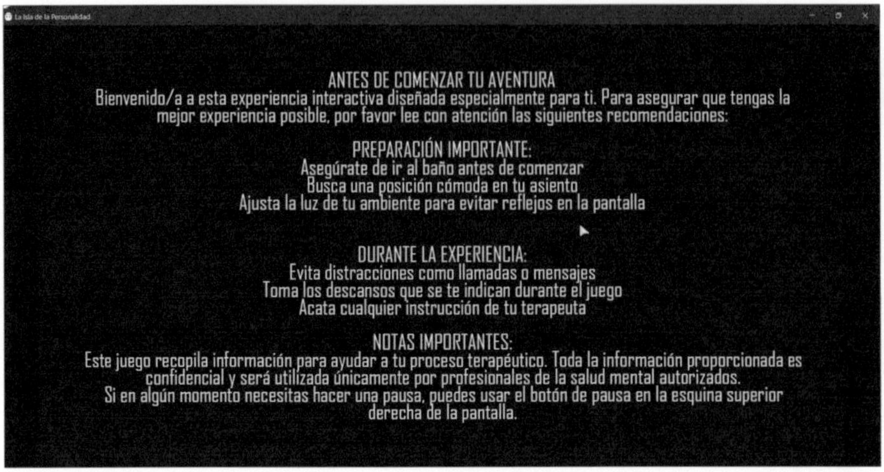

Fig. 3 Main screen of the gamified application showing the interface presented to users for psychometric assessment completion. Translation of the instructions originally presented in Spanish: before starting the interactive experience, users are welcomed and provided with preparation guidelines, including using the restroom, finding a comfortable position, and adjusting the ambient lighting to reduce screen glare. During the experience, users are advised to avoid distractions, take suggested breaks, and follow any therapeutic instructions. Additionally, it is clarified that all collected information is confidential and will be used solely by authorized mental health professionals

5.2 Machine Learning Module Performance

The C4.5 classification engine was tested using structured datasets derived from psychometric assessments collected through the gamified module. The model was trained and evaluated locally using available data to assess its ability to classify psychological profiles as demonstrated in Fig. 4.

Performance Results

- Accuracy: The classification engine achieved high accuracy in categorizing profiles into defined psychological risk levels.
- Interpretability: The resulting decision trees provided clear rule paths, showing which attributes were most relevant for each classification. This transparency added value for future clinical interpretation.
- Consistency: The algorithm produced stable and repeatable outputs when applied to similar input conditions, supporting its reliability within the prototyping environment.

CBR Recommendation Validation

- Clinical validation of the CBR outputs showed that 90% of treatment suggestions generated were rated as clinically appropriate or aligned with expert standards.

```
Matriz de confusión:
[[10432     0   835]
 [    0 13274   773]
 [  732   843 73111]]

Reporte de clasificación:
                precision    recall  f1-score   support

    Riesgo_Alto      0.93      0.93      0.93     11267
    Riesgo_Bajo      0.94      0.94      0.94     14047
Riesgo_Moderado      0.98      0.98      0.98     74686

       accuracy                          0.97    100000
      macro avg      0.95      0.95      0.95    100000
   weighted avg      0.97      0.97      0.97    100000
```

Fig. 4 Result of C4.5 algorithm. Translation of the confusion matrix and classification report originally shown in Spanish. The model evaluates three risk categories (high, moderate, and low), showing precision, recall, and F1-score values. The algorithm achieved an overall accuracy of 97%, confirming its effectiveness in classifying clinical profiles for decision support

```
=== Matriz de Confusión ===
[[ 1067     0  1203]
 [    1  1687  1171]
 [  622   793 13457]]
=== Exactitud ===
90.68%
```

Fig. 5 Results of CBR. CBR recommendation examples and expert validation. Translation of the confusion matrix originally displayed in Spanish. The Case-Based Reasoning system shows its recommendation performance across risk levels, with a global accuracy of 90.68%. These results were validated through expert evaluation, supporting the system's applicability in clinical contexts

- The system was particularly effective in handling typical case patterns, while more complex cases still require refinement in similar metrics and adaptation rules.

Figure 5 presents the confusion matrix results of the Case-Based Reasoning system, which confirm its recommendation accuracy.

5.3 ISPS Module Integration

A visual overview of ISPS components was created to demonstrate the structural interaction between modules and the decision-making flow. To illustrate the system's architecture, Fig. 6 presents the structure and data flow between ISPS components.

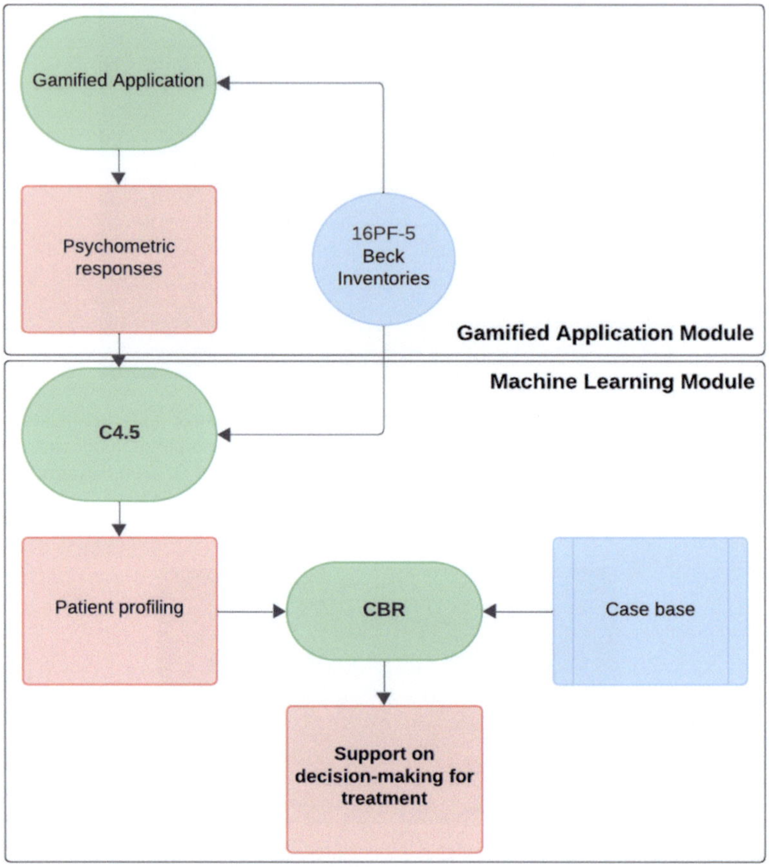

Fig. 6 ISPS module structure: integration of data collection, classification, and recommendation

5.4 Domain Expert Feedback

Feedback gathered from participating clinical psychologists included the following points:

- Assessment Quality: The psychologist appreciated the fidelity of the assessment items and how the interface preserved the original structure of the psychological instruments.
- Visual Appeal: The engaging design of the application was considered a strength for maintaining patient motivation during longer assessments. The visual aesthetics and professional design increased trust in the system's outputs. Suggestions were made to include dashboards and dynamic feedback mechanisms in future iterations.

- Clinical Utility: The concept of balancing engagement and clinical rigor was well received. Experts highlighted the benefit of having a tool that could reduce patient fatigue while retaining diagnostic validity.

5.5 *Summary of Key Findings*

Area	Outcome
User engagement	High engagement and satisfaction with the interactive assessment format
Classification accuracy	Strong profile consistency and rule-based outputs
Recommendation validity	Strong alignment with expert opinion
Feedback	Positive expert validation with suggestions for future enhancements

6 Conclusion

The initial implementation of the Integrated System for Psychologists Support (ISPS) represents an important step toward enhancing clinical decision-making through the integration of interactive assessment and intelligent analysis. The system was designed under the principles of the KMoS-SSA methodological framework, which provided a structured foundation for addressing the complexity of the clinical psychology domain.

By working within a complex and dynamic context, ISPS was conceptualized to align with real-world needs while remaining adaptable to future developments. The methodological approach emphasized the importance of capturing expert insights, structuring relevant knowledge, and translating it into actionable system components.

The Gamified Application Module successfully demonstrated the feasibility of using interactive technologies to administer standardized psychometric instruments. Participants and specialists reported positive experiences with the interface, noting improved engagement and usability.

The Machine Learning Module, built in Python, incorporates the C4.5 decision tree algorithm for patient profiling and a Case-Based Reasoning (CBR) engine for treatment recommendation., delivered consistent and interpretable classifications of psychological profiles. The clarity of the model's output supports clinical interpretation and lays the foundation for data-informed decision support.

The results obtained confirm the technical and conceptual viability of the ISPS architecture. The separation between data acquisition and analysis, supported by a modular structure, enables future scalability and integration with more advanced features, including personalized recommendations and adaptive interfaces.

The ISPS prototype validates the core idea of enhancing psychological assessment and decision-making through digital tools grounded in systemic thinking. By combining human-centered design with interpretable machine learning, ISPS offers a promising direction for the development of intelligent support systems in clinical psychology.

AI Use Statement
Some portions of this manuscript were enhanced with the assistance of generative AI tools. The authors carefully reviewed, edited and approved all AI-generated content. The authors take full responsibility for the accuracy and integrity of the entire chapter.

References

1. World Health Organization.: World Mental Health Report: Transforming Mental Health for All, Geneva (2022)
2. Patel, V., Saxena, S., Lund, C., et al.: The Lancet commission on global mental health and sustainable development. The Lancet **392**(10157), 1553–1598 (2018)
3. Bujold, M., Pluye, P., Légaré, F., et al.: Decision-making and related outcomes of patients with complex care needs in primary care settings: a systematic literature review with a case-based qualitative synthesis. BMC Primary Care (2022)
4. Simpson, K., Nham, W., Thariath, J., et al.: How health systems facilitate patient-centered care and care coordination: a case series analysis to identify best practices. BMC Health Serv. Res. (2022)
5. American Psychological Association.: Psychologists Struggle to Meet Demand Amid Mental Health Crisis. APA (Nov 2022)
6. Freedman, R.: The initial field trials of DSM-5: new blooms and old thorns. Am. J. Psychiatry **170**(1), 1–5 (2013)
7. Pade, H.: The evolution of psychological testing: embarking on the age of digital assessment. In: The Changing Faces of Therapy. Lulu Publishing Services (2016)
8. Olmos-Sánchez, K., Rodas-Osollo, J., Maldonado-Macías, A., Jiménez-Galina, A.: Harmonization of knowledge representation: integrating systems thinking ideas with appropriate domain representation strategies. J. Syst. Comput. (2024)
9. Nebeker, S. et al.: Building research integrity and capacity (BRIC): an educational initiative to increase research literacy among clinical staff. J. Empir. Res. Hum. Res. Ethics. **13**(3) (2018)
10. Rodas-Osollo, J.: An interesting adventure accompanied by CMCg.I model supporting digital transformation in the cognitive era. Universidad Autónoma de Ciudad Juárez, Ciudad Juárez, México (2023)

New Rainfall Missing Values Estimation Proposal: Normal Ratio with Additive Correlation

Edgar Omar Ruiz-DelAngel, Rocío del Carmen Vargas-Castilleja, Julio César Rolón-Aguilar, Mirna Patricia Ponce-Flores, and Jesús David Terán-Villanueva

Abstract Rainfall is one of the most essential climatological variables for developing hydrological balances, as with other studies. In Mexico, there is a lack of reliable and spatially homogeneous historical information. This study evaluates a proposal to estimate missing data in rainfall time series incorporating a correlation between rain gauge stations using the Normal Ratio as a basis. We used records from eight rain gauge stations over the Chicayan River basin in the Panuco River basin in northern Veracruz. We compared the results of the proposed imputation technique against those obtained with the Normal Ratio, and we found several similarities. Having for both techniques in all rain gauge stations values above 0.88 for R^2 and values above 0.94 for r. The *RMSE* and *MAE* were also analysed, and we found that both yield very close values.

Keywords Imputation techniques · Missing data · Rain gauge stations · Rainfall time series · Correlation matrix

---

E. O. Ruiz-DelAngel · R. C. Vargas-Castilleja · J. C. Rolón-Aguilar ·
M. P. Ponce-Flores (✉) · J. D. Terán-Villanueva
Universidad Autónoma de Tamaulipas (UAT), Centro Universitario Tampico Madero, Tampico, Tamaulipas, Mexico
e-mail: mirna_poncef@hotmail.com

R. C. Vargas-Castilleja
e-mail: rocvargas@docentes.uat.edu.mx

J. C. Rolón-Aguilar
e-mail: jrolon@docentes.uat.edu.mx

J. D. Terán-Villanueva
e-mail: jdteran@docentes.uat.edu.mx

W. Pedrycz et al. (eds.), *Decision-Making in Computational Intelligence-Based Systems*, Studies in Systems, Decision and Control 628,
https://doi.org/10.1007/978-3-032-04056-5_22

1 Introduction

In this paper, we study basins as essential components for environmental, social, and economic development. To conduct a thorough assessment of a basin, adequate climatological records are necessary. These records are crucial for understanding weather patterns, hydrological processes and broader climatic trends.

One of the most significant challenges encountered in analyzing climatological time series data is the presence of missing values, which can significantly hinder accurate assessments of climate trends. These gaps can surface for many reasons, including human error during data entry, breakdowns in recording equipment, and extreme weather events that may disrupt the standard data collection processes, such as hurricanes, floods, or severe storms. Such interruptions not only compromise the continuity of the dataset but can also lead to distorted analyses and unreliable conclusions about long-term climate patterns. Canchala-Nastar et al. [6] and Gallegos-Cedillo et al. [14] mention that it is common for rain gauge stations to contain periods of missing data. Gyau-Boakye and Schultz [15] highlight that insufficient information over an extended period of time can significantly hinder water supply systems' effective planning, design, and operation. This lack of data can lead to misguided decisions and inefficient resource allocation, ultimately affecting the reliability and sustainability of water services. The scarcity of reliable data is a prevalent challenge in operational hydrology, often hindering the ability to accurately assess and manage water resources. This lack of comprehensive information can impact flood forecasting, water quality monitoring, and the overall understanding of hydrological cycles, making it difficult for professionals to make informed decisions and implement effective strategies [24].

Wangwongchai et al. [27] highlighted the importance of rainfall as a key variable in hydrological studies. They mentioned that developing efficient techniques for estimating missing rainfall data becomes essential to ensure that time series datasets are complete and reliable. Their analysis employed various statistical methods, including Multiple Linear Regression and the Normal Ratio method, alongside cutting-edge artificial intelligence techniques for data imputation. This research focused on multiple rain gauge stations located in the northern regions of Thailand, evaluating the effectiveness of these methods through metrics such as Mean Absolute Error (MAE), which reflects the average absolute differences between estimated and actual data points, Root Mean Square Error ($RMSE$), which quantifies the average error between predicted and observed values, Coefficient of determination (R^2) measures the proportion of the total variance in the observed data that is predictable from the independent variables in the model and Correlation Coefficient (r), which measures the strength and direction of the relationship between variables.

Zulfaqar et al. [21] noted that the lack of rainfall records negatively impacts water management, flood forecasting, and irrigation programs. They developed an imputation study by analyzing twenty-one rain gauge stations in the Johor River Basin, Malaysia. They used daily data and multivariate imputation techniques available in R. They employed MAE, $RMSE$ and R^2 as evaluation metrics. Finally, they emphasized the importance of selecting appropriate imputation methods based on the spatial variability of rainfall.

Burhanuddin et al. [4] emphasize that missing rainfall data can result from observation errors, measuring instrument failures, extreme weather conditions, or human error. They conducted a study in the Malaysian peninsula addressing missing data for 13 rain gauge stations using the Normal Ratio and its variations, yielding satisfactory results.

In several regions of Mexico, the lack of sufficient climatological information poses considerable challenges for effective basin management. It is noteworthy that only about 30% of the annual rainfall is considered renewable [9], which makes accuracy in hydrological balance calculations imperative. This study presents a proposal for estimating missing rainfall data by leveraging correlations between rain gauge stations, utilizing the Normal Ratio method as a foundational approach.

The reminder of the paper is structured as follows: In Sect. 2 we provide the necessary context to support our proposal by describing some previous work and some technical reference documents. Section 3 contains the location and data description. Section 4 presents a baseline method for estimating the missing rainfall values along with our proposal. In Sect. 5 we discuss the results obtained through the applied imputation techniques. Finally, Sect. 6 includes our conclusions.

2 Background

To effectively tackle the missing data problem, it is vital to implement methods and techniques for estimating missing values; this often involves drawing on data from nearby rain gauge stations with similar climatic conditions. By examining the relationships and correlations between an incomplete dataset and those from adjacent rain gauge stations, researchers can apply statistical interpolation methods. Examples of such methods are linear interpolation, and more advanced approaches like time series forecasting and machine learning algorithms.

These methods enable climatologists to generate estimated values that reflect the underlying patterns of the dataset, ensuring continuity and integrity in their analyses. Moreover, improving the accuracy of these estimations is crucial for maintaining reliable climate models, which are essential for predicting future climate scenarios and informing policymakers about potential environmental impacts. Addressing the challenge of missing data enhances the robustness of climatological research and contributes to a more comprehensive understanding of the Earth's changing climate.

In their study, Paulhus and Kohler [19] introduced the Normal Ratio (NR) method as an alternative approach to estimating precipitation at a target station, referred to as Station x. This method involves weighting the precipitation data collected from nearby index stations strategically selected based on their proximity and the climatic characteristics of the area. The weighting is determined by comparing the target station's normal annual precipitation with that of the index stations, allowing for a more tailored and accurate estimate. The researchers found that the NR method significantly outperformed the conventional three-station average method commonly used at that time, particularly in regions characterized by challenging mountainous terrain. Across all five sets of stations examined, the NR method consistently produced more reliable results. Notably, the average estimation error when using the NR method was less than half that of the three-station average method in four out of the five tested sets, illustrating its superior accuracy. Furthermore, the study explored the use of ratios based on normal monthly precipitation. However, these tests revealed that the results obtained were not significantly different from those using the annual ratios, underscoring the stability of the NR approach. As a result of these findings, the U.S. Weather Bureau adopted the NR method for operational use whenever the normal annual precipitation at any index station varied by more than 10% from that of the target station. In cases where the difference fell within 10%, the more straightforward three-station average method could still be applied, providing a practical solution tailored to specific circumstances.

Choosing the proper method for addressing missing data presents quite a challenge, as researchers have developed and refined various techniques over the years. One notable study by Young [32] sought to fill in gaps in a monthly precipitation time series from 1860 to 1990, specifically for rain gauge stations in Arizona, New Mexico, and California. This research concentrated on three imputation methods: Multiple Discriminant Analysis, Multiple Linear Regression, and the NR method. In this comprehensive analysis, he found that the NR method is particularly suitable in situations where a substantial volume of data is missing. This method allows for greater flexibility and tolerance regarding the imputation process, making it a practical choice for researchers dealing with incomplete datasets in climatological studies. In this comprehensive analysis, he found the NR method suitable, notably when a substantial volume of data is missing. This method allows for greater flexibility and tolerance regarding the imputation process, making it a practical choice for researchers dealing with incomplete datasets in climatological studies. This study significantly advanced the understanding of station data by integrating the correlation between different measurement stations into the NR method. This integration suggests that more straightforward analytical techniques, such as calculating the arithmetic average or applying the NR, can adequately interpret data when stations exhibit a high degree of correlation. On the other hand, Young's modification presents a more nuanced approach, instrumental in situations where the correlation between stations fluctuates widely, allowing for more accurate and reliable results in complex datasets.

Eischeid et al. in [13] also highlighted the importance of considering the correlation between rain gauge stations in the NR method. Their research discovered that the variability in the locations and distributions of rain gauges significantly influences the degree of correlation between these stations. This finding emphasizes that differences in geographical factors can affect rainfall measurements and the associated data interpretations.

Alfaro and Pacheco [1] undertook a comprehensive study to evaluate the reliability of various simple methods for filling in missing annual rainfall data. The methods examined were Simple Regression, Multiple Regression, and the NR method. This study emphasizes the implementation of the NR method, particularly in areas where rainfall data is incomplete or missing.

Yozgatligil et al. in [33] conducted an in-depth study to evaluate the accuracy, robustness, and efficiency of various imputation methods for addressing missing meteorological data. Their analysis focused on extensive monthly time series data related to precipitation and temperature in Turkey, spanning the years from 1965 to 2006. In their research, the authors employed a range of straightforward imputation methods, including the Simple Arithmetic Average, NR, and Correlation Weighted NR. The Correlation Weighted approach computes a weighted average using data from neighboring rain gauge stations. Their study incorporated data from seven distinct geographic regions across Turkey, comprehensively evaluating how imputation techniques perform in different climatic contexts. Additionally, they introduced the concept of using correlation as a valuable alternative metric for assessing the effectiveness of missing data imputation, broadening the analytical tools available for researchers in the field.

Amin Burhanuddin et al. [2] developed a refined approach that modifies the NR method by incorporating elements from the geographic coordinate method. Focusing their study on the diverse climate of Peninsular Malaysia, the researchers analyzed an extensive dataset spanning 36 years, from 1975 to 2010, collected from 18 strategically located rain gauge stations throughout the region. They selected three key target stations for a detailed investigation: Batu Kurau, Sg. Bernam and Genting Klang also identify their neighboring stations for comparative analysis. The study's primary aim is to propose enhancements to the techniques for estimating missing precipitation data. By merging the robust NR method with geographic coordinate elements, the researchers seek to improve the accuracy and reliability of precipitation data. They highlighed the critical role of high-quality precipitation data in ensuring the accuracy of hydrological and meteorological analyses, and they pointed out that poor data quality can significantly skew results, leading to misleading conclusions.

Another study by Herrera-Olivia et al. [16] focused on filling in missing data from annual rainfall records collected at 13 rain gauge stations within the Guadalupe Basin in Baja California, Mexico. The researchers utilized the Linear Regression method to develop a complete and dependable database that would support future hydrological, agronomic, and climatological studies in this vital region. To evaluate the effectiveness of their statistical approach, they calculated the Linear Correlation Coefficient, which indicates the strength of relationships between variables. Remarkably, the study yielded consistently high correlation values, with an impressive average of 0.85, demonstrating the robustness and reliability of their data analysis.

One study that considers correlation was conducted by Sanusi et al. [23]; they explored the significance of correlation in estimating missing values within monthly precipitation data. They comprehensively compared several estimation methods, including the Arithmetic Mean (AM), NR, Modified NR, and Correlation Coefficient Weighting methods. Their research aimed to determine the most effective estimation technique for addressing gaps in precipitation data collected from four rain gauge stations situated in the diverse climate of Makassar City, Indonesia. To rigorously assess the performance of each method, the researchers employed a range of statistical indicators. These included r, $RMSE$, and MAE.

Abu Romman et al. [20] conducted an extensive study to compare ten distinct methods for imputing missing monthly precipitation data in a dry Mediterranean region. This area, characterized by its unique climate and water scarcity challenges, requires precise precipitation data for effective hydrological modeling and efficient water resource management. The researchers sought to identify the most accurate method for filling in these crucial gaps in data. To achieve this, they employed several straightforward techniques, including the AM, NR, and Inverse Distance Weighting (IDW), assessing each method's effectiveness in providing reliable estimates for the missing precipitation values.

Chinasho et al. [7] evaluated seven gap-filling techniques on daily precipitation data at five rain gauge stations in and around Wolaita Zone in southern Ethiopia. Their main objective was to determinate the most accurate technique for estimating missing values in daily precipitation time series. They used methods such as Simple AM (SAM), NR, Correlation Coefficient Weighting (CCW) among others. The data used in the study were daily rainfall and observed maximum and minimum temperatures from five stations for the periods (1987–2017). They mention that SAM, NR, and CCW methods did not differ statistically, and they used MAE, $RMSE$, and r as evaluation criteria.

Valencia-Gonzales et al. [26] developed a study focused on the estimation and verification of methods to complete missing daily data in climatological series records, specifically in the rice production area in the department of Valle del Cauca, Colombia, their main objective is to determine the method with the lowest error to fill in the missing data of variables such as precipitation, sunlight, evaporation, maximum temperature, minimum temperature, and relative humidity. Nine rain gauge stations with distances no greater than 50 km and altitudinal differences of less than 750 m^3 were analyzed. Three main methods were evaluated: NR, Fourier Series (SF), and IDW. The methods were evaluated using the statistical indices of $RMSE$ and R^2. They point out that the NR method performed well in imputing missing data.

According to the previously mentioned works, the NR method is a simple technique with good results for filling in rainfall data when incomplete or missing. Although the method has had some variants, in this case, we highlight the importance of the correlation between the stations as a factor that can improve the quality of the imputed data and thus improve the quality of the time series.

2.1 World Meteorologic Organization Recomendations

The World Meteorological Organization (WMO) is a United Nations organization that is responsible for coordinating and promoting international cooperation in meteorology, climatology, operational hydrology, and related geophysical sciences. Its main objective is to improve the collection and use of meteorological and climatic data for the security and well-being of humanity and its main functions: Coordinate global climate and atmospheric observation networks; Develop standards and methodologies in meteorology and climatology; Provide alerts on extreme phenomena such as hurricanes, heat waves, and droughts; Support research on climate change and natural disasters; Promote open access to climate and meteorological data [31].

The WMO provides a diverse and extensive collection of documents related to climatology, meteorological data processing, and analysis techniques. These resources include guides, guidelines, manuals, and technical reports, which serve as valuable references for researchers, meteorologists, and climate scientists worldwide. For the development of our study, we have consulted several of these documents, which offer methodological frameworks, best practices, and standardized procedures for handling and analyzing climatological data, such as the WMO Climatological Practices Guide [18, 30], which mentions that the presence of missing data can significantly influence the accuracy of climatological normals; if they are derived from incomplete datasets, the results can become distorted, leading to misleading conclusions. These normals serve two essential functions: they act as a benchmark against which current observations can be compared and facilitate the identification of anomalies that may arise within various data series. The guide recommends the NR method for estimating missing data. It also highlights the importance of correlation, which examines the degree of similarity in behavior between two or more data series recorded at different rain gauge stations. This correlation can reveal valuable information about broader climatological patterns and trends, ultimately improving the overall understanding of climate dynamics. The guide recommends the use of cross-validation as a verification tool when applying methods for missing data estimation.

The WMO guidelines for calculating climatic normals [29] emphasize the importance of utilizing 30-year periods as a standard reference timeframe. This long-term approach aims to enable effective comparisons of climatic records collected from different regions around the globe. Climatic normals are defined as systematic measurements calculated over these extended periods, with each normal derived from a minimum of three decades of data. They can cover an array of climatic parameters, including temperature, precipitation, and vapor pressure. In this study, we will concentrate specifically on rainfall.

Trewin [25] describes the history of climatological normals, several of the metrics used for their evaluation, and a critical perspective on their applicability in the context of climate change. Another antecedent of the calculation of the normals is the document WMO 341 Calculation of Monthly and Annual 30-Year Standard Normals [28], a document prepared by a group of experts in Washington D.C. in March 1989. Its purpose was to establish general procedures for the calculation of the standard normals of 30 years of the WMO (such as the period 1961–1990 and later periods), as well as the provisional normals, and to suggest other climatic descriptors. This document mentions the criterion of rule 3/5 to determine the validity of the monthly data and mentions that the month's records should not be considered for the normal calculation if it presents three consecutive missing days or five days in total in the whole month. We took this criterion for our study.

3 Location and Data Description

3.1 Study Area

Mexico boasts an extensive network of 633,000 kilometers of water currents, which play a vital role in the country's ecological and hydrological systems. Among these, 50 major river basins stand out, encompassing 65% of Mexico's total surface area. These basins not only contribute to the nation's water supply but also support diverse ecosystems and numerous communities throughout the region [22].Across the landscape, there are 731 distinct hydrological basins, each playing a crucial role in managing water resources. These basins are organized into 13 administrative regions, which oversee the sustainable coordination of water supplies to meet the diverse needs of various sectors that depend on this vital resource and 37 hydrological regions. Each region ensures that water is effectively allocated and preserved for agricultural, industrial, domestic, and ecological purposes [8].

The study area is located within the hydrological administrative region IX, referred to as Gulf North, in the country's northeastern part. This region comprises the states of San Luis Potosí, Tamaulipas, Veracruz, Querétaro, Guanajuato, and Hidalgo. It showcases a diverse landscape characterized by mountainous areas that rise sharply against the skyline and serene coastal plains that stretch toward the horizon. Combining these geographical features creates a rich and varied environment for exploration and study.

The administrative region is structured around two primary basin councils: San Fernando-Soto La Marina and Panuco. Our research is focused on the Panuco basin, particularly within the Chicayan River basin (Fig. 1). This area is classified as part of hydrological region No. 26, which encompasses the Panuco River, a significant waterway that flows through the lush northern landscapes of Veracruz state. The Chicayan River itself meanders through diverse ecosystems, making it a crucial area for our study of hydrology and the regional environment.

The Chicayan River basin encompasses a substantial drainage area of 3488 km^2 [12], which is divided into two main sections: Rio Chicayan 1 to the south and Rio Chicayan 2 to the north. This division is marked by a red line in Fig. 1, visually representing the basin's layout and structure.

The basin predominantly covers the localities of Pánuco, Ozuluama de Mascareñas, Tempoal, and Tantoyuca. Additionally, it includes the smaller areas of municipalities of Tampico Alto, El Higo, Ixcatepec, Chontla, and Citlatepetl. These communities broadly are engaged in agricultural and livestock activities. With agriculture as the cornerstone of their economies, these municipalities exhibit a limited industrial presence. This is primarily due to the substantial water requirements for farming and livestock management, emphasizing the importance of water resources in sustaining their livelihoods and rural traditions.

This basin contains one of the eleven irrigation districts within the Gulf North administrative hydrological region. The irrigation district, designated 092 Río Pánuco Unidad Chicayán, encompasses an extensive 21,415 ha, with 21,256 ha (97% of the

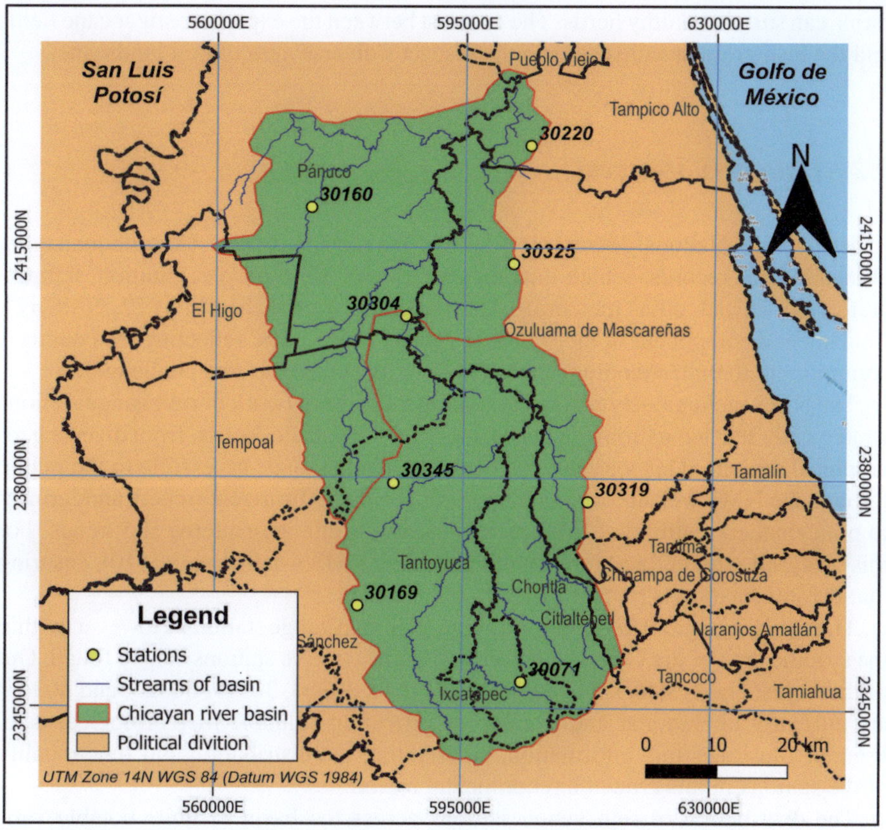

Fig. 1 Localization of rain gauge stations in Chicayan river basin northern Veracruz

total) designated as irrigable land. This irrigable expanse is meticulously managed by four civil associations, each dedicated to ensuring a reliable water supply for 1746 users. The district benefits from two primary water sources that are crucial for maintaining its irrigation needs: the Paso de Piedras dam and the La Tortuga pumping plant [9].

The Paso de Piedras dam, a vital component of the district's water infrastructure, boasts a substantial storage capacity of 180.05 cubic hectometers at its Maximum Ordinary Level (MOL) [11]. This large volume of water is essential for supporting agricultural activities, sustaining the livelihoods of local farmers, and contributing to the region's overall agricultural productivity.

Most of the irrigation district is devoted to agricultural activities, with a particular emphasis on sugar cane cultivation. This crop thrives in this region's fertile soil and abundant water supply. In addition, a smaller portion of the area is set aside for growing grass, which serves as essential forage for livestock, ensuring that local farms can sustain healthy herds. The contrast between the expansive sugar cane fields and the lush grass pastures adds to the district's diverse agricultural landscape.

3.2 Data and Sources

In Mexico, the National Meteorological Service (NMS) is entrusted with managing climatological records, which encompass essential data on precipitation, temperature, and various other meteorological variables. Established in 1877, NMS is a fundamental component of the National Water Commission, reflecting the country's commitment to understanding and addressing its climate-related challenges.

The NMS meticulously manages a comprehensive network of rain gauge stations strategically located nationwide, ensuring accurate data collection from diverse geographical regions. This wealth of information is then made accessible to the public through the NMS website, empowering users (ranging from researchers and farmers to policymakers) with vital insights into Mexico's climate patterns and trends. For this study, all rainfall data were sourced from the NMS's online portal [10], ensuring that the information is both accessible and up to date.

The northern region of Veracruz is home to 54 rain gauge stations. However, within the specific area of the Chicayan River basin, only twelve stations can be found. Out of these, just eight (30071, 30160, 30169, 30220, 30304, 30319, 30325, and 30345) are currently operational and capable of providing valuable meteorological data. Figure 1 illustrates this information, highlighting the distribution and functionality of the stations in this crucial environmental region.

The distribution of rain gauge stations across the basin exhibits notable variability, reflecting the diverse topography of the region. Station 30220 Quebrache is positioned in the northern part of the basin, specifically at a modest elevation of 16 m above sea level (masl). This area represents the lower elevation section of the basin, characterized by lush landscapes that thrive in the abundant moisture. In contrast, station 30071 Ixcatepec is found in the southern part of the basin, perched at a significant elevation of 173 masl, and considered the highest station in the analysis. The terrain is generally steeper, and the vegetation tends to be different due to the higher altitude. Table 1 provides a comprehensive overview of rain gauge stations.

Additionally, there are several stations at intermediate elevations, such as station 30345 Rancho Santa Rosa, which sits at an elevation of 98 masl, and station 30169 Tantoyuca, located at 77 masl. These intermediate sites contribute to a comprehensive understanding of precipitation patterns across the basin, each reflecting unique ecological conditions shaped by their respective heights.

Rain gauge stations are identified by a unique code or ID that reflects their geographical location within the republic. This code consists of a number indicating the state and three digits that signify the specific station number. The states are arranged alphabetically; for Veracruz, this designated number is 30. Consequently, all rain gauge stations located in Veracruz are identifiable by codes that begin with this number.

Table 1 Description of rain gauge stations

Station	Station name	Lat.	Long.	Elev.	Period
30071	Ixcatepec	21.24	−98.00	173	1961–2023
30160	Santa Trinidad	21.89	−98.29	18	1961–2019
30169	Tantoyuca	21.35	−98.23	77	1925–2023
30220	Quebrache	21.98	−97.99	16	1972–2023
30304	Presa Paso de Piedras	21.74	−98.16	40	1976–2019
30319	El Mercado	21.49	−97.91	72	1978–2024
30325	La Puente	21.81	−98.01	25	1978–2023
30345	Rancho Santa Rosa	21.51	−98.18	98	1979–2018

This table details essential properties such as the number of each rain gauge station, the name of each station, geographic coordinates (latitude and longitude) in decimal degrees, elevation above sea level in meters (masl), and the period of years where the rainfall was recorded

We analyzed stations 30071 Ixcatepec and 30160 Santa Trinidad, which began their operations in 1961. The next station in chronological order is station 30220 Quebrache, which started recording rainfall in 1972. In 1978, we welcomed two more stations into the network: 30319 El Mercado and 30325 La Puente. Lastly, the most recent addition, 30345 Rancho Santa Rosa, began recording in 1979. While the data collection periods for these stations are extensive and diverse, a significant amount of data remains missing, which could impact the comprehensiveness of our analysis.

In our detailed analysis of daily rainfall records, we began by assessing the percentage of missing data for each rain gauge station, considering the total days of recording. This information is visually represented in Fig. 2, where we can observe a monthly breakdown of the data. Notably, the pluviometer 30169 in Tantoyuca stands out, showing the highest percentage of missing data, hovering around 40% from January through August. In September and October, this percentage experiences a slight decline to 38%. However, as we approach the end of the year, the missing data percentage climbs back to nearly 40% for both November and December.

Most stations have a percentage of missing data that remains below 30%. For example, the rain gauge station at Rancho Santa Rosa experiences notable variability: in February, March, and April, it records around 20% of missing data, indicating some inconsistencies in data collection during those periods. This percentage swells to approximately 25% in July before gradually decreasing in the following months, ultimately dropping below 20% after September, suggesting improved data reliability as the year progresses.

Other stations, such as La Puente, display a more stable pattern with missing values fluctuating between 10% and 20%. La Puente hits a peak of nearly 20% only during May and July. Another station, Presa Paso de Piedras, shows similar trends, reaching close to 20% in April and July as well. In contrast, the Santa Trinidad station maintains lower percentages throughout most of the year. However, it experiences a

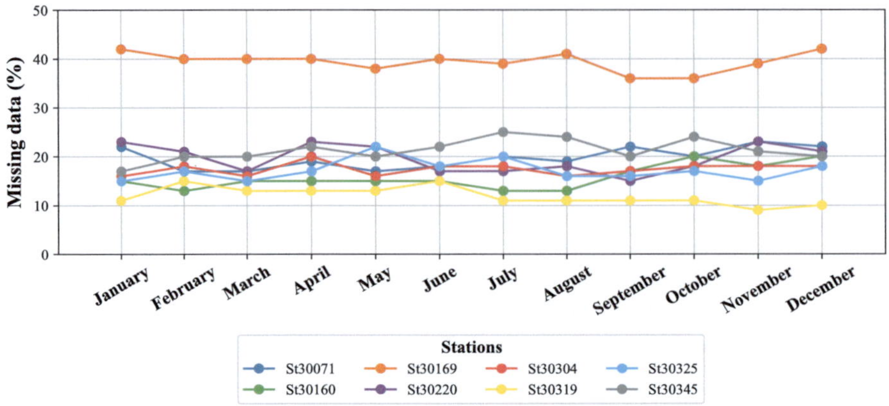

Fig. 2 Monthly percentage of missing data at rain gauge stations

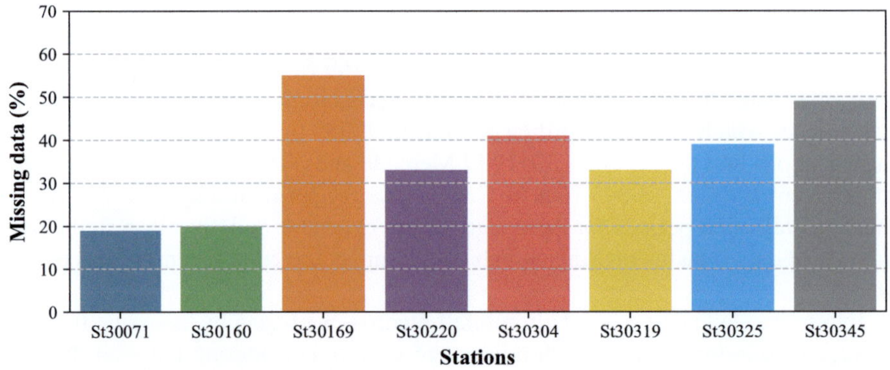

Fig. 3 Total percentage of missing data at rain gauge stations

notable increase in October, where missing data spikes to 20% and then stabilizes at that level through November and December, possibly indicating challenges in data collection during the latter part of the year.

The standout station regarding data completeness is El Mercado, which consistently has the lowest percentages of missing data. It records approximately 15% of missing values from May to June. However, it shines with exceptionally high data reliability during January, July, August, September, October, November, and December, where missing data hovers around 10%. This reliable performance lends confidence to the data provided by this station throughout the year.

Figure 3 illustrates the percentage of missing data for each rain gauge station throughout its entire recording period, focusing exclusively on rainfall data. This analysis adheres to the World Meteorological Organization (WMO) [28] criterion for missing data, which stipulates that if there are gaps of more than three consecutive days or a total of five days a month, those records should be classified as lost. For our calculations, we require 26 days of valid records for months with 31 days, 25 days for months with 30 days, and 23 days for February.

Based on this criterion, the findings reveal that station 30169 Tantoyuca experiences a strikingly high percentage of missing data, exceeding 50%. Following closely is station 30345 Rancho Santa Rosa, which shows approximately 49% missing data. Station 30304 Presa Paso de Piedras also reflects a notable extent of missing data at 40%. Meanwhile, station 30220 registers nearly 39% of its records as missing.

Stations 30319 Quebrache and 30071 El Mercado each report missing data around the 30% mark. In contrast, the stations with the least missing data (30160 Ixcatepec and 30160 Santa Trinidad) show significantly lower percentages at around 20%. These readings highlight the varying reliability of rainfall records across different stations.

4 Estimation of Missing Rainfall Values

4.1 Normal Ratio (NR)

The Normal Ratio is one of the procedures to estimate missing data; in this method, we use the mean rainfall of stations to obtain rainfall weights and, with them, impute values. The ratio is established between the station under analysis and all the other stations. The impute value results from the sum of the weights multiplied by rainfall of the respective neighboring station on the analysis date divided by the total number of stations analyzed. According to Campos-Aranda (1998) [5] and Aparicio, (1992) [3], it can be described with Eq. 1

$$RF_x = \frac{1}{n} \cdot \left[\frac{\overline{RF_x}}{\overline{RF_1}} RF_1 + \frac{\overline{RF_x}}{\overline{RF_2}} RF_2 + \frac{\overline{RF_x}}{\overline{RF_3}} RF_3 + \cdots + \frac{\overline{RF_x}}{\overline{RF_n}} RF_n \right] \tag{1}$$

As we can see, NR is a summation, so we can simplify Eq. 1 according to McCuen (1998) [17] as follows:

$$RF_x = \frac{1}{n} \cdot \sum_{i=1}^{n} \left(\frac{\overline{RF_x}}{\overline{RF_i}} RF_i \right) \tag{2}$$

where RF_x is the missing value of monthly rainfall; RF_i is the value of monthly rainfall at station i; $\overline{RF_x}$ is the average of rainfall at station in which value is missing; $\overline{RF_i}$ is the average of rainfall at station i and n is the number of stations analyzed [1, 16].

4.2 Proposal Normal Ratio with Additive Correlation (NRAC)

In this case, we propose a technique that takes the NR as a starting point and then involves the correlation between the station under analysis and the neighboring stations. In other words, it involves the weightings between rainfall averages and the correlations that could occur with the same stations. Our proposal is described in Eq. 3.

$$RF_x = \frac{1}{div} \cdot \sum_{i=1}^{n} \left(\frac{\overline{RF_x}}{\overline{RF_i}} RF_i \cdot C_{ix} \right) \tag{3}$$

where:

$$div = \sum_{i=1}^{n} C_{ix} | C_{ix} = \begin{cases} C_{ix} & if\ RF_i \neq 0 \\ 0 & otherwise \end{cases}$$

where RF_x is the missing value of monthly rainfall; RF_i is the value of monthly rainfall at station i; $\overline{RF_x}$ is the average of rainfall at station in which value is missing; $\overline{RF_i}$ is the average of rainfall at station i; C_{ix} is the coefficient of correlation between station x with the missing value and the station i; and n is the number of stations analyzed.

4.3 Experimental Design

To evaluate the effectiveness of the imputation techniques, we analyzed rainfall records gathered from various rain gauge stations, as detailed in Sects. 4.1 and 4.2. This dataset underwent a comprehensive processing and analytical approach to extract pertinent information and rigorously assess the overall quality of the data. The steps involved in this analytical process are visually depicted in Fig. 4.

Our study was conducted through a comprehensive process consisting of four steps. Each step is detailed below:

Step 1: The analysis began with thoroughly examining daily rainfall records collected from various selected stations. We meticulously processed this information by the standards set by the World Meteorological Organization, allowing us to derive finely tuned monthly rainfall values suitable for further analysis. Our focus was narrowed to the time frame from 2000 to 2014, during which we observed convergence in records across all stations. This careful selection yielded a comprehensive data matrix encompassing monthly rainfall measurements from eight rain gauge stations. It is important to note that the monthly values are derived from the cumulative totals of daily rainfall recordings. Step 2: Once we established the matrix of monthly rainfall records, we determined the most effective method for testing our imputation

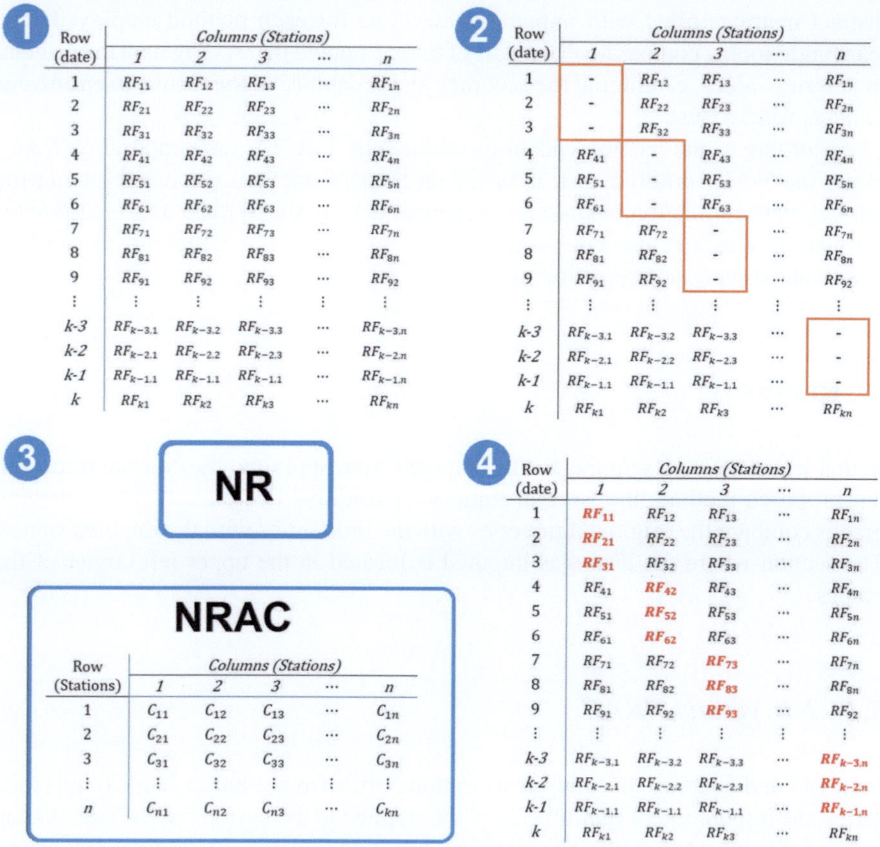

Fig. 4 Experimental design diagram

techniques. Following the recommendations of the WMO, we applied eight-fold cross-validation, and based on the amount of available data, we opted to remove ten data points per rain gauge station in a staggered manner, equivalent to ten months of data for each station. Specific sections of the records were systematically removed to ensure a balanced and representative validation. This strategic omission ensured that each station had data gaps while allowing other stations in the same period to maintain usable information for the imputation process. The removed areas were designated as "NaN" (not a number), marking them clearly for further analysis. Step 3: In the next phase, we implemented the imputation techniques described in Sect. 4, focusing specifically on the NR and the $NRAC$. For the $NRAC$ technique, we initiated the process by calculating a detailed correlation matrix that illustrated the relationships between the different stations. This essential step provided a foundation for applying both imputation techniques effectively, allowing the data to be filled in accurately. Step 4: Finally, after applying the chosen imputation techniques, we generated two distinct matrices filled with imputed values, one for each method employed. The team undertook a comparative analysis of these imputed datasets against the original time series values, examining the accuracy and reliability of the results to ensure our findings were robust.

According to the recommendations outlined in Sect. 2.1, our proposal ($NRAC$) meets the WMO criteria, as it is based on the NR method (included among the missing data estimation techniques recommended by the WMO) and incorporates correlation, which is also considered by the organization. Additionally, it employs cross-validation as a verification tool.

5 Results

In this section, we present the results obtained from applying: the Normal Ratio and our proposed method, the Normal Ratio with Additive Correlation. The following graphs compare the original time series with the time series with the imputed values. The section where the data was imputed is located in the upper left corner of the graphs.

5.1 NR Versus NRAC

Figures 5 and 6 shows the analysis of station 30071 from 2000 to 2014. In this time series, the imputation of rainfall values was applied to the period 2000–2002. As can be seen, the imputed values shown with an orange line are greater than real values (blue line).

For simplicity only the period where the value filling was performed is shown the following figures. As can be seen in Fig. 7, the imputed values tend to be higher than the real values. For station 30160, the period during which the imputation of values was carried out was from 2002 to 2004 (Fig. 8). In this case, the imputed values were lower than the real values.

In the case of station 30169, the period during which values were computationally calculated was from 2003 to 2005 (Fig. 9). In this case, the imputed values were higher than the real values.

The analysis period for station 30220 was from 2005 to 2007, which shows a correspondence between the imputed data and the real values (Fig. 10). As shown in Fig. 11, the data imputed at station 30304 were higher than the real values, and the analysis period was from 2006 to 2008. For station 30319, the imputation results show values larger than the real values (Fig. 12). Here, we can see an increase in the imputed values with respect to the original time series.

As we can see in the case of station 30325 (Fig. 13), the behavior of imputed values tends to be greater than the real values; it was analyzed from 2008 to 2011. There is a more remarkable similarity between 2009 and 2010, unlike 2008–2009.

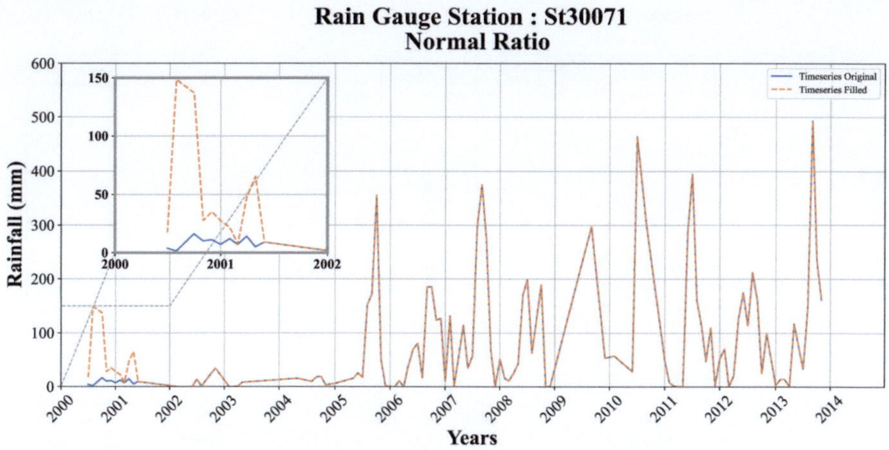

Fig. 5 Comparison of timeseries in Station 30071

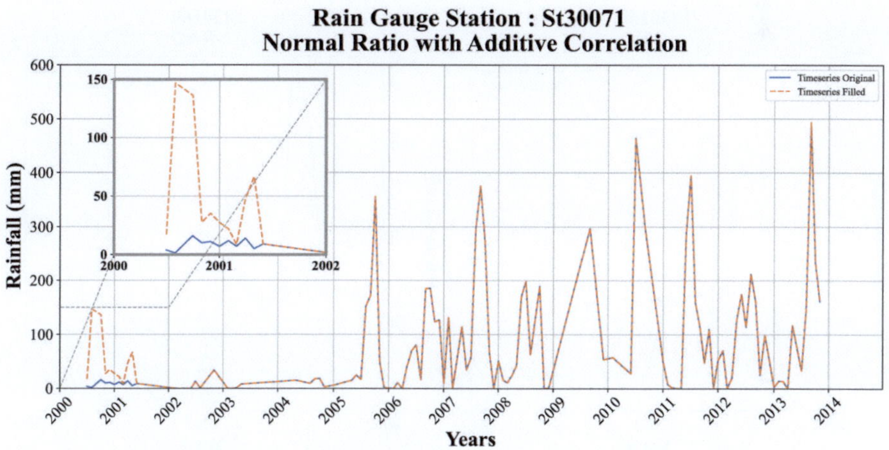

Fig. 6 Comparison of timeseries in Station 30071

Finally, in Station 30345, a good correspondence is observed between the two sets of values(Fig. 14). For this case, we analyzed the period from 2010 to 2012, where similarity is presented between the imputed values and the original values recorded.

According to the results, the imputed values are higher than the original ones in some cases and lower in others. This variation appears to be related to the absence of rainfall (distinct from missing data) in one or more neighboring stations. When rainfall values at neighboring stations are zero, the imputed values tend to be lower than the originals; conversely, when those stations report rainfall, the imputed values tend to be higher.

In our proposal, the similarity of the imputed and real values depends on the correlation between stations. As can be seen in Fig. 15, most of them are above

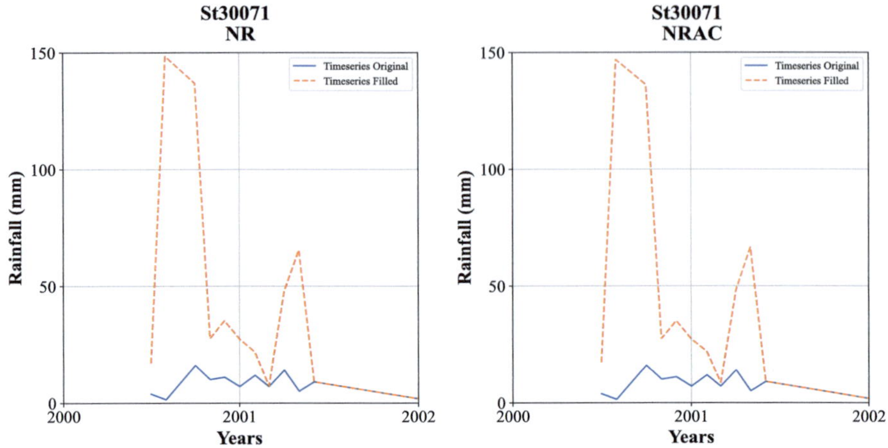

Fig. 7 Comparison of timeseries in Station 30071

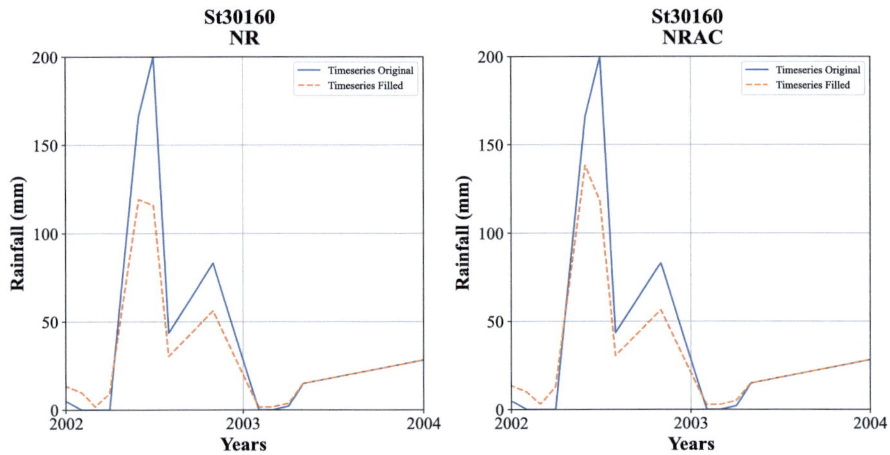

Fig. 8 Comparison of timeseries in Station 30160

0.5. The lowest correlation was observed between stations 30169 and 30325, with a coefficient of 0.62, while the highest occurred between stations 30160 and 30304, with a coefficient of 0.87. Station 30345 showed consistently strong correlations, exceeding 0.8 with stations 30319, 30304, 30169, and 30325 and ranging from 0.7 to 0.8 with stations 30160, 30220, and 30071. In contrast, station 30169 had the weakest correlations overall—above 0.8 only with station 30345, between 0.7 and 0.8 with stations 30071, 30160, and 30304, and from 0.6 to 0.7 with stations 30319, 30220, and 30325. Overall, the stations exhibit high correlation values.

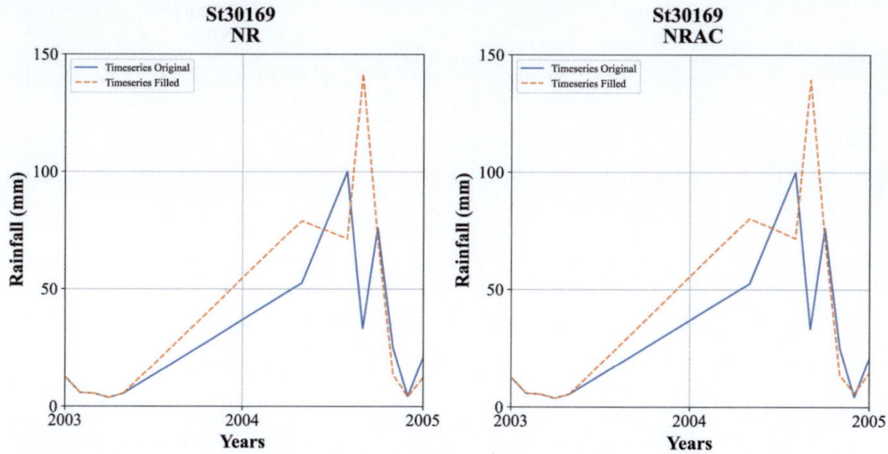

Fig. 9 Comparison of timeseries in Station 30169

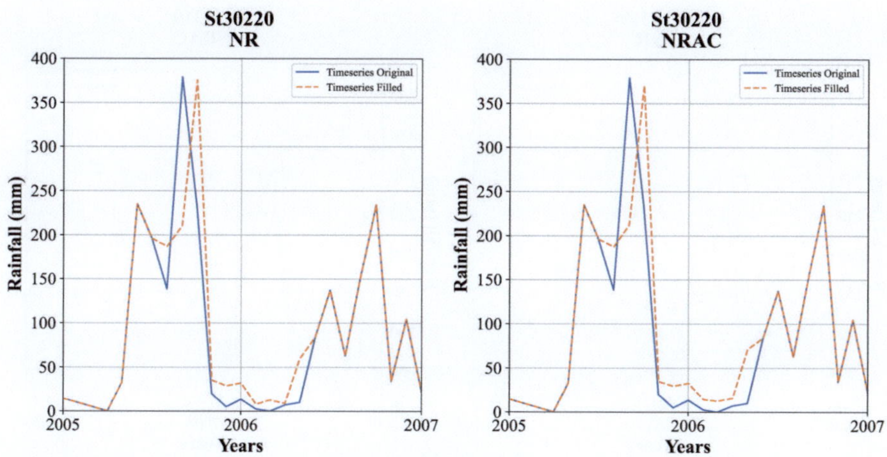

Fig. 10 Comparison of timeseries in Station 30220

5.2 Evaluation Metrics

Each metric concerning the original values was obtained once the analysis was carried out using the two imputation techniques. As we can see in Table 2, the coefficient of determination (R^2) values for both NR and $NRAC$ for all stations are above 0.80, with the highest being stations 30160, 30169 and 30345 with 0.99, followed by 30304 with 0.98, 30071 and 30220 with 0.97, 30319 with 0.96 and finally 30325 with 0.88. These results indicate a good fit with the original values for both methods, with identical R^2 values across all rain gauge stations. For $RMSE$ we observed that there are some differences between the NR and $NRAC$. The values for stations

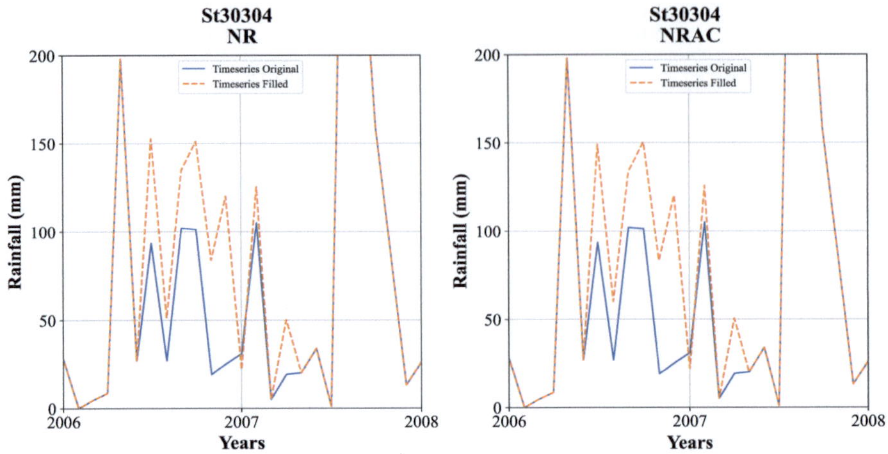

Fig. 11 Comparison of timeseries in Station 30304

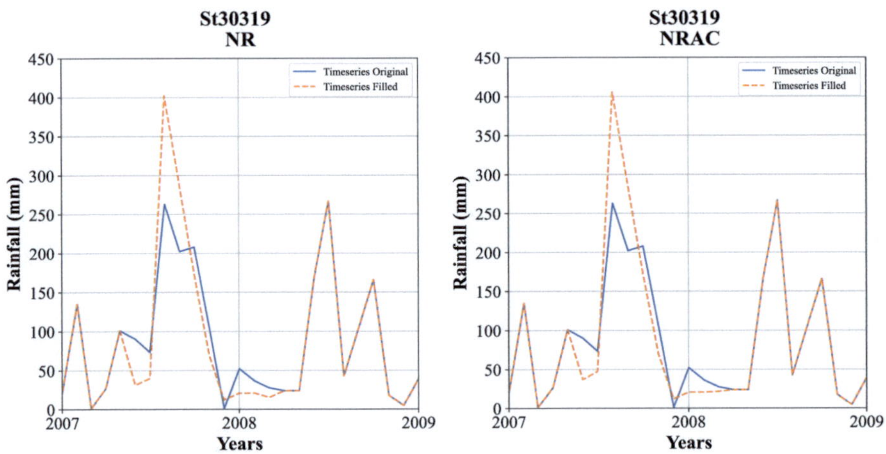

Fig. 12 Comparison of timeseries in Station 30319

30325 and 30345 were lower for NR, indicating smaller prediction errors, while for stations 30071, 30160, 30169, 30220, 30304 and 30319 $NRAC$ showed better performance. In general, the differences between both methods were small. For MAE, NR performs better for stations 30220, 30304, 30325 and 30345 with smaller errors; $NRAC$ showed smaller errors for stations 30160, 30169, and 30319. Station 30071 showed the same value in both cases. The correlation coefficients (r) for both imputation methods were above 0.90, indicating an excellent correspondence between the imputed and original values. Rain gauge station 30160 showed better performance with $NRAC$, while all other stations showed identical correlation values for both NR and $NRAC$.

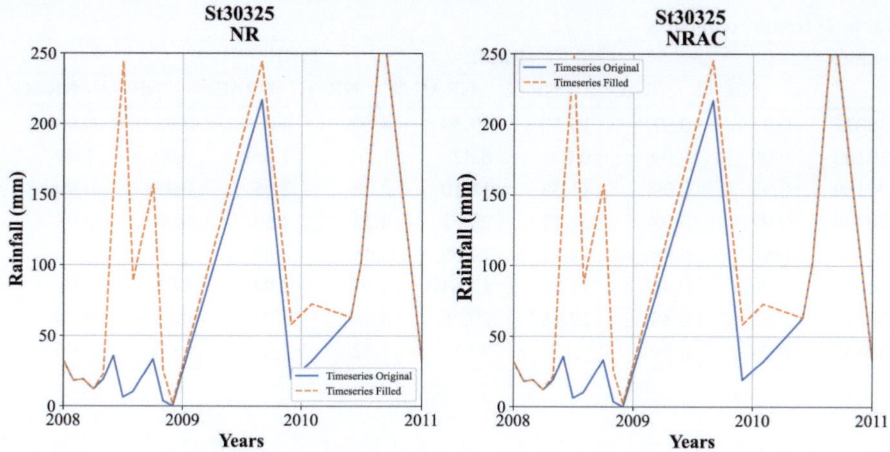

Fig. 13 Comparison of timeseries in Station 30325

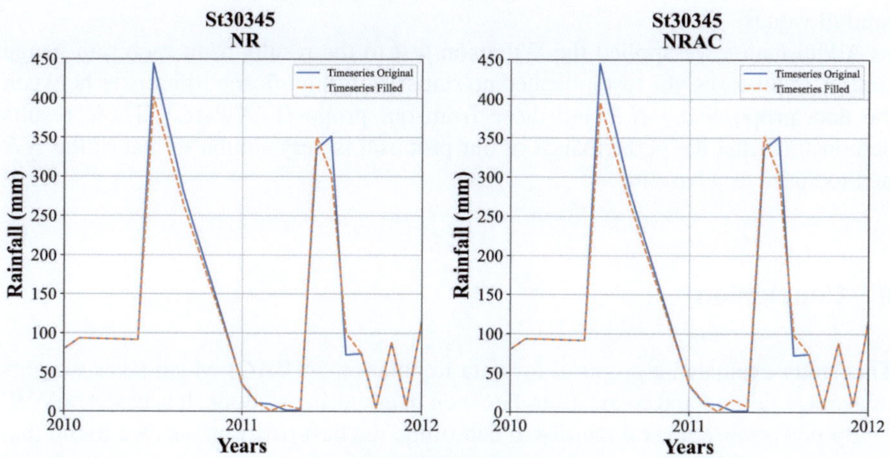

Fig. 14 Comparison of timeseries in Station 30345

Fig. 15 Matrix of rainfall correlation between stations

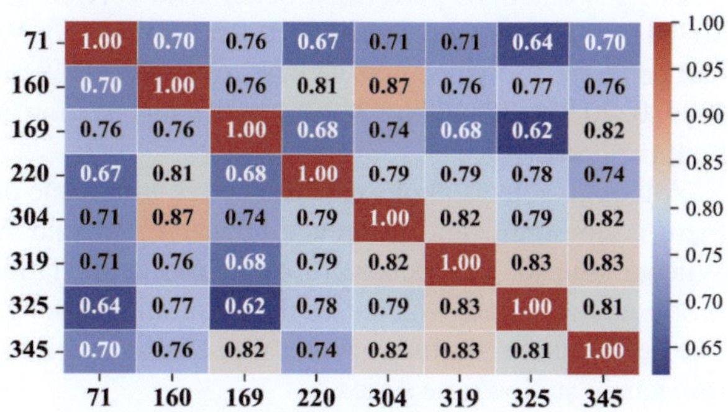

Table 2 Results of metrics

Station	R^2		RMSE		MAE		r	
	NR	NRAC	NR	NRAC	NR	NRAC	NR	NRAC
30071	0.97	0.97	19.50	**19.39**	4.00	4.00	0.98	0.98
30160	0.99	0.99	9.63	**8.82**	1.76	**1.65**	0.99	**1.00**
30169	0.99	0.99	11.33	**11.10**	2.08	**2.05**	0.99	0.99
30220	0.97	0.97	22.17	**22.11**	**4.35**	4.53	0.98	0.98
30304	0.98	0.98	14.12	**14.08**	**3.45**	3.48	0.99	0.99
30319	0.96	0.96	17.70	**17.61**	4.07	**3.94**	0.98	0.98
30325	0.88	0.88	**29.41**	29.76	**6.20**	6.25	0.94	0.94
30345	0.99	0.99	**7.12**	7.51	**1.52**	1.66	1.00	1.00

The metrics values obtained with our proposal ($NRAC$) and the NR method showed small differences, indicating that $NRAC$ provides a good fit for estimating rainfall values.

Additionally, we applied the Wilcoxon test to the results from each rain gauge station. In all cases, the test indicated no statistically significant differences between the data proposed by NR and those from our proposal ($NRAC$). These results demonstrate that the performance of our proposal is very similar to that of the NR method used as a baseline.

6 Conclusion

This study evaluated a proposal for data imputation ($NRAC$) which takes weights of average rainfall and correlations between neighboring stations. It is based on NR.

By performing several metrics to determine the best performance, we found that for the case of R^2, both techniques obtained the same values, all above 0.88. For r, only station 30160 had a different behavior, all of which were above 0.94. In the case $RMSE$, six stations performed better with $NRAC$ (30071, 30160, 30169, 30220, 30304 and 30319), leaving (30325 and 30345) to NR. For MAE, three stations had better performance with $NRAC$ (30160, 30169 and 30319), and four with NR (30220, 30304, 30325 and 30345), rain gauge station 30071 had the same values for both methods.

Although some stations perform better with some of the metrics in several cases, there is no significant difference between NR and $NRAC$ (as confirmed by the Wilcoxon tests). The values are very close in all stations, so we considered that the proposed technique, which uses the correlation among stations, has a state of the art like performance. As future work, we consider applying $NRAC$ to a more extensive rain gauge station time series.

AI Use Statement: Some portions of this manuscript were enhanced with the assistance of generative AI tools. The authors carefully reviewed, edited and approved all IA-generated content. The authors take full responsibility for the accuracy and integrity of the entire chapter.

References

1. Alfaro, R., Pacheco, R.: Aplicación de algunos métodos de relleno a series anuales de lluvia de diferentes regiones de costa rica. Top. Meteor. Oceanogr **7**, 1–11 (2000)
2. Amin-Burhanuddin, S.N.Z., Mohd-Deni, S., Mohamed-Ramli, N.: Revised normal ratio methods for imputation of missing rainfall data. Sci. Res. J. **13** (2016)
3. Aparicio-Mijares, F.: Fundamentos de hidrología de superficie. Limusa S.A de C.V. (1992)
4. Burhanuddin, S.N.Z.A., Deni, S.M., Shaadan, N.: Controlled Sampling Approach in Improving Multiple Imputation for Missing Seasonal Rainfall Data (2021). https://doi.org/10.21203/rs.3.rs-679692/v1. https://www.researchsquare.com/article/rs-679692/v1
5. Campos-Aranda, D.: Procesos del Ciclo Hidrológico, 3 edn (1998)
6. Canchala-Nastar, T., Carvajal-Escobar, Y., Alfonso-Morales, W., Cerón, W.L., Caicedo, E.: Estimation of missing data of monthly rainfall in Southwestern Colombia using artificial neural networks. Data in Brief **26** (2019). https://doi.org/10.1016/j.dib.2019.104517
7. Chinasho, A., Bedadi, B., Lemma, T., Tana, T., Hordofa, T., Elias, B.: Evaluation of seven gap-filling techniques for daily station-based rainfall datasets in South Ethiopia. Adv. Meteorol. (2021). https://doi.org/10.1155/2021/9657460
8. CONAGUA: Programa hídrico regional 2014-2018 de la región hidrológico -administrativa IX Golfo Norte. Comisión Nacional del Agua (2015). https://www.gob.mx/conagua
9. CONAGUA: Estadísticas del agua en méxico 2023 (2024)
10. CONAGUA: Servicio meteorologico nacional (2024). https://smn.conagua.gob.mx/es/
11. CONAGUA: Sistema de seguridad de presas (2024). https://presas.conagua.gob.mx/inventario
12. DOF: Acuerdo por el que se actualiza la disponibilidad media anual de las aguas superficiales. (2011). https://www.dof.gob.mx
13. Eischeid, J.K., Pasteris, P.A., Diaz, H.F., Plantico, M.S., Lott, N.J.: Creating a serially complete, national daily time series of temperature and precipitation for the Western United States. J. Appl. Meteorol. **39**, 1580–1591 (2000)
14. Gallegos-Cedillo, J., Arteaga-Ramírez, R., Vázquez-Peña, M.A., Juárez-Méndez, J.: Estimation of missing daily precipitation and maximum and minimum temperature records in san luis potosí. Ingeniería Agrícola y Biosistemas **8**, 3–16 (2016). https://doi.org/10.5154/r.inagbi.2015.11.008. https://revistas.chapingo.mx
15. Gyau-Boakye, P., Schultz, G.A.: Filling gaps in runoff time series in West Africa. Hydrol. Sci. J. **39**, 621–636 (1994). https://doi.org/10.1080/02626669409492784
16. Herrera-Oliva, C., Campos-Gaytán, J., Carrillo-Gonzalez, F.M.: Estimación de datos faltantes de precipitación por el método de regresión lineal: Caso de estudio cuenca guadalupe, baja california, méxico. Investigación y Ciencia **25**, 34–44 (2017). http://www.redalyc.org/articulo.oa?id=67452917005
17. McCuen, R.H.: Hydrologic Analysis and Design, 2nd edn. Prentice-Hall, Inc (1998)
18. OMM: Guía de prácticas climatológicas. Organización Meteorológica Mundial (2018). https://public.wmo.int/es/recursos/meteoterm
19. Paulhus, J., Kohler, M.: Interpolation of missing precipitation records. Monthly Weather Rev. **80**, 129–133 (1952)
20. Romman, Z.A., Al-Bakri, J., Kuisi, M.A.: Comparison of methods for filling in gaps in monthly rainfall series in arid regions. Int. J. Climatol. **41**, 6674–6689 (2021). https://doi.org/10.1002/joc.7219

21. Sa'adi, Z., Yusop, Z., Alias, N.E., Chow, M.F., Muhammad, M.K.I., Ramli, M.W.A., Iqbal, Z., Shiru, M.S., Rohmat, F.I.W., Mohamad, N.A., Ahmad, M.F.: Evaluating imputation methods for rainfall data under high variability in Johor river basin, Malaysia. Appl. Comput. Geosciences **20** (2023). https://doi.org/10.1016/j.acags.2023.100145
22. SAGARPA, UAZ: Asistencia técnica para el desarrollo de procesos de purificación y recuperación de acuíferos, manantiales y cuerpos de agua en los estados de la región norte. durango, sonora, chihuahua, coahuila, nuevo león, tamaulipas y zacatecas, que permitan rescatar el dinamismo de la producción de ganado, leche y granos, al amparo del programa regional de desarrollo previsto en el pnd (2015)
23. Sanusi, W., Zawiah, W., Zin, W., Mulbar, U., Side, S.: Comparison of the methods to estimate missing values in monthly precipitation data. Int. J. Adv. Sci. Eng. Information Technol. **7** (2017)
24. Simonovic, S.P.: Synthesizing Missing Streamflow Records on Several Manitoba Streams Using Multiple Nonlinear Standardized Correlation Analysis (1995)
25. Trewin, B.C.: Función de las normales climatológicas en un clima cambiante. World Meteorological Organization (2007)
26. Valencia-Gonzalez, J.N., Arteaga-Ramirez, R., Vasquez-Peña, M.A., Quevedo-Nolasco, A.: Relleno de datos diarios faltantes en registros de series climatológicas temporales. Revista Mexicana de Ciencias Agricolas **13** (2022)
27. Wangwongchai, A., Waqas, M., Dechpichai, P., Hlaing, P.T., Ahmad, S., Humphries, U.W.: Imputation of missing daily rainfall data; A comparison between artificial intelligence and statistical techniques. MethodsX **11**, 102459 (2023). https://doi.org/10.1016/j.mex.2023.102459
28. WMO: Calculation of Monthly and Annual 30-Year Standard Normals (1989). https://posmet.ufv.br/wp-content/uploads/2017/04/MET-481-WMO-341.pdf
29. WMO: WMO Guidelines on the Calculation of Climate Normals. World Meteorological Organization (2017). https://www.agroorbi.pt/
30. WMO: Guide to Climatological Practices. World Meteorological Organization (2023)
31. WMO: About World Meteorological Organizartion (2024). https://wmo.int/about
32. Young, K.: A three-way model for interpolating for monthly precipitation values. Monthly Weather Rev. **120**, 1561–2569 (1992)
33. Yozgatligil, C., Aslan, S., Iyigun, C., Batmaz, I.: Comparison of missing value imputation methods in time series: the case of Turkish meteorological data. Theoretical Appl. Climatol. **112**, 143–167 (2013). https://doi.org/10.1007/s00704-012-0723-x

MIX
Papier aus verantwortungsvollen Quellen
Paper from responsible sources
FSC® C105338

If you have any concerns about our products,
you can contact us on
ProductSafety@springernature.com

In case Publisher is established outside the EU,
the EU authorized representative is:
Springer Nature Customer Service Center GmbH
Europaplatz 3, 69115 Heidelberg, Germany

Printed by Libri Plureos GmbH
in Hamburg, Germany